SELECTED ASTM STANDARDS ON PACKAGING

Sponsored by ASTM Committee D-10 on Packaging

Third Edition
1991

ASTM Publication Code Number (PCN): 03-410091-11

ASTM
1916 Race St., Philadelphia, PA 19103

Editorial Staff

Library of Congress Cataloging-in-Publication Data

American Society for Testing and Materials.
Selected ASTM standards on packaging.—3rd ed.
"Sponsored by ASTM Committee D10 on Packaging"—Foreword.
"ASTM publication code number (PCN) 03-410091-11"—T.p. verso.
Includes bibliographical references and index.
ISBN 0-8031-1246-7
1. Packaging—Standards—United States. 2. Packaging—Materials—Standards—United States. 3. Packaging—Testing—Standards—United States. I. ASTM Committee D-10 on Packaging. II. Title.
TS195.6.A54 1991
688.8'09218'73—dc20 91-15892
CIP

Library of Congress Catalog Card Number: 91-15892
ISBN 0-8031-1246-7

Printed in Baltimore, MD U.S.A.
August 1991

FOREWORD

This compilation is sponsored by ASTM Committee D-10 on Packaging. It contains test methods, specifications, practices, and guides for the evaluation and selection of packaging materials and complete packages, based on their physical properties and performance, and in their proper application. As they are applied to packaging, the standards cover wood, fiberboard and other paper products; plastics, glass, metals, rubber, and composite structures in the form of containers, cushioning, wrapping, barriers, and closures; and sealing, unitizing, reinforcing, and handling aids. Definitions of terms related to the complete packaging and distribution environment are also provided.

The major source for the standards in this compilation is Volume 15.09 of the *Annual Book of ASTM Standards*. This volume contains standards on Paper; Packaging; Flexible Barrier Materials; and Business Copy Products.

The standards contained within this compilation are not limited to those developed by Committee D-10. Included are standards from ASTM committees C-16 on Thermal Insulation, D-6 on Paper and Paper Products, D-20 on Plastics, and F-2 on Flexible Barrier Materials. Nine standards from the Technical Association of the Pulp and Paper Industry (TAPPI) covering paper, paperboard, containerboard, and corrugated and solid fiberboard are also included.

This compilation is intended as a technical resource for students in universities that offer a packaging curriculum; for packaging engineers and designers of packaging for industry and government, and for manufacturers of packaging materials and processes.

Suggestions and comments regarding the compilation are welcome and should be addressed to Director, Editorial Services, Standards Department, ASTM, 1916 Race Street, Philadelphia, PA 19103.

ASTM

ASTM, founded in 1898, is a developer and publisher of technical information designed to promote the understanding and development of technology and to ensure the quality of commodities and services and safety of products.

ASTM's primary mission is to develop voluntary full consensus standards for materials, products, systems, and services. It provides a forum for producers, users, ultimate consumers, and those having a general interest (representatives of government and academia) to meet on common ground to write standards that best meet their needs.

COMMITTEE D-10

ASTM Committee D-10 on Packaging is one of many technical, standards-writing committees within ASTM. The scope of the committee is the promotion of knowledge and the development of standards for packages. Standards include terminology classifications (including dimensions), practices, test methods, and specifications. The committee's work also includes defining or generating closely related packaging design criteria and developing related material handling standards.

Tests on basic materials used in package fabrication are excluded from the work of D-10 in cases where the work is being carried forward by other committees.

The committee was organized in 1914. Today it has over 400 members organized into three divisions, an executive subcommittee, sixteen technical subcommittees, and three administrative subcommittees. A total of 125 standards have been developed.

For additional information on committee activities and membership, see the application at the end of this book, or contact: ASTM, Member Services, 1916 Race Street, Philadelphia, PA 19103.

RELATED PUBLICATIONS

ASTM has issued several publications which may be of interest to the users of this compilation of standards. These are listed as follows:

Special Technical Publications (STPs):
STP 1113 Food Packaging Technology (1991)
STP 994 Current Trends in Protective Packaging of Computers and Electronic Components (1988)
STP 912 Current Technologies in Flexible Packaging (1986)

Compilation:
Performance Testing of Shipping Containers (1987)

For further information on these and other ASTM publications, please contact: ASTM, Department 8408, 1916 Race Street, Philadelphia, PA 19103 (215-299-5585).

CONTENTS

In the serial designations prefixed to the following titles, the number following the dash indicates the year of original adoption as standard or, in the case of revision, the year of last revision. Thus, standards adopted or revised during the year 1991 have as their final number, 91. A letter following this number indicates more than one revision during the year, that is, 91a indicates the second revision in 1991, 91b the third revision, etc. Standards that have been reapproved without change are indicated by the year of last reapproval in parentheses as part of the designation number, for example, (1991). A superscript epsilon (ϵ) indicates an editorial change since the last revision or reapproval—ϵ1 for the first change, ϵ2 for the second change, etc.

ASTM Standards:

D 4728 – 87	Test Method for Random Vibration Testing of Shipping Containers
D 4919 – 89$^{\epsilon 1}$	Specification for Testing of Hazardous Materials Packagings
D 4991 – 89	Test Method for Leakage Testing of Empty Containers by Vacuum Method
D 5077 – 90	Terminology Relating to Electrostatic Discharge (ESD) Packaging Materials
D 5094 – 90	Test Method for Gross Leakage of Liquids from Containers With Threaded or Lug Style Closures
D 5112 – 90	Test Method for Vibration (Horizontal Linear Sinusoidal Motion) Test of Products
D 5118/D 5118M – 90	Practice for Fabrication of Fiberboard Shipping Boxes
E 96 – 90	Test Methods for Water Vapor Transmission of Materials
G 26 – 90	Practice for Operating Light-Exposure Apparatus (Xenon-Arc Type) With and Without Water for Exposure of Nonmetallic Materials
G 53 – 88	Practice for Operating Light- and Water-Exposure Apparatus (Fluorescent UV-Condensation Type) for Exposure of Nonmetallic Materials

TAPPI Test Methods:

T 208	Moisture in Wood, Pulp, Paper and Paperboard by Toluene Distillation
T 410	Grammage of Paper and Paperboard (Weight Per Unit Area)
T 411	Thickness (Caliper) of Paper, Paperboard, and Combined Board
T 414	Internal Tearing Resistance of Paper
T 423	Folding Endurance of Paper (Schopper Type Tester)
T 441	Water Absorptiveness of Sized (Non-Bibulous) Paper and Paperboard (Cobb Test)
T 803	Puncture Test of Containerboard
T 808	Flat Crush Test of Corrugated Board
T 810	Bursting Strength of Corrugated and Solid Fiberboard

RELATED MATERIAL

The following list identifies other significant ASTM standards relating to packaging that may be found in the indicated volumes of the *Annual Book of ASTM Standards*:

A 623 – 90	Specification for Tin Mill Products, General Requirements (Volume 01.06)
A 623M – 87a$^{\epsilon 1}$	Specification for Tin Mill Products, General Requirements (Metric) (Volume 01.06)
A 624 – 86	Specification for Tin Mill Products, Electrolytic Tin Plate, Single-Reduced (Volume 01.06)
A 624M – 87	Specification for Tin Mill Products, Electrolytic Tin Plate, Single-Reduced (Metric) (Volume 01.06)
B 117 – 90	Test Method of Salt Spray (Fog) Testing (Volume 03.02)
C 149 – 86	Method of Thermal Shock Test on Glass Containers (Volume 15.02)
D 646 – 86	Test Method for Grammage of Paper and Paperboard (Weight Per Unit Area) (Volume 15.09)
D 1251 – 79 (1987)$^{\epsilon 1}$	Test Method for Water Vapor Permeability of Packages by Cycle Method (Volume 15.09)
D 3066 – 79 (1984)	Practice for Labeling Aerosol Packages (Volume 15.09)
D 3071 – 79 (1984)	Method of Drop Testing of Glass Aerosol Bottles (Volume 15.09)
D 3473 – 88	Test Methods for Lifting Force Required to Remove Certain Child-Resistant Snap Caps (Volume 15.09)
D 3480 – 88	Test Methods for Downward Force Required to Open or Activate Child-Resistant Snap-Engagement Packages (Volume 15.09)
D 3611 – 89	Practice for Accelerated Aging of Pressure-Sensitive Tapes (Volume 15.09)
D 3652 – 83 (1988)$^{\epsilon 1}$	Test Method for Thickness of Pressure-Sensitive and Gummed Tapes (Volume 15.09)
D 3652M – 83 (1988)$^{\epsilon 1}$	Test Method for Thickness of Pressure-Sensitive and Gummed Tapes (Metric) (Volume 15.09)
D 3715 – 88	Practice for Quality Assurance of Pressure-Sensitive Tapes (Volume 15.09)
D 3811 – 88$^{\epsilon 1}$	Test Method for Unwind Force of Pressure-Sensitive Tapes (Volume 15.09)
D 3811M – 88$^{\epsilon 1}$	Test Method for Unwind Force of Pressure-Sensitive Tapes (Metric) (Volume 15.09)
D 3813 – 88$^{\epsilon 1}$	Test Method for Curling and Twisting on Unwinding of Pressure-Sensitive Tapes (Volume 15.09)
D 3813M – 88$^{\epsilon 1}$	Test Method for Curling and Twisting on Unwinding of Pressure-Sensitive Tapes (Metric) (Volume 15.09)
D 3815 – 89	Practice for Accelerated Aging of Pressure-Sensitive Tapes by Carbon-Arc Exposure Apparatus (Volume 15.09)
D 3815M – 89	Practice for Accelerated Aging of Pressure-Sensitive Tapes by Carbon-Arc Exposure Apparatus (Metric) (Volume 15.09)
D 3816 – 88$^{\epsilon 1}$	Test Method for Water Penetration Rate of Pressure-Sensitive Tapes (Volume 15.09)
D 3816M – 88$^{\epsilon 1}$	Test Method for Water Penetration Rate of Pressure-Sensitive Tapes (Metric) (Volume 15.09)
D 3833 – 88$^{\epsilon 1}$	Test Method for Water Vapor Transmission of Pressure-Sensitive Tapes (Volume 15.09)
D 3833M – 88$^{\epsilon 1}$	Test Method for Water Vapor Transmission of Pressure-Sensitive Tapes (Metric) (Volume 15.09)
D 4504 – 85	Specification for Molded Polyethylene Open-Head-Pails for Industrial Shipping (Volume 15.09)
D 4649 – 87	Guide for Selection of Stretch, Shrink, and Net Wrap Materials (Volume 15.09)
D 4675 – 87	Guide for Selection and Use of Flat Strapping Materials (Volume 15.09)
D 5105 – 90	Practice for Performing Accelerated Outdoor Weathering of Pressure-Sensitive Tapes Using Concentrated Natural Sunlight (Volume 15.09)
E 171 – 87	Specification for Standard Atmospheres for Conditioning and Testing Materials (Volume 15.09)
E 685 – 79	Practice for Testing Fixed-Wavelength Photometric Detectors Used in Liquid Chromatography (Volume 14.01)
F 88 – 85	Test Method for Seal Strength of Flexible Barrier Materials (Volume 15.09)
F 372 – 73 (1984)$^{\epsilon 2}$	Test Method for Water Vapor Transmission of Flexible Barrier Materials Using an Infrared Detection Technique (Volume 15.09)
F 392 – 74 (1987)	Test Method for Flex Durability of Flexible Barrier Materials (Volume 15.09)
F 1140 – 88	Test Method for Failure Resistance of Unrestrained and Nonrigid Packages for Medical Applications (Volume 15.09)
F 1306 – 90	Test Method for Slow Rate Penetration Resistance of Flexible Barrier Films and Laminates (Volume 15.09)

Standard Test Method for Machine Direction of Paper and Paperboard[1]

This standard is issued under the fixed designation D 528; the number immediately following the designation indicates the year of original adoption or, in the case of revision, the year of last revision. A number in parentheses indicates the year of last reapproval. A superscript epsilon (ε) indicates an editorial change since the last revision or reapproval.

1. Scope

1.1 This test method covers four procedures for determining the machine direction of most grades of paper and paperboard, embodying the principle that machine direction alignment of fibers results in:

1.1.1 Cross-direction shrinkage to produce curl with axis in machine direction,

1.1.2 Higher cross-direction tear,

1.1.3 Higher tensile strength in machine direction, and

1.1.4 Higher stiffness in machine direction.

1.2 Application of the procedures in this method to certain grades of paper, such as sheets laminated to film, creped papers, extensible papers (where it is not unusual for the machine direction tensile to be relatively low and the stretch to be relatively high) and papers reinforced with textile materials, may result in unreliable determinations.

1.3 *This standard does not purport to address the safety problems associated with its use. It is the responsibility of the user of this standard to establish appropriate safety and health practices and determine the applicability of regulatory limitations prior to use.*

2. Referenced Documents

2.1 *ASTM Standards:*

D 585 Method for Sampling and Accepting a Single Lot of Paper, Paperboard, Fiberboard, or Related Product[2]

D 774 Test Method for Bursting Strength of Paper[2]

D 828 Test Method for Tensile Breaking Strength of Paper and Paperboard[2]

2.2 *TAPPI Standard:*

T 409 Machine Direction of Paper[3]

3. Definitions

3.1 The two major directions of paper or paperboard are defined as:

3.1.1 *machine direction*—the direction of a paper or paperboard corresponding or parallel to the direction of flow of the stuff on the paper machine.

3.1.2 *cross direction*—the direction of a paper or paperboard at right angles to the machine direction.

4. Summary of Test Method

4.1 This test method describes four physical procedures for determining the machine direction of paper and paperboard. Two of the procedures employ no special apparatus, the one using the difference in stiffness between machine and cross directions, and the other utilizing the tendency of paper to curl when preferentially wetted on one side. The remaining two procedures use standard test equipment, a tensile tester in one case and a bursting tester in the other.

5. Significance and Use

5.1 From the standpoint of use requirements, the determination of the machine direction of paper is essential where creasing is required, such as the folding of pages in books and pamphlets, or when scoring or creasing is performed as for cut-outs and folders.

5.2 Determination of machine direction is necessary, in many instances, before further testing can be done. For example, machine direction must be known when determining brightness, gloss, tear, tensile and folding endurance.

6. Apparatus

6.1 *Bursting Tester*, meeting the requirements set forth in Test Method D 774, to be used for Procedure C.

6.2 *Tensile Tester*, meeting the requirements set forth in Test Method D 828, to be used for Procedure D.

6.3 *Other Apparatus*—Indelible pencil (optional for 9.1), shallow glass or metal pan.

7. Sampling and Test Specimens

7.1 Obtain a sample of the paper in accordance with Methods D 585. From each test unit sheet of the sample prepare specimens as follows:

7.1.1 *Procedure A*—For purposes of identification, draw a line through adjacent parts of the paper and the specimen to be cut. Cut circular specimens approximately 50 mm in diameter or square specimens approximately 50 mm on a side. For square specimens, the sides of the specimen must be cut parallel to the sides of the test unit sheet.

7.1.2 *Procedures B and D*—Cut two test specimen strips 15 by 250 mm. Cut them at right angles to each other and parallel to the edges of the test unit sheet.

7.1.3 *Procedure C*—Use the test unit sheet as received.

8. Procedure

8.1 *Procedure A*—Float a specimen on tap water in a pan and note or mark with an indelible pencil, the final axis of curl. Observe the curl before water penetrates completely through the specimen.

8.2 *Procedure B—Bend*—Place two specimen strips together, one on top of the other, making sure they are aligned at one end. Grasp the two between the thumb and forefinger

---

[1] This test method is under the jurisdiction of ASTM Committee D-6 on Paper and Paper Products and is the direct responsibility of Subcommittee D06.92 on Test Methods.

Current edition approved Aug. 19, 1987. Published October 1987. Originally published as D 528-39T. Last previous edition D 528-76.

[2] *Annual Book of ASTM Standards*, Vol 15.09.

[3] Available from the Technical Association of the Pulp and Paper Industry, P.O. Box 105113, Atlanta, GA 30348.

and hold them so that they are free to bend of their own weight. Repeat, placing the bottom specimen on top. Note which specimen bends more when it is placed on the bottom.

8.3 *Procedure C—Bursting Test*—Perform a bursting test on the test unit sheet in accordance with Test Method D 774. Remove the test unit sheet from the bursting tester and observe the principal line of rupture.

8.4 *Procedure D—Tensile Test*—Perform tensile tests on the specimens in accordance with Test Method D 828.

8.5 *Recommendations for Procedures:*

8.5.1 Procedures A and C are recommended when the test unit is not in square cut sheet form or where it is not certain that the edges of the test unit are parallel to the machine and cross directions.

8.5.2 Procedure A may not be applicable to unsized papers.

8.5.3 A modification of Procedure D can be also used where the test unit is not square cut or where it is not certain that the edges of the test unit are parallel to the machine and cross directions. Cut consecutive test specimens at 0, 30, 60, 90, 120, and 150° from an arbitrarily selected reference line. The specimen having the greatest tensile strength can be considered to have its length parallel to the machine direction. This will not apply to extensible papers.

9. Interpretation of Results

9.1 The axis of curl in Procedure A will be parallel to the machine direction of paper. Papers with a high degree of dried-in strain may first exhibit an axis of curl in the cross direction. After strain relaxation, the curl axis changes and parallels the machine direction.

9.2 The specimen in Procedure B cut with its length parallel to the cross direction, will bend more because of the lesser cross-direction stiffness and will, when on the bottom, fall away from the specimen cut with its length parallel to the machine direction.

9.3 The principal line of rupture in Procedure C (with approximately perpendicular fractures at either end) will be perpendicular to the machine direction.

NOTE—The bursting test is convenient for papers with a normal distribution of tensile and stretch characteristics; however, there are numerous exceptions to this. The principal line of rupture is parallel to the direction with the higher stretch. In those papers where there is no significant difference in the stretch for the two directions, the rupture tends to be more random and less positive.

9.4 The specimen in Procedure D, cut with its length in the machine direction, will normally have the greater tensile strength and the lesser stretch.

10. Precision and Bias

10.1 This test method is considered reliable for the identification of machine direction except for papers specified in 1.2.

Designation: D 585 – 86

Standard Method for Sampling and Accepting a Single Lot of Paper, Paperboard, Fiberboard, or Related Product[1]

This standard is issued under the fixed designation D 585; the number immediately following the designation indicates the year of original adoption or, in the case of revision, the year of last revision. A number in parentheses indicates the year of last reapproval. A superscript epsilon (ϵ) indicates an editorial change since the last revision or reapproval.

1. Scope

1.1 This method covers a procedure for obtaining a sample to represent a lot of paper or paperboard, fiberboard, or related product, including converted paper products (all hereafter referred to as *paper*).

1.2 Prior to purchase, there should be agreement between buyer and seller on the size of the lot to be sampled (5.1), on details of the sampling procedure, the required physical and chemical properties, dimensional tolerances, etc., and the test methods to be employed.

1.3 Appendixes X1 and X2, which provide useful, statistical criteria for accepting individual lots of paper on the basis of the number of defective test units, can assist the buyer and the seller in selecting an agreed-upon sampling and acceptance procedure.

1.4 *This standard does not purport to address the safety problems associated with its use. It is the responsibility of the user of this standard to establish appropriate safety and health practices and determine the applicability of regulatory limitations prior to use.*

2. Referenced Documents

2.1 *ASTM Standards:*

D 528 Test Method for Machine Direction of Paper and Paperboard[2]

D 725 Method for Identification of Wire Side of Paper[3]

2.2 *ISO Standard:*

ISO 186 Paper and Board–Sampling for Testing[4]

2.3 *Military Standards:*

MIL-STD-105D Sampling Procedures and Tables for Inspection by Attributes[5]

MIL-STD-414 Sampling Procedure and Table for Inspection by Variables for Percent Defective[5]

3. Description of Terms Specific to This Standard

3.1 *lot,* a quantity of paper of a single type, grade, grammage, thickness, and composition, about which it is desired to make a judgment (usually as to conformance to specification) by examining or testing a small fraction called the sample.

3.2 *sample,* a specified number of test units selected in accordance with a prescribed procedure to represent the lot.

3.3 *test unit,* an area of paper sufficient to obtain a single adequate set of test results for all the properties to be measured.

3.4 *test specimen,* a test unit, or a portion of a test unit, upon which (for a specified property) a single test determination is to be made.

NOTE 1—A detailed description of the appropriate specimen should be found in the applicable test method or the specification. A specimen may consist of several pieces of paper, as in a multiple-ply tear test, or a single area of paper may be used in whole or part consecutively for several tests. For example, a grammage specimen, which may have been used previously for detecting spots or a dirt count, may subsequently be cut for use as a tension specimen. This last may be used simultaneously for tensile breaking strength, elongation at break, and tensile energy absorption. A "multiple observation" may be involved when the uniformity of a specimen is of interest.

3.5 *test determination,* (*a*) the process of carrying out the series of operations specified in the test method whereby one or more readings (observations) are made on a test specimen and the observations combined to obtain the value of a property of the test specimen, or (*b*) the value obtained by the process.

3.6 *test result,* the value of a property associated with one test unit of the sample by carrying out the complete protocol of the test method, the value being (as specified in the test method) either a single test determination or the average or other specified combination of a specified number of test determinations.

4. Significance and Use

4.1 If a lot of paper is to be accepted or rejected on the basis of a series of tests made on the paper, it is important to sample the paper in a way that will give a test result that is representative of the lot of paper.

5. Establishing the Lot

5.1 *General Considerations*—Avoid grouping together as a lot batches of paper likely to differ significantly from each other in raw materials or manufacture. If the shipment is small or consists of a large number of batches, it may be uneconomical or impracticable to form lots conforming with the definition (3.1). If this is the case, divide the shipment into portions, to be designated "sublots," in which each portion conforms to the definition of *lot.* Before starting the sampling, obtain a complete understanding of the paper to

[1] This method is under the jurisdiction of ASTM Committee D-6 on Paper and Paper Products and is the direct responsibility of Subcommittee D06.04 on Precision and Sampling.

Current edition approved Aug. 29, 1986. Published October 1986. Originally published as D 585 – 40 T. Last previous edition D 585 – 74 (1982)$^{\epsilon 1}$.

[2] *Annual Book of ASTM Standards*, Vol 15.09.

[3] Discontinued, see *1979 Annual Book of ASTM Standards*, Part 20.

[4] Available from the American National Standards Institute, 1430 Broadway, New York, NY 10018.

[5] Available from Naval Publications and Forms Center, 5801 Tabor Ave., Philadelphia, PA 19120.

be sampled, including the composition and size of lots and sublots, rolls, skids, etc.

5.2 *Location of Sampling:*

5.2.1 *Sampling at the Paper Mill*—When paper is to be delivered in large rolls or skids, if agreed, have the sample taken at the paper mill and delivered to the purchaser for examination and testing. Include in the purchase agreement a definition of the number of rolls or skids constituting a lot and details of the sampling procedure to ensure compliance with the rules in selecting the sample in accordance with Section 6. Especially indicate whether the sample was taken from mill reels or from the finished rolls or skids.

5.2.2 *Sampling in Transit or Warehouse*—It may be necessary to take a sample at a dock or warehouse en route between the supplier and the purchaser, especially for international shipments or in referee sampling for testing. Sampling in transit or warehouse can be quite expensive and should be avoided whenever possible, because of the problem of resealing. Arranging to inspect the shipment prior to unloading is recommended on inspection in transit. Establish the size of each lot and determine which rolls or skids will be sampled, and arrange to sample and reseal the rolls or skids as they are loaded or unloaded to minimize additional handling. Obtain the sample in accordance with Section 6.

5.2.3 *Sampling After Receipt*—This type of sampling has the obvious disadvantage of delaying evaluation of the shipment and of providing the least favorable conditions for disposal of a rejected lot. Other considerations are the same as above for sampling in transit.

5.3 *Representativeness of the Sample*—The test units of the sample usually cannot be taken at random from anywhere within a large roll or skid, but must be taken near the outer layers so as not to destroy the roll or skid. If such test units are regularly taken at the paper mill (5.2.1) and a reasonably uniform quality of production is being maintained, then the test units will, in the long run, be quite representative of average production, even though not randomly selected and therefore not "representative" from a statistical viewpoint. Such test units, however, taken when the lot is in transit or warehouse (5.2.2) or after receipt (5.2.3), cannot be said with certainty to be representative of the shipment; for example, rolls shipped to a customer may have been selected because their outer layers met the customer's specification, with no knowledge as to whether properties vary from the outer layers to the core of the roll. Such test units therefore may be neither random nor "representative," but they are all that is available "to represent" the lot, which experience shows they do fairly well unless nonuniform manufacture or aging has occurred.

6. Selecting the Sample

6.1 Determine the area of paper required for each test unit (see 3.3) and the number of test units required. Take a sample consisting of the required number of test units in accordance with a predetermined procedure, designed to eliminate deliberate selection of any particular area of paper (see X1.5), and as far as practicable, conforming to the following rules:

6.1.1 *Rule 1: Take test units in proportion to the sizes of the sublots*—When a lot divides naturally into sublots consisting of discrete quantities such as carloads, pallet loads, rolls, etc., or when certain portions of the lot differ from one another and are segregated as "sublots" (5.1), take the test units in such a way that the number of test units taken from a given sublot is proportional to the size of the sublot.

6.1.2 *Rule 2: Take the test units so that each area of paper in the lot or sublot has an equal probability of being selected*—It is recognized that difficulties of handling may not permit the taking of test units from locations scattered throughout the lot, but this should be done to the greatest practicable extent.

6.1.3 *Rule 3: Take test units indiscriminately*—Take the units without regard to their condition or quality but do not sample the outermost sheets of a roll or skid. Also do not sample from inner sheets that have been damaged by handling, abrasion, etc., unless it is evident that similarly damaged areas occur throughout the lot.

6.2 *Compliance with Rules 2 and 3* may be assured in accordance with the following procedure: Divide the lot or sublot into "locations" (carloads, skids, cartons, etc.) in such a way that each location contains an equal quantity of paper. Assign a number to each location and select the locations to be sampled by drawing numbers at random from a hat or from a table or random numbers. Next assign and select the sublocations or areas within a location by the same procedure. In order to minimize handling during sampling, observe the precautions of Section 5.

6.2.1 When test units are to be taken from *small rolls, reams, bundles, cartons or shipping containers* of completed paper products, divide the lot or sublot into "locations," etc., and proceed as described above.

6.2.2 When test units are to be taken from *large rolls or skids*, divide the lot or sublot into "locations" and "sublocations" no smaller than a roll or unopenable skid (see 6.3.2), assign and select numbers as described above, then proceed as described in 6.3.

NOTE 2—In ISO and British Standards, the word *reel* is used for a continuous sheet or board wound on a core and the word *roll* for the same when wound without a core. In the United States, the distinction is between the "reel" at the end of the paper machine and the "rolls" (with or without the cores) made therefrom.

6.3 *Large Rolls or Skids:*

6.3.1 When taking a test unit from a *large roll* remove all damaged layers from the outside of the roll and, in addition, discard at least three undamaged layers for sheets having a nominal grammage of less than 250 g/m^2 or at least one undamaged layer for heavier sheets. Cut the roll across its full width and to a sufficient depth to enable the requisite number of sheets to be taken. Let the cut sheets fall to each side and remove the roll.

6.3.2 When taking a test unit from a *skid or roll that may not be opened* at the time of sampling, cut a window 300 by 450 mm or larger if larger test specimens are required. Cut the window with its longer side in the machine direction when this is known, or if not known, mark "Direction Unknown" or determine the machine direction using Test Method D 528 and mark whether the machine direction is the short dimension or long dimension. For paper that has been sheeted simultaneously from several rolls, take a sufficient number of adjacent sheets to assure representation of all the rolls used. Cut the sheets to sufficient depth to enable the requisite number of sheets to be taken and remove them. Remove all outside sheets that are damaged and, in addition, discard at least the three outermost undamaged sheets for paper having a nominal grammage less than 250 g/m^2 or at least one outmost undamaged sheet of heavier paper. Vary the position of the window at random among the skids or rolls from which test units are to be drawn, making sure that the sides of the window are parallel to those of the skid or to the roll edge and axis.

6.3.3 In either of the above, if a test for *moisture content* is to be made, it may be necessary to discard considerably more than three undamaged sheets of paper and one undamaged sheet of board. In one case reported in the literature,[6] it was necessary to discard 40 sheets of manila board to obtain the moisture content of the major portion of the roll.

7. Care of Samples

7.1 Keep the test units smooth and flat, except for transporting, when it may be better to ship the test unit in a tube. Protect the sample from exposure to direct sunlight, moisture of the hands, contact with liquids or other harmful influences such as extremes of temperature or humidities above 58 % relative humidity. Consult the product specification and the test methods for directions as to any precautions to be taken or special handling necessary.

[6] Yezek, M., "Some Aspects of Moisture Measurement in Paper and Paperboard," *TAPPI* 41 (8): 193A (1958).

8. Cutting and Marking

8.1 Trim test units with their edges parallel to the machine and cross directions. Avoid watermarks or creases for other than grammage determinations; also avoid any unusual flaws or blemishes that might subsequently affect the test results. Mark test units for identification, for example, the locations from which they were taken, and, if needed, their machine direction (see Test Method D 528) and top side (see Method D 725).

9. Sampling Report

9.1 When required, give a brief description of the shipment or lot and the sampling including:

9.1.1 Type and grade of paper or paperboard with a reference to the specification, if available.

9.1.2 Form in which purchased (that is, dimensions of rolls or sheets; packaging; etc., if not covered in the product specification).

9.1.3 Total quantity (usually weight or area), or purchase order number, or both.

9.1.4 Lot number or other identification of specific lot sampled (if divided into sublots, identification or description of sublots).

9.1.5 Date of sampling.

9.1.6 Location of sampling (mill, warehouse, in transit, etc.).

9.1.7 Description and enumeration of any portions of the shipment excluded from sampling because of damage.

9.1.8 Deviations from the specified sampling procedure if it was found to be not reasonably possible to follow the prescribed directions.

9.1.9 Authority requesting the sample (if appropriate).

APPENDIXES

(Nonmandatory Information)

X1. Useful Acceptance Plans

X1.1 *Individual Lot Acceptance*—The appropriate sampling-acceptance plan provided in Appendix X2 is to be used only for individual lot acceptance when no other plan has been specified and this method has been referenced. A plan that is tailored to the specific paper product, grade, type of defects, and use, etc., is preferable to a general plan. However, the general plans provided in Appendix X2 are considered to be a good compromise between the costs of testing and the risks of wrong decisions, and are applicable to a wide range of paper products.

X1.2 The acceptance plans given in Appendix X2 are called "attribute" acceptance plans because a test unit fails if it fails to conform to one or more of the requirements for which it is tested and acceptance of the lot is based on an acceptably low number of units failing.

X1.3 ISO 186 uses an engineered sampling plan to determine the lot average, with 5 % of the shipping units sampled but with a minimum of 5 and a maximum of 20.

X1.4 Maltenfort and Boedecker[7] describe a "variables" acceptance plan in which acceptance is based on the average of the values obtained for the test units and the variability among these values.

X1.5 Deliberate nonrepresentative or semirepresentative selection of rolls or skids for sampling may be desirable to emphasize expected stock problems. The plan uses the bills of lading to select skids or rolls at the beginning and near the end of the run, then randomly through the run but stressing front and back positions. These are the areas from which production problems may be expected to arise. This approach is nonrandom, not statistical, but may be more economical when extremes and not averages are sought.

X1.6 When a continuing series of lots is being obtained from the same manufacturer, other plans (such as, MIL-STD 105D and MIL-STD 414) are likely to be more efficient as they can provide for increasing or reducing inspection as experience with that manufacturer indicates.

[7] Maltenfort, G. G., and Boedeker, R. E., "Sampling of Paper and Paper Products," *Industrial Quality Control* 14(11): 19(1958).

TABLE X2.1 Plan I for Large Rolls or Skids

Lot Size, *N*	Sample Size		Acceptance and Rejection Numbers			
	n	n_t[A]	*Ac*	*Re*	Ac_t[B]	Re_t[B]
1	1		0	1		
2	2		0	1		
3 to 5	3		0	1		
6 to 10	4		0	1		
11 to 100	5		0	1		
101 to 300	8	16	0	2	1	2
301 to 1200	13	26	0	3	2	3
1201 and over	20	40	0	3	3	4

[A] n_t = total sample size, that is, sum of test units in first and second part, of double sample.
[B] Ac_t and Re_t = acceptance and rejection numbers for double sample.

X2. Attribute Plans for Single-Lot Acceptance

X2.1 *Assumptions:*

X2.1.1 Since a lot is presumed to be reasonably homogeneous (3.1), the plans presented below assume homogeneity and therefore provide only for acceptance or rejection of the lot as a whole. If the lot is not homogeneous, a test on one test unit of the sample might be so far off specification as to make at least the corresponding part of the lot unacceptable even though the proportion of off-specification test units would be so small as to indicate the whole lot should be accepted. Provision should be made for this situation in advance, for example, by calling for rejection of the lot as a whole because of its excess nonhomogeneity or by requiring complete screening (that is, testing of each part and rejecting substandard parts) if nonhomogeneity is found.

X2.1.2 The plans are based also on the assumption that the properties of a test unit drawn from the outer layers or sheets (6.3) or at random (in 6.2.1) are identical with the properties throughout the roll, skid, ream, carton, etc., from which the test unit was taken. While this assumption is obviously not true, if the lot or sublot is reasonably uniform and the rules for selecting the sample are carefully followed, the acceptance plans will generally provide a satisfactory level of protection.

X2.2 *Application of Plans:*

X2.2.1 *Plan I*—For individual lots composed of large rolls that cannot be unwound or skids (pallets) that cannot be opened at the time of sampling (6.3.2).

X2.2.2 *Plan II*—For individual lots composed of small rolls (as tapes, toweling), reams, bundles, cartons, or shipping containers from which test units may be selected at random.

TABLE X2.2 Plan II for Small Rolls, Reams, Bundles, Etc.

Lot Size, *N*	Sample Size		Acceptance and Rejection Numbers			
	n	n_t[A]	*Ac*	*Re*	Ac_t[B]	Re_t[B]
2 to 25	2		0	1		
26 to 150	3		0	1		
151 to 1 200	5		0	1		
1 201 to 35 000	8	16	0	2	1	2
35 001 and over	13	26	0	3	2	3

[A] n_t = total sample size, that is, sum of test units in first and second part, of double sample.
[B] Ac_t and Re_t = acceptance and rejection numbers for double sample.

TABLE X2.3 Plan III for Noncritical Chemical Tests

Lot Size, *N*	Sample Size, *n′*	Acceptance and Rejection Numbers	
		Ac	*Re*
2 to 100	2	0	1
101 to 1 200	3	0	1
1 201 to 35 000	4	0	1
35 001 and over	5	0	1

X2.2.3 *Plan III*—For "noncritical" chemical tests for individual lots composed as in X2.3.1 or X2.3.2.

NOTE X1—As used herein, *noncritical* means tests for which variability within the lot is of little or no significance and therefore the amount of testing called for by Plan I or II could not be justified.

X2.3 *Acceptance Definitions:*

X2.3.1 *Lot Size:*

X2.3.1.1 *Plan I*—The lot size (*N*) is the number of rolls or skids of which the lot (3.1) is composed.

X2.3.1.2 *Plan II*—The lot size (*N*) is the number of small rolls, reams, bundles, cartons, or shipping containers of which the lot is composed, as appropriate for the test to be made. When a choice is possible (as in testing a property of the paper in a roll when an equal number of rolls are packed in each carton), select the unit (roll or carton) on which lot size is based so that the lot size will be greater than 25.

X2.3.1.3 *Plan III*—As for Plan I or II, as appropriate.

X2.3.2 *Sample Size*—The number of test units (3.3) of which the sample (3.2) for test is composed, determined from lot size in accordance with Tables X2.1, X2.2, or X2.3.

X2.3.3 *Acceptance*—A lot is either accepted or rejected as a whole on the basis of tests carried out on the sample. Each test unit of the sample is separately evaluated.

X2.3.4 *Defective Test Unit*—A test unit that fails to conform to one or more of the requirements for which it has been tested.

X2.3.5 *Acceptance Number*—A number used in connection with a sampling plan (Tables X2.1, X2.2, or X2.3), such that if the number of defective test units in the sample is less than or equal to this number, the lot should be passed for the properties tested.

X2.3.6 *Rejection Number*—A number used in connection with a sampling plan (Tables X2.1, X2.2, or X2.3), such that if the number of defective test units in the sample is greater than or equal to this number, the lot should be rejected.

X2.4 *Level of Protection*—If the above assumptions are true, the plans provide the following:

X2.4.1 *Protection to Producer*—The lot will be accepted with a probability of at least 95 % if the proportion of the defective items contained in it does not exceed 2.5 %.

X2.4.2 *Protection to Consumer*—The lot will be rejected with a probability of at least 90 % if the proportion of the defective items contained in it reaches 16 to 32 % in Plan I, 19 to 68 % in Plan II, and 37 to 68 % in Plan III, the smaller figure in each case applying to the largest lot size.

X2.5 *Instructions for The Use of Tables X2.1, X2.2, and X2.3.*

X2.5.1 Select Table X1 or Table X2 according to the composition of the lot (X2.3.1 and X2.3.2).

X2.5.2 Locate the lot size (X2.4.1) in the first column of the selected table. The sampling and acceptance criteria to be used are those given on the line corresponding to this lot size.

X2.5.3 Take a first sample consisting of the number of test units equal to the number in the column headed *n*. Sample according to the rules for selecting a sample to represent a lot (6.1), including in each test unit enough additional material for noncritical chemical tests (X2.5.10 to X2.5.13).

X2.5.4 Subject each of the *n* test units of this sample to the appropriate number of test determinations for each of the required test properties. For noncritical chemical tests follow instructions in X2.5.10 to X2.5.13.

X2.5.5 Record the number of defective units thus found.

X2.5.6 *First Sample Criteria:*

X2.5.6.1 If the number from X2.5.5 does not exceed the number in column *Ac*, the lot should be considered to meet the requirements relating to the properties tested.

X2.5.6.2 If the above number equals or exceeds the number in column *Re*, the lot should be considered as having failed to meet the requirements of the detail specification.

X2.5.6.3 If the above number exceeds the acceptance number (*Ac*) but is less than the rejection number (*Re*), proceed to the next step.

X2.5.7 Take a second sample equal in size to the first so that the total number of test units in the first and second sample is n_t. Again take this sample in accordance with the rules for selecting a sample to represent a lot.

X2.5.8 Follow X2.5.4 and X2.5.5, and compute the total number of defective test units in the two samples.

X2.5.9 *Total Sample Criteria:*

X2.5.9.1 If the number in X2.5.8 does not exceed the number in Ac_t, the lot is considered to meet the requirements relating to the properties tested.

X2.5.9.2 If the above number equals or exceeds the number in column Re_t, the lot shall be considered as having failed to meet the requirements of the detail specification.

X2.5.10 If the lot satisfies X2.5.6.1 or X2.5.9.1 and noncritical chemical tests are specified, determine the sample size (n') for these tests in accordance with Table X3.

X2.5.11 Select the above n' test units at random from the *n* test units obtained in accordance with 5.3.

X2.5.12 Subject each of the n' test units of this sample to all the required noncritical chemical tests, and record the number of defective test units thus found.

X2.5.13 *Chemical Test Criteria:*

X2.5.13.1 If the number of defective test units (X2.5.12) is zero, the lot is considered to meet the requirements relating to noncritical chemical tests.

X2.5.13.2 If the number is greater than zero, the lot shall be considered as having failed to meet the requirements of the detail specification.

X2.6 *Acceptance Report*—Report lot size, sample size, number of defective test units found, and the nature of the defects.

Standard Test Method for Determining Compressive Resistance of Shipping Containers, Components, and Unit Loads[1]

This standard is issued under the fixed designation D 642; the number immediately following the designation indicates the year of original adoption or, in the case of revision, the year of last revision. A number in parentheses indicates the year of last reapproval. A superscript epsilon (ε) indicates an editorial change since the last revision or reapproval.

1. Scope

1.1 This test method covers compression tests on shipping containers, components or both. Shipping containers may be tested with or without contents. The procedure can be used for measuring the ability of the container to resist external compressive loads applied to its faces, to diagonally opposite edges, or to corners (Figs. 1 and 2).

1.1.1 This test method covers testing of multiple containers or unit loads, in addition to individual shipping containers, components or both.

1.2 The test method of applying load may be used to compare the characteristics of a given design of container with a standard, or to compare the characteristics of containers differing in construction.

1.3 *This standard does not purport to address the safety problems associated with its use. It is the responsibility of the user of this standard to establish appropriate safety and health practices and determine the applicability of regulatory limitations prior to use.*

2. Referenced Documents

2.1 *ASTM Standards:*

D 644 Test Method for Moisture Content of Paper and Paperboard by Oven Drying[2]

D 996 Terminology of Packaging and Distribution Environments[2]

D 2016 Test Method for the Moisture Content of Wood[2]

D 4332 Practice for Conditioning Containers, Packages or Packaging Components for Testing[2]

D 4577 Test Method for Compression Resistance of a Container Under Constant Load[2]

E 4 Practice for Load Verification of Testing Machines[3]

E 122 Practice for Choice of Sample Size to Estimate a Measure of Average Quality for a Lot or Process[2]

3. Terminology

3.1 *Definitions*—For definitions of terms, see Terminology D 996.

3.2 *Descriptions of Terms Specific to This Standard:*

3.2.1 *fixed platen testing machine*—a testing machine equipped with two platens which are both restrained from tilting.

3.2.2 *swiveled platen testing machine*—a testing machine equipped with two platens, one rigidly restrained from tilting while the other platen is universally mounted and allowed to tilt freely.

4. Significance and Use

4.1 Compressive resistance is one of the properties used to evaluate the ability of shipping containers, components, and unit loads to successfully survive the compressive forces to which they are subjected during storage and distribution (see Note 1).

4.2 Compressive resistance may be determined with either fixed or swiveled platen type testing machines. However, a fixed head compression machine is required to perform edgewise and corner-to-corner orientations on test specimens (see Note 2).

NOTE 1—For constant load test, refer to Test Method D 4577.

NOTE 2—Fixed platen machines generally cause failure to occur at the specimen's strongest point, while swivel platen machines cause failure at the specimen's weakest point.

5. Apparatus

5.1 *Compression Testing Machines:*

5.1.1 *Fixed Platen Testing Machine*, Two parallel platens, flat to within 0.02 in. (0.5 mm), one of which is moveable in the vertical direction so as to compress the specimen between the platens. One is the load measuring platen, and both are of sufficient size so that the test specimen does not extend beyond the edges of the platens. The platens are fixed in the horizontal directions so as to have no lateral movement greater than 0.05 in. (1.3 mm) and are held parallel throughout the test to within 0.04 in./1 ft (1 mm/300 mm) in the length and width dimensions.

5.1.2 *Swivel Platen Testing Machine*, two parallel platens, flat to within 0.02 in. (0.5 mm), one of which is moveable in the vertical direction so as to compress the specimen between the platens. One is the load measuring platen, and both are of sufficient size so that the test container does not extend beyond the edges of the platens. One platen is fixed in the horizontal direction so as to have no lateral movement greater than 0.05 in. (1.3 mm). The second platen is attached to the machine by a swivel or universal joint to a point directly centered on the platen thus allowing the platen to tilt freely.

5.1.3 *Common Features of the Machines:*

5.1.3.1 Each machine must have a means of driving the moveable platen at a uniform speed of 0.5 ± 0.10 in./min (12.7 ± 2.5 mm/min).

5.1.3.2 Each machine should have a means of recording

[1] This test method is under the jurisdiction of ASTM Committee D-10 on Packaging, and is the direct responsibility of Subcommittee D10.23 on Storage Test Methods.

Current edition approved June 29, 1990. Published August 1990. Originally published as D 642 – 41 T. Last previous edition D 642 – 76 (1983)$^{\epsilon 1}$.

[2] *Annual Book of ASTM Standards*, Vol 15.09.

[3] *Annual Book of ASTM Standards*, Vol 04.09.

[4] *Annual Book of ASTM Standards*, Vol 03.01.

[5] *Annual Book of ASTM Standards*, Vol 14.02.

FIG. 1 Compression Applied Top-to-Bottom

FIG. 2 Compression Applied Diagonally Opposite Edges

or indicating the applied load to within ± 0.5 % of the full scale capacity of the load sensor being used; for example, 30 000 ± 150 lb or 3 000 ± 15 lb.

5.1.3.3 Each machine should have a means of recording or indicating the resultant deformation to within ± 0.025 in. (±0.64 mm).

5.2 *Suitable Closure Apparatus*—see Appendix A.

5.3 *Conditioning Apparatus*—Adequate facilities for conditioning test specimens at proper relative humidity and temperature prior to test in accordance with the requirements of the specifications covering the containers to be tested. Use of Practice D 4332 is recommended in the absence of other requirements.

5.4 *Miscellaneous Equipment*—Drying oven, scales, knife, saws, direct moisture reading devices, etc., for use in determination of the moisture content, or for making other supplementary tests of the materials from which the test specimens are made.

6. Sampling, Test Specimens, and Test Units

6.1 Choose test specimens and sample quantities to provide an adequate determination of representative performance. For large production runs, lot sampling is advised. Application of Practice E 122 is suggested.

6.2 Whenever sufficient containers and contents are available, it is recommended that five or more replicate tests be conducted to improve the statistical reliability of the data obtained.

6.3 The specimens being tested shall be complete in all respects. Depending on the purpose of the test, interior components may or may not be included. Tests shall be made on specimens with or without contents as prescribed.

6.4 The test specimen shall be closed and secured in the same manner as will be used in preparing them for shipment unless otherwise specified. The method of flap securement for corrugated containers may affect test results (see Appendix X1).

7. Conditioning

7.1 Condition test specimens prior to test or during test, or both, in accordance with the requirements of the applicable specification. When no conditioning requirements are given, and container materials are moisture sensitive, a standard conditioning atmosphere is recommended in accordance with Practice D 4332.

7.1.1 Determination of test specimen moisture content, for moisture sensitive materials, at the time of the test will confirm test effects of conditioning (see Appendix X2).

7.2 For special applications, and depending on the purpose of the tests, the test specimens may be conditioned prior to the compression test by water immersion, exposure to water spray, or other specific conditions.

8. Procedure

8.1 Center the specimen on the lower platen of the testing machine in the desired orientation, so as not to incur eccentric loading. Bring the platens into contact with the specimen applying an initial pressure or pre-load (see Note 3). Both platens must be fixed for tests where the compressive loads are applied on test specimen edges, or on diagonal corners (Fig. 1). Either fixed platen or swivel platen machines may be used for face-to-face compressive tests (Fig. 2).

NOTE 3—For single-wall corrugated containers, an initial pressure or preload, of 50 lbf (222 N) on the specimen is recommended. For doublewall and triplewall boxes, preloads of 100 lbf (445 N) and 500 lbf (2220 N) respectively, are recommended. For other types of test specimens a suitable pre-load may or may not be selected.

8.2 Set the load-deformation recorder pen at zero deformation. If the testing machine is not fitted with a load-deformation recorder, record the test load for every 0.1 in. (2.5 mm) deformation of the container (see Note 4).

8.3 Apply the load with a continuous motion of the moveable platen of the testing machine at a speed of 0.5 ± 0.1 in. (12.7 ± 2.5 mm)/min until failure, maximum load, or both, has been reached.

8.4 Prior to testing for each type of loading, critical points shall be established where applicable. Record the compressive load at these critical deformations, together with the maximum load and deformation.

NOTE 4—When testing full containers, and the load sensing device is located under bottom platen, be sure to zero the test machine with the product on it, or subtract the container weight from peak load readings.

9. Report

9.1 Report the following information:

9.1.1 Container structural and physical specifications, including:

9.1.1.1 Inside dimensions for all corrugated and solid fiberboard containers,

9.1.1.2 Description of the contents of the container and gross weight, as tested,

9.1.1.3 Type of material, style of container, printing, access holes, and double scores,

9.1.1.4 Description and specifications for interior packaging and contents, if used, and

9.1.1.5 Spacing, size, and type of fasteners, and method of securement.

9.1.2 Detailed data documentation for each test specimen, including:

9.1.2.1 Damage to the container and contents,

9.1.2.2 Any observations that may assist in correctly interpreting the results or aid in improving the design of the container,

9.1.2.3 Nature and cause of failure,

9.1.2.4 Any tests performed on the test specimen prior to compression testing,

9.1.2.5 Printing amount and location on container,

9.1.2.6 A tabulation of individual maximum load and deformation results,

9.1.2.7 Graph or table showing the load-deformation relationship for each test,

9.1.2.8 Number of specimens tested, and

9.1.2.9 Mean and standard deviation calculations of all specimens tested.

9.1.3 Identification of test apparatus and instrumentation used, including:

9.1.3.1 Manufacturer's names and model numbers,

9.1.3.2 Type of test machine used: fixed or swiveled platen, including details of any known modifications,

9.1.3.3 Orientation in which the specimen was tested: platen speed, and preload applied to test specimen, if any,

9.1.3.4 Date of last calibration of apparatus and recording instrumentation,

9.1.5 Method, if any, of conditioning the container, including the moisture content of the wood, plywood, or fiberboard, if determined,

9.1.6 Description of test set-up used such as pallets, fixtures, or simulation devices, and

9.1.7 The results of any supplementary tests of the materials from which the container is made.

10. Precision and Bias[6]

10.1 The precision and bias statement is based on data developed from a round robin compression test[7] conducted by eleven laboratories. The test specimens consisted of 200-lb test C-flute regular slotted containers measuring 16 by 12 by 10 in. When testing empty RSC style corrugated containers for top to bottom compression strength using a fixed platen compression tester, the precision of the test method is as follows:

10.1.1 *Repeatability*—The difference between successive results obtained by the same operator with the same apparatus under constant operating conditions on identical test materials would in the long run, in normal and correct operation of the test method, exceed the following values only one time in twenty:

Repeatability = 8.5 % (standard deviation)

10.1.2 *Reproducibility*—The difference between two single and independent results obtained by different laboratories on identical material would, in the long run, in normal and correct operation of the test method, exceed the following values only one time in twenty:

Reproducibility = 11.3 % (standard deviation)

10.2 *Bias*—Since there is no accepted reference material suitable for determining the bias for the procedure in this test method, no statement on bias is being made.

[6] Miles, J. G. "Compression Strength of Corrugated Containers: An Interlaboratory Study," *Materials Research Standards*, Vol 6, No. 3, March 1966, pp 142–146.

[7] Supporting data are available from ASTM Headquarters. Request RR:D10-1859.

APPENDIXES

(Nonmandatory Information)

X1. CLOSURE APPARATUS FOR CORRUGATED CONTAINERS

X1.1 Suitable facilities are required, including sealing boards, clip or rods and proper adhesives, for sealing both the top and bottom flaps of box specimens without bracing the material within the boxes causing reinforcing of the sidewalls of the container or vertical reinforcement that will give false results as to compressive strength.

X1.2 Other closure apparatus may be used for taping, stapling or strapping.

X1.3 When testing empty top loaded containers, restraining of inner flaps may affect test results.

X1.4 When testing side and end loaded containers, sealing should be performed in a comparable manner to what is done in actual practice.

X2. MOISTURE CONTENT

X2.1 *Fiberboard*—Determine the moisture content of fiberboard at the time of test in accordance with Method D 644 or equivalent.

X2.2 *Wood*—Determine the moisture content of wood at the time of test by selecting duplicate samples from different parts of the container, immediately following the test, of minimum 2 in.^{3} (32.8 cm^3) each, weighing to the nearest 0.003 oz (0.1 g), drying to constant weight between 212 and 221°F (100 and 105°C), and reweighing. Express the moisture content as a percentage of the oven-dry weight of the specimen. Refer to Test Method D 2016.

Standard Test Method for Moisture Content of Paper and Paperboard by Oven Drying[1]

This standard is issued under the fixed designation D 644; the number immediately following the designation indicates the year of original adoption or, in the case of revision, the year of last revision. A number in parentheses indicates the year of last reapproval. A superscript epsilon (ε) indicates an editorial change since the last revision or reapproval.

1. Scope

1.1 This test method covers the determination of the moisture in all papers, paperboards, and paperboard and fiberboard containers, except those containing matter other than water that is volatile at 105°C.

1.2 *This standard does not purport to address the safety problems associated with its use. It is the responsibility of the user of this standard to establish appropriate safety and health practices and determine the applicability of regulatory limitations prior to use.*

NOTE—This test method is technically equivalent to TAPPI T 412 - 83.

2. Referenced Document

2.1 *ASTM Standard:*

D 585 Method for Sampling and Accepting a Single Lot of Paper, Paperboard, Fiberboard, or Related Product[2]

3. Apparatus

3.1 *Weighing Container*, airtight, for weighing the specimen before and after drying. A glass weighing bottle with a ground stopper is suitable for specimens of the order of 2 g. The weighing bottle may be of either high or low form; a volume of about 100 mL is appropriate. An airtight metal container, preferably containing a light-weight large-mesh wire basket, is suitable for larger specimens, particularly those over 10 g.

3.2 *Thermometer*, to indicate the temperature of the drying oven. This thermometer shall include the temperature range from 100 to 110°C, and its scale shall be divided in 1° intervals.

3.3 *Drying Oven*, constant-temperature, equipped with means for ensuring adequate temperature control and air circulation, and preferably equipped with means for drying the air entering the oven.

3.4 *Chemical Balance*, sensitive to 1 mg, for weighing specimens of 2 g and under, and sensitive to 0.05 % of the original weight of the specimen for larger specimens.

3.5 *Desiccator*, in which weighing containers and specimens are cooled after drying. It is recommended that anhydrous alumina (indicator grade) be used as the desiccant.

4. Test Specimens

4.1 When moisture is determined for the purpose of calculating the results of a chemical analysis of paper or board on a moisture-free basis, the test specimens shall weigh not less than 1 g, and preferably not less than 2 g each. At the time of initial weighing, these specimens shall be in moisture equilibrium with the samples being analyzed.

4.2 When moisture is determined for the purpose of calculating the amount of moisture in a shipment, the test specimens shall weigh not less than 50 g each, and shall be taken from samples obtained in accordance with Method D 585.

4.3 When moisture is determined on combined board or containers that are to be tested for other physical properties, test specimens shall weigh approximately 50 g, and shall be taken so as to be representative of the material being tested. In the case of containers, specimens shall be cut from unsealed sections, and preferably from unprinted sections, of the container, and shall be taken from each type of container being tested.

5. Procedure

5.1 Determine the tare weight of the oven-dried weighing container as follows: Heat the open container (container, stopper or cover, and wire basket, if any) in the oven at 105 ± 3°C for 1 h. At the end of that period, quickly close the container, remove it from the oven, and place it in the desiccator to cool for 1 h. Remove the closed container from the desiccator, momentarily loosen the cover to equalize the pressure, and then weigh the container.

5.2 Transfer the test specimen to the tared weighing container as soon as the specimen is withdrawn from the sample of material under test, and close the container immediately. Great care must be taken to avoid change in moisture content during the transfer of the specimen from the sample to the container. Handle the specimen with tweezers or clean, dry rubber gloves. If a delay of over a second or two in transferring the specimen to the container is unavoidable, keep the specimen covered on both sides with several adjacent layers of the paper or board from which it is withdrawn, until it can be placed in the container. Then discard the protecting layers of paper or board. Unless the specimen is later to be spread out in the oven, avoid filling the container tightly. Weigh the closed container holding the test specimen, to allow calculation of the original weight of the specimen. The weighing container should not be touched with the fingers during these manipulations.

5.3 Remove large specimens from the container. If the container has a removable basket, leave the specimen in the basket, and place the basket and open container in the oven. If the container does not have a removable basket, spread the

[1] This test method is under the jurisdiction of ASTM Committee D-6 on Paper and Paper Products and is the direct responsibility of Subcommittee D06.92 on Test Methods.

Current edition approved Sept. 29, 1989. Published November 1989. Originally published as D 644 – 55. Last previous edition D 644 – 88.

[2] *Annual Book of ASTM Standards*, Vol 15.09.

specimen in a basket or tray that will permit free circulation of air around the specimen, and place the basket or tray and the open container in the oven. Heat for 2 h at 105 ± 3°C. Replace the specimen in its container and close the container, doing this, if possible, without removing the specimen from the oven. Place the closed container in the desiccator, and allow it to cool for 1 h. Loosen the top of the container momentarily to equalize the pressure, and then weigh the container to allow calculation of the oven-dry weight of the specimen.

5.4 Place small specimens in the drying oven without removing them from the weighing bottle, remove the stopper of the bottle, heat for 1 h at 105 ± 3°C, close the bottle in the oven, cool to room temperature in the desiccator, and weigh. Remove the stopper momentarily just before weighing to equalize the pressure.

5.5 Repeat the required drying and weighing operations until the difference in weight between two successive weighings is not more than 0.1 % of the weight of the specimen.

5.6 Make all weighings to the nearest 1 mg for specimens of the order of 2 g, and weigh to within 0.05 % of the weight of the original specimen for specimens over 2 g.

6. Calculation

6.1 When the percentage of the moisture based on the original weight is required, calculate it as follows:

$$\text{Moisture, \%} = [(W_1 - W_2)/W_1] \times 100$$

where:

W_1 = original weight of the specimen, and
W_2 = weight of the specimen after oven drying.

6.2 When the percentage of moisture based on the oven-dry weight is required, calculate it as follows:

$$\text{Moisture, \%} = [(W_1 - W_2)/W_2] \times 100$$

where:

W_1 = original weight of the specimen, and
W_2 = weight of the specimen after oven drying.

7. Report

7.1 Report the moisture as the percent loss in weight of the specimen to the nearest 0.1 %, on the basis of the original weight (6.1) or on the oven-dry basis (6.2).

8. Precision and Bias

8.1 *Repeatability*—0.35 %

8.2 *Reproducibility*—To be determined. See Appendix X1.

8.3 *Bias*—Unknown.

APPENDIX

(Nonmandatory Information)

X1. Reproducibility

X1.1 Reproducibility or between laboratory variability is a function of how well the technician follows the procedure, especially in 5.2. Some laboratories have reported reproducibility figures as good as the repeatability figures while others have experienced differences as great as 1.5 %.

X1.2 The ambient temperature and humidity will affect the humidity in the oven and, therefore, the moisture determination. For very critical work, the oven should be in a controlled environment.

Standard Method of Conditioning Paper and Paper Products for Testing[1]

This standard is issued under the fixed designation D 685; the number immediately following the designation indicates the year of original adoption or, in the case of revision, the year of last revision. A number in parentheses indicates the year of last reapproval. A superscript epsilon (ε) indicates an editorial change since the last revision or reapproval.

1. Scope

1.1 This method defines the standard atmospheres for normal preconditioning, conditioning, and testing of paper and paper products, paperboard, fiberboard, and containers made from them. It also specifies procedures for handling these materials in order that they may reach equilibrium with the respective atmosphere.

1.2 This method does not include special conditioning and testing atmospheres, such as those that attempt to simulate tropical or arctic environments.

1.3 *This standard does not purport to address the safety problems associated with its use. It is the responsibility of the user of this standard to establish appropriate safety and health practices and determine the applicability of regulatory limitations prior to use.*

2. Referenced Documents

2.1 *ASTM Standards:*

D 585 Method for Sampling and Accepting a Single Lot of Paper, Paperboard, Fiberboard, or Related Product[2]

E 171 Specification for Standard Atmospheres for Conditioning and Testing Materials[2]

2.2 *ISO Standard:*

ISO/TC 125 Enclosures and Conditions for Testing[3]

3. Significance and Use

3.1 The preconditioning requirement is important because physical properties of a sample at 50 % relative humidity depend upon whether the sample is brought to 50 % from a higher or lower relative humidity. A detailed discussion of the importance of preconditioning may be found in Annex A1.

3.2 The conditioning and testing atmospheres are important because both temperature and relative humidity have significant effects on the physical properties of paper and board. A more detailed discussion of the importance may be found in Annex A2.

4. Standard Atmospheres

4.1 *Preconditioning Atmosphere*—10 to 35 % relative humidity and 22 to 40°C (see Annexes).

4.2 *Conditioning Atmosphere*—50.0 ± 2.0 % relative humidity and 23.0 ± 1.0°C.

4.3 *Testing Atmosphere*—Same as 3.2.

NOTE 1—It is important to distinguish between the overall limits of the temperatures within which conditioning and testing may be carried out and the limits within which the temperature must be maintained in order to maintain the specified relative humidity limits; that is, the close temperature tolerance of ±1°C required in 4.2 and 4.3 will not in itself ensure the close relative humidity requirement of ±2 % relative humidity, as a sudden change of 1°C when at 23°C and 50 % relative humidity will change the relative humidity about 3 %.

5. Apparatus

5.1 *Preconditioning Chamber*—A room or cabinet in which sample sheets or specimens may be individually exposed to circulating air at the preconditioning relative humidity and temperature.

NOTE 2—For smaller sheets of specimens, the required preconditioning may be achieved easily with a simple cabinet, if no other means are available. If this cabinet is operated in a room maintained at 50 % relative humidity and 23°C and so designed that room air is drawn through it, and if the air entering and in the cabinet is heated to a temperature of 39 ± 1°C, the relative humidity in the cabinet will be in the required range (that is, 20 ± 3 % relative humidity). Commercially available forced-ventilation "ovens" should prove satisfactory. Input air to the oven should be drawn from the standard room, output should be vented *outside* of the standard room.

5.1.1 For many papers and boards, approximately the same preconditioning moisture content obtained by the above procedure may be achieved by using a sealed cabinet operated in a room maintained at 23 ± 1°C and using a saturated solution of lithium chloride (LiCl) to obtain a relative humidity of about 12 to 13 %.

5.1.2 For larger rooms needed for preconditioning sealed containers and large sheets, the required low relative humidity may be achieved by drawing air over a refrigerator expansion coil operating at a few degrees above freezing temperature.

5.2 *Conditioning and Testing Chamber*—One or more rooms or cabinets in which sample sheets or specimens may be individually exposed to circulating air at the conditioning relative humidity and temperature, and then tested under the same conditions.

NOTE 3—The required relative humidity (50.0 ± 2.0 %) and temperature (23.0 ± 1.0°C) are difficult to achieve and therefore careful attention must be given to the design, evaluation, and maintenance of the "standard room."

5.3 *Hygrometer*—Any instrument that can indicate directly or indirectly the relative humidity of the air with a calibrated accuracy of ±0.5 % relative humidity.

5.4 *Thermometer*—Either the dry bulb of a psychrometer (that is, a hygrometer of the wet- and dry-bulb type) or a

[1] This method is under the jurisdiction of ASTM Committee D-6 on Paper and Paper Products and is the direct responsibility of Subcommittee D06.92 on Test Methods.

Current edition approved Oct. 30, 1987. Published December 1987. Originally published as D 685 – 42 T. Last previous edition D 685 – 73 (1980).

[2] *Annual Book of ASTM Standards,* Vol 15.09.

[3] Available from the American National Standards Institute, 1430 Broadway, New York, NY 10018.

separate thermometer of any convenient design. If a separate thermometer, it should be graduated to 0.2°C or closer with scale errors not exceeding these values. (Psychrometer thermometers must be graduated and correct to 0.1°C or closer and carefully matched to each other to give the required accuracy and repeatability in the measurement of relative humidity.)

6. Procedure

6.1 Obtain and preserve the sample in accordance with Method D 585. In particular, avoid exposure of the sample to direct sunlight, to extremes of temperature, and, to relative humidities above 58 % (see Annex A1). If the sample is to be held for some time before testing, preserve it at a temperature below 25°C and relative humidity below 58 %, but not below 10 % as the paper may curl or cockle and change in other respects.

NOTE 4—It has been reported that low-temperature storage below 10°C slows aging effects. Samples should be protected against condensation.

6.2 Cut the specimens from a sample where critical dimensions are involved only after conditioning the sample. Where there are no critical dimensions, the specimens may be cut before preconditioning or conditioning.

6.3 If shipping containers are to be sealed with aqueous adhesives, make the seal prior to preconditioning.

6.4 Expose the sample sheets or specimens to the preconditioning atmosphere so that both surfaces of single sheets and exterior surfaces of laminated products or sealed containers are freely exposed. This is best achieved by suspending them from overhead or supporting them on a wire grid or rack.

6.5 Precondition the sample sheets or specimens by exposing them as specified in 6.4 to the preconditioning atmosphere. Precondition for a minimum of 24 h, unless a lesser time has been found to give satisfactory results.

NOTE 5—For preconditioning apparatus of ample capacity and air circulation, the following preconditioning times have usually been found satisfactory: single sheets of paper, less than 1 h; liners corrugating medium, chip board, box board, 1 to 2 h; corrugated and solid fiber board in sheet form, 5 to 10 h; sealed boxes and shipping containers, 12 to 16 h; specially treated water vapor resistant papers and boards, 24 h and more.

NOTE 6—If the sample is to be stored for some time after preconditioning, store at a temperature below 25°C and a relative humidity below 40 % but not below 10 %.

6.6 Condition the sample sheets or specimens by exposing them as specified in 6.4 to the standard conditioning atmosphere for a sufficient time for them to come into equilibrium with the atmosphere. Determine that equilibrium has been obtained by weighing the sheets or specimens at time intervals which increase roughly geometrically. Plot the weight against log time (that is, on semilogarithmic paper); the desired equilibrium exists when the plotted curve becomes essentially parallel to the time axis.

NOTE 7—With good air circulation and ample capacity, a conditioning period of 4 h is usually sufficient for paper of ordinary weight and composition. A minimum time of exposure of 5 to 8 h will be required for boards, and unsealed boxes, with a minimum time of 16 h or more for sealed fiberboard boxes, and much longer periods for boards of heavy substances and specially treated water-vapor resistant papers (for example, a minimum of 72 h for wax treated containers).

6.7 Handle the preconditioned and conditioned sheets or specimens as little as possible; especially avoid touching or breathing on test areas.

6.8 Test the specimens in the standard testing atmosphere.

7. Report

7.1 The report shall include the following:

7.1.1 Temperature, relative humidity, and total time of:

7.1.1.1 Preconditioning,

7.1.1.2 Conditioning before testing,

7.1.2 Minimum and maximum temperature and relative humidity during testing, and

7.1.3 Any variations in procedure from this.

8. Precision and Bias

8.1 No statement is made about either the precision or the bias of the method since it merely states the environmental conditions under which paper tests are to be carried out. They could affect the precision and bias of the test methods, but such effects are reflected in the precision and bias statement appearing in the individual test method.

ANNEXES

(Mandatory Information)

A1. IMPORTANCE OF PRECONDITIONING

A1.1 The physical properties of a sample at 50 % relative humidity depend on whether the sample was brought to 50 % from higher or lower relative humidities. This "humidity hysteresis effect" is 5 to 25 % of the test value for many physical properties. For example, a hysteresis effect of 1.5 % moisture content (or 25 % of the test value of 6 % moisture content) is typical. Preconditioning on the dry side within the range specified will avoid most of the hysteresis effect and result in the moisture content of a given sample being established within 0.15 % when the sample is later conditioned to 50 % relative humidity and 23 °C. Conditioning *down* to 50 % gives most papers a moisture content very nearly the same as conditioning *up* to 60 %.

A1.2 For the sake of obtaining close interlaboratory agreement, especially on physical properties, a specified preconditioning procedure is necessary but not always sufficient. While preconditioning practically eliminates the hysteresis effect, it has little influence on strain relaxation effects. The latter depend on the entire previous moisture history of the sample, especially on the conditions of initial drying and tension, and on the duration and degrees of subsequent excursions to high humidities (that is, above about 58 % relative humidity). Consequently, for very close interlaboratory agreement a standardized procedure for handling the sample from manufacture to testing is required.

A1.3 For mill control and similar intra-laboratory purposes, the preconditioning step may often be eliminated. For some properties and materials, preconditioning may not be necessary, either because of the smallness of the humidity hysteresis effect or because of lower test accuracy requirements.

A2. IMPORTANCE OF TEMPERATURE

A2.1 A tolerance of 1.0°C has been adopted in a number of countries. It is recommended along with ±2 % relative humidity, by ASTM (see Specification E 171) and by ISO/TC 125 on Enclosures and Conditions for Testing whenever close tolerances are required, as in paper and board testing. Both temperature and relative humidity have significant effects on the physical properties of paper and board. For some properties of paper and board (for example, MD tensile and CD stretch) a change of 1°C may have nearly as much effect as a change of 2 % relative humidity. For synthetic fibers and plastic laminates the temperature effect may be greater than the relative humidity effect.

Standard Test Method for Bursting Strength of Paper[1]

This standard is issued under the fixed designation D 774; the number immediately following the designation indicates the year of original adoption or, in the case of revision, the year of last revision. A number in parentheses indicates the year of last reapproval. A superscript epsilon (ϵ) indicates an editorial change since the last revision or reapproval.

This test method has been approved for use by agencies of the Department of Defense and for listing in the DoD Index of Specifications and Standards.

1. Scope

1.1 This test method covers measurements of the bursting strength of paper and paper products having a bursting strength of not over 200 points and occurring as single or laminated flat sheets not over 0.025 in. (0.6 mm) in thickness.

1.2 This test method is not intended for use in testing corrugated boxboard, liner board, or hardboards that tend to cut the thin rubber diaphragm.

NOTE 1—A related method is TAPPI Standard T 403 m, as revised, (**1**),[2] Bursting Strength of Paper.

1.3 *This standard does not purport to address the safety problems associated with its use. It is the responsibility of the user of this standard to establish appropriate safety and health practices and determine the applicability of regulatory limitations prior to use.*

2. Referenced Documents

2.1 *ASTM Standards:*

D 585 Method for Sampling and Accepting a Single Lot of Paper, Paperboard, Fiberboard, or Related Product[3]

D 685 Method of Conditioning Paper and Paper Products for Testing[3]

3. Definition

3.1 *bursting strength*—the hydrostatic pressure, in pounds per square inch (or pascals), required to produce rupture of the material when the pressure is applied at a controlled increasing rate through a rubber diaphragm to a circular area 1.200 in. (30.48 mm) in diameter, the area of the material under test being initially flat and held rigidly at the circumference but free to bulge under the increasing pressure during test. To avoid confusion this hydrostatic pressure is referred to as "points bursting strength."

4. Significance and Use

4.1 Bursting strength is widely used as a measure of the resistance to rupture in many kinds of paper. The test is relatively easy and inexpensive to make and appears to simulate some use requirements.

5. Apparatus

5.1 A bursting tester having the following features and auxiliary equipment:

5.1.1 *Clamps*, for firmly and uniformly securing the test specimen between two annular, plane, parallel, and preferably stainless steel surfaces, without slippage during the test. Use sufficient pressure to effect the practicable minimization of slippage.

5.1.1.1 The upper clamping surface (the clamping ring) shall have a circular opening 1.200 ± 0.001 in. (30.48 ± 0.02 mm) in diameter. On a new clamp, the surface that is in contact with the paper during testing shall have a continuous spiral 60° V-groove, not less than 0.010 in. (0.25 mm) deep and of 1⁄32-in. (0.8-mm) pitch, the groove starting at 1⁄8 in. (3.2 mm) from the edge of the opening (Note 2). The circular edge of the opening that is in contact with the paper during testing shall be relieved of sharpness by means of a very fine abrasive cloth (such as crocus cloth), but not rounded off enough to alter effectively the diameter of the opening.

5.1.1.2 The lower clamping surface (the diaphragm plate) shall have a thickness of 0.128 in. (3.25 mm) and an opening 1.302 ± 0.003 in. (33.07 ± 0.08 mm) in diameter. On a new clamp, the surface shall have a series of concentric 60° V-grooves 0.01 in. (0.3 mm) deep, 1⁄32 in. (0.8 mm) apart, the center of the first groove being 1⁄8 in. (3.2 mm) from the edge of the opening (Note 2). The thickness of the plate at the opening is 0.025 in. (0.64 mm). The lower edge which is in contact with the rubber diaphragm shall be rounded to a 1⁄4-in. (6.4-mm) radius, to prevent cutting of the diaphragm when pressure is applied.

5.1.1.3 The clamping ring shall be connected to a clamping mechanism through a swivel joint or other means to ensure an even clamping pressure. During tests, the circular edges of the openings in the two clamping plates shall be concentric to within 0.1 in. (0.3 mm).

NOTE 2—Since the clamping mechanism and clamping surfaces are subject to considerable wear or distortion, they should be examined periodically and repaired or replaced when necessary (**2**). The effectiveness of grooving the clamping surfaces in the manner specified has not been determined.

5.1.2 *Rubber Diaphragm*—A diaphragm of pure gum rubber, 0.034 ± 0.002 in. (0.86 ± 0.05 mm) in thickness, clamped between the lower clamping plate and the rest of the apparatus so that before the diaphragm is stretched by pressure underneath it the center of its upper surface is below the plane of the clamping surface. The pressure required to

---

[1] This test method is under the jurisdiction of ASTM Committee D-6 on Paper and Paper Products and is the direct responsibility of Subcommittee D06.92 on Test Methods.

Current edition approved Aug. 19, 1987. Published October 1987. Originally published as D 774 – 44 T. Last previous edition D 774 – 67 (1971)$^{\epsilon 1}$.

[2] The **boldface** numbers in parentheses refer to the list of references at the end of this test method.

[3] *Annual Book of ASTM Standards*, Vol 15.09.

raise the free surface of the diaphragm 3/8 in. (9.5 mm) above the top surface of the diaphragm plate shall be 4.3 ± 0.8 psi (30 ± 5 kPa). This pressure shall be checked at least once a month. To test, a bridge gage may be used, the test being carried out with the clamping ring removed. The diaphragm should be inspected frequently for permanent distortion and renewed if necessary.

5.1.3 *Hydraulic Pressure System*—A means of applying controlled increasing hydrostatic pressure to the underside of the diaphragm until the specimen bursts through a fluid displaced at the rate of 95 ± 5 mL/min. The fluid is displaced by a piston in the pressure chamber of the apparatus. The recommended chamber fluid is USP chemically pure 96 % glycerin. The hydraulic system, including the gages, shall be mounted so as to be free from externally induced vibrations.

NOTE 3—Ethylene glycol may be substituted for the glycerin if desired **(2)**.

NOTE 4—Since the bursting resistance of paper increases with increased rate of loading, the rate of strain must be maintained effectively constant to obtain reproducible results. Air must be substantially removed, since air present in the hydraulic system of the tester will lower the rate of distortion of the specimen. Air is most commonly trapped under the rubber diaphragm and in the tubes of the gages. A simple method of testing for the presence of allowable quantities of air is given in Section 7. It has also been found important to connect the gage directly to the gage line and not through a manifold.

5.1.4 *Pressure Gage*—A maximum-reading pressure gage of the bourdon type of appropriate capacity and with a graduate circular scale 3¾ in. (95 mm) or more in diameter. The choice and characteristics of the gage are given in Table 1.

6. Calibration of Gage

6.1 Calibrate the gage while it is inclined at the same angle at which it is used, by means of a dead-weight tester of the piston type, or by means of a column of mercury. Such calibration is preferably carried out with the gage in its normal position in the instrument.

6.2 Calibrate gages in frequent use at least once a month. If a gage is accidentally used beyond its capacity, recalibrate it before using it again.

TABLE 1 Gages

NOTE—An appreciable flow of liquid into the gage occurs from the start of the test to the instant of burst. The gage therefore reduces the rate of distension of the specimen by an amount depending upon the expansibility of the gage. When a number of gages are mounted on a single apparatus, care must be taken that only the gage on which the measurement is being made is open to the hydraulic system; otherwise, erroneously low burst pressures will be recorded.

Range of Bursting Strength, points	Range of Gage, psi (kPa)	Scale Graduation Intervals, psi (kPa)	Expansibility[B] of Gage, mL/psi (kPa)
5 to 25	0 to 30 (207)	0.2 (1.4)	0.011 (2)
25 to 50	0 to 60 (414)	0.5 (3.4)	0.007 (1)
50 to 110	0 to 120[A] (827)	1.0 (6.9)	0.004 (0.6)
110 to 200	0 to 300 (2068)	1.0 (6.9)	0.001 (0.1)

[A] The 0 to 120-range gage may be used for any test within its capacity, if so noted in the report.

[B] The expansibility of a gage is defined as the volume of liquid entering the gage tube per unit increase in pressure, when air is absent. It can be determined most conveniently by means of a dilatometer device described in Ref **(4)**. The gage is expansibility is required to be within 15 % of the specified value.

NOTE 5—Maximum-reading pressure gages are subject to dynamic errors in addition to the ordinary static calibration errors. A suitable method of dynamic calibration is described in Ref **(3)**.

6.3 For rough checking purposes, bursting tests on aluminum foil may be used.[4] An alternative procedure is to carry out comparison bursts with a standardized gage.

7. Checking for Air in System

7.1 When first assembled or subsequent to changing gage or diaphragms or adding liquid, check the tester as follows:

7.1.1 Fill the opening from the rubber diaphragm to the level of the clamping surface with water and clamp a rigid metal plate with a sheet of thin rubber to prevent leakage in the position normally occupied by the sample. Turn the shaft of the pump of the hydraulic system by hand until a pressure is just registered on the gage. Then turn the shaft so that 0.4 mL of fluid is displaced.

NOTE 6—In the Mullen tester where a pump shaft rotates at 115 rpm, a displacement of 0.4 mL corresponds to a half turn (180°) of the shaft.

7.1.2 Observe the pressure developed. Pressures less than the following indicate the presence of excessive quantities of air in the system or erroneous gage expansibilities:

Gage Range	Pressure Developed, psi (kPa)
0 to 30	12 (83)
0 to 60	20 (138)
0 to 120	35 (241)
0 to 300	90 (621)

7.1.3 A gradual loss in the pressure obtained indicates a leak in the system; this should be corrected.

8. Sampling

8.1 The sample to be tested shall be obtained in accordance with Method D 585.

9. Test Specimens

9.1 Not less than ten specimens, each at least 2.5 by 2.5 in. (64 by 64 mm), shall be obtained from each test unit of the sample, so as to be representative of the test unit.

10. Conditioning

10.1 Condition the sample in accordance with Method D 685, and make all tests in the same standard atmosphere.

11. Procedure

11.1 Unless the approximate strength of the paper under test is known, make a preliminary test to determine the required capacity of the gage (Table 1). To avoid overloading and possible damage to a gage, begin the preliminary bursting tests with a high-capacity gage.

11.2 Clamp the specimen securely in position, apply the hydrostatic pressure as specified in 5.1.3 until the specimen ruptures, and record the maximum registered by the pressure gage. Watch carefully for any movement (buckling) of the

[4] Standardized foils for this purpose, bursting over the range from 51 to 150 psi (352 to 1034 kPa) may be obtained from the Pulp and Paper Research Institute of Canada, 3420 University St., Montreal, Canada; also from Testing Foil Service, 304 N. Stevens St., Rhinelander, WI 54501.

unclamped margin of the specimen. If slippage is indicated, discard the test and increase the clamping pressure for subsequent tests.

11.3 Make at least ten acceptable tests, applying an equal number to each side of the paper. Make no tests on areas containing watermarks, creases, imperfections, or visible damage.

11.4 After each test, return the indicator needle of the gage gently to zero.

12. Report

12.1 The report shall include the following:

12.1.1 Bursting strength in points as the arithmetical mean of the tests, corrected for any gage error, to three significant figures,

12.1.2 Number of tests made, and

12.1.3 Maximum and minimum values of acceptable tests.

13. Precision and Bias

13.1 *Repeatability* (*Within a Laboratory*)—5.4 %.

13.2 *Comparability* (*Between Materials*)—9.5 %.

13.3 *Reproducibility* (*Between Laboratories*)—14.3 %.

13.4 *Bias*—This test method has no bias because the property "bursting strength" can be defined only in terms of a test method.

REFERENCES

(1) "Bursting Strength of Paper—T 403 m," *Tappi*, TAPPA, Vol 45, No. 10, October 1962, p. 144A.

(2) Aldrich, L. C., "Standardization of Model C Mullen Testers in a Multi-Mill Company," *Tappi*, TAPPA, Vol 43, No. 5, May 1960.

(3) Tuck, N. G. M., Faichney, L. M., and Mason, S. G., *The Dynamic Calibration of Maximum-Reading Pressure Gages*, (to be published).

Standard Test Method for Drop Test for Loaded Boxes[1]

This standard is issued under the fixed designation D 775; the number immediately following the designation indicates the year of original adoption or, in the case of revision, the year of last revision. A number in parentheses indicates the year of last reapproval. A superscript epsilon (ε) indicates an editorial change since the last revision or reapproval.

1. Scope

1.1 This test method covers four procedures for the drop testing of loaded boxes.

1.2 *This standard does not purport to address the safety problems associated with its use. It is the responsibility of whoever uses this standard to consult and establish appropriate safety and health practices and determine the applicability of regulatory limitations prior to use.*

2. Referenced Documents

2.1 *ASTM Standards:*

D 685 Method of Conditioning Paper and Paper Products for Testing[2]

D 880 Method of Incline Impact Test for Shipping Containers[2]

D 997 Test Method for Drop Test for Loaded Cylindrical Containers[2]

D 1083 Test Methods for Mechanical Handling of Unitized Loads and Large Shipping Cases and Crates[2]

E 122 Practice for Choice of Sample Size to Estimate a Measure of Quality of a Lot or Process[3]

3. Significance and Use

3.1 This test method is used to provide an indication of the ability of a box to withstand the damage which can be caused by the sudden shock induced by dropping a package. It simulates those shocks that are likely to occur in the handling of packages through a variety of distributions cycles and the ability of the box to protect the contents against these shocks. The test may also permit observation of progressive failure of the box and damage to the contents.

3.2 Drop tests are performed with either of the two following objectives (see Annex A1), either objective being of value for comparing the characteristics of a package design to a standard or for comparing differences in package construction:

3.2.1 *Objective A*—To measure the ability of the shipping container to withstand rough handling.

3.2.2 *Objective B*—To measure the ability of the package (shipping container and interior packing materials) to protect its contents.

3.3 These procedures are also particularly suitable for boxes that are normally handled manually. Boxes of such bulk or mass that they cannot be handled manually may be more satisfactorily tested in accordance with Method D 880 and Test Method D 1083.

4. Apparatus

4.1 *Drop Test Apparatus*—The drop test equipment shall conform to the following requirements:

4.1.1 Permit the box to be placed in a position, prior to release, that will assure correct orientation, within 2° for flat-face drops and 10° for edge and corner drops upon impact.

4.1.2 Permit accurate control of the drop from specified heights.

4.1.3 Utilize lifting devices that will not damage the test member.

4.1.4 Provide a release mechanism that does not impart rotational or side-wise forces to the test package. If drop leaves are used, the apparatus shall provide a spring or other mechanism so that the leaves do not interfere with a free, unobstructed fall.

4.1.5 Provide a rigid and level dropping surface that is integral with a mass that shall be at least 50 times the weight of the dropping container. Neither the depth nor width of the mass shall be less than half the length. The dropping surface, firmly anchored to the mass, shall be a steel plate not less than ½ in. (13 mm) thick.

4.1.6 A test unit for calibration purposes should be constructed. It shall be of the same general size and mass as maximum size and mass packages to be tested, and shall be dropped from heights in the expected range. The center or gravity of the unit shall be at its geometric center. When this unit is dropped, it must strike the impact surface within the specified limits. Compliance with the specifications may conveniently be determined with a camera.

4.1.7 For very large packages it may be necessary to use a hoist, sling, and tripping devices or solenoid operated drop test mechanism and suspension devices as shown in the Appendix of Test Method D 997.

4.2 *Conditioning Apparatus*—Adequate facilities shall be provided for conditioning test specimens at proper humidity and temperature prior to test in accordance with the specifications covering the containers to be tested.

5. Sampling

5.1 Determine the number of specimens to be tested in accordance with Recommended Practice E 122.

---

[1] This test method is under the jurisdiction of ASTM Committee D-10 on Packaging and is the direct responsibility of Subcommittee D10.22 on Handling and Transportation.

Current edition approved March 3, 1980. Published May 1980. Originally published as D 775 – 44 T. Last previous edition D 775 – 61 (1973).

[2] *Annual Book of ASTM Standards*, Vol 15.09.

[3] *Annual Book of ASTM Standards*, Vol 14.02.

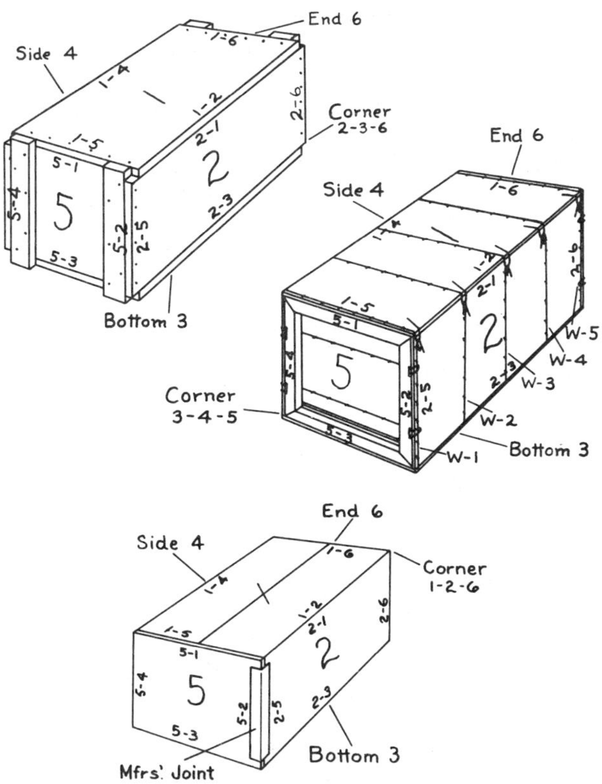

FIG. 1 Diagram Showing Identification of Faces, Edges, and Corners of Containers

6. Test Specimens

6.1 When the protective ability of the package is to be evaluated, pack the package with the actual contents for which it was designed (Note 1). When the ability of the box to withstand rough handling is to be evaluated, pack the box with either the actual contents or with a load simulating the contents. Regardless of which procedure is used, close and strap the container in the same manner that will be used in preparing it for shipment.

NOTE 1—Where the use of actual contents is prohibitive because of excessive cost or danger, a "dummy" load simulating the contents with respect to dimensions, center of gravity, moment of inertia, etc., may be used with accelerometers or other indicating mechanisms.

6.2 The procedure for identification of the faces, edges, and corners of box shall be as follows (see Fig. 1): Facing one end of the box with the manufacturer's joint on the observer's right if the container is fiberboard, designate the top of the box as *1*, the right side as *2*, the bottom as *3*, the left side as *4*, the near end as *5*, and the far end as *6*. Identify the edges by the numbers of the two faces that form that edge: for example, *1-2* identifies the edged formed by the top and the right side, and *2-5* the edge formed by the right side and near end. (The latter is the edge having the manufacturer's joint, in fiberboard container.) Identify the corners by the numbers of the three faces that meet to form that corner; for example, *1-2-6* identifies the corner where the top, the right side, and the far end meet.

7. Conditioning

7.1 Depending on the purpose of the tests, the boxes may be conditioned prior to the drop test by either a different physical test, water immersion, exposure to water spray, or exposure to standard or other fixed conditions of air temperature or humidity. Fiberboard boxes shall be conditioned in accordance with Method D 685.

8. Procedure

8.1 Perform the tests in the conditioned atmosphere or immediately upon removal from that atmosphere.

8.2 Drop test procedures may be conducted by dropping the package either on one member of the box or on several different members in a prescribed sequence (cycle testing). A member is defined as a face, corner, or edge of the box.

8.2.1 When the box is to be dropped on a face, position it so that upon impact, there is no more than a 2° angle between the plane of this face and the dropping surface.

8.2.2 When the box is to be dropped on an edge, position it so that upon impact, this edge makes no more than 2° angle with the dropping surface, and the plane containing this edge and the center of gravity of the box makes no more than a 10° angle with the vertical.

8.2.3 When the box is to be dropped on a corner, position it so that upon impact, the line containing this corner and the center of gravity of the box makes no more than a 10° angle with the vertical.

8.3 *Choice of End Point*—In conducting drop tests, use either of two types of end points: (*1*) continue the test until obvious damage occurs, or (*2*) conduct the test at a specified number of drops or cycles followed by examination of the resulting damage to the contents or the box (depending upon which objective, A or B, is being considered (see Annex A1)).

8.4 *Height of Drop*—The height from which the box shall be dropped will depend upon the purpose of the test. In some instances the height and number of drops will be prescribed in the specifications for the box. For end points based upon obvious damage, choose the height of drop as a height judged to be sufficient to cause failure to any of the boxes of a comparison series within a reasonable and convenient number of falls, but also that will not cause failure to any of the samples in less than five falls. If the approximate drop resistance of the boxes is unknown, determine this height by preliminary tests. If the end point is a prescribed number of drops, then choose the height of drop by consideration of the expected shipment hazards. In the single zone drop procedure (8.10) the height of drop is determined, as the test progresses, from previous drops.

8.5 In a comparison series where a number of different samples are to be compared use the following procedure: Test one of the specimens of the first sample, chosen at random, followed by the testing of a randomly chosen specimen of each of the other samples until one specimen from each sample has been tested. Follow the same procedure with the remaining specimens of the various samples.

8.6 After the purpose of the test has been designated, select the procedure from those in 8.7, 8.8, 8.9, or 8.10 and after consulting the outline in Annex A1 to see which procedures are applicable. The outline will serve as a standard code for completely describing the method chosen.

8.7 *Constant-Height Drop Test*—The procedure for the constant-height drop test consists of repeatedly dropping the test package from one higher until the end point of the test is reached.

8.7.1 Properly position the first specimen in the drop tester at the chosen height 8.4 and drop upon one member or upon several members in a cycle depending upon previous choice (8.3) (Annex A1, Note A1.1).

8.8 *Progressive–Height Drop Test*—The procedure for the progressive-height drop test consists of dropping the test package a specified number of drops (or a specified cycle) from several progressively increasing heights until the end point is reached.

8.8.1 Properly position the first specimen in the drop tester at the chosen first height of drop (8.4). Drop the specimen repeatedly from this height for the required number of drops, or through the required cycle of drops (8.3) prior to increasing the height to the second level. Then increase the height of drop by an increment previously chosen, at which height the specimen shall be subjected to the same number of drops or the same cycle of drops as given at the first height. Continue the drops or cycles at progressively increasing heights until the end point is reached. (Annex A1, Notes A1.2 and A1.3).

8.9 *Safe-Drop Height-Infinite Number of Drops*—This procedure consists essentially of repeated applications of the constant-height drop test procedure (8.7) with the difference that a different height of drop shall be chosen for each specimen of the sample, which height shall be dependent on the results of the testing of the previous specimens. The height of drop shall be determined that will not cause failure until the specimen package has been subjected to a very large number of drops. Then the data obtained in this manner shall be plotted on a graph with the number-of-drops (or cycles) to failure as abscissa and corresponding height-of-drop as ordinate. This plot will give a curve that can be considered to approximate a hyperbola. This ordinate of the horizontal asymptote of this curve can be interpreted as representing the height of drop below which failure will not occur after an infinite number of drops. In practice, this can be estimated from inspection of the curve. This procedure may be used with many methods outlined in Annex A1 which have obvious damage end points.

8.10 *Single-Drop Test*—The single-drop test consists of finding the failure zone determined by the lowest height of drop that will cause failure in a single drop and the greatest height of drop that will not cause failure in a single drop. The average of these two heights is defined as the mean drop height of the failure zone. The safe single-drop height is the height immediately below the failure zone. This procedure may be used whenever an end point of obvious damage is applicable.

8.10.1 To determine the mean drop height of the failure zone and the safe single-drop height, properly position the first specimen in the drop tester at a height estimated to cause failure on the first drop (8.4). If the specimen does not fail as a result of the drop from the selected height or fall, drop the second specimen from an arbitrarily chosen greater height. If the second specimen fails as a result of the single drop from the greater height, select an intermediate height for the drop of specimen No. 3. Depending on whether the third specimen does or does not fail when dropped from the intermediate height, either decrease or increase the intermediate height for the drop of the fourth specimen. Likewise, the choice of the height of drop for each subsequent test depends upon the results of the preceding drop. Thus, with a few specimens, a narrow zone of height can be established from which some specimens of a specific sample will fail while others will not fail. The width of the failure zone depends to a large extent on the total variability. Determine the mean drop height of the failure zone by taking the average of the lowest height of drop that caused failure in a single drop and the greatest height of drop that did not cause failure in a single drop. The safe single-drop height is the height immediately below the failure zone.

8.10.2 This procedure may be used to measure the protective ability of a package, in which case it is coded (see A1.2.1.1 (4)) or, it may be used to measure the ability of the container to withstand rough handling in which case it will be coded as item A1.1.1.1 (4).

9. Report

9.1 The report shall include the following:

9.1.1 The dimensions of the box under test; its complete structural specifications; kinds of materials; description and specifications for blocking and cushioning, if used; spacing, size and kind of fasteners; method of closing and strapping, if any; and the tare and gross masses.

9.1.2 A description of the contents of the box under test,

9.1.3 Description of apparatus and special instrumentation, if used,

9.1.4 The procedure used (coded as in Annex A1),

9.1.5 A description of prescribed sequence, if used,

9.1.6 A height of drop, if applicable,

9.1.7 Size of the increment if the progressive height drop test procedure (8.8) is used,

9.1.8 Number of drops (at each incremental height, if applicable),

9.1.9 Number of specimens tested per sample,

9.1.10 Detailed record of test on each box including damage to the box and contents, together with any other observation which may assist in correctly interpreting the results or aid in improving the design of the box or the method of packing, blocking, or bracing,

9.1.11 Method, if any, of conditioning the box the moisture content, if determined, of the wood, plywood, or fiberboard; and the results of supplementary tests of the materials from which the box is made, and

9.1.12 Statement to the effect that all tests were made in full compliance with this test method.

10. Precision and Bias

10.1 The drop test is subject to a high degree of variation, the variation being due to the nonhomogeneous nature of the box and possibly the contents as well (glass bottles for example). Therefore, the reproducibility may not be expected to lie between generally accepted limits.

10.2 In addition to generally indicating the ability of a package to withstand the rigors of a shock environment, the test has great value in comparing the performance of similar packages and is therefore a valuable tool for package design.

ANNEXES

(Mandatory Information)

A1. DROP TEST METHODS AND PROCEDURES BY CODE

A1.1 *Objective A—To Measure Ability of a Box to Withstand Rough Handling*

A1.1.1 *Dropping On Only One Member:*

A1.1.1.1 *End Point No. 1*, Obvious Damage of Box:

(*1*) Constant-height drop test procedure (8.7),

(*2*) Progressive-height drop test procedure (8.8),

(*3*) Safe-drop height-infinite number of drops procedure (8.9), and

(*4*) Single-drop test procedure (8.10).

A1.1.1.2 *End Point No. 2*, Prescribed Number of Drops:

(*1*) Constant-height drop test procedure (8.7) and

(*2*) Progressive-height drop test procedure (8.8).

A1.1.2 *Dropping on Several Members in Prescribed Sequence:*

A1.1.2.1 *End Point No. 1*, Obvious Damage of Box:

(*1*) Constant-height drop test procedure (8.7),

(*2*) Progressive-height drop test procedure (8.8) (Note A1.1), and

(*3*) Safe-drop height-infinite number of drops procedure (9).

A1.1.2.2 *End Point No. 2*, Prescribed Number of Drops:

(*1*) Constant-height drop test procedure (8.7) and

(*2*) Progressive height drop test procedure (8.8).

A1.2 *Objective B—To Measure Protective Ability*

A1.2.1 *Dropping On Only One Member:*

A1.2.1.1 *End Point No. 1*, Obvious Damage of Contents:

(*1*) Constant-height drop test procedures (8.7),

(*2*) Progressive-height drop test procedures (8.8) (Note A1.2),

(*3*) Safe-drop height-infinite number of drops procedure (8.9), and

(*4*) Single-drop test procedure (8.10).

A1.2.1.2 *End Point No. 2*, Prescribed number of drops:

(*1*) Constant-height drop test procedure (8.7) and

(*2*) Progressive-height drop test procedure (8.8).

A1.2.2 *Dropping On Several Members in Prescribed Sequence:*

A1.2.2.1 *End Point No. 1*, Obvious Damage of Contents:

(*1*) Constant-height drop test procedure (8.7),

(*2*) Progressive-height drop test procedure (8.8), and

(*3*) Safe-drop height-infinite number of drops procedure (8.9).

A1.2.2.2 *End Point No. 2*, Prescribed Number of Drops:

(*1*) Constant-height drop test procedure (8.7) (Note A1.3) and

(*2*) Progressive-height drop test procedure (8.8).

NOTE A1.1—Another method for the progressive-height drop test procedure, A1.1.2.1.(*2*) is as follows: In this procedure, several members of the package are under text in that the package is dropped upon all corners. Drop the package cornerwise from a height of 12 in. (305 mm). In the following sequence of drops: *5-1-2, 6-3-4, 5-2-3, 6-1-4, 5-3-4, 6-1-2, 5-1-4*, and *6-2-3*. These eight cornerwise drops shall constitute one cycle. Make additional cycles at heights increasing in 3-in. (76-mm) increments until failure of the box results. Failure, depending upon the purpose of the test, may be considered to have occurred when: (*1*) the contents are exposed, or (*2*) the contents spill from the box, or (*3*) one edge breaks open along its entire length, or (*4*) some other predetermined type of damage has occurred.

NOTE A1.2—One of the methods following the progressively increasing height of drop test procedure A1.2.1.1.(*2*) is as follows: Drop the packages upon their bottom surfaces for ten consecutive drops from a height of 1 ft (0.3 m) in the drop tester. Then increase the height by an increment of 3 in. (76 mm) to give a height of drop of 15 in. (381 mm). After ten drops at this height, increase the height by an increment of 3 in., where ten additional drops shall be given, etc., until obvious damage to the contents results. An alternative of this method is to drop the package once (instead of ten times) from each increment height.

NOTE A1.3—An example of the constant height procedure for A1.2.2.2.(*1*) is as follows: Drop the packages from a height of 12 in. (305 mm) in the following sequence, which constitutes the first cycle of ten drops:

(*1*) A corner drop on the *5-1-2* corner,

(*2*) An edge drop on the shortest edge radiating from that corner,

(*3*) An edge drop on the next shortest edge radiating from that corner,

(*4*) An edge drop on the longest edge radiating from that corner,

(*5*) A flatwise drop on one of the smallest faces,

(*6*) A flatwise drop on the opposite smallest face,

(*7*) A flatwise drop on one of the medium faces,

(*8*) A flatwise drop on the opposite medium faces,

(*9*) A flatwise drop on one of the largest faces, and

(*10*) A flatwise drop on the opposite large face.

Start the second cycle of ten drops with a drop on the *6-3-4* corner, which is diagonally opposite to the corner on which the first drop was made, and complete in the sequence prescribed above. Start the succeeding cycles of ten drops each at the following corners and complete in the same sequence:

Cycle	Corner
Third	*5-2-3*
Fourth	*6-1-4*
Fifth	*5-3-4*
Sixth	*6-1-2*
Seventh	*5-1-4*
Eighth	*6-2-3*

Standard Test Method for Tensile Breaking Strength of Paper and Paperboard[1]

This standard is issued under the fixed designation D 828; the number immediately following the designation indicates the year of original adoption or, in the case of revision, the year of last revision. A number in parentheses indicates the year of last reapproval. A superscript epsilon (ϵ) indicates an editorial change since the last revision or reapproval.

1. Scope

1.1 This test method covers the determination of the tensile breaking strength of papers and paperboards. Related methods include Test Methods D 829 and D 987.

1.2 *This standard does not purport to address the safety problems associated with its use. It is the responsibility of the user of this standard to establish appropriate safety and health practices and determine the applicability of regulatory limitations prior to use.*

2. Referenced Documents

2.1 *ASTM Standards:*

D 585 Method for Sampling and Accepting a Single Lot of Paper, Paperboard, Fiberboard, of Related Product[2]

D 685 Method of Conditioning Paper and Paper Products for Testing[2]

D 829 Test Method for Wet Tensile Breaking Strength of Paper and Paper Products[3]

D 987 Test Method for Stretch of Paper and Paper Products Under Tension[4]

3. Significance and Use

3.1 Tensile strength is a fundamental property and belongs to that group of properties associated with paper manufacture since it is influenced by the kind, quality, and treatment of the fiber constituents and by the way the sheet has been formed on the paper machine. This test method is indicative of the serviceability of many grades such as wrapping, bag, gummed tape, and cable wrapping, which are subjected to direct tensile stresses. Tensile strength measurements indicate the potential resistance to breaking when the paper web is subjected to strains during travel from the roll through the web press mechanism in the process of printing.

4. Apparatus

4.1 *Tensile Testing Machine,* having the following characteristics:

4.1.1 Two clamps from ½ to 2.0 in. (12.7 to 51 mm) wide (see Section 6 for preferred widths) with clamping surfaces in the same plane parallel to the direction of motion of the applied stress and so aligned that they hold the test specimen in that plane throughout the test without slippage. At the start of the test the edges of the jaws of the clamps are set apart at $7.1^{+1.0}_{-0.2}$ in. (180^{+25}_{-5} mm). (See Explanatory Notes at end of this test method.)

NOTE 1—Although many machines are arranged for an 8-in. (203-mm) jaw separation distance, especially for paperboard and coarse papers, the preferred jaw separation at the start of the test is 180 mm in all cases. If necessary, it is usually not difficult to have the machine changed accordingly.

4.1.2 Means for applying a smoothly increasing load to the test specimen until it breaks, the increase being such that additional load applied each second is not different by more than 10 % from the additional load applied in the previous second.

NOTE 2—This condition is fulfilled by most motor-driven tensile breaking strength testers, including the incline plane testers, temperature-compensated spring testers, the usual pendulum type of apparatus, and the constant-rate-of-elongation testers, provided that the slope of the stress-strain curve does not change abruptly.

4.1.3 Means of indicating the applied load at the instant of rupture, to within ±1 %.

5. Calibration

5.1 Check the apparatus for cleanliness. Then level it accurately in each of the principal directions. Check all the moving parts to ensure that they move freely. Apply various dead weights to the clamp actuating the indicating mechanism and note the scale readings when the weights and mechanism come gently into an equilibrium position from the direction and at the approximate speed used in an actual test. In general, each range of the machine should be verified at three or four widely spaced points, and if errors greater than 1 % are found, a correction curve should be constructed. Verify the apparatus within 1 month prior to any test.

NOTE 3—When calibrating the pendulum type of tester, first determine the effect of the friction of the pawls, as follows:

Procedure A—With the pawls engaged as in normal use, allow the calibration weight to exert its force by lowering it gradually by hand at about the speed used in an actual test. Do not drop it nor lower it too slowly.

Procedure B—With the pawls wedged up with a small piece of paper bent double, gently apply the same calibration weight at the speed to be used and allow the pendulum to come to equilibrium from the same direction as in an actual test.

If the reading by Procedure A is measurably less than by Procedure B, then the friction of the pawls is the more important; therefore calibrate the scale by Procedure A, but first remove any noticeable friction in the pawls. If the reading by Procedure A is greater than by Procedure B, the friction is less significant than the inertia; therefore calibrate by Procedure B.

---

[1] This test method is under the jurisdiction of ASTM Committee D-6 on Paper and Paper Products and is the direct responsibility of Subcommittee D06.92 on Test Methods.

Current edition approved Oct. 30, 1987. Published December 1987. Originally published as D 828 – 45 T. Last previous edition D 828 – 60 (1971)$^{\epsilon 1}$.

[2] *Annual Book of ASTM Standards,* Vol 15.09.

[3] Discontinued, see *1983 Annual Book of ASTM Standards,* Vol 15.09.

[4] Discontinued, see *1968 Annual Book of ASTM Standards,* Part 15.

With other types of testers, follow the manufacturer's instructions and the general instructions given above.

6. Test Specimen

6.1 Obtain the samples and condition them in accordance with Methods D 585 and D 685, respectively. For each principal direction to be tested, cut at least ten specimens with clean and parallel edges to a width within 0.1 mm of the nominal width, and over 200 mm, preferably 9.9 in. (250 mm) long. Avoid abnormalities, water-marks, creases, and wrinkles.

6.2 For paper, cut the specimens preferably 0.59 in. (15.0 mm) wide. For coarse papers, such as building papers and paperboards over 0.12 in. (3 mm) thick, cut them preferably 1.00 in. (25.4 mm) wide. In either case, do not cut the test specimens less than 0.5 in. (13 mm) nor more than 2.0 in. (51 mm) wide.

NOTE 4—Varying the width of the test specimen between 0.5 and 2 in. (13 and 51 mm) does not, in general, make much difference in the proportionate test results, except for unbeaten long-fiber papers when the difference may be appreciable.

7. Procedure

7.1 Make the test in an atmosphere conditioned in accordance with Method D 685.

7.2 Avoid touching the portions of the specimens that will be between the jaws. Tightly clamp one end of a test specimen in the upper jaw after placing the specimen loosely in the lower jaw and checking its alignment. Then tightly clamp the lower end of the specimen and apply the load.

NOTE 5—Except for a referee test, time may be saved by aligning and clamping the ends of the ten specimens together in the upper jaw and breaking them one at a time, after clamping the lower end of each in the lower jaw.

7.3 Operate the tester so that the *average* time for the completion of the test will be within 10 ± 5 s. The appropriate setting of the tester may be determined from a trial test specimen, or a table may be prepared giving the approximate settings for a particular type of tester for each width of specimen and its expected breaking strength and elongation.

NOTE 6—In cases of dispute, the average time to fracture should be 10 ± 2 s. Normally the rate of loading is not a very important factor; doubling it usually results in an increase of only about 2 %.

7.4 Reject readings from individual specimens if the specimen slips or breaks in or at the edge of the clamps.

7.5 Record the result of each individual breaking load to the nearest two or three significant figures.

7.6 Test at least ten specimens cut in both principal directions of the paper, unless the strength in one direction only is required.

7.7 Compute the average breaking load and, preferably, also the standard deviation for each direction tested.

8. Report

8.1 Report the results obtained on specimens cut in the machine direction as the tensile strength, machine direction, and on specimens cut in the cross direction as the tensile strength, cross direction.

8.2 Report the average value of the breaking load calculated either to kilograms-force per 15 mm or to pounds-force per inch to three significant figures (kgf/15 mm × 3.73 = lbf/in.).

8.3 A complete report requires in addition, statements of the following:

8.3.1 Width of the test specimen,

8.3.2 Distance between the jaws at the start of the test,

8.3.3 Average value of the "time to break" for each principal direction tested,

8.3.4 Number of specimens tested in each direction,

8.3.5 Standard deviation for each direction tested, and

8.3.6 Type of testing machine and speed of operation of the driven head.

NOTE 7—In a pendulum tester, the speed depends on the scale used and the leverage of the pendulum as well as the rate of movement of the stressing jaw. The speed of operation may be expressed as the average time to break, or as the average increase of load per second, but preferably, if available, as the average extension per second.

9. Precision and Bias

9.1 *Repeatability (Within-Laboratory)*—4.1 %.

9.2 *Reproducibility (Between Laboratories)*—12.8 %.

9.3 *Bias*—Unknown.

9.4 These values are based on data and reports 25 through 70 of the TAPPI CRP. Tensile strength precision is based on data from 42 laboratories and 5 different materials.

REFERENCES

(1) Carson, F. T., "Some Notes on Revision of Methods for Measuring the Strength of Paper," *Paper Trade Journal,* PTJOA, Vol 102, No. 19, May 7, 1936, pp. 39–42; *Technical Association Papers,* TAPAA, Vol 19, 1936, pp. 172–175.

(2) Clark, J. d'A. "Some Observations on Burst, Tensile, and Stretch Tests," *Paper Trade Journal,* PTJOA, Vol 102, No. 2, Jan 9, 1936, p. 40-A; *Technical Association Papers,* TAPAA, Vol 19, 1936, pp. 264–266; discussion, *ibid,* Vol 19, 1936, pp. 90–94.

(3) Griffin, R. C., and McKinley, R. W., "Studies of Tensile and Bursting Tests," *Paper Trade Journal,* PTJOA, Vol 102, No. 2, Jan 9, 1936, pp. 34–35; *Technical Association Papers,* TAPAA, Vol 19, 1936, pp. 222–223.

(4) Institute of Paper Chemistry, "Instrumentation Studies XL. Measurement of the Tensile Strength of Paper. XLI. The Scott Serigraph, Model J-3. XLII. The Amthor Universal Tensile Strength Tester. XLIII. The Schopper electrically driven strength tester," *Paper Trade Journal,* PTJOA:
Vol 115, No. 5, July 30, 1942, pp. 12–18
Vol 115, No. 7, Aug. 13, 1942, pp. 13–22
Vol 115, No. 19, Nov. 5, 1942, pp. 18–22
Vol 115, No. 26, Dec. 24, 1942, pp. 14–21
Vol 116, No. 6, Feb. 11, 1943, pp. 10–12
Vol 116, No. 23, June 10, 1943, pp. 13–14

(5) Oliver, C. V., "Variability of Results Met With in Paper Strength Testing," *Proceedings Technical Section,* PMAIA, Paper Markers' Assoc. Gt. Britain and Ireland, Vol 11, 1930, pp. 53–84.

(6) Paper Makers' Assoc. Gt. Britain and Ireland, Tech. Section, Paper Testing Committee, First Report, London, the Association, 1937, 85 pages: *Proceedings Technical Section,* PMAIA, Vol 18, Part 1A, 1937; *World's Paper Trade Review,* WPTRA, Technical Convention Number, March 1937, pp. 4–72.

(7) Hindman, H., and Burr, G. S., "The Instron Tensile Tester," *Transactions*, ASME, TASMA, October 1949, p. 798; G. S. Burr, "Servo-Controlled Tensile Strength Tester," *Electronics*, ELECA, Vol 22, No. 5, May 1949, pp. 101–105.

(8) Van den Akker, J. A., and Hardacker, K. W., "Instrument Studies. LXXXI. The Automatic Recording of the Load Elongation Characteristic of Paper. III. The Table Model Instron (Universal Testing Instrument)," *Tappi*, TAPPA, Vol 41, No. 8, August 1958, pp. 224-A to 231-A.

EXPLANATORY NOTES

NOTE 1—If a lesser distance between the clamps is employed, a higher test result is usual. For papers that have a poor formation, the difference between 100 mm and the standard distance of 180 mm, may amount to over 10 %. Where a length of specimen of 180 mm is not possible, the machine may be adapted to 6 in. (150 mm), 4 in. (100 mm), or 2 in. (50 mm). If the specimens are too short for the tester, they may be extended by means of strips of strong, gummed kraft. In any of these deviations from standard, however, the fact should be recorded in a prominent position in the report.

NOTE 2—The jaw separation distance of 100 mm is standard for pulp test sheets. The results are calculated to breaking length in metres, based on the moisture-free weight of the test sheets. See Standard T 202 m-53 of the Technical Association of Pulp and Paper Industry, entitled "Physical Testing of Pulp Handsheets."

NOTE 3—The main change in this revision is to replace the requirement for the rate of loading from a rather complicated procedure to the same time of 10 ± 5 s for all papers and paperboards.

Standard Method of Incline Impact Test for Shipping Containers[1]

This standard is issued under the fixed designation D 880; the number immediately following the designation indicates the year of original adoption or, in the case of revision, the year of last revision. A number in parentheses indicates the year of last reapproval. A superscript epsilon (ϵ) indicates an editorial change since the last revision or reapproval.

1. Scope

1.1 This method covers three procedures for making incline impact tests on loaded shipping containers, as follows:

1.1.1 *Procedure A*, to determine the ability of a container to withstand impact stresses, and

1.1.2 *Procedure B*, to determine the ability of a container or interior packing, or both, to provide protection to the contents, when subjected to impact stresses, and

1.1.3 *Procedure C*, to simulate the types of shock pulses experienced by lading in railcar switching, with the use of a rigid bulkhead on the front edge of the dolly, and programming techniques. Procedure C may also be used to simulate compressive forces experienced by lading during railcar switching with the use of backloading.

1.2 Any of the three procedures may be used to obtain average results permitting a comparison of different designs of containers of the same size and carrying the same load. These procedures are suitable for testing various types of containers such as boxes, crates, barrels, drums, kegs, or pails made of metal, wood, fiberboard, or combinations of these materials.

1.3 The procedures prescribed in this method are particularly suitable for testing large or heavily loaded containers.

1.4 *This standard does not purport to address the safety problems associated with its use. It is the responsibility of the user of this standard to establish appropriate safety and health practices and determine the applicability of regulatory limitations prior to use.*

2. Referenced Documents

2.1 *ASTM Standards:*

D 644 Test Method for Moisture Content of Paper and Paperboard by Oven Drying[2]

D 4003 Methods of Controlled Horizontal Impact Test for Shipping Containers[2]

3. Significance and Use

3.1 The incline impact test has been found useful in laboratory simulation of shocks such as might be encountered in handling and transportation. The test results within a given laboratory using this method have been found to be valid, particularly in situations where one package is being compared with another. However, it has been found that interlaboratory correlation of test results has been poor. This is due in large part, to variation in design and construction of the test apparatus, particularly the backstop, which has often not complied with the description in this method. Therefore, the apparatus and instruments must be carefully checked against each other when interlaboratory data are to be compared. Although the test does not provide fundamental data, it is useful for evaluating and comparing packages.

4. Apparatus

4.1 *Inclined Track, Backstop, and Dolly*—The principal apparatus consists of a two-rail steel track inclined 10° from the horizontal, a rolling carriage or dolly, and a rigid backstop. The backstop shall be a barrier with a face made of group 4 woods,[3] of sufficient size to permit full contact with the shipping container, constructed at the bottom of the incline, with the plane of the face perpendicular to the direction of movement of the dolly. The backstop may, if desired, be equipped with removable hazards such as boards or timbers, which can be so placed as to contact the container at the time of impact at any desired position between top and bottom edges of the container. The track shall accommodate the flatbed rolling carriage or dolly which is equipped with steel wheels and a renewable face made of group 4 woods or group 4 plywood. The wood faces of the carriage and the backstop shall be maintained free of prominent projections which may affect the test results, such as bolt or nail heads, scores, abrasions, and splits. The track shall be clean and the wheels well lubricated. The incline shall be graduated in suitable length increments of 6 in. (150 mm) and may be equipped with an automatic counting device which records each passage of the dolly to the lower end of the track. The apparatus may also have a cable and winch, or pulleys, to aid in bringing the dolly to the elevated end of the track and an automatic tripping device for releasing the dolly from a predetermined point on the incline. A typical apparatus[4] is shown in Figs. 1, 2, 3 and 4. The wheels shall be not less than 3 in. (76 mm) in diameter.

4.2 *Dolly Bulkhead* (*used for Procedure C only*)—Dolly shall be fitted with a bulkhead at a 90° angle to the dolly top surface. Bulkhead shall be so fitted that upon release the bulkhead impacts first on the programming material, or the bulkhead and dolly impact at the same time. The bulkhead shall be securely attached, and shall be a rigid structure.

---

[1] This method is under the jurisdiction of ASTM Committee D-10 on Packaging and is the direct responsibility of Subcommittee D10.22 on Handling and Transportation.

Current edition approved Feb. 28, 1986. Published April 1986. Originally published as D 880 – 68 (1973). Last previous edition D 880 – 79.

[2] *Annual Book of ASTM Standards*, Vol 15.09.

[3] Grouping as established by the U. S. Forest Products Laboratory.

[4] A detailed description of this apparatus with construction drawings appears in *General Information Series 511*, Freight Loading and Container Section, Assn. of Am. Railroads, 59 E. Van Buren St., Chicago, IL 60605.

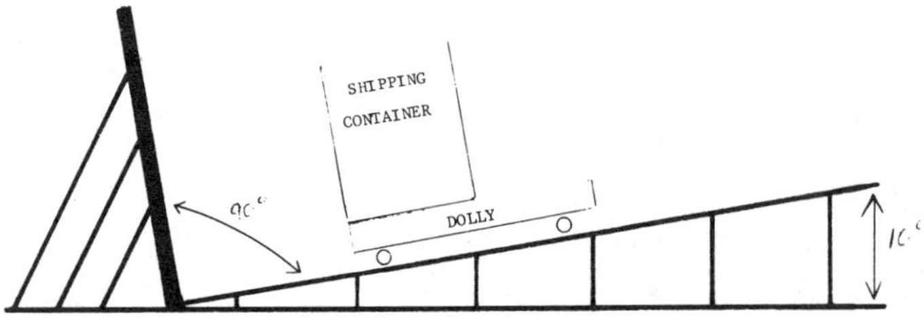

FIG. 1 Incline Impact Test Apparatus for Procedures A and B

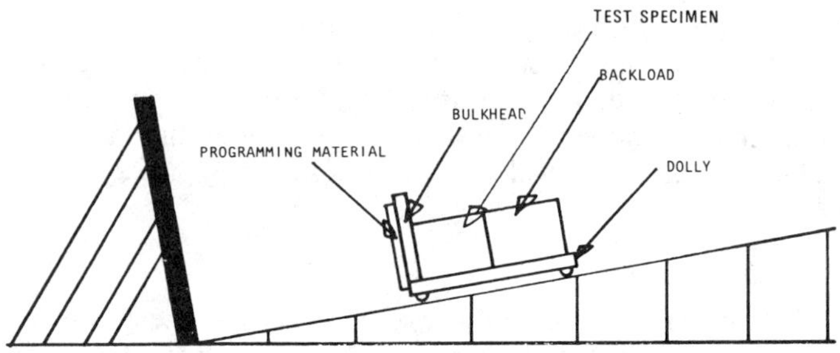

FIG. 2 Incline Impact Test Apparatus for Procedure C

4.3 *Programming Material* (*used for Procedure C only*), is required to shape and control the pulse seen by the specimen. This material may be fastened to the backstop or the bulkhead, and may be any material that gives a repeatable, controllable test, such as package cushioning material, or special programmers.

4.4 *Conditioning Apparatus*—Proper facilities for conditioning test containers at humidity and temperature in accordance with specification requirements.

4.5 *Miscellaneous Equipment*—Drying oven, scales, sharp knife, saw, etc., for taking samples for determination of the moisture content of the containers.

NOTE 1—The test equipment described in Method D 4003 may be used to conduct Procedure C of this method.

5. Test Specimens

5.1 When the container is being tested by Procedure A to determine the ability of the container to withstand impact stresses, it may be packed with either the actual contents or with a dummy load simulating such contents. When the container is being tested by Procedure B to determine the ability of the container or interior packing, or both, to provide protection to the contents, it shall be packed with the actual contents (Note 2) for which it was designed. In either procedure the container shall be closed, strapped, or sealed in the same manner as will be used in actual shipment.

NOTE 2—Certain contents or ingredients of the objects under test may be replaced by inert materials of approximately the same weight and characteristics.

5.2 The faces, edges, and corners of the box shall be identified, as follows: Facing one end of the box (in the case of a fiberboard container, with the manufacturer's joint on the right), the top of the container shall be designated as *1*, the right side as *2*, the bottom as *3*, the left side as *4*, the near end as *5*, and the far end as *6*. The edges shall be identified by the numbers of the two faces that form that edge; for example, *1-2* identifies the edge where the top and right side meet, and *2-5* the edge formed by the right side and the near end. (In a fiberboard container, this edge will have the manufacturer's joint.) The corners shall be identified by the numbers of the three faces that meet to form that corner; for example, *1-2-6* identifies the corner where the top, the right side, and the far end meet. This identification is illustrated in Fig. 5.

5.3 In the case of other than rectangular containers, any convenient and adequate system of identification may be used.

FIG. 3 Impact on Edge of Container

FIG. 4 Impact on End of Container with Optional Timber Hazard in Position to Contact Lower Edge

6. Number of Tests

6.1 A minimum of five identical specimens, or more depending on the objectives of the test, shall be tested to obtain an average result.

7. Calibration

7.1 The impact test apparatus shall be fully calibrated so that the velocity attained at point of impact is known for each increment of height (or length of travel). A chart or graph shall be prepared from these values from which, for any desired velocity at impact, the height of the release point (or length of travel) may be selected.

7.2 The calibration shall be made on the empty dolly and may be made by any suitable apparatus such as an electric timing device by which the velocity may be determined in units of 0.01 ft/s (0.003 m/s). If a timing device is used, the average velocity shall be calculated from the known length of roll and the elapsed time. The velocity at impact shall be taken as twice the average velocity.

8. Conditioning

8.1 Depending on the purpose of the test, the containers may be conditioned prior to the incline impact test by either a different physical test, water immersion, exposure of water spray or exposure to fixed conditions of air temperature or humidity, or other anticipated service conditions.

9. Moisture Content

9.1 When the moisture content must be determined the following procedures shall be used:

9.1.1 *Fiberboard*—Determine the moisture content of fiberboard at the time of test in accordance with Test Method D 644.

9.1.2 *Wood*—Determine the moisture content of wood by selecting a minimum of eight samples from different parts of the box, immediately following the test, of approximately 2 in.^{3} (33 cm^3) each. Weigh them to the nearest 0.1 g, dry to constant mass at 100 to 105°C (212 to 221°F), and reweigh. Express the moisture content as a percentage of the oven-dry mass of the specimen.

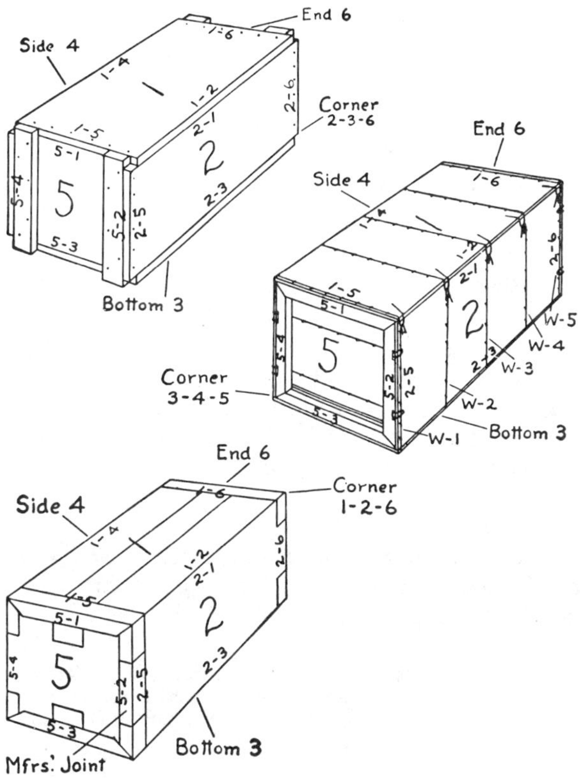

FIG. 5 Identification of the Faces, Edges and Corners of Test Specimen

10. Distance of Travel

10.1 The distance from the bumper at which the dolly and container shall be released will depend upon the purpose of the test. The cycle of impacts may be started at any suitable distance, and succeeding cycles run from: (*1*) the same distance, or (*2*) from distances that are increased by equal increments for each cycle of impacts.

11. Procedure A

11.1 *Determination of the Ability of a Container to Withstand Impact Stresses*—Place the container to be tested on the dolly with the face or edge that is to receive the impact located either: (*1*) coincident with the forward end of the dolly or (*2*) projecting 2 in. (51 mm) beyond the forward end of the dolly. Bring the dolly to the predetermined position on the incline and release. If it is desired to concentrate the impact at any particular position on the container, attach the optional board or timber hazard to the backstop in the desired position before the test. Allow no part of the hazard to be struck by the dolly. Then draw the dolly and container up the incline to the selected position and release. The position of the container on the dolly and the sequence in which the faces and edges are subjected to impacts may be at the option of the operator and will depend on the objective of the tests. Continue the test until failure of the container occurs. Depending on the purpose of the test, failure shall be considered to have occurred when:

11.1.1 The contents have spilled,

11.1.2 Any part of the contents may be removed from the container without damaging it further, or

11.1.3 Some predetermined type of types of damage have occurred to the container.

NOTE 3—It is presumed that if the contents are damaged, as determined by leakage of liquid contents or by other evidence, the test will be discontinued.

12. Procedure B

12.1 *Determination of the Ability of a Container or Interior Packing to Provide Protection to the Contents*—Conduct this test in accordance with Section 11 until a predetermined number of impacts of prescribed intensity have been applied, whereupon open the container for inspection, unless previous evidence of failure is apparent.

13. Procedure C

13.1 *Determination of the Performance of Shipping Unit in Simulated Rail Switching Impact*—Conduct this test by placing the specimen(s) on the dolly and against the bulkhead. Use impact velocity and programming material to produce the pulse(s) required, with simulated backload in place. The backload may be a duplicate specimen, similar to the test package or weights, to simulate the other lading in the railcar. When required, the bulkhead shall be monitored adjacent to test specimen(s) and deceleration level and time duration recorded.

14. Report

14.1 The report shall include the following:

14.1.1 Dimensions of the container under test; its complete structural specifications, kind of wood, plywood or fiberboard, description and specifications for blocking and cushioning; spacing, size, and kind of fasteners; method of closing and strapping, if any; and the net and gross masses.

14.1.2 Description of the contents of the container under test.

14.1.3 Detailed record of test on each container, including damage of the container and contents, together with any other observation that may assist in correctly interpreting the results or aid in improving the design of the container or the method of packing, blocking, or cushioning.

14.1.4 Method, if any, of conditioning the containers; the moisture content, if determined, of the wood, plywood, or fiberboard, and the results of any supplementary tests of the materials from which the container is made,

14.1.5 Record the calibration of the impact test apparatus, and

14.1.6 Statement to the effect that all tests were made in full compliance with this method, including a description of the location of specimens on the dolly, and an account as to the use of the optional hazard.

15. Precision and Bias

15.1 It has been found that interlaboratory correlation of test results has been poor. This is due, in large part, to variation in design and construction of the test apparatus, particularly the backstop, which has often not complied with the description in this method (see 3.1).

15.2 In addition to generally indicating the ability of a large package to withstand the rigors of a shock environment, the test has great value in comparing the performance of similar packages and is therefore a valuable tool for package design.

Standard Test Methods for Tensile Properties of Thin Plastic Sheeting[1]

This standard is issued under the fixed designation D 882; the number immediately following the designation indicates the year of original adoption or, in the case of revision, the year of last revision. A number in parentheses indicates the year of last reapproval. A superscript epsilon (ε) indicates an editorial change since the last revision or reapproval.

These test methods have been approved for use by agencies of the Department of Defense to replace Method 1013 of Federal Test Method Standard 406. Consult the DoD Index of Specifications and Standards for the specific year of issue which has been adopted by the Department of Defense.

1. Scope

1.1 These test methods cover the determination of tensile properties of plastics in the form of thin sheeting, including film (less than 1.0 mm (0.04 in.) in thickness).

NOTE 1—Film has been arbitrarily defined as sheeting having nominal thickness not greater than 0.25 mm (0.010 in.).

NOTE 2—Tensile properties of plastics 1.0 mm (0.04 in.) or greater in thickness shall be determined according to Test Method D 638.

1.2 Two types of tension tests are described in these test methods, differing basically only in manner of load application. These test methods may be used to test all plastics within the thickness range described and the capacity of the machine employed.

1.2.1 *Test Method A. Static Weighing—Constant-Rate-of-Grip Separation Test*—This test method employs a constant rate of separation of the grips holding the ends of the test specimen.

1.2.2 *Test Method B. Pendulum Weighing—Constant-Rate-of-Power-Grip Motion Test*—This test method employs a constant rate of motion of one grip and a variable rate of motion of the second grip. The variable-rate grip is attached to a pendulum weighing head, and its movement is dependent on the load-deformation behavior of the material under test.

1.3 Specimen extension may be measured in these methods by grip separation, extension indicators, or displacement of gage marks.

1.4 A procedure for determining the tensile modulus of elasticity is included, using Test Method A at one strain rate.

1.5 The values stated in SI units are to be regarded as the standard. The values in parentheses are provided for information only.

NOTE 3—This modulus determination procedure is based on the use of grip separation as a measure of extension; however, the desirability of using extension indicators accurate to ±1.0 % or better as specified in Test Method D 638 is recognized, and provision for the use of such instrumentation is incorporated in the procedure.

1.6 *This standard does not purport to address all of the safety problems associated with its use. It is the responsibility of the user of this standard to establish appropriate safety and health practices and determine the applicability of regulatory limitations prior to use.*

2. Referenced Documents

2.1 *ASTM Standards:*

D 374 Test Methods for Thickness of Solid Electrical Insulation[2]

D 618 Methods of Conditioning Plastics and Electrical Insulating Materials for Testing[2]

D 638 Test Method for Tensile Properties of Plastics[2]

E 691 Practice for Conducting an Interlaboratory Test Program to Determine the Precision of a Test Method[3]

3. Terminology

3.1 *Definitions*—Definitions of terms and symbols relating to tension testing of plastics appear in the Annex to Test Method D 638.

3.1.1 *tear failure*—a tensile failure characterized by fracture initiating at one edge of the specimen and progressing across the specimen at a rate slow enough to produce an anomalous load-deformation curve.

3.1.2 *line grips*—grips having faces designed to concentrate the entire gripping force along a single line perpendicular to the direction of testing stress. This is usually done by combining one standard flat face and an opposing face from which protrudes a half-round.

4. Significance and Use

4.1 Tensile properties determined by these test methods are of value for the identification and characterization of materials for control and specification purposes. Tensile properties may vary with specimen thickness, method of preparation, speed of testing, type of grips used, and manner of measuring extension. Consequently, where precise comparative results are desired, these factors must be carefully controlled. Since the actual loading rates vary between Test Methods A and B, the results obtained using these two methods cannot be directly compared. Test Method A is preferred and shall be used for referee purposes, unless otherwise indicated in particular material specifications.

4.2 Tensile properties may be utilized to provide data for research and development and engineering design as well as quality control and specification. However, data from such tests cannot be considered significant for applications differing widely from the load-time scale of the test employed.

4.3 The tensile modulus of elasticity is an index of the stiffness of thin plastic sheeting. The reproducibility of test

---

[1] These test methods are under the jurisdiction of ASTM Committee D-20 on Plastics and are the direct responsibility of Subcommittee D20.10 on Mechanical Properties.

Current edition approved October 26, 1990. Published December 1990. Originally published as D 882 – 46 T. Last previous edition D 882 – 88.

[2] *Annual Book of ASTM Standards*, Vol 08.01.

[3] *Annual Book of ASTM Standards*, Vols 08.03 and 14.02.

results is good when precise control is maintained over all test conditions. When different materials are being compared for stiffness, specimens of identical dimensions must be employed.

4.4 The tensile energy to break (TEB) is the total energy absorbed per unit volume of the specimen up to the point of rupture. In some texts this property has been referred to as *toughness*. It is used to evaluate materials that may be subjected to heavy abuse or that might stall web transport equipment in the event of a machine malfunction in end-use applications. However, the rate of strain, specimen parameters, and especially flaws may cause large variations in the results. In that sense, caution is advised in utilizing TEB test results for end-use design applications.

4.5 Materials that fail by tearing give anomalous data which cannot be compared with those from normal failure.

5. Apparatus

5.1 *Grips*—A gripping system that minimizes both slippage and uneven stress distribution.

NOTE 4—Grips lined with thin rubber, crocus-cloth, or pressure-sensitive tape as well as file-faced or serrated grips have been successfully used for many materials. The choice of grip surface will depend on the material tested, thickness, etc. More recently, line grips padded on the round face with 1.0-mm (40-mil) blotting paper have been found superior. Air-actuated grips have been found advantageous, particularly in the case of materials that tend to "neck" into the grips, since pressure is maintained at all times. In cases where samples frequently fail at the edge of the grips, it may be advantageous to increase slightly the radius of curvature of the edges where the grips come in contact with the test area of the specimen.

5.2 *Thickness Gage*—A dead-weight dial micrometer as prescribed in Method C of Test Methods D 374, reading to 0.0025 mm (0.0001 in.) or less.

5.3 *Width-Measuring Devices*—Suitable test scales or other width measuring devices capable of measuring 0.25 mm (0.010 in.) or less.

5.4 *Specimen Cutter*—Razor blades, fixtures incorporating razor blades, suitable paper cutters, or other devices capable of cutting the specimens to the proper width and producing straight, clean, parallel edges with no visible imperfections, shall be used. Devices that use razor blades have proved especially suitable for materials having an elongation-at-fracture above 10 to 20 %. A device consisting of two parallel knives mounted firmly against a precision-ground base shear block (similar to a paper cutter) has also proved satisfactory. The use of striking dies is not recommended because of poor and inconsistent specimen edges which may be produced. It is imperative that the cutting edges be kept sharp and free from visible scratches or nicks.

5.5 *Extension Indicators* (if employed) shall conform to requirements specified in Test Method D 638. In addition, such apparatus shall be so designed as to minimize stress on the specimen at the contact points of the specimen and the indicator (see 8.3).

5.6 *Testing Machines:*

5.6.1 *For Test Method A*—A testing machine of the constant rate-of-jaw-separation type. The machine shall be equipped with a weighing system that moves a maximum distance of 2 % of the specimen extension within the range being measured. The machine shall be equipped with a device for recording the tensile load and the amount of separation of the grips; both of these measuring systems shall be accurate to ±2 %. The rate of separation of the jaws shall be uniform and capable of adjustment from approximately 1.3 to 500 mm (0.05 to 20 in.)/min in increments necessary to produce the strain rates specified in 9.3 and 9.4. This test method (A) shall be used for tensile modulus of elasticity measurements (Note 5).

5.6.2 *For Test Method B*—A testing machine of the pendulum type. This machine shall be equipped with a pendulum weighing head to measure the load applied to the test specimen and a device for indicating or recording the tensile load carried by the specimen with an accuracy of ±2 %. The rate of travel of the power-activated grip shall be uniform and capable of adjustment to 50.8 and 508 mm (2 and 20 in.)/min.

NOTE 5—A high response speed in the recording system is desirable, particularly when relatively high strain rates are employed for rigid materials. The speed of pen response for recorders is supplied by manufacturers of this equipment. Care must be taken to conduct tests at conditions such that response time (ability of recorder to follow actual load) will produce less than 2 % error.

6. Test Specimens

6.1 The test specimens shall consist of strips of uniform width and thickness at least 50 mm (2 in.) longer than the grip separation used.

6.2 The nominal width of the specimens shall be not less than 5.0 mm (0.20 in.) or greater than 25.4 mm (1.0 in.).

6.3 A width-thickness ratio of at least eight shall be used. Narrow specimens magnify effects of edge strains or flaws, or both.

6.4 The utmost care shall be exercised in cutting specimens to prevent nicks and tears which are likely to cause premature failures (Note 6). The edges shall be parallel to within 5 % of the width over the length of the specimen between the grips.

NOTE 6—Microscopical examination of specimens may be used to detect flaws due to sample or specimen preparation.

6.5 Wherever possible, the test specimens shall be selected so that thickness is uniform to within 10 % of the thickness over the length of the specimen between the grips in the case of materials 0.25 mm (0.010 in.) or less in thickness and to within 5 % in the case of materials greater than 0.25 mm (0.010 in.) in thickness but less than 1.00 mm (0.040 in.) in thickness.

NOTE 7—In cases where thickness variations are in excess of those recommended in 6.5, results may not be characteristic of the material under test.

6.6 If the material is suspected of being anisotropic, two sets of test specimens shall be prepared having their long axes respectively parallel with and normal to the suspected direction of anisotropy.

6.7 For tensile modulus of elasticity determinations, a specimen gage length of 250 mm (10 in.) shall be considered as standard. This length is used in order to minimize the effects of grip slippage on test results. When this length is not feasible, test sections as short as 100 mm (4 in.) may be used if it has been shown that results are not appreciably affected. However, the 250-mm gage length shall be used for referee purposes. The speed of testing of shorter specimens must be adjusted in order for the strain rate to be equivalent to that of the standard specimen.

NOTE 8—Two round robins[4] have shown that, for materials of less than 0.25-mm (10-mil) thickness, line grips padded on the round side with 1.0-mm (40-mil) blotting paper give the same results with a 100-mm test section as a 250-mm test section produces with flat-face grips.

NOTE 9—Excessive jaw slippage becomes increasingly difficult to overcome in cases where high modulus materials are tested in thicknesses greater than 0.25 mm (0.010 in.).

7. Conditioning

7.1 *Conditioning*—Condition the test specimens at 23 ± 2°C (73.4 ± 3.6°F) and 50 ± 5 % relative humidity for not less than 40 h prior to test in accordance with Procedure A of Methods D 618 for those tests where conditioning is required. In cases of disagreement, the tolerances shall be 1°C (1.8°F) and ±2 % relative humidity.

7.2 *Test Conditions*—Conduct tests in the Standard Laboratory Atmosphere of 23 ± 2°C (73.4 ± 3.6°F) and 50 ± 5 % relative humidity, unless otherwise specified in the test methods or in this specification. In cases of disagreements, the tolerances shall be ±1°C (±1.8°F) and ±2 % relative humidity.

8. Number of Test Specimens

8.1 In the case of isotropic materials, at least five specimens shall be tested from each sample.

8.2 In the case of anisotropic materials, at least ten specimens, five normal and five parallel with the principal axis of anisotropy, shall be tested from each sample.

8.3 Specimens that fail at some obvious flaw or that fail outside the gage length shall be discarded and retests made, unless such flaws or conditions constitute a variable whose effect is being studied. However, jaw breaks (failures at the grip contact point) are acceptable if it has been shown that results from such tests are in essential agreement with values obtained from breaks occurring within the gage length.

NOTE 10—In the case of some materials, examination of specimens, prior to and following testing, under crossed optical polarizers (polarizing films) provides a useful means of detecting flaws which may be, or are, responsible for premature failure.

9. Speed of Testing

9.1 The speed of testing is the rate of separation of the two members (or grips) of the testing machine when running idle (under no load). This rate of separation shall be maintained within 5 % of no-load value when running under full-capacity load.

9.2 The speed of testing shall be calculated from the required initial strain rate as specified in Table 1. The rate of grip separation may be determined for the purpose of this test method from the initial strain rate as follows:

$$A = BC$$

where:
A = rate of grip separation, mm (or in.)/min
B = initial distance between grips, mm (or in.), and
C = initial strain rate, mm/mm·min (or in./in.·min).

[4] Supporting data are available from ASTM Headquarters. Request RR:D20-1058.

TABLE 1 Speed of Testing

Method	Percentage Elongation at Break	Initial Strain Rate, mm/mm·min (in./in.·min)
	Modulus of Elasticity Determination	
A		0.1
	Determinations other than Elastic Modulus	
A	Less than 20	0.1
	20 to 100	0.5
	Greater than 100	10.0
B	Less than 100	0.5
	Greater than 100	10.0

9.3 The initial strain rate shall be as in Table 1 unless otherwise indicated by the specification for the material being tested.

NOTE 11—Results obtained at different initial strain rates are not comparable; consequently, where direct comparisons between materials in various elongation classes are required, a single initial strain rate should be used. For some materials it may be advisable to select the strain rates on the basis of percentage elongation at yield.

9.4 In cases where conflicting material classification, as determined by percentage elongation at break values, results in a choice of strain rates, the lower rate shall be used.

9.5 If modulus values are being determined, separate specimens shall be used whenever strain rates and specimen dimensions are not the same as those employed in the test for other tensile properties.

10. Procedure

10.1 Select a load range such that specimen failure occurs within its upper two thirds. A few trial runs may be necessary to select a proper combination of load range and specimen width.

10.2 Measure the cross-sectional area of the specimen at several points along its length. Measure the width to an accuracy of 0.25 mm (0.010 in.) or better. Measure the thickness to an accuracy of 0.0025 mm (0.0001 in.) or better for films less than 0.25 mm (0.010 in.) in thickness and to an accuracy of 1 % or better for films greater than 0.25 mm (0.010 in.) but less than 1.0 mm (0.040 in.) in thickness.

10.3 The initial grip separation shall be at least 50 mm (2 in.) for materials having a total elongation at break of 100 % or more, and at least 100 mm (4 in.) for materials having a total elongation at break of less than 100 %.

NOTE 12—Since slippage is a potential problem in these tests, as great an initial distance between grips as possible should be employed.

10.4 Set the rate of grip separation to give the desired strain rate based on the initial distance between the grips (Note 13). Balance, zero, and calibrate the load weighing and recording system.

NOTE 13—Suggested crosshead speeds and initial grip separation to give the desired initial strain rate described in Table 1 are shown in Table 2.

10.5 In cases where it is desired to measure a test section other than the total length between the grips, mark the ends of the desired test section with a soft, fine wax crayon or with ink. Do not scratch these marks onto the surface since such scratches may act as stress raisers and cause premature specimen failure. Extensometers may be used if available; in

TABLE 2 Crosshead Speeds and Initial Grip Separation

Method	Percentage Elongation at Break	Initial Strain Rate, mm/mm·min (in./in.·min)	Initial Grip Separation mm	in.	Rate of Grip Separation mm/min	in./min
		Modulus of Elasticity Determination				
A		0.1	250	10	25	1.0
		Determinations other than Elastic Modulus				
A	Less than 20	0.1	125	5	12.5	0.5
	20 to 100	0.5	100	4	50	2.0
	Greater than 100	10.0	50	2	500	20.0
B	Less than 100	0.5	100	4	50	2.0
	Greater than 100	10.0	50	2	500	20.0

this case, the test section will be defined by the contact points of the extensometer.

NOTE 14—Measurement of a specific test section is necessary with some materials having high elongation. As the specimen elongates, the accompanying reduction in area results in a loosening of material at the inside edge of the grips. This reduction and loosening moves back into the grips as further elongation and reduction in area takes place. In effect this causes problems similar to grip slippage, that is, exaggerates measured extension.

10.6 Place the test specimen in the grips of the testing machine, taking care to align the long axis of the specimen with an imaginary line joining the points of attachment of the grips to the machine. Tighten the grips evenly and firmly to the degree necessary to minimize slipping of the specimen during test.

10.7 Start the machine and record load versus extension.

10.7.1 When the total length between the grips is used as the test area, record load versus grip separation.

10.7.2 When a specific test area has been marked on the specimen, follow the displacement of the edge boundary lines with respect to each other with dividers or some other suitable device. If a load-extension curve is desired, plot various extensions versus corresponding loads sustained, as measured by the load indicator.

10.7.3 When an extensometer is used, record load versus extension of the test area measured by the extensometer.

10.8 If modulus values are being determined, select a load range and chart rate to produce a load—extension curve of between 30 and 60° to the *X* axis. For maximum accuracy, use the most sensitive load scale for which this condition can be met. The test may be discontinued when the load—extension curve deviates from linearity.

10.9 In the case of materials being evaluated for secant modulus, the test may be discontinued when the specified extension has been reached.

10.10 If tensile energy to break is being determined, some provision must be made for integration of the stress-strain curve. This may be either an electronic integration during the test or a subsequent determination from the area of the finished stress-strain curve (see Annex A2).

11. Calculation

11.1 *Breaking Factor* (nominal) shall be calculated by dividing the maximum load by the original minimum width of the specimen. The result shall be expressed in force per unit of width, usually newtons per metre (or pounds per inch) of width, and reported to three significant figures. The thickness of the film shall always be stated to the nearest 0.0025 mm (0.0001 in.).

Example—Breaking Factor = 1.75 kN/m (10.0 lbf/in.) of width for 0.1300-mm (0.0051-in.) thickness.

NOTE 15—This method of reporting is useful for very thin films (0.13 mm (0.005 in.) and less) for which breaking load may not be proportional to cross-sectional area and whose thickness may be difficult to determine with precision. Furthermore, films which are in effect laminar due to orientation, skin effects, nonuniform crystallinity, etc., have tensile properties disproportionate to cross-sectional area.

11.2 *Tensile Strength* (nominal) shall be calculated by dividing the maximum load by the original minimum cross-sectional area of the specimen. The result shall be expressed in force per unit area, usually megapascals (or pounds-force per square inch). This value shall be reported to three significant figures.

NOTE 16—When tear failure occurs, so indicate and calculate results based on load and elongation at which tear initiates, as reflected in the load-deformation curve.

11.3 *Tensile Strength at Break* (nominal) shall be calculated in the same way as the tensile strength except that the load at break shall be used in place of the maximum load (Notes 16 and 17).

NOTE 17—In many cases tensile strength and tensile strength at break are identical.

11.4 *Percentage Elongation at Break* shall be calculated by dividing the extension at the moment of rupture of the specimen by the initial gage length of the specimen and multiplying by 100. When gage marks or extensometers are used to define a specific test section, only this length shall be used in the calculation, otherwise the distance between the grips shall be used. The result shall be expressed in percent and reported to two significant figures (Note 16).

11.5 *Yield Strength*, where applicable, shall be calculated by dividing the load at the yield point by the original minimum cross-sectional area of the specimen. The result shall be expressed in force per unit area, usually megapascals (or pounds-force per square inch). This value shall be reported to three significant figures. Alternatively, for materials that exhibit Hookean behavior in the initial part of the curve, an offset yield strength may be obtained as described in the Appendix of Test Method D 638. In this case the value should be given as "yield strength at—% offset."

11.6 *Percentage Elongation at Yield* where applicable shall be calculated by dividing the extension at the yield point by the initial gage length of specimen and multiplying by 100. When gage marks or extensometers are used to define a specific test section, only this length shall be used in the calculation. Before calculating, correct the extension for "toe compensation" as described in Annex A1. The results shall be expressed in percent and reported to two significant figures. When offset yield strength is used, the elongation at the offset yield strength may be calculated.

11.7 *Elastic Modulus* shall be calculated by drawing a tangent to the initial linear portion of the load—extension curve, selecting any point on this tangent, and dividing the tensile stress by the corresponding strain. Before calculating, correct the extension for "toe compensation" as described in Annex A1. For purposes of this determination, the tensile stress shall be calculated by dividing the load by the average original cross section of the test section. The result shall be expressed in force per unit area, usually megapascals (or pounds-force per square inch), and reported to three significant figures.

11.8 *Secant Modulus* at a designated strain shall be calculated by dividing the corresponding stress (nominal) by the designated strain. Elastic modulus values are preferable and shall be calculated whenever possible. However, for materials where no proportionality is evident, the secant value shall be calculated. Draw the tangent as directed in A1.3 and Fig. A1.2 of Annex A1, and mark off the designated strain from the yield point where the tangent line goes through zero stress. The stress to be used in the calculation is then determined by dividing the load at the designated strain on the load—extension curve by the original average cross-sectional area of the specimen.

11.9 *Tensile Energy to Break* where applicable shall be calculated by integrating the energy per unit volume under the stress-strain curve or by integrating the total energy absorbed and dividing it by the volume of the original gage region of the specimen. As indicated in Annex A2, this may be done directly during the test by an electronic integrator, or subsequently by computation from the area of the plotted curve. The result shall be expressed in energy per unit volume, usually in megajoules per cubic metre (or inch-pounds-force per cubic inch). This value shall be reported to two significant figures.

11.10 For each series of tests, the arithmetic mean of all values obtained shall be calculated to the proper number of significant figures.

11.11 The standard deviation (estimated) shall be calculated as follows and reported to two significant figures:

$$s = \sqrt{(\Sigma X^2 - n\bar{X}^2)/(n-1)}$$

where:

s = estimated standard deviation,
X = value of a single observation,
n = number of observations, and
$\bar{X}$ = arithmetic mean of the set of observations.

12. Report

12.1 Report the following information:

12.1.1 Complete identification of the material tested, including type, source, manufacturer's code number, form, principal dimensions, previous history, and orientation of samples with respect to anisotropy (if any),

12.1.2 Method of preparing test specimens,

12.1.3 Thickness, width, and length of test specimens,

12.1.4 Number of specimens tested,

12.1.5 Strain rate employed,

12.1.6 Grip separation (initial),

12.1.7 Crosshead speed (rate of grip separation),

12.1.8 Gage length (if different from grip separation),

12.1.9 Type of grips used, including facing (if any),

12.1.10 Test method (A or B),

12.1.11 Conditioning procedure (test conditions, temperature, and relative humidity if nonstandard),

12.1.12 Anomalous behavior such as tear failure and failure at a grip,

12.1.13 Average breaking factor and standard deviation,

12.1.14 Average tensile strength (nominal) and standard deviation,

12.1.15 Average tensile strength at break (nominal) and standard deviation,

12.1.16 Average percentage elongation at break and standard deviation,

12.1.17 Where applicable: Average tensile energy to break and standard deviation,

12.1.18 In the case of materials exhibiting "yield" phenomenon:

Average yield strength and standard deviation, and

Average percentage elongation at yield and standard deviation,

12.1.19 For materials which do not exhibit a yield point:

Average—% offset yield strength and standard deviation, and

Average percentage elongation at—% offset yield strength and standard deviation,

12.1.20 Average modulus of elasticity and standard deviation (if secant modulus is used, so indicate and report strain at which calculated), and

12.1.21 When an extensometer is employed, so indicate.

13. Precision and Bias

13.1 Two interlaboratory tests have been run for these tensile properties. The first was run for modulus, only in 1977 in which randomly drawn samples of four thin (~ 0.025 mm (0.001-in.)) materials were tested with five specimens in each laboratory. Elastic (tangent) modulus measurements were made by six laboratories, and secant (1 %) modulus measurements were taken by five laboratories. The relative precision obtained in this interlaboratory study is in Table 3.

13.1.1 In deriving the estimates in Table 3, statistical

TABLE 3 Precision Data for Modulus

Material	Thickness, mils	Average, 10^3 psi	S_r, 10^3 psi	S_R, 10^3 psi	I_r, 10^3 psi	I_R, 10^3 psi
Tangent Modulus						
LDPE	1.4	53.9	1.81	8.81	5.12	24.9
HDPE	1.6	191	5.47	16.2	15.5	45.9
PP	1.1	425	10.3	31.5	29.0	89.1
PET	0.9	672	13.8	55.5	39.1	157.1
Secant Modulus						
LDPE	1.4	45.0	2.11	3.43	5.98	9.70
HDPE	1.6	150	3.29	9.58	9.30	27.1
PP	1.1	372	4.66	26.5	13.2	74.9
PET	0.9	640	10.0	27.5	28.4	77.8

TABLE 4 Precision Data for Yield Stress

Material	Thickness, mils	Average, 10^3 psi	$(S_r)^A$ 10^3 psi	$(S_R)^B$ 10^3 psi	$I(r)^C$ 10^3 psi	$I(R)^D$ 10^3 psi
LDPE	1.0	1.49	0.051	0.13	0.14	0.37
HDPE	1.0	4.33	0.084	0.16	0.24	0.44
PP	0.75	6.40	0.13	0.52	0.37	1.46
PC	4.0	8.59	0.072	0.29	0.20	0.82
CTA	5.3	11.4	0.12	0.50	0.34	1.43
PET	4.0	14.3	0.12	0.23	0.34	0.66
PET	2.5	14.4	0.14	0.54	0.40	1.52
PET	7.0	14.4	0.13	0.36	0.37	1.03

[A] S_r is the within-laboratory standard deviation of the average.
[B] S_R is the between-laboratories standard deviation of the average.
[C] $I_r = 2.83\ S_r$.
[D] $I_R = 2.83\ S_R$.

TABLE 5 Precision Data for Yield Elongation

Material	Thickness, mils	Average, %	$(S_r)^A$, %	$(S_R)^B$, %	$I(r)^C$, %	$I(R)^D$, %
PP	0.75	3.5	0.15	0.41	0.42	1.2
PET	2.5	5.2	0.26	0.92	0.74	2.6
PET	4.0	5.3	0.25	0.60	0.71	1.7
PET	7.0	5.4	0.14	1.05	0.40	3.0
CTA	5.3	5.4	0.19	0.99	0.54	2.8
PC	4.0	6.9	0.24	0.98	0.68	2.8
HDPE	1.0	8.8	0.32	1.82	0.91	5.2
LDPE	1.0	10.0	0.55	3.41	1.56	9.6

NOTE—See Table 4 for footnote explanation.

outliers were not removed, in keeping with Practice E 691.[5]

13.1.2 The within-lab standard deviation of a mean value, $S_{\bar{x}}$, in each case was determined from the standard deviation, S_x, of the five individual specimens as follows: $S_{\bar{x}} = S_x/(5)^{1/2}$. The $S_{\bar{x}}$ values were pooled among laboratories for a given material to obtain the within-lab standard deviation, S_r, of a test result (mean of 5 specimens). See 13.3 through 13.3.2 for definitions of terms in the tables.

13.2 An interlaboratory test was run for all the other tensile properties except modulus in 1981 in which randomly drawn samples of six materials (one of these in three thicknesses) ranging in thickness from 0.019 to 0.178 mm (0.00075 to 0.007 in.) were tested in seven laboratories. A test result was defined as the mean of five specimen determinations. However, each laboratory tested 8 specimens, and the $S_{\bar{x}}$ was determined from: $S_{\bar{x}} = S_x/(5)^{1/2}$ as above. This was done to improve the quality of the statistics while maintaining their applicability to a five-specimen test result. The materials and their thicknesses are identified in Tables 4 to 8, each of which contain data for one of the following properties: tensile yield stress, yield elongation, tensile strength, tensile elongation at break, and tensile energy at break (see Note 18).[6]

NOTE 18—Subsequent to filing the research report, examination of the LDPE used in this study between crossed polarizers revealed lengthwise lines representing substantial widthwise variation in molecular orientation that probably was not successfully randomized out of the between-labs component of variance.

NOTE 19: **CAUTION**—The following explanations of I_r and I_R (13.3 through 13.3.3) are only intended to present a meaningful way of considering the *Approximate* precision of this test method. The data in Table(s) 3 to 8 should not be rigorously applied to the acceptance or rejection of material, as those data are specific to the round robin and may not be representative of other lots, conditions, materials, or laboratories. Users of this test method should apply the principles outlined in Practice E 691 to generate data specific to their laboratory and materials, or between specific laboratories. The principles of 13.3 through 13.3.3 would then be valid for such data.

13.3 *Concept of I_r and I_R*—If S_r and S_R have been calculated from a *large enough body of data*, and for test results that were averages (medians/other function) from testing five specimens:

13.3.1 *Repeatability, I_r*, (Comparing two test results for the same material, obtained by the same operator using the same equipment on the same day)—The two test results should be judged not equivalent if they differ by more than the I_r value for that material.

13.3.2 *Reproducibility, I_R*, (Comparing two test results for the same material, obtained by different operators using different equipment on different days)—The two test results should be judged not equivalent if they differ by more than the I_R value for that material.

13.3.3 Any judgment made in accordance with 13.3.1 and 13.3.2 would have an approximate 95 % (0.95) probability of being correct.

13.4 *Bias*—The systematic error which contributes to the difference between a test result and a true (or reference) value. There are no recognized standards on which to base an estimate of bias for this test method.

[5] Supporting data are available from ASTM Headquarters. Request RR:D20-1084.

[6] Supporting data are available from ASTM Headquarters. Request RR:D20-1101.

TABLE 6 Precision Data for Tensile Strength

Material	Thickness, mils	Average, 10^3 psi	$(S_r)^A$ 10^3 psi	$(S_R)^B$ 10^3 psi	$I(r)^C$ 10^3 psi	$I(R)^D$ 10^3 psi
LDPE	1.0	3.42	0.14	0.53	0.40	1.5
HDPE	1.0	6.87	0.27	0.81	0.76	2.3
PC	4.0	12.0	0.34	0.93	0.96	2.6
CTA	5.3	14.6	0.20	1.37	0.57	3.9
PP	0.75	28.4	1.57	4.56	4.4	12.9
PET	4.0	28.9	0.65	1.27	1.8	3.6
PET	7.0	30.3	0.83	1.32	2.3	3.7
PET	2.5	30.6	1.22	2.64	3.4	7.5

NOTE—See Table 4 for footnote explanation.

TABLE 7 Precision Data for Elongation at Break

Material	Thickness, mils	Average, %	$(S_r)^A$, %	$(S_R)^B$, %	$I(r)^C$, %	$I(R)^D$, %
CTA	5.3	26.4	1.0	4.3	3	12
PP	0.75	57.8	4.4	12.7	12	36
PET	2.5	120	8.0	14.6	23	41
PET	7.0	132	5.8	10.6	16	30
PET	4.0	134	4.4	12.2	12	35
PC	4.0	155	5.4	17.1	15	48
LDPE	1.0	205	24.4	73.3	69	210
HDPE	1.0	570	26.0	91.7	74	260

NOTE—See Table 4 for footnote explanation.

TABLE 8 Precision Data for Tensile Energy to Break

Material	Thickness, mils	Average, 10^3 $\frac{\text{in./lb}}{\text{in.}^3}$	$(S_r)^A$ 10^3 $\frac{\text{in./lb}}{\text{in.}^3}$	$(S_R)^B$ 10^3 $\frac{\text{in./lb}}{\text{in.}^3}$	$I(r)^C$ 10^3 $\frac{\text{in./lb}}{\text{in.}^3}$	$I(R)^D$ 10^3 $\frac{\text{in./lb}}{\text{in.}^3}$
CTA	5.0	3.14	0.14	0.70	0.4	2.0
LDPE	1.0	5.55	0.84	2.47	2.4	7.0
PP	0.75	11.3	1.19	3.11	3.4	8.8
PC	4.0	12.9	0.59	1.55	1.7	4.4
HDPE	1.0	26.0	1.87	5.02	5.3	14.2
PET	2.5	26.1	2.13	4.20	6.0	11.9
PET	4.0	27.1	1.42	2.75	4.0	7.8
PET	7.0	28.4	1.71	2.72	4.8	7.7

NOTE—See Table 4 for footnote explanation.

ANNEXES

(Mandatory Information)

A1. TOE COMPENSATION

A1.1 In a typical stress-strain curve (Fig. A1.1) there is a toe region, *AC*, which does not represent a property of the material. It is an artifact caused by a takeup of slack, and alignment or seating of the specimen. In order to obtain correct values of such parameters as modulus, strain, and offset yield point, this artifact must be compensated for to give the corrected zero point on the strain or extension axis.

A1.2 In the case of a material exhibiting a region of Hookean (linear) behavior (Fig. A1.1), a continuation of the linear (*CD*) region of the curve is constructed through the zero-stress axis. This intersection (*B*) is the corrected zero-strain point from which all extensions or strains must be measured, including the yield offset (*BE*), if applicable. The elastic modulus can be determined by dividing the stress at any point along the line *CD* (or its extension) by the strain at the same point (measured from point *B*, defined as zero-strain).

A1.3 In the case of a material that does not exhibit any linear region (Fig. A1.2), the same kind of toe correction of the zero-strain point can be made by constructing a tangent to the maximum slope at the inflection point (*H′*). This is extended to intersect the strain axis at point *B′*, the corrected zero-strain point. Using point *B′* as zero strain, the stress at any point (*G′*) on the curve can be divided by the strain at that point to obtain a secant modulus (slope of line *B′ G′*). For those materials with no linear region, any attempt to use the tangent through the inflection point as a basis for determination of an offset yield point may result in unacceptable error.

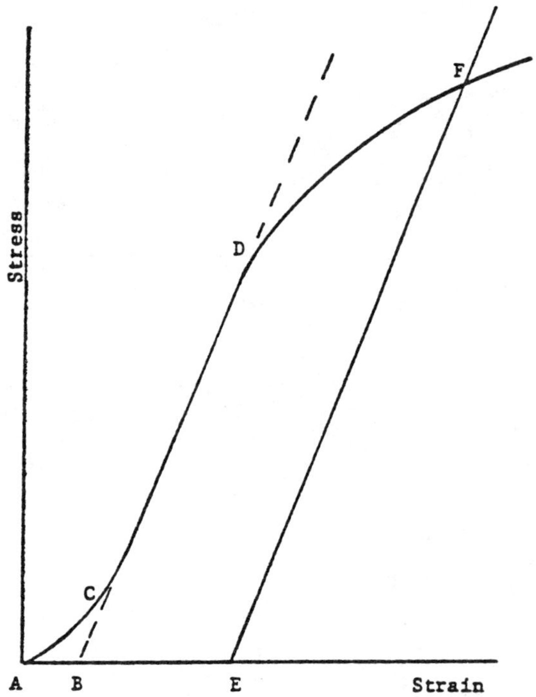

NOTE—Some chart recorders plot the mirror image of this graph.

FIG. A1.1 Material with Hookean Region

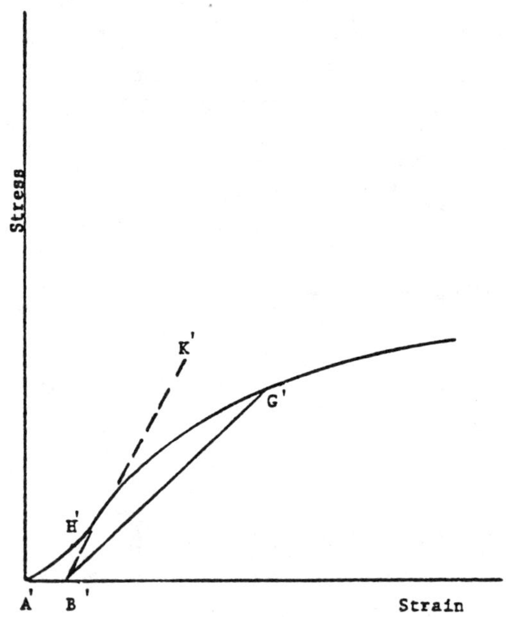

NOTE—Some chart recorders plot the mirror image of this graph.

FIG. A1.2 Material with No Hookean Region

A2. DETERMINATION OF TENSILE ENERGY TO BREAK

A2.1 Tensile energy to break (TEB) is defined by the area under the stress-strain curve, or

$$TEB = \int_0^{\epsilon_T} S\,d\epsilon$$

where S is the stress at any strain, ϵ; and ϵ_T is the strain at rupture. The value is in units of energy per unit volume of the specimen's initial gage region. TEB is most conveniently and accurately measured with a tension tester equipped with an integrator. The calculation is then:

$$TEB = (I/K)\,\frac{\text{(full scale load) (chart speed) (crosshead speed/chart speed)}}{\text{(mean caliper) (specimen width) (gage length)}}$$

where I is the integrator count reading and K is the maximum possible count per unit time for a constant full scale load. This whole calculation is typically done electronically. The results are best expressed in megajoules per cubic metre (or inch-pounds-force per cubic inch).

A2.2 Without an integrator, the area under the recorded stress-strain curve can be measured by planimeter, counting squares, or weighing the cut-out curve. These techniques are time-consuming and likely to be less accurate, since the load scale on some chart paper is not in round-number dimensions. Moreover, if the curve coordinates are in terms of force and extension instead of stress and strain, the calculated energy, corresponding to the measured area, must be divided by the product of gage length, specimen width, and mean caliper:

$$TEB = \frac{\text{(curve area) (force per unit chart scale)}\ \text{(extension per unit chart travel)}}{\text{(mean caliper) (specimen width) (gage length)}}$$

A2.3 For example, if the area under a force-extension curve is 60 000 mm^2, the load coordinate is 2.0 N/mm of chart scale, the extension coordinate is 0.25 mm of extension per mm of chart travel, and the specimen dimensions are 0.1 mm caliper, 15 mm width and 100 mm gage length, then the calculation for tensile energy to break is:

$$TEB = \frac{(60\ 000\ mm^2)\ (2.0\ N/mm)\ (0.25 \times 10^{-3}\ m/mm)}{(0.1 \times 10^{-3} m)\ (15 \times 10^{-3} m)\ (100 \times 10^{-3} m)}$$

$$TEB = 200\ MJ/m^3$$

Designation: D 895 – 79 (Reapproved 1984)

Standard Test Method for Water Vapor Permeability of Packages[1]

This standard is issued under the fixed designation D 895; the number immediately following the designation indicates the year of original adoption or, in the case of revision, the year of last revision. A number in parentheses indicates the year of last reapproval. A superscript epsilon (ε) indicates an editorial change since the last revision or reapproval.

1. Scope

1.1 This test method covers the determination of the water vapor permeability of finished packages when closed and sealed in the conventional manner. This test method is intended for "shelf-size packages" which are shipped in outside containers.

1.2 This test method may be applied to the package as packed or after one or more performance tests such as drum, vibration, drop, or actual shipping tests have been completed, as required.

1.3 Packages with body walls having the property of absorbing water are not suitable for testing under this method.

1.4 *This standard does not purport to address the safety problems associated with its use. It is the responsibility of the user of this standard to establish appropriate safety and health practices and determine the applicability of regulatory limitations prior to use.*

2. Significance and Use

2.1 This test method is normally used to:

2.1.1 Evaluate materials and constructions for a specific type of package,

2.1.2 Compare performance of different types of packages,

2.1.3 Determine adequacy of protection for a specific product or application, and

2.1.4 Maintain quality control.

2.2 When the test is conducted to determine the water vapor permeability of the packages for research, development, manufacturing control, specification acceptance, etc., a desiccant is used as the water absorbing medium.

2.3 When the test is used to determine the suitability of the package with respect to water vapor resistance for a particular product, the product is used instead of the desiccant.

3. Definition

3.1 *water vapor permeability, package*—for this test method, the rate at which water is transmitted into the package from the test atmosphere (normally 90 ± 2 % relative humidity and 100 ± 2°F (37.8 ± 1.1°C)) surrounding it while a desiccant is sealed within.

4. Apparatus

4.1 *Weighing Balance*—A suitable laboratory balance sensitive to 1 mg or one ten-thousandth of the mass of the package, whichever is larger.

4.2 *Test Chamber*—A test room or cabinet provided with conditioned air which is continuously circulated around the specimens under test. The conditions in the chamber shall be such that no condensation occurs on the specimens.

5. Materials

5.1 *Desiccant*—A desiccant shall be used that has a high affinity for water vapor and a high drying efficiency; that is, having a low water vapor pressure even after absorbing a large amount of water.

NOTE—A suitable desiccant is anhydrous calcium chloride ($CaCl_2$) in the form of small lumps that will pass a No. 8 (2.36-mm) sieve, and free of fines that will pass a No. 30 (600-μm) sieve.

5.2 *Product*—When the test is made to determine the suitability of a specific package for a particular product, that product may be used inside the test specimen instead of the desiccant, in which case the specimen shall be filled to normal capacity.

6. Test Specimens

6.1 Test specimens shall be representative of the packages being tested, and shall be closed and sealed in the normal manner.

6.2 Performance shall be based on tests of a sufficient number of representative specimens of a given size and type of package to give valid results.

6.3 A sufficient quantity of desiccant (Section 5) shall be used inside the test specimen so that it does not become noticeably moist during the test. The minimum amount used in any package shall be 80 to 100 g except in instances where the actual capacity is less than 80 g, in which event not less than one half the capacity of the package shall be filled. When the product is used, the package shall be filled to its normal capacity and the moisture content of the product shall be determined immediately before the start of the test.

7. Procedure

7.1 Place the selected quantity of the desiccant or product inside the test specimen with no other contents, but with a suitable liner, if the package normally requires a liner, so that the package can be handled with the regular automatic packaging machinery. Close and seal in exactly the same manner that packages with normal contents are closed and sealed. This method may be applied to hand-sealed packages, but take great care to ensure a complete seal. Note use of the hand seal in the report.

---

[1] This test method is under the jurisdiction of ASTM Committee D-10 on Packaging and is the direct responsibility of Subcommittee D 10.32 on Consumer Packages.

Current edition approved July 27, 1979. Published September 1979. Originally published as D 895 – 47 T. Last previous edition D 895 – 68 (1973).

7.2 Prewarm the sealed package slowly to a temperature of approximately 100°F (37.8°C).

7.3 Place the warmed specimen inside the test room or cabinet in a position where free access of the conditioned circulating air is provided on all surfaces of the package.

7.4 Make successive weighings of the package at suitable intervals and plot the weight gained against time. Accuracy of the test is adversely affected by too frequent weighings. For highly permeable packages a minimum weighing frequency of 3 days is recommended. For packages having a low rate of permeability, a weighing frequency of biweekly to monthly is recommended. All weighings shall preferably be made without removing the specimen from the test chamber. In instances where this is not possible, remove the specimens one at a time, weigh immediately, and return to the test chamber in the same sequence each time they are weighed. Continue until a constant rate of gain is attained as indicated by at least three successive points in a straight line. The slope of this portion of the curve will furnish a measure of water vapor permeability.

7.5 In instances where tests are made in which specific products are used instead of the standard desiccant, a constant rate of gain may not be attainable. Therefore, for product tests, fill the package to its normal capacity and determine the moisture content from a composite sample taken at the start of the test. Place the specimen in the test room for a predetermined period (usually one month), remove, and determine the average moisture content from a composite sample taken at the end of this period. Continue the test by immediately closing and resealing, after which return the specimen without delay to the test room or cabinet.

8. Report

8.1 The report shall include the following:

8.1.1 Identification of the package, including data on closure, liners, etc.,

8.1.2 Desiccant used and quantity, or identification of contents if a product is used,

8.1.3 Temperature and relative humidity at which the test was conducted. If the tolerance of ±2 % relative humidity or ±2°F (1.1°C) temperature is exceeded, the variation shall be specified.

8.1.4 Number of specimens tested,

8.1.5 Weighing frequency,

8.1.6 Water vapor permeability reported as grams of water per 30 days for the package as a unit. The calculation shall be made for the period of constant rate of gain. If a product is used instead of a desiccant, the moisture content at the start and completion of the test shall be given, together with the net weight of contents, from which data the actual amount of water pickup can be calculated,

8.1.7 Data on any previous tests to which the specimens might have been subjected, and

8.1.8 A statement to the effect that all tests were made in accordance with this test method.

9. Precision

9.1 Duplicate determinations on uniform, undamaged specimens should agree within 10 % within a given laboratory. However, the reproducibility between different laboratories has been found to vary as much as 25 %. This must be recognized when making interlaboratory comparisons of data.

Standard Test Method for Water Resistance of Shipping Containers by Spray Method[1]

This standard is issued under the fixed designation D 951; the number immediately following the designation indicates the year of original adoption or, in the case of revision, the year of last revision. A number in parentheses indicates the year of last reapproval. A superscript epsilon (ϵ) indicates an editorial change since the last revision or reapproval.

1. Scope

1.1 This test method covers the determination of the water resistance of shipping containers.

1.2 This test method is frequently used in conjunction with other tests made prior to or after the spray test, such as the drop test, vibration test, inclined impact test, or compression test.

1.3 *This standard does not purport to address the safety problems associated with its use. It is the responsibility of the user of this standard to establish appropriate safety and health practices and determine the applicability of regulatory limitations prior to use.*

2. Referenced Document

2.1 *ASTM Standard:*

D 685 Method of Conditioning Paper and Paper Products for Testing[2]

3. Significance and Use

3.1 This test method is used to determine the water resistance of shipping containers. It can be used to determine the ability of the container to resist deterioration caused by water or the ability of the container to protect the contents from water. It is frequently used in conjunction with other tests made prior to or after the spray test, such as the drop test, inclined impact test, vibration test or compression test.

4. Apparatus

4.1 The apparatus, illustrated schematically in Fig. 1, shall consist of the components described in 4.2 to 4.8. Modifications are permissible, such as the use of fresh tap water instead of a recirculating system, as long as the specified temperature and spray intensity are achieved.

4.2 *Test Room*—A test room or cabinet shall be of water-resistant construction, insulated and heated when necessary so that proper temperature control can be maintained. The bottom shall be covered with a false floor of slats and have an outlet drain.

4.3 *Sprays*—Spray nozzles shall be of such size and so spaced that the specified intensity of spray falls uniformly distributed over the floor area. The spray nozzles shall be so located that the droplets are falling from gravitational force only when they strike the specimens.

4.4 *Flow Control Valves*—Flow control valves, to control the intensity of the spray, are required.

4.5 *Circulating System*—A circulating system shall, where used, consist of a fine-mesh strainer, pump, relief by-pass, gages, and pressure regulating valve, together with the necessary piping between the flow control valves, spray nozzles, and water reservoir.

4.6 *Reservoir*—A reservoir for storage and conditioning of the spray water shall, where used, be equipped with an overflow to a sewer and with a drain to facilitate changing the water at the start of each test. Make-up water, regulated by a float control, shall also discharge into this tank.

4.7 *Water Temperature Control*—A thermostatically controlled means of heating for maintaining the supply water at the desired temperature is required.

4.8 *Conditioning Apparatus*—Adequate facilities shall be provided for conditioning test specimens at proper humidity and temperature prior to test.

5. Test Specimens

5.1 Test at least three containers, closed and sealed in the normal manner. Select as test specimens samples that are representative of the containers being tested. For some purposes (such as for checking leakage alone)' the containers may be tested empty. Normally, however, the containers shall be packed with the contents designed to be shipped in them or with a material of similar character.

5.2 Condition the test specimens sufficiently, after closing and sealing, so that any adhesives, protective coatings, glued flaps, and the like, will have reached their final normal condition, and shall be at the same temperature as the test room before being placed in it. Condition fibreboard containers in accordance with Method D 685.

6. Procedure

6.1 Weigh each sample after conditioning and prior to test.

6.2 Adjust spray nozzles and flow control valves to the required spray intensity. See Appendix X1.

6.3 Start and operate the water sprays until the entire system, including the test room, has reached equilibrium. Unless otherwise specified, the standard temperature of the spray water and test room shall be 23 ± 2°C (73.4 ± 3.6°F), maintained by thermostatic control.

6.4 Place each sample to be tested on the false floor of the test room in normal stacking position or in any other position specifically called for in the test. Be certain the distance between adjacent samples is not less than 150 mm (6 in.). Operate the water sprays continuously at the specified intensity for the duration of the test.

---

[1] This test method is under the jurisdiction of ASTM Committee D-10 on Packaging and is the direct responsibility of Subcommittee D10.23 on Storage Test Methods.

Current edition approved Feb. 26, 1988. Published May 1988. Originally published as D 951 – 47 T. Last previous edition D 951 – 86.

[2] *Annual Book of ASTM Standards*, Vol 15.09.

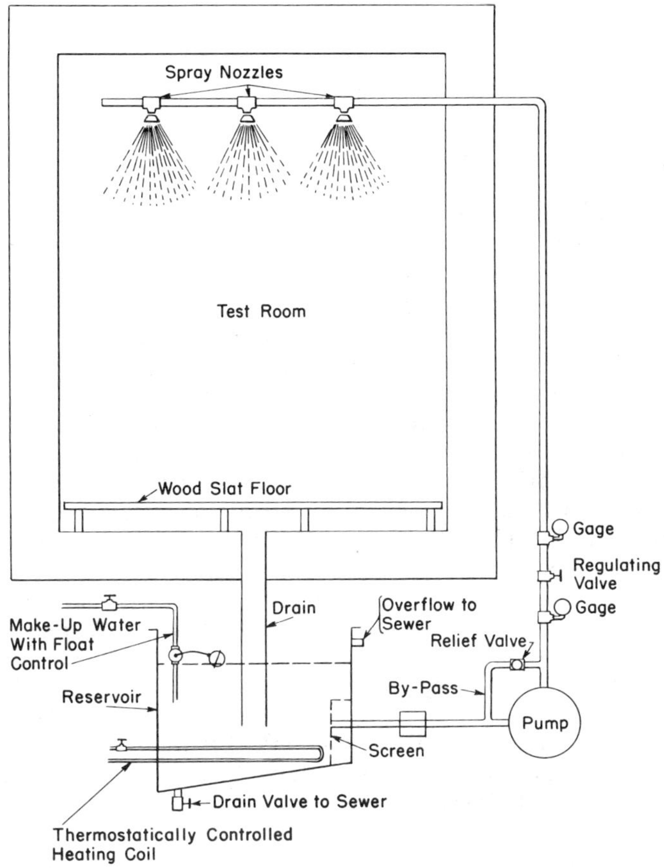

FIG. 1 Schematic Diagram of Spray Apparatus

6.5 Immediately at the conclusion of the spray test, remove any liquid water from the exterior surfaces of each sample using a squeegee or other suitable means, and weigh the samples.

6.6 Immediately following the weighing of the container, make any required examination to determine the condition of the containers and their contents. Observe location and extent of any leakage into the container where the test is being performed to determine the ability of the container to protect the contents from water. Immediately follow this examination with any subsequent tests that are required.

7. Report

7.1 The report shall include the following:

7.1.1 Identification of container, including data on closure, liners, etc.

7.1.2 Identification of contents.

7.1.3 Temperature at which the test was conducted.

7.1.4 Intensity of the spray.

7.1.5 Duration of test.

7.1.6 Position of containers during test.

7.1.7 Data on condition of the container liners, and contents at the completion of the spray test. These comparative data shall be as detailed and specific as the nature of the container and its contents will permit and shall include, when possible, changes in mass, water leakage determined by measurement or estimated by visual examination, amount of product spoilage, etc.

7.1.8 Initial and final weight.

7.1.9 Data on any previous or subsequent tests to which the test containers are submitted.

8. Precision and Bias

8.1 No statement is made about either the precision or bias of this test method since the result merely states whether there is conformance to the criteria for success specified in the procedure, and because the results depend on the precision and bias of other tests carried out following this test.

APPENDIX

(Nonmandatory Information)

X1. SPRAY INTENSITIES

X1.1 The following intensities have been used in carrying out the water spray test.

X1.1.1 *High Intensity*, 100 ± 25 mm/h (4 ± 1 in./h)—This intensity is useful to simulate tropical or subtropical conditions.

X1.1.2 *Medium Intensity*, 50 ± 10 mm/h (2 ± ½ in./h)—This intensity is useful to simulate temperate zone or protected storage conditions. The International Atomic Energy Agency (IAEA) uses this intensity for package testing.

X1.1.3 *Low Intensity*, 25 ± 10 mm/h (1 ± ½ in./h)—This intensity is useful to simulate protected storage conditions where incidental exposure to weather may occur during loading and unloading from covered transport into the covered storage.

Designation: D 959 – 80 (Reapproved 1986)

Standard Method of Drop Test for Filled Bags[1]

This standard is issued under the fixed designation D 959; the number immediately following the designation indicates the year of original adoption or, in the case of revision, the year of last revision. A number in parentheses indicates the year of last reapproval. A superscript epsilon (ϵ) indicates an editorial change since the last revision or reapproval.

1. Scope

1.1 This method covers drop tests on loaded bags for measuring the ability of the bag to withstand handling or for comparative evaluation of various bag constructions.

1.2 This procedure is suitable for testing all types of bags.

1.3 *This standard does not purport to address the safety problems associated with its use. It is the responsibility of whoever uses this standard to consult and establish appropriate safety and health practices and determine the applicability of regulatory limitations prior to use.*

2. Referenced Documents

2.1 *ASTM Standards:*

D 585 Method for Sampling and Accepting a Single Lot of Paper, Paperboard, Fiberboard, or Related Product[2]

D 685 Method of Conditioning Paper and Paper Products for Testing[2]

3. Significance and Use

3.1 In the life cycle of many bags, there is a point at which the bag is handled either manually or mechanically in a manner such that there is an increased probability that the bag will fall and possibly rupture. Since bags are most often used to contain products in the form of small particles (powders, granules, etc.) which are flowable to some extent, the rupture of a bag will result in the product spilling from the rupture. This not only results in an economic loss but may additionally be the source of a health hazard and a difficult clean-up. This test method is used to indicate the ability of the bag to withstand the shock produced when it is dropped. This test method also is an aid in the design and improvement of bags.

4. Apparatus

4.1 *Drop Test Apparatus*—Any suitable apparatus may be used that conforms to the following requirements:

4.1.1 Permits accurate prepositioning of the bag to assure a true fall and impact at the exact places and in the direction desired,

4.1.2 Permits accurate and convenient control of the height of drop,

4.1.3 Facilitates handling and elevation of the bags, particularly when weights are in higher brackets,

4.1.4 Utilizes lifting devices that will not damage the bags,

4.1.5 Permits an absolutely free, unobstructed fall,

4.1.6 Provides for variations in height of drops within limits of anticipated requirements,

4.1.7 Provides a solid surface of concrete, stone, or steel of sufficient mass to absorb all shock without deflection, and

4.1.8 Provides a 4 by 4-in. (102 by 102-mm) timber (about 4 ft (1.2 m) in length) having the edges rounded to a radius of not more than ¼ in. (6.4 mm), firmly positioned on a drop surface.

4.2 *Conditioning Apparatus*—Adequate facilities for conditioning test specimens at proper humidity and temperature prior to test in accordance with the requirements of the specifications covering the bags to be tested.

4.3 *Miscellaneous Equipment*—Drying oven, scales, knife, etc., for use in determination of the moisture content or for making other supplementary tests of the materials from which the bags are made.

5. Sampling

5.1 Any convenient or specified sampling plan may be used. When the test is for the purpose of accepting a lot, Method D 585 shall be used.

5.2 In the absence of any specific plan, no fewer than five representative specimens of a given size or construction shall be selected for evaluating performance.

6. Test Specimens

6.1 Pack the bag to be tested with the actual contents for which it was designed or a dummy load simulating these contents. Close and seal the bag in the same manner as will be used in preparing it for shipment.

6.2 The procedure for the identification of the faces, edges, and butts of bags shall be as follows: Facing the front of the bag with top up, designate the front of the bag as *1*, the right side edge as *2*, the rear side as *3*, the left side edge as *4*, the bottom as *5*, and the top as *6*.

7. Conditioning

7.1 Depending on the purpose of the tests, bags filled and sealed may be conditioned prior to the drop test by either water immersion, exposure to water spray, or exposure to fixed conditions of air temperature or humidity.

8. Procedure

8.1 *Procedure A*—Drop the bags on any face, butt, or side in accordance with 8.3.

8.2 *Procedure B*—Drop the bags on faces, butts, and sides in a specified sequence of drops as follows: *(1)* front, *(2)* back, *(3)* right side, *(4)* left side, and optionally, *(5)* bottom, *(6)* top,

---

[1] This test method is under the jurisdiction of ASTM Committee D-10 on Packaging and is the direct responsibility of Subcommittee D10.22 on Handling and Transportation.

Current edition approved April 28, 1980. Published June 1980. Originally published as D 959 – 48 T. Last previous edition D 959 – 50 (1973).

[2] *Annual Book of ASTM Standards,* Vol 15.09.

and in accordance with 8.3. Before each drop uniformly distribute the contents throughout the bag.

8.3 Position the bag with the center of gravity of the contents over the point of impact, and the face, side, or butt on which the bag is to fall parallel to the floor. Make certain the drops are vertical, free, and unobstructed.

8.4 *Height of Drop*—Drops on faces, butts, and sides may be from the same height, or drops on butts and sides may be from lesser heights, in a definite relationship. Unless otherwise specified, the height of face drops shall be 4 ft (1.2 m).

8.5 The use of the 4 by 4-in. (102 by 102-mm) hazard is optional.

8.6 Continue the test until a specified predetermined number of drops, estimated to be equivalent to handling in actual service, have been given the bag or until failure occurs. Failure may be considered to have occurred when the contents are exposed, the contents spill from the bag, or the bag breaks open.

9. Report

9.1 The report shall include the following:

9.1.1 Dimensions of bag under test, its complete specifications, kind of material, method of closing and sealing, net and gross masses,

9.1.2 Description of the contents of the bag under test and load density in pounds per cubic foot (or kilograms per cubic metre),

9.1.3 Method, if any, of conditioning the bag and the results of any supplementary tests of the material from which the bag is made,

9.1.4 Type of apparatus and procedure used,

9.1.5 Number and height of drops to completion of test or to failure of bag,

9.1.6 Detailed record of test on each bag including damage to the bag together with any other observation that may assist in correctly interpreting the results or improving the design of the bag, and

9.1.7 Statement to the effect that all tests were made in full compliance with this method.

10. Precision and Bias

10.1 The drop test for filled bags produces a high degree of variation. The variation may be caused by the lack of homogeneity of both the material from which the bag is made and the contents as well. Therefore, reproducibility will not generally lie between acceptable limits.

Designation: D 996 – 90a

Standard Terminology of Packaging and Distribution Environments[1]

This standard is issued under the fixed designation D 996; the number immediately following the designation indicates the year of original adoption or, in the case of revision, the year of last revision. A number in parentheses indicates the year of last reapproval. A superscript epsilon (ε) indicates an editorial change since the last revision or reapproval.

INTRODUCTION

The terms and definitions in this standard are grouped into related areas under principal concepts. The broad descriptor term for each group is followed in alphabetical order by narrower terms and related terms. Cross-references are included where the concept group is not obvious.

absorbent packing—See **packing.**

adhesive, *n*—a substance capable of holding materials together by surface attachment.

DISCUSSION—Adhesive is the general term and includes among others, cement, glue, mucilage, and paste. All of these terms are loosely used interchangeably. Various descriptive adjectives are applied to the term adhesive to indicate certain characteristics as follows: (*1*) Physical form, that is, liquid adhesive, tape adhesive, (*2*) Chemical type, that is, silicate adhesive, resin adhesive, (*3*) Materials bonded, that is, paper adhesive, metal-plastic adhesive, can label adhesive, and (*4*) Conditions of use, that is, hot-setting adhesive (D 907, D-14)[2].

contact adhesive, n—an adhesive that is apparently dry to the touch and that will adhere to itself instantaneously upon contact; also called contact bond adhesive or dry bond adhesive (D 907, D-14)[2].

aerosol package—See **package.**

ampoule, *n*—a hermetically sealed, small bulbous glass or plastic vessel. Opening is achieved by breaking the stem. (Also *ampule* or *ampul.*)

anchor, *v*—to secure firmly (*Webster*).

anti-skid plate—See **loading.**

bag, *n*—a preformed **container** of tubular construction made of flexible material, generally enclosed on all sides except one forming an opening that may or may not be sealed after filling. (See also **pouch.**)

DISCUSSION—A bag may be made of any flexible material, or multiple plies of the same, or combination of various flexible materials. The term bag is used as a synonym for sack, but the term sack generally refers to the heavier duty or shipping sacks. It is made in various standard styles and may be open-mouth or valve type. The five basic standard types of bags are: (*1*) grocery bag, (*2*) merchandise paper, (*3*) industrial, (*4*) textile, and (*5*) paper shipping sack.

paper multiwall-sack—a flexible **container** made of several plies, usually of kraft paper. The various plies may be specially treated, such as waxed paper, glassine, greaseproof, polyethylene, **wet strength paper,** or other specialty sheets. The particular nature of the sack depends upon the material to be packed and the type of transportation to be employed.

bag liner—See **liner.**

bail, *n*—the usually arched handle of a pail or can.

bale, *n*—in packaging a shaped unit, bound with cord or metal ties under tension, and containing compressed articles or materials. It may be wrapped.

banding—Use **strapping.**

barrel, *n*—a bulged cylindrical **container** of greater length than breadth, made of wooden **staves** bound together with hoops and having two flat ends of equal diameter. (Compare **drum.**)

cask, n—a term used synonymously with **barrel** but usually of large size or capacity.

keg, n—a small slack or tight **barrel** of 30-gal capacity or less.

barrier material:

grease-resistant barrier—a material that prevents or retards the transmission of grease or oils.

water-resistant barrier—a material that retards the transmission of liquid water.

water-vapor-resistant barrier—a material that retards the transmission of water vapor.

basket, *n*—a semirigid **container** usually open at the top and provided with one or two handles for carrying. (Compare **hamper.**)

DISCUSSION—A **basket** is sometimes made of thin strips of wood, woven or stapled, or otherwise bound together, or it may be made of fiberboard or combinations of wood and fiber, or plastic. (See **stave.** (*2*))

batten—See **box.**

blister pack—See **pack.**

blocking—See **loading.**

body—See **container.**

bottle—See **container.**

bottom—See **box.**

box, *n*—a rigid **container** having closed faces and completely enclosing the contents. When this term is used in connection with fiberboard boxes, such fiber boxes must comply with all the requirements of the carrier rules. (See **carton.**)

bottom, n—the **face** of a **box** on which it usually rests while filling.

[1] These terms are under the jurisdiction of ASTM Committee D-10 on Packaging and is the direct responsibility of Subcommittee D10.11 on Terminology.

Current edition approved Nov. 30, 1990. Published January 1991. Originally published as D 996 – 48 T. Last previous edition D 996 – 90.

[2] *Annual Book of ASTM Standards*, Vol 15.06.

DISCUSSION—In terms of fiberboard shipping boxes, the face created by the flaps of regular (or similar style) slotted boxes are the top or bottom, regardless of loading or stacking.

box batten, n—a reinforcing member, (*1*) for a **wood box** internally or externally applied to the sides, top and bottoms. When applied externally it should be applied in pairs; (*2*) in a **wirebound box,** a batten is a reinforcement used on the ends of the container only.

cleated fiberboard box—a rigid **container** having five or six **panel faces** with wood strips fastened to them, the panels being made of **solid** or **corrugated fiberboard.**

cleated plywood box—a rigid **container** having five or six **panel faces** with wood strips fastened to them, the panels being made of plywood.

flange, n—in **fiberboard boxes,** an extension to a panel similar to a short flap that may be folded in or out, usually at angles of 90 or 180° to the panel.

nailed wood box—a rigid **container** constructed of wood in several standard styles, assembled by fastening sides, top and bottom to the ends with nails or other suitable fasteners.

skid box—a metal, wooden, or fiber **box** fastened to a platform raised on skid members or legs; it may or may not be collapsible. (See also **skid.**)

wirebound box—a rigid **container** whose sides, top, and bottom are of rotary-cut lumber, sliced lumber, **resawn lumber, fiberboard,** or combinations thereof, usually 3/8 in. (9.5 mm) or less in thickness, fastened to **cleats** and to each other by means of binding wires and staples; and ends of similar material, plain or stapled to **battens** or **liners,** fastened in place by means of nails or staples or wires stapled thereto.

DISCUSSION—The **closure** is made by twisting or looping together the ends of the binding wires.

boxboard—See **paperboard.**

bracing—See **loading.**

bubble packaging material—a material consisting of a flexible plastic film having uniformly spaced bubbles integrally molded therein.

DISCUSSION—These bubbles may or may not be permanently affixed to a separate backing film to either seal the air within the bubbles or to add dimensional stability to the structure. Bubble packaging is primarily used as a cushioning material.

buffer, *n*—a material or device, such as folded up **corrugated fiberboard,** placed in a container to position and protect the contents from the forces of impact.

DISCUSSION—A **buffer** is usually made of a cushioning, or compressible material. It may be made in a variety of styles such as spring buffer, rolled-up buffer, die-cut, etc. (See **cushioning material.**)

bulk packaging—see **packaging.**

bundle, *n*—two or more articles held together with rope, wire, or **strapping** so as to form a shipping unit; it may be wrapped.

bung hole—*in packaging,* an opening in a **barrel** or **drum** through which material can be poured to fill, empty or vent.

bursting strength—See **package testing.**

bursting strength test—See **package testing.**

caliper—See **package testing.**

can, *n—in packaging,* a receptacle generally of 10-gal capacity or less, normally not used as a **shipping container.**

DISCUSSION—The body is made of lightweight metal or is a composite of paperboard and other materials having the ends made of paperboard, metal, plastic, or a combination thereof.

Cady test—See **package testing.**

cap—See **container (cover).**

carboy, *n*—a **container** made of glass, ceramic, plastic, or metal, having a capacity of 5 to 15 gal (19 to 57 L) with the pouring and filling opening at the top.

DISCUSSION—For shipment, carboys are generally encased in a protective rigid outer container.

carton, *n*—a folding **box,** generally made from **boxboard** for merchandising consumer quantities of products (for example, shelf packages or prime packages).

case—See **container.**

case liner—See **liner.**

cask—See **barrel.**

child-resistant packaging—See **packaging.**

chime (chine), *n—in packaging,* the rim of a **container,** such as a **drum, barrel,** or **can.**

chipboard—See **paperboard.**

cleat, *n*—a wood or metal strip attached along the edge of a **panel** of a **container** for the attaching of an adjacent panel, or fastened to the panel between the edges, or to barrel heads, for reinforcement and stiffening.

cleated fiberboard box—See **box.**

cleated plywood box—See **box.**

closure, *n—in packaging,* a means of closing a **container** to retain the contents.

plug, n—in packaging, a type of **closure** that is designed to be inserted into a **container** opening. It may be held by friction or by screw threads. (See **cap.**)

cocoon, *v—in packaging,* to employ strippable, usually plastic, sometimes multi-layered films to encapsulate an item.

code, *v*—to assign numbers, letters, words, or symbols as identifying marks to **containers,** packaged materials, or articles to convey information concerning the qualities of the container or its contents, date, place of manufacture, or other significant identification. (Compare **marking.**)

collapsible tube—See **tube.**

composite tube—See **tube.**

Conbur test—See **package testing.**

constant load—See **load.**

consumer package—a primary container (which may be enclosed in a secondary container) designed to contain, store, and protect from the point of manufacture to the point of use, a product intended for household or individual use. (See *unit package.*)

DISCUSSION—The primary or secondary container may also be used as the shipping container and it may be used to communicate information relative to the product or its use.

contact adhesive—see **adhesive.**

container—a nonspecific term for a receptacle capable of **closure** (See also: **bag, barrel, basket, box, can, carton, crate, cylinder, drum, envelope, hamper, pail, tube.**)

body, n—in packaging, the principal part of a **container,** usually the largest part in one piece containing the sides.

bottle, n—a rigid or semirigid **container** typically of glass or plastic, having a comparatively narrow neck or mouth, and usually no handle (*Webster*).

case, n—a nonspecific term for a **shipping container.** In domestic commerce, case usually refers to a box made from **corrugated** or **solid fiberboard** wood, or metal.

cover, n—in packaging, the top or bottom, or both of a **container,** usually the part that closes the filling and dispensing opening. It is often called a cap when used with **fiberboard containers.** (See also **shroud.**)

cylinder, n—a rigid cylindrical metal **container** designed as a portable vessel for the storage and transportation of compressed gases. Generally equipped with protected valve closure and suitable pressure-relief safety device.

die-cut, adj.—(*1*) a method of preparation in which a part or **container** has been cut, slotted, and scored or any combination of these by custom-made dies; (*2*) *n*, a part so made.

expendable container—a **container** for shipping or storage, or both, intended primarily for a single trip.

face, n—in packaging, any one of the plane surfaces of a **container.**

fast pack container—a standard size, **reusable container** with foam cushion **inserts.**

DISCUSSION—Some designs permit shipment of a large variety of items within certain limits of size, weight, configuration and fragility.

fiberboard container—a **box, package,** or **drum** made of **fiberboard.** When the term box is used for classification purposes, the structure must comply with all requirements of the carrier rules.

flat, n—one of the closing members of a **fiberboard container.**

glass container—any glass receptacle capable of holding a **seal** or **closure** for retention of contents.

intermodal container—a reusable **shipping container** manufactured to standard dimensions intended to unitize cargo or freight for shipping by one or more modes of transportation without the need for intermediate handling of the contents.

jar, n—a widemouthed container made typically of glass, plastic, or earthenware.

jug, n—a large, deep, usually glass, plastic or earthenware container with a narrow mouth and a handle.

manufacturer's joint—that part of a **fiberboard container** where the ends of the box blank are joined together in the manufacturing process by taping, stitching, or gluing.

modular container—a family of **containers** designed to be assembled into a **unit load.**

returnable container—a **shipping container** of any material designed to be used for more than one shipment.

reusable container—a shipping and storage **container** designed for reuse without impairment of its protective function.

DISCUSSION—It may be repaired or refitted to prolong its life, or to adapt it for items other than originally intended.

seam, n (when referring to a **fiberboard container**)—the lines of junction created by any free edge of a container flap or wall where it abuts or overlaps another portion of the **container** (except the **manufacturer's joint**).

DISCUSSION—A seam may be fastened by tape, stitches, or adhesive in the process of closing a fiberboard container.

shipping container—a container that is sufficiently strong to be used in commerce for **packing,** storing, and shipping commodities. (See also **barrel, crate, drum.**)

containerboard—any **paperboard** made specifically for the manufacture of **corrugated** and **solid fiberboard shipping containers.** Basis weight is expressed in pounds per 1000 ft^2 (or grams per square metre). It is customarily shipped in rolls.

cylinder kraft—**containerboard** made from kraft pulp on a cylinder machine.

Fourdrinier kraft—**containerboard** made from kraft pulp on a Fourdrinier machine, basically of single-ply formation, although possibly with supplementary second-ply, with less prominent grain direction. The sheet is formed on a traveling endless-wire screen which may also be vibrated to obtain more random orientation of fibers.

solid fiberboard—a solid board made by laminating two or more plies of **containerboard.**

containerization, *n*—(*1*) a shipping method in which material (such as merchandise) is packaged together in one **container.** (*2*) the use of transport **containers** to unitize cargo for transportation, supply, and storage.

core, *n—in packaging,* a cylindrical structure used as a carrier of flexible material that is wound around it.

corrosion, *n*—the chemical or electrochemical reaction between a material, usually a metal, and its environment that produces a deterioration of the material and its properties (G 15, G-1)[3].

corrosion, inhibitor, n—a chemical substance or combination of substances that, when present in the proper concentration and form in the environment, prevents or reduces corrosion.

volatile corrosion inhibitor (*VCI*)—a material that slowly releases vapor to inhibit corrosion within a **package** by neutralizing the effects of moisture-laden air.

corrugated box—See **box.**

corrugated fiberboard:

(*1*) *single face*—the structure formed by one corrugated member glued to the flat facing;

(*2*) *single wall*—the structure formed by one corrugated inner member glued between two flat facings; also known as double face;

(*3*) *double wall*—the structure formed by three flat facings and two intermediate corrugated members;

(*4*) *triple wall*—the structure formed by four flat facings and three intermediate corrugated members. (See also **containerboard.**)

corrugating medium—**paperboard** used in forming the fluted portion of the **corrugated board.**

corrugation flute—one of the wave shapes formed in the inner member, that is, the corrugating medium, of corrugated fiberboard. Flutes most commonly used are:

[3] *Annual Book of ASTM Standards,* Vol 03.02.

	Number, per Linear ft	Span Between Adjacent Flutes, mm	Height, in. (mm)[A,B]
A-flute	36 ± 3	7.9 to 9.1	3/16 (4.7)
B-flute	50 ± 3	4.7 to 6.6	3/32 (2.4)
C-flute	42 ± 3	6.8 to 7.8	9/64 (3.6)
E-flute	94 ± 4	3.0 to 3.5	3/64 (1.2)

[A] The values are approximate.
[B] Height does not include thickness of facing.

V-board—a term adopted from the grade symbol of **corrugated** or **solid fiberboard** made to comply with the weather-resistant class as defined in Federal Specification PPP-F-320, "Fiberboard, Corrugated and Solid. Sheet Stock (Container Grade), and Cut Shapes,"[4] made of wet strength **paperboard.** Components are especially made to exhibit high strength (against bursting, tearing, or rupturing) when wet.

W-board—same as V-board except the "W" grades are of lower test requirements and primarily for use as interior or intermediate containers.

cover—See **container.**

crate, *n*—a rigid **shipping container** of framed construction joined together with nails, bolts or any equivalent method of fastening. The framework may or may not be enclosed with sheathing. It may be demountable (reusable) or nondemountable. (See also **rubbing strip, strut.**)

open crate—a **crate** with exposed frame members and not enclosed by **sheathing.**

sheathed crate—a **crate** that is enclosed by having the frame members completely covered with **sheathing** boards or material.

creped duplex paper—See **paper.**

creped paper—See **paper.**

cube, *n—in packaging,* the volume of space occupied by the unit under consideration, computed by multiplying overall exterior length, width, and height. For shipping purposes cube is expressed to the nearest 0.1 ft^3 (2830 cm^3). (Compare **displacement.**)

cushion, *v*—to use **cushioning material** to reduce shock and vibration transmitted to a packaged product from an externally applied force.

cushioning material—a material used to isolate or reduce the effect of externally applied shock or vibration forces, or both. (See also **buffer, divider, molded shape.**)

creped cellulose wadding—a material consisting of cellulose fibers produced by the sulfite process, loosely matted into sheet form and then creped. It is available in single or multiple sheet form, either plain or embossed, and may be backed with various papers. It is also available impregnated with asphalt for water resistance.

filler pad—in packaging, a **pad** used to fill space; sometimes applied to a soft flexible pad made with various loose filling materials to provide cushioning effects. (Compare **buffer.**)

foam-in-place cushioning material, n—one formed by dispensing, usually into a box or mold, reactive chemical components that expand to envelop items packaged or occupy void areas.

macerated paper—generally, waste paper torn up mechanically for use as a **cushioning material.**

cylinder—See **container.**

cylinder kraft—See **containerboard.**

dead load—See **load.**

density, *n*—mass per unit volume (D 3288, D-9)[5].

desiccant, *n*—a hygroscopic substance used to absorb water vapor from the air to maintain a low relative humidity in a container.

diagonal bracing—See **loading.**

die cut—See **container.**

dimensions, *n—in packaging,* the measurement of length, width (or diameter), and depth of **containers,** expressed in that order; it should be stated as "inside" or "outside."

DISCUSSION—For fiberboard and most other types of boxes, length is the larger of the two dimensions of the open face, width is the lesser of the two dimensions of the open face, depth is the distance between the innermost surfaces of the box measured perpendicular to the length and width, and are given as inside dimensions.

displacement, *n—in packaging,* the volume occupied by a **container,** calculated from its outside dimensions. (Compare **cube.**)

divider, *n*—a device, made of various materials, that separates the space within a **container** into two or more spaces, cells, compartments, or layers.

DISCUSSION—A divider may be plain, interlocking, scored, horizontal, vertical, or diagonal. The primary purpose of a divider is to separate the articles, or to furnish cushioning, or both. Also, it frequently adds stacking strength (Compare **buffer, fiberboard partition, liner, pad, separator, spacer.**)

dolly, *n*—a low platform or structure mounted on wheels or casters, designed primarily for moving bulky loads for short distances. (Compare **pallet.**)

drop test—See **package testing.**

drum, *n*—(*1*) a cylindrical **shipping container** having straight sides, and flat, convex or embossed ends, designed for storage and shipment as an unsupported outer package that may be shipped without boxing or crating. It may be made of metal, or of plywood, or of fiber with wooden, metal or fiber ends. Drums are also made of rubber or plastics (Compare **barrel.**); (*2*) in set-up paper boxes, a shell or tube with paper or cellophane head, used for powder box. (See **divider.**)

dunnage—See **loading.**

dynamic load—See **load.**

edge protector—See **loading.**

Elmendorf test—See **package testing.**

end-grain nailing—nailing in such a way that the point of the nail follows the grain of the wood so that the shank is parallel, or nearly parallel to the grain in that member holding the nail point. It is weaker than side-grain nailing in direct withdrawal.

envelope, *n—in packaging,* a container of flexible material having only two **faces** and joined at three edges to form a partial enclosure. The nonjoined edge provides a filling opening which later may be closed or sealed.

expendable container—See **container.**

expendable pallet—See **pallet.**

[4] Available from Superintendent of Documents, US Government Printing Office, Washington, DC 20402.

[5] *Annual Book of ASTM Standards,* Vol 10.02.

exterior pack—See **pack.**
face—See **container.**
facing, *n*—a form of *linerboard* used as a flat member of **corrugated fiberboard** (sometimes erroneously called a **liner**).
fastener, *n—in packaging,* a device that serves to secure one part to another; for example, nail, screw, **staple, strapping, stitch** or adhesive.
fast pack container—See **container.**
fiberboard—See **containerboard.**
fiberboard container—See **container.**
fiberboard partitions—a set of **corrugated** or **solid fiberboard** or **paperboard** pieces slotted so they will interlock when assembled to form a number of cells into which articles may be placed for shipment. (See also **divider.**)
fiberboard tube—See **tube.**
filler pad—See **cushioning material.**
finish, n—See **paperboard.**
fire-retardant, *adj.*—having or providing comparatively low flammability or flame spread properties (E 176, E-5)[6].
flange, n—See **box.**
flat—See **container.**
floating controlled load—See **load.**
flute—See **corrugated fiberboard.**
foam-in place cushioning material—See **cushioning material.**
Fourdrinier kraft—See **containerboard.**
friction, *n*—resistance to relative motion between two bodies in contact (*Websters*).
coefficient of friction—the ratio of the force required to move one surface over another, to the total force applied normal to those surfaces.
kinetic coefficient of friction—the ratio of the force required to move one surface over another, to the total force applied normal to those surfaces, once that motion is in progress.
static coefficient of friction—the ratio of the force required to move one surface over another, to the total force applied normal to those surfaces, at the instant motion starts.
G—symbol for the dimensionless ratio between an acceleration in length per time-squared units, and the acceleration of gravity in the same units (D 1596)[7].
g—symbol for the acceleration of gravity at the earth's surface.

DISCUSSION—Its value differs slightly at different points on the earth, but the standardized value of 32.2 ft/s^2 or 9.806 m/s^2 is usually used.

glass container—See **container.**
grease-resistant barrier—See **barrier material.**
gummed paper tape—See **tape.**
hamper, *n*—a **container** (commonly used for shipping fruits and vegetables) circular, elliptical or polygonal in horizontal cross section, the tube dimensions being usually greater than the bottom. It has slotted sides and a bottom that may be loose, stapled, or nailed in place. The top may or may not be open. (Compare **basket.**)
heat seal—See **seal.**
hermetic seal—See **seal.**
humidity indicator—an instrument or device that displays the approximate humidity condition within a **package.**
ID—inside dimensions or inside diameter.
incline impact (Conbur) test—See **package testing.**
inner packing—See **packing.**
insert, *n*—usually a thin filler or frame of wood, **fiberboard,** plastic, or other suitable material used to take up space, or separate articles within a **package.** (Compare **buffer, divider, separator.**)
interior packing—See **packing.**
intermediate pack—See **pack.**
intermediate package—See **package.**
intermodal container—See **container.**
keg—See **barrel.**
kraft, n—See **paperboard.**
label, *n*—a piece of paper or other material to be affixed to a container or article, on which is printed a legend, information concerning the product, or addresses. It may also be printed directly on the container. (Compare **tag.**)
liner, *n—in packaging,* (*1*) generally, any linear material that separates a product within a **container** from the basic walls of the **container,** (*2*) in **fiberboard containers,** a creased fiberboard sheet inserted in a container and usually fitting against the side and end panels (*liner* is sometimes erroneously used for *linerboard* or **facing**). (Compare **divider.**)
case liner (*bag liner*)—a lining, usually paper, or treated materials placed inside a shipping container for the purpose of preventing sifting, or entrance of moisture, dust, or dirt.
linerboard—See **paperboard.**
live load—See **load.**
load, *n*—(*1*) the force in weight units applied to a body; (*2*) the weight of the contents of a **container** or transportation device; (*3*) a qualitative term denoting the contents of a container.
constant load, n—a load that is invariable or unchanging.
dead load, n—a constant load that, in structures (as a bridge, building, or machines) is due to the weight of the members, the supporting structure, and permanent attachments or accessories (*Webster*).
dynamic load, n—an imposed force in motion; that is, one that may vary in magnitude, sense, and direction.
floating load—(*1*) a shipment, usually a *unitized load* (or loads), or a large individual article, so prepared that it may move in the carrying vehicle, the movement being restricted or retarded by friction between the load and the vehicle; (*2*) a method of packing in which the contents of the **container** are supported within the container by **cushioning** devices or **materials**; (*3*) *floating controlled load*—a **floating load** within a vehicle in which the movement is retarded, snubbed, or restrained by suitable devices.
live load, n—a moving load on a structure.
palletized load—a load made up of articles, loose or in **containers** placed on **pallets** or skids.
palletized unit load—a unitized **load** fixed to a pallet.
static load, n—an imposed stationary force, constant in magnitude, sense and direction.

[6] *Annual Book of ASTM Standards,* Vol 04.07.
[7] *Annual Book of ASTM Standards,* Vol 15.09.

unitized load—in packaging, a type of unit load consisting of articles or containers secured together so as to be handled as an entity.

unit load—in distribution, an item or assembly of items assembled or restrained for handling and transportation as a single entity.

loading, *n*—the act of placing a load on or in; to load a car, a vessel, or a test specimen. (See also **unitization.**)

anti-skid plate—a device, generally metal, about 4 by 6 in. (100 by 150 mm), with sharp projections on each face, placed under, against, or between containers and car floor, to retard shifting of load in transit.

blocking, n—material used to prevent or control movement of the unit or *load* or to facilitate handling (A 700, A-1).

bracing, n—material or devices used to hold articles or sections of loads in position and prevent shifting within a transportation vehicle or within a **container.** (See **molded shape, strapping.**)

diagonal bracing—(*1*) a member reaching at an angle from a gate or other structure to the wall or floor of a freight car or truck to strengthen and reinforce the gate or structure; (*2*) a member, single or multiple, of a **crate** or **box** attached diagonally or at an angle to add reinforcement to the container.

dunnage, n—(*1*) in a carrying vehicle, the temporary **blocking,** flooring or lining, racks, standards, strips, stakes, or similar **bracing,** or supports not constituting a part of the carrying vehicle, used to protect and make freight secure in, or on a carrying vehicle (Classification, Rule 30);[8] (*2*) in a **container,** materials not constituting a part of the container, frequently by-product or scrap, used for filling space, for **blocking** or **bracing,** or otherwise to protect and secure the contents.

dynamic compression loading—in packaging, the application of a force in motion that usually occurs in 10 s or less.

DISCUSSION—These forces result most often from random impacts, vibration, or shocks in handling and transit.

edge protector—a right-angle piece placed over the edge of **boxes, crates, bundles** and gates, or **bracing** to distribute the pressure from metal bands or ties in order to prevent cutting.

spreader, n—(*1*) a horizontal wooden member placed between two gates in a car to hold the gates in position against the load; (*2*) a rigid device placed between sling legs, lifting cables, or ropes to prevent them from damaging cargo being loaded or unloaded.

static compression loading—in packaging, the application of a force for an extended period, often days, weeks, or months.

DISCUSSION—These forces usually are associated with stacked loads. Many factors, such as material fatigue, environmental conditions, and handling affect the rate and duration of loading.

load limit—(*1*) the maximum force, in weight units, a body can withstand without damage; (*2*) the greatest allowable weight that may be placed in a **container** or vehicle. For containers, load limit is often expressed in terms of the gross weight of the container and its contents.

load type, *n*—a qualitative description of the contents of a **container** with respect to **density,** fragility, and degree of **blocking, bracing,** and **cushioning** necessary. Load types are further divided into the following categories:

Type I load (easy)—a qualitative term denoting contents of low or moderate **density** conforming to the shape of the **container** and lending support to all faces of the container.

Type II load (average)—a qualitative term denoting contents of low or moderate density providing, when packed directly into a **shipping container,** nonshifting support at several points on the face of the container.

Type III load (difficult)—a qualitative term denoting contents characterized by irregular shape not lending support to the container, or by great density or extreme fragility.

lumber, *n*—the product of the saw and planing wood mill, not further manufactured than by sawing, resawing, planing, crosscutting to length, and matching.

matched lumber—**lumber** that is shaped to make a close-tongued and grooved joint at its edges or ends

plank, n—a broad board usually more than 1 in. thick, laid with its wide dimension horizontal, and used as a bearing surface.

resawn lumber—rough or surfaced **lumber** that has been divided into two or more thicknesses by sawing.

rough lumber—undressed **lumber** as it comes from the saw.

shiplapped lumber—**lumber** that is shaped at its edges to make a close-rabbeted, or lapped joint.

surfaced lumber—**lumber** that is dressed on one or more faces.

macerated paper—See **cushioning material.**

mailing tube—See **tube.**

manufacturer's joint—See **container.**

marking, *n—in packaging,* the application of numbers, letters, **labels, tags,** symbols or colors to provide identification and to expedite handling during shipment and storage. (Compare **code.**)

matched lumber—See **lumber.**

modular container—See **container.**

molded pulp—See **packaging.**

molded shape—*in packaging,* a material formed to specific contours for **cushioning,** or **blocking** or **bracing** purposes.

Mullen test—See **package testing.**

nest, *n*—a group of articles, **cans, baskets,** or **boxes** of various sizes, that fit one within the other.

OD—the outside dimension of a **package** or part, or outside diameter of a cylindrical container or wire or rod.

open crate—See **crate.**

overpacking, *n*—(*1*) **packaging** that exceeds minimum requirements, (*2*) additional packaging used to increase protection.

pack, *n*—the final configuration of material with necessary protection afforded for the distribution system.

blister pack, n—encapsulation of an item inside a preformed semi-rigid transparent, or opaque plastic bubble that is affixed to a semi-rigid backing.

[8] *Uniform Freight Classification,* Uniform Classification Committee, 222 South Riverside Plaza, Chicago, IL 60606.

exterior pack—a **container, bundle,** or assembly that is sufficient by reason of material, design, and construction to protect its contents during shipment or storage.

intermediate pack—*in packaging,* a **wrap, box,** or **bundle** that contains two or more unit packs of identical items, and that is in turn enclosed by an outer **shipping container.**

pack, *v*—to place material into a container for handling, storage, and transportation.

package—(*1*) *n,* a container providing means of protection and handling to a product; (*2*) *v,* to design, manufacture, or provide protection to a product; (*3*) *when referring to a fiber container,* a container not necessarily complying with all the requirements for a "box" in accordance with the Uniform Freight Classification[8] and National Motor Freight Classification[9]; (*4*) when referring to a fiber container, one of the special authorized containers described in detail in the Classifications in the section titled "Authorized Packages or Shipping Containers," of the above Freight Classifications.

aerosol package—a package that contains in addition to its contents, a gas under pressure. It is fitted with a valve which enables the contents to be dispensed.

intermediate package—a wrap, box, or bundle (that is, a container) that contains two or more unit packages of identical items.

unit package—the first tie, wrap, or container applied to a single item, a quantity of the same item, a set, or an item with all its component parts, that constitutes a complete and identifiable package containing the unit of issue of a product for ultimate use.

package testing:

bursting strength—the strength of a material in pounds-force per square inch (or kilopascals), measured by the Cady or Mullen tester. (See **Cady test,** and **Mullen test.**)

bursting strength test—a test for measuring the resistance of a material to bursting measured in pounds per square inch (or kilopascals). (See **Mullen tester.**)

Cady test—a test for **bursting strength** made on a specific type of machine.

caliper—(*1*) *n,* thickness (as related to **paperboard**) of a sheet measured under specified procedures expressed in thousandths of an inch (or millimetres). Thousandths of an inch are sometimes termed "points"; (*2*) *n,* the precision instrument used in the paperboard industry to measure thickness; (*3*) *v,* to measure with a caliper.

compression test—a test for measuring resistance to external compressive forces. (See **loading.**)

drop test—a test for measuring the durability of an article, or the protection, or the retention properties of a container, or all three by subjecting the packaged product to a free fall from predetermined heights onto a surface with prescribed characteristics.

Elmendorf test—a test for measuring the tearing resistance of tape, paperboard, or other materials.

[9] *National Motor Freight Classification,* National Classification Board, 2200 Mill Road, Alexandria, VA 22314.

incline impact (Conbur) test—a test for determining principally the resistance of a packaged product to damage from impacts.

Mullen test—a test for bursting strength made on a specific type of machine.

puncture test—(*1*) a test for measuring the resistance of board to puncture; (*2*) the strength of a material, primarily involving tear and stiffness, expressed in units as measured by the puncture tester.

DISCUSSION—1 unit = 0.265 in.·lbf (0.0298 J) of energy.

revolving drum test—a test for measuring the protection to contents, or the retention properties of a container, or both, by subjecting the packaged products to rough handling in a standard revolving drum.

testing machine compression—*in package testing,* an imposed force in motion applied at a predetermined rate and condition until a predetermined end point is achieved.

packaging, *n*—(*1*) the technique of preparing goods for distribution;

(*2*) the design criteria, processes, and procedures used to protect material from deterioration and damage from the time manufacturing is completed until ultimate use or disposal;

DISCUSSION—It includes cleaning, drying, preserving, packing, unitization, and marking.

(*3*) the processes and procedures used to protect an item in a unit package.

bulk packaging—(*1*) a method of containing loose or granular materials for shipping or storage, (*2*) a method of assembling many items into a **container** for shipment or storage.

child-resistant packaging—packaging designed or constructed to be significantly difficult for most children under 5 years of age to open, or to obtain a toxic or harmful amount of the contents within a reasonable time; and not difficult for normal adults to use properly.

commercial packaging—the methods and materials used by a supplier to satisfy the requirements of the distribution system.

DISCUSSION—Commercial packaging includes industrial packaging, and consumer packaging; and may be applicable for certain levels of military packaging.

foam-in-place packaging—See **cushioning material.**

industrial packaging—packaging of partially manufactured or finished goods for distribution from manufacturer to manufacturer, and to users other than retail customers.

molded pulp—a type of **packaging** which has been formed from wet paper pulp to predetermined shape to generally fit one or more individual items. Used as apple trays, egg flats, egg carton. (See **molded shape.**)

skin packaging—a process whereby a product placed on a backing material is covered by a closely fitting thermoplastic film bonded to the backing material.

DISCUSSION—The article usually is placed on a porous, rigid backing sheet, heated film is draped over it, and vacuum is applied to draw the film tightly over the article.

vacuum packaging—the technique of **packaging** in **containers** from which substantially all air has been removed prior to sealing.

packing, *n*—the selection or construction of the **shipping container** and assembling of items or **packages** therein, including any necessary **blocking, bracing,** or **cushioning,** weatherproofing, exterior **strapping,** and **marking** of shipping container for identification of contents. (Compare **packaging**).

absorbent packing, n—a material included within a **package** to soak up liquids resulting from leakage or liquefaction of contents.

interior (inner) packing—material or parts used in supporting, positioning, or cushioning an item in its outer shipping container. (See also **buffer, cushioning material, divider, liner, separator.**)

saddle, n—a form of **interior packing** shaped to fit the article in order to increase the area of contact between the **blocking** and the article.

shell, (when referring to a *fiberboard packaging*), *n*—a specific form of inner **packing** consisting of a sheet of **corrugated** or **solid fiberboard** scored and folded to form a joined or unjoined **tube** open at both ends.

pad, *n*—(when referring to a **fiberboard container**) a **corrugated** or **solid fiberboard** sheet or other authorized material used for extra protection or for separating tiers or layers of articles when packed for shipment. (Compare **buffer, divider, separator**). See **cushioning material.**

pail, *n*—a **container,** usually cylindrical, with a handle, available in sizes of about 1 to 12 gal (3 to 50 L).

pallet, *n*—horizontal platform device used as a base for assembling, storing, handling, and transporting materials and products in a unit load. (Compare **skid, dolly.**)

expendable pallet—a **pallet** intended to be discarded after a single use.

reusable pallet—a **pallet** intended for more than one shipment (reusable for economic life).

stringer, n—a wooden member to which the face or faces of a **pallet** or the deck of a platform are fastened and running at right angles to the members making up the face or the deck.

palletized load—See **load.**

palletized unit load—See **load.**

panel, n—(*1*) a **face** or *side* of a **box** or **crate,** (*2*) any flat area between folds, or edges of a **box, carton,** or interior part.

paper:

creped duplex paper—a double sheet composed of two layers of **creped paper** united with asphalt, latex, or other adhesives, generally with 20, 25, or 30-lb paper and 20, 25, or 30 lb of laminating material.

creped paper—paper, usually **kraft** (bleached sulfate) in various basis weights, that has been machined to resemble crepe and thus has stretch. It is used for wrapping purposes and for bag and barrel **liners.**

water-resistant paper—paper that is treated by the addition of materials to provide a degree of resistance to damage or deterioration by water in liquid form.

wet-strength paper—paper that has been treated with chemical additives to aid in the retention of resistance to bursting, tearing, or rupturing when wet.

paperboard, *n*—one of the two broad subdivisions of paper (general term, the other being paper as specific term). Paperboard is, in general, heavier and thicker than paper and is constructed primarily for strength properties such as stiffness, tearing resistance and the like. Sheets with thickness greater than 0.010 in. (0.25 mm) (with some exceptions) fall in this class. (See also **containerboard.**)

DISCUSSION—The broad classes of paperboard are (*1*) **containerboard,** (*2*) **boxboard,** and (*3*) other special types.

boxboard, n—a general term designating the grades of **paperboard** used for fabrication of folding and set-up **boxes (cartons).** Customarily shipped in sheets.

chipboard, n—a **paperboard** generally made from reclaimed paper stock.

DISCUSSION—It is used for many purposes, including facings for partitions, center plies of solid **fiberboard pads,** and other forms of paperboard that may or may not have specifications of strength, color, or other characteristics. The board is of relatively low density in thickness of 0.006 in. (0.15 mm) and up.

finish, n—in packaging, a term descriptive of the surface of **paperboard** depending on the final machining in manufacture:

(*1*) *dry finish*—not dampened before going through the calender rolls. The surface is not as dense nor as smooth as water-finished board;

(*2*) *water finish*—dampened on one (or both) sides in the calendering opening to provide a relatively hard and glossy finish.

kraft, n—a term (derived from a German word meaning strength) applied to pulp, paper, or **paperboard** produced from virgin wood fibers by the sulphate process.

linerboard, n—(*1*) **paperboard** used for the flat facings in **corrugated fiberboard;** also paperboard used as the outer plies of solid fiberboard, (*2*) **containerboard** made of blends of reworked paper fibers usually made on cylinder machines but can be made on Fourdrinier or Inverform machines. It is sometimes called jute—a misnomer, since jute fibers are no longer used (See **facing.**)

paperboard blank—a flat piece of **paperboard** that has been cut to size, die-cut, corner cut, scored, printed, or otherwise prepared and ready to be formed or folded and joined, or the flat pieces of parts thereof.

paperboard step—a term applied to trays, shells, or platforms positioned inside a **container** to raise one portion of contents higher than others. It may be attached to a lid for the purpose of holding certain articles positioned when the lid is in closed position.

paperboard stop—pieces of **boxboard** attached to the inside surface of the ends of a set-up paper box lid to hold the base in proper position when the height of the base is less than the height of the lid.

paperboard tube—See **tube.**

paper multiwall-sack—See **bag.**

partition—See **fiberboard partition.**

plank—See **lumber.**

plug—See **closure.**

ply, *n*—(*1*) any of the several layers of solid **fiberboard;** (*2*) any of the layers in plywood or laminated **paperboard;** (*3*) any of the walls of a **multiwall bag.** (See also **veneer.**)

plywood box—See **box.**

point, *n*—a term used to describe the thickness of **paperboard,** a point being one thousandth of an inch. (See **caliper.**)

pouch, *n*—a small or moderate-sized bag, sack, or receptacle for carrying or containing parts.

preservation, *n*—*in packaging,* the application of protective measures such as cleaning, drying, and the use of preservatives, **barrier materials, cushioning,** and **containers.**

pressure-sensitive-adhesive tape—See **tape.**

primary container—a container which is in contact with the contents.

DISCUSSION—The primary container may be used as the shipping container and may be used to communicate information relative to the product or its use. (See *unit package* under **package.**)

puncture test—See **package testing.**

reinforced gummed tape—See **tape.**

resawn lumber—See **lumber.**

returnable container—See **container.**

reusable container—See **container.**

reusable pallet—See **pallet.**

revolving drum test—See **package testing.**

rough lumber—See **lumber.**

rubbing strips—boards nailed to the underside of **skids,** or floor boards of **crates** or **boxes** to provide a surface upon which to slide the containers. (See also **skid.**)

sack—See **bag.**

saddle—See **packing.**

score, *n*—an impression or crease in **paperboard, corrugated** or **solid fiberboard** to locate and facilitate folding.

DISCUSSION—Set-up boxes made of nonbending paperboard are generally partially cut through at score lines and may be reinforced by gummed paper stays.

score, *v*—to form a score.

seal:

heat seal—*in packaging,* a method of bonding mating surfaces under controlled application of heat, pressure, and dwell time.

hermetic seal—*in packaging,* a seal that prevents passage of air and other gases.

seam—See **container.**

secondary container—a container which encloses one or more primary containers.

DISCUSSION—The secondary container may be used as the shipping container and may be used to communicate information relative to the product or its use. (See *intermediate package* under **package.**)

separator, *n*—any material interposed between articles or packing components. (Compare **buffer, divider, insert, liner.**)

sheathed crate—See **crate.**

sheathing, *n*—**lumber** or sheet materials, such as plywood, **fiberboard** or other suitable **panel** or sheet material fastened to the frame across the openings of a crate to deter pilferage, or entry of dirt or water or of other articles.

shelf life—the length of time packaged materials can be stored under specified conditions and still remain effective in use (C 717, C-24)[6].

shell—See **packing.**

shiplapped lumber—See **lumber.**

shipping container—See **container.**

shroud, *n*—*in packaging,* a protective **cover** placed over a load unit, or **package** to cover the top and four sides.

skid, *n*—one of a pair or series of parallel runners usually wood, affixed to the underside of **boxes, crates,** or an item to allow entry of truck forks, or to facilitate sliding. (Compare **rubbing strip.**)

skid box—See **box.**

skid platform—a single platform of wood, metal, plastic, or combination of these materials, elevated a short distance above the floor by runners or legs to facilitate mechanical handling. (Compare **pallet.**)

skin packaging—See **packaging.**

slat, *n*—a thin flat strip of material used as a member of a **box, crate, pallet, basket,** or other similar structure. A basket **stave.**

sleeve, *n*—*in packaging,* a form, made of various materials and in various shapes, usually open at both ends, that is slipped over an item.

solid fiberboard—See **containerboard.**

spacer, *n*—a device made of any material that serves to maintain a predetermined distance between any two points in a **shipping container,** or between any interior part and articles contained therein. (Compare **divider.**)

spreader—See **loading.**

staple, *n*—a U-shaped round or flat wire fastener that is: (*1*) used to secure parts together. (It may or may not be clinched); (*2*) driven over wires and bands to hold parts in position; (*3*) used to fasten together the flaps of **fiberboard** boxes. (Compare **stitch.**)

staple, *v*—to fasten with staples.

static load—See **load.**

stave, *n*—(*1*) one of the shaped vertical units of a coopered wooden **container** which, when placed edge-to-edge and bound with hoops around a circular bottom (sometimes top and bottom), forms a **barrel** or other staved container; (*2*) one of the thin, wood or veneer **slats** forming the **body,** sides, or **cover** of a **basket** or **hamper.**

stay—paper or metal material used to join parts of a **box** together at a corner, or to reinforce a cut-score on **solid fiberboard.** It is most commonly used for set-up boxes.

stitch, *n*—*in packaging,* a metal fastener. Stitches are machine-formed using wire drawn from a spool. (Compare **staple. Staples** are preformed.)

stitch, *v*—(*1*) to unite, by means of **staples** or stitching wire, the joints or parts of containers, or the flaps for **closure;** (*2*) to unite by sewing with thread, as a **sack.**

strapping, *n*—a flexible strip material, generally of flat or round cross section, used to: (*1*) fasten merchandise within a **container,** (*2*) hold together a **bundle** or **bale,** (*3*) reinforce a **shipping container,** (*4*) secure a shipping container or an article to a pallet. (Compare **bracing.**)

stress retention—the residual load expressed as a percentage of the original load on a material after the test specimen has been maintained at a constant elongation for a specified time.

stringer—See **pallet.**

strut, *n*—a short member of a **crate** or **bracing** used as reinforcement and designed to resist compressive forces in the direction of its length.

surfaced lumber—See **lumber.**

tag, *n*—a **paperboard** card, metal plate, or plastic card on which shipping or identification data are given. (Compare **label.**)

DISCUSSION—It is fastened to an article or container, by wires, staples, tacks, or an adhesive.

tape:

gummed paper tape—a **kraft** paper in a wide range of basic weights, usually 35, 60 and 90 lb (24 by 35–500) that is gummed on one side and slit into rolls of various widths.

DISCUSSION—*in packaging,* it is used largely for sealing packages, bundles and fiberboard boxes.

pressure-sensitive-adhesive tape—in packaging, a backing, as tape, coated so as to permit it to adhere instantaneously to most surfaces with the application of slight pressure.

reinforced gummed tape—an adhesive-coated tape that contains glass, rayon, or sisal fibers embedded in latex, resin or asphaltic laminants between two sheets of **kraft.**

DISCUSSION—It is used for **manufacturer's joints,** or **closure** of corrugated fiber boxes. It is manufactured with the fibers running in a cross-machine direction (most commonly used), or with the fibers running in both directions, (bi-directional), or in a diamond-shaped pattern. (See carrier regulations.)

tare weight, *n*—the weight of a **container** or of packaging materials excluding the contents.

tongue-and-grooved, *adj*—type of **lumber** joint consisting of a tongue and a groove on opposite edges to provide close fitting into an adjacent grooved or tongued piece.

tube, *n*—a cylindrical paperboard **container,** convolutely or spirally wound, having paper, wood, metal, or combination ends.

collapsible tube—in packaging, a flexible **container** having a dispensing tip and cap at one end, and closed at the opposite end, serving as both package and dispenser.

composite tube—in packaging, a **tube** consisting usually of multiple layers of **fiberboard,** film, or foil providing barrier properties, and an outer layer of paper.

fiberboard tube—in packaging, a cylindrical or multisided form.

DISCUSSION—It may be an element of a box style, or a unit of packaging to add protection or compressive strength, or both. It may or may not be joined. (See **drum** (*2*), **insert, liner, spacer.**)

paperboard (mailing) tube—in packaging, a cylindrical **container** wherein the length does not exceed ten times the girth with a minimum ⅛-in. (3-mm) wall thickness.

DISCUSSION—Closure method should meet U.S. Postal Service requirements.

unitization, *n*—the assembled group of containers or items in a single load that can be handled as a unit throughout the distribution system.

DISCUSSION—Unitization encompasses, but is not limited to, consolidation in a container, placement on a pallet or load base, or securely binding together.

unitized load—See **load.**

unit load—See **load.**

unit package—See **package.**

vacuum packaging—See **packaging.**

V-board—see **corrugated fiberboard.**

veneer, *n*—thin sheets of wood, produced on a rotary lathe, or by slicing, or sawing.

vial, *n*—a small, usually cylindrical, vessel capable of closure, such as for medications, perfumes, essences, and samples.

volatile corrosion inhibitor (*VCI*)—See **corrosion.**

wadding—See **cushioning material.**

warp, *n*—a flexible sheet material used to protect items in **packaging.**

waterproof, *adj*—deprecated term.

DISCUSSION—ASTM policy does not permit inclusion of absolute terms in its standards unless actually used and defined in their absolute sense (*Form and Style for ASTM Standards* ("Blue Book"), Section E1).[10]

water-resistance, *n*—measured ability to retard both penetration and wetting by water in liquid form.

DISCUSSION—The method of test must be stated, since the degree of resistance depends on the way it is determined or measured.

water-resistant barrier—See **barrier material.**

water-vapor-resistance—measured ability to retard penetration and permeation by water-vapor.

DISCUSSION—The method of test must be stated, since the degree of resistance depends on the way it is determined or measured.

water-vapor-resistant barrier—See **barrier material.**

W-board—See **corrugated fiberboard.**

weather-resistance—measured ability of a material, assembly, or package to maintain designated physical properties and appearance under specified weathering conditions.

DISCUSSION—The method of test must be stated, since the degree of resistance depends on the way it is determined or measured

wet strength paper—See **paper.**

wirebound box—See **box.**

[10] Available from ASTM Headquarters.

This standard is subject to revision at any time by the responsible technical committee and must be reviewed every five years and if not revised, either reapproved or withdrawn. Your comments are invited either for revision of this standard or for additional standards and should be addressed to ASTM Headquarters. Your comments will receive careful consideration at a meeting of the responsible technical committee, which you may attend. If you feel that your comments have not received a fair hearing you should make your views known to the ASTM Committee on Standards, 1916 Race St., Philadelphia, PA 19103.

Standard Test Method for Drop Test for Loaded Cylindrical Containers[1]

This standard is issued under the fixed designation D 997; the number immediately following the designation indicates the year of original adoption or, in the case of revision, the year of last revision. A number in parentheses indicates the year of last reapproval. A superscript epsilon (ε) indicates an editorial change since the last revision or reapproval.

ε1 NOTE—Figure 1, which was inadvertently omitted in 1981, was reinstated in December 1989.

1. Scope

1.1 This test method covers drop tests on loaded, substantial cylindrical containers such as barrels, drums, kegs, or pails (metal, fiber, plastic, or wood) to measure their ability to withstand rough handling.

1.2 *This standard does not purport to address the safety problems associated with its use. It is the responsibility of whoever uses this standard to consult and establish appropriate safety and health practices and determine the applicability of regulatory limitations prior to use.*

2. Referenced Documents

2.1 *ASTM Standards:*

D 585 Method for Sampling and Accepting a Single Lot of Paper, Paperboard, Fiberboard, or Related Product[2]

D 644 Test Method for Moisture Content of Paper and Paperboard by Oven Drying[2]

D 685 Method of Conditioning Paper and Paper Products for Testing[2]

D 775 Test Method for Drop Test for Loaded Boxes[2]

3. Significance and Use

3.1 Materials shipped in cylindrical containers vary from solids of varied particle size and shape (powders, granules, etc.) to liquids of all densities. Even the solids are to some degree capable of flowing. The rupture of any of these containers will cause material to escape easily from the container resulting not only in an economic loss but in the possible creation of hazards such as health, fire, housekeeping, etc. This test method is to evaluate the ability of the container to withstand sudden shock resulting from a free fall. This test method may also be used to design or redesign suitable containers.

4. Apparatus

4.1 *Drop Test Apparatus*, consisting of:

4.1.1 Chain hoist or other appropriate lifting device,

4.1.2 Rope or belt sling for placing around the container for lifting,

4.1.3 Tripping devices similar to that shown in Figs. X1 through X4 of X1.

4.1.4 Solid surface of concrete, stone, or steel of sufficient mass to absorb all shock without deflection, and

4.1.5 A 4 by 4-in. (102 by 102-mm) timber (about 4 ft (1.2 m) in length) having the edges rounded to a radius of not more than 1/4 in. (6.4 mm).

4.1.6 For small containers, the apparatus described in Section 4 of Test Method D 775 may be used.

4.2 *Conditioning Apparatus*—Adequate facilities for conditioning test specimens at proper humidity and temperature prior to test, in accordance with the requirements of the specifications covering the container to be tested.

4.3 *Miscellaneous Equipment*—Drying oven, scales, knife, saw, micrometer caliper, etc., for use in the determination of the moisture content or for making other supplementary tests of the materials from which the containers are made.

5. Sampling

5.1 Any convenient or specified sampling plan may be used. For accepting fiberboard drums, samples should be selected in accordance with Method D 585.

5.2 In the absence of any specific sampling plan, performance shall be based on not less than three representative specimens of a given size and type of a container.

6. Test Specimens

6.1 Pack the container to be tested with the actual contents for which it was designed or with a dummy load simulating these contents. Close and seal the container in the normal manner and dry and age sufficiently so that any adhesives, protective coatings, sealing tape, etc., will have reached their final normal condition.

6.2 The procedure for the identification of the component parts as well as point of impact upon drop testing shall be as follows (See Fig. 1): The quarter points around the top of the container shall be designated as points *1*, *3*, *5*, and *7* for diagonal cover chime drops. The quarter points around the bottom of the container shall be designated as *2*, *4*, *6*, and *8* for diagonal bottom chime drops. The lines between *1* and *2*, *3* and *4*, *5* and *6*, and *7* and *8* shall be designated *12*, *34*, *56*, and *78*, respectively, for flatwise sidewall drops.

7. Conditioning

7.1 Depending on the purpose of the test, the containers may be conditioned prior to the drop test by either water

---

[1] This test method is under the jurisdiction of ASTM Committee D-10 on Packaging and is the direct responsibility of Subcommittee D10.22 on Handling and Transportation.

Current edition approved March 3, 1980. Published May 1980. Originally published as D 997 – 48 T. Last previous edition D 997 – 50 (1973).

[2] *Annual Book of ASTM Standards*, Vol 15.09.

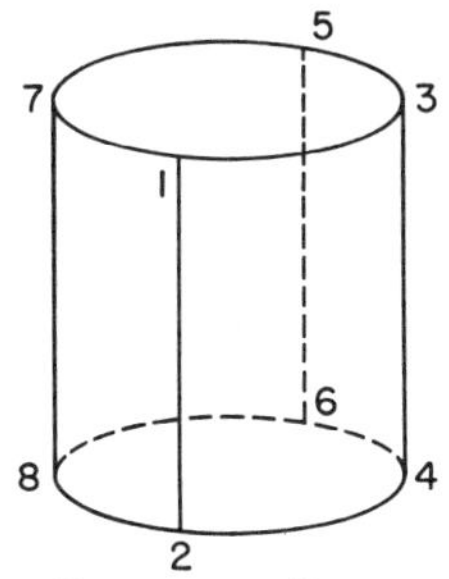

FIG. 1 Container Component Parts and Quarter Points

immersion, exposure to water spray, or exposure to fixed conditions of air temperature or humidity.

8. Moisture Content

8.1 When the moisture content of fiberboard at the time of test is required, it shall be determined in accordance with Method D 644.

9. Procedure

9.1 For a diagonal chime drop, suspend the container by means of the sling connected to the tripping device and hoist so that the center of gravity of the container and contents is vertically above the point of impact. This correct position can be ascertained by gently lowering the specimen until it touches the floor in order to see whether it is balanced properly.

9.2 For a flatwise drop, position the face on which the container is to fall in a plane parallel with the floor.

9.3 For determining the puncture resistance of the sidewall, make a flatwise drop, as described in 9.2, so that the container will strike across a 4 by 4-in. timber resting on the floor at right angles to the container.

9.4 The height from which containers shall be dropped will depend upon the purpose of the test, but normally will be 4 ft (1.2 m). In some instances the height as well as number and position of drops will be prescribed in the specifications for the container.

9.5 Failure, depending on the purpose of the test, may be considered to have occurred when:

9.5.1 The contents are exposed,

9.5.2 The contents spill from the container, or

9.5.3 Some other predetermined type of damage has occurred.

10. Report

10.1 The report shall include the following:

10.1.1 Identification of container, including data on closure, liners, etc.,

10.1.2 Identification of contents, including net and gross masses,

10.1.3 Detailed record of test on each container including height of drop or drops; position of drop, as for example a cover chime drop at *1*, bottom chime drop at *2*, flat sidewall drop at *12*, etc.; and damage to the container upon each drop, together with any other observations that may assist in correctly interpreting the results or aid in improving the design of the container or the method of testing.

10.1.4 Method, if any, of conditioning the container; the moisture content, if determined, of the wood, plywood, or fiberboard; and the results of supplementary tests of the materials from which the container is made, and

10.1.5 Statement to the effect that all tests were made in full compliance with this method.

11. Precision and Bias

11.1 The variables associated with many of the materials from which these containers are made and the variables associated with the test are such that reproducibility will not be obtained within generally acceptable limits.

APPENDIX

(Nonmandatory Information)

X1. DROP TEST APPARATUS

X1.1 *Hoist with Suitable Slings, Tripping Device, and Hooks*[3]—When this equipment is used, it shall conform substantially to the requirements shown in Figs. X1.1 and X1.2.

X1.2 *Solenoid Operated Drop Test Mechanism*[3] *and Suspension Device*—See Figs. X1.3 and X1.4.

[3] Detailed drawings of this apparatus are available at a nominal cost from ASTM Headquarters, 1916 Race St., Philadelphia, PA 19103. Request Adjunct No. 12-409970-00.

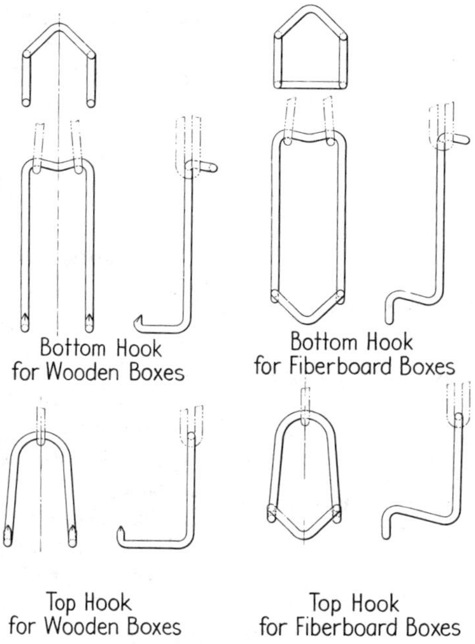

FIG. X1.1 Hooks for Hoist-Sling Drop Test Apparatus

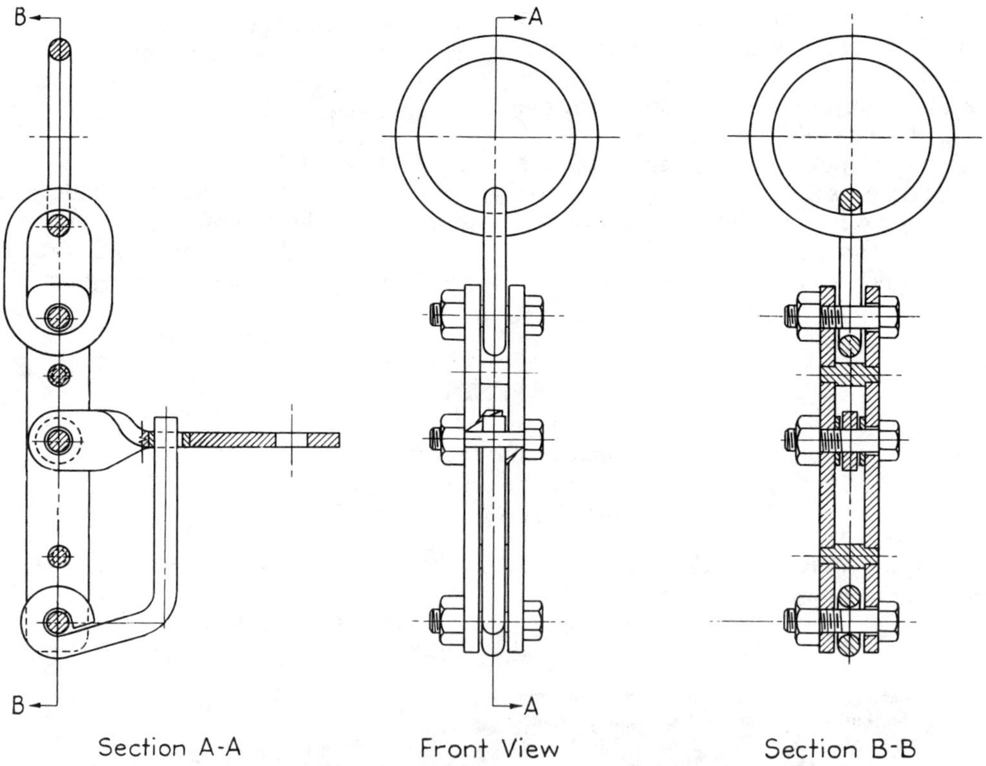

FIG. X1.2 Tripping Device for Hoist-Sling Drop Test Apparatus

FIG. X1.3 Solenoid Release Mechanism With Face Plate Removed and Holding Jaws in Open Position

FIG. X1.4 Corner Suspension Device Attached to a Conventional Trunk Corner that is Sealed Inside Box to be Tested

Standard Methods for Vibration Testing of Shipping Containers[1]

This standard is issued under the fixed designation D 999; the number immediately following the designation indicates the year of original adoption or, in the case of revision, the year of last revision. A number in parentheses indicates the year of last reapproval. A superscript epsilon (ε) indicates an editorial change since the last revision or reapproval.

1. Scope

1.1 These methods cover vibration tests of filled shipping containers. Such tests may be used to assess the performance of a container, with its interior packing and means of closure, both in terms of its strength and of the protection it provides its contents when it is subjected to vibration such as it experiences in transportation. These procedures are suitable for testing containers of any form, material, kind, design of interior packing, means of closure, and any size and weight. They are not intended for determining the response of products to vibration for product design purposes, nor are they intended for tests of products in their operational configuration as other more suitable procedures are available for these purposes.[2,3]

1.2 The following methods appear:

Method A1—Repetitive Shock Test (Vertical Motion).

Method A2—Repetitive Shock Test (Rotary Motion).

Method B—Single Container Resonance Test.

Method C—Palletized Load, Unitized Load, or Vertical Stack Resonance Test.

1.3 *This standard does not purport to address all of the safety problems associated with its use. It is the responsibility of the user of this standard to establish appropriate safety and health practices and determine the applicability of regulatory limitations prior to use.* Specific precautionary statements are given in Section 6.

2. Referenced Documents

2.1 *ASTM Standards:*

D 4169 Practice for Performance Testing of Shipping Containers and Systems[4]

D 4332 Practice for Conditioning Containers, Packages, or Packaging Components for Testing[4]

E 122 Practice for Choice of Sample Size to Estimate a Measure of Quality of a Lot or Process[5]

[1] These test methods are under the jurisdiction of ASTM Committee D-10 on Packaging and are the direct responsibility of Subcommittee D10.22 on Handling and Transportation.

Current edition approved Dec. 26, 1986. Published February 1987. Originally published as D 999 – 48 T. Last previous edition D 999 – 75 (1981).

[2] Military Standard Environmental Test Methods, MIL-STD-810D, Method 514, Vibration, available from Naval Publications and Forms Center, 5801 Tabor Ave., Philadelphia, PA 19120.

[3] International Electrotechnical Commission Recommendation, Publication 68-2-6, Part 2, Test F: Vibration, Basic Environmental Testing Procedures for Electronic Components and Electrical Equipment, available from American National Standards Institute, Inc., 1430 Broadway, New York, NY 10018.

[4] *Annual Book of ASTM Standards*, Vol 15.09.

[5] *Annual Book of ASTM Standards*, Vol 14.02.

3. Terminology

3.1 *Definition:*

3.1.1 *resonance*—a vibration of large amplitude in a mechanical system caused by a relatively small periodic stimulus of the same or nearly the same period as the natural vibration period of the system.

3.2 *Descriptions of Terms Specific to This Standard:*

3.2.1 *double amplitude*—the maximum value of a sinusoidal quantity (peak-to-peak).

3.2.2 *octave*—the interval between two frequencies having a ratio of two (2).

3.2.3 *repetitive shock*—impacts of a package on a test platform which occur cyclically from input vibration.

4. Significance and Use

4.1 Shipping containers are exposed to complex dynamic stresses when subjected to vibration present in all transportation vehicles. Approximating the actual damage, or lack of damage, experienced in shipping may require subjecting the container(s) and contents to vibration inputs.

4.2 Resonant responses during shipment can be severe and may lead to package or product failure. Identification of critical frequencies, and the nature of package stresses can aid in minimizing the effect of these occurrences.

4.3 Vibration tests should be based on representative field data. When possible, the confidence level may be improved by comparing laboratory test results with actual field shipment data.

4.4 Exposure to vibration can affect the shipping container, its interior packaging, means of closure, and contents. These tests allow analysis of the interaction of these components. Design modification to one or more of these components may be utilized to achieve optimum performance in the shipping environment.

4.5 *Methods A1 and A2, Repetitive Shock Tests*, are suitable for tests of individual containers that are transported unrestrained on the bed of a vehicle and may be suitable for tests of containers that might be subjected to repetitive shocks due to magnification of vibrations in unit loads or stacks.

NOTE 1—Methods A1 and A2 produce different vibration motions and therefore will generate different forces which may result in different damage modes and intensities. Results from these two methods may not correlate with one another.

4.6 *Method B, Single Container Resonance Test*, covers the determination of the ability of an individual container and its interior packaging to protect the contents from transportation vibration, particularly when the container and its contents might exhibit resonant responses.

NOTE 2—Individual products that are palletized might be better tested using Method C.

4.7 *Method C, Palletized Load, Unitized Load or Vertical Stack Resonance Test*, covers the determination of the presence and the effects of resonance in palletized loads and multiple-unit stacked loads, and whether or not the strength of the containers is sufficient to withstand dynamic loads when stacked.

4.8 Any or all of these test methods may be employed, as determined by the appropriate performance specification, with test intensities, frequency ranges, and test durations as called for in the specification. Although these tests do not simulate the shipping environment, they are intended to create the damage-producing potential of the shipping environment.

5. Apparatus

5.1 *Method A1, Repetitive Shock Test (Vertical Motion):*

5.1.1 *Vibration Test Machine*, with a platform having a horizontal surface of sufficient strength and rigidity so that the applied vibrations are essentially uniform over the entire test surface when loaded with the test specimen. The platform shall be supported by a mechanism that vibrates it so the motion is approximately a vertical sinusoidal input. (A rotary motion of the platform is not acceptable.) The double amplitude displacement of the vibration shall be fixed at or controlled to 1 in. (25 mm), and the frequency shall be variable within the range from 2 to at least 5 Hz (cycles per second). The vibration test machine shall be equipped with fences, barracades, or other restraints to keep the test specimen from falling off of the platform without restricting its vertical motion.

5.2 *Method A2—Repetitive Shock Test (Rotary Motion):*

5.2.1 *Vibration Test Machine*, with a platform having a horizontal surface of sufficient strength and rigidity so that the applied vibrations are essentially uniform over the entire test surface when loaded with the test specimen. The platform shall be supported by a mechanism that vibrates it so that the motion is a rotational input with the vertical component approximately sinusoidal. The double amplitude displacement of the vibration shall be fixed at 1 in. (25 mm), and frequency shall be variable from 2 to at least 5 Hz (cycles per second). The vibration test machine shall be equipped with fences, barricades, or other restraints to keep the test specimen from falling off the platform without restricting the vertical motion.

5.3 *Methods B and C—Resonance Tests* (also suitable for Method A1):

5.3.1 *Vibration Test Machine*, with a platform having a horizontal surface of sufficient strength and rigidity so that the applied vibrations are essentially uniform over the entire test surface when loaded with the test specimen. The platform shall be supported by a mechanism capable of producing a sinusoidal vibration in the vertical linear plane at controlled accelerations or displacements, or both, over a controlled continuously variable range of frequencies. (A rotary motion of the platform is not acceptable.) Suitable fixtures and attachment points shall be provided to rigidly attach the test container to the platform for Method B. Restraints shall be provided to restrain the horizontal motion of the test specimens on the platform without restricting the vertical motion of the specimen(s), for Method C.

5.4 *Instrumentation*—Accelerometers, signal conditioners, and data display or storage devices are required to measure and control the accelerations at the test surface in Methods B and C. Instrumentation may also be desirable for monitoring the response of the containers and packaged items. The instrumentation system shall have a response accurate to within ±5 % over the range specified for the test. Detailed information on suitable instrumentation may be found in the *Shock and Vibration Handbook.*[6]

5.5 *Conditioning Apparatus*—Adequate facilities shall be provided for conditioning test specimens at selected humidity and temperature prior to or during the test, or both, in accordance with the requirements of the applicable specification.

6. Safety Precautions

6.1 These test methods may produce severe mechanical responses of the test specimens. Therefore, fences, barricades, and other restraints must have sufficient strength and must be adequately secured. Operating personnel must remain alert to potential hazards and take necessary precautions for their safety. Stop the test immediately if a dangerous condition should develop.

7. Test Specimens

7.1 The test specimen shall consist of the container, as intended for shipment, loaded with the interior packaging and the actual contents for which it was designed. Blemished or rejected products may be used, if the defect is recorded prior to the test. Dummy test items should be used for developmental testing when necessary, but may not be used for final acceptance testing. Sensors and transducers may be applied with the minimum possible alteration of the test specimen, to obtain data on the container or packaged item. When it is necessary to observe the contents during the test, holes may be cut in noncritical areas of the container.

7.2 Whenever sufficient containers and contents are available, it is highly desirable that five or more replicate tests be conducted to improve the statistical reliability of the data obtained (see Practice E 122).

8. Conditioning

8.1 Condition test specimens prior to the test or during the test, or both, in accordance with the requirements of the applicable specification. When no conditioning requirements are given, and the container materials are climatically sensitive, a conditioning atmosphere is recommended (see Practice D 4332 for standard and special conditions).

9. Procedure

9.1 *Methods A1 and A2—Repetitive Shock Tests:*

9.1.1 Place the test specimen on the test machine platform in its normal shipping orientation. Attach restraining devices to the platform to prevent the specimen from moving horizontally off the platform and to prevent excessive rocking, without restricting the vertical movement. Adjust

[6] Harris, C. M., and Crede, C. E., *Shock and Vibration Handbook*, McGraw-Hill, New York, NY, 1976, Chapter 16.

the restraining devices to permit free movement of the specimen of approximately 10 mm (0.4 in.) in any horizontal direction from its center position.

9.1.2 Start the vibration of the platform at a frequency of about 2 Hz, and steadily increase the frequency until some portion of the test specimen repeatedly leaves the test surface. To ensure that the test specimen receives a continuing series of repetitive shocks, a shim with a 1/16-in. thickness and a width of 2 in. shall be used to determine when the test specimen is leaving the test platform by inserting it under the package, a minimum of 4 in., and moving it intermittently along one entire length of the package.

9.1.3 Continue the test at this frequency for a length of time stated in the applicable specification, if any, or for a predetermined period, or until a predetermined amount of damage may be detected. The test may be stopped momentarily to inspect for damage.

9.1.4 If the container might possibly be transported in any other orientations, test at least one container in each possible orientation for the full specified test duration.

NOTE 3—When no test duration is specified, a test duration of 1 h is recommended. Practice D 4169 may also be referred to for test durations.

9.1.5 Inspect the container and its contents and record any damage or deterioration resulting from the test.

9.2 *Method B—Single Container Resonance Test:*

9.2.1 Fasten the test specimen, in its normal shipping orientation, securely to the platform of the vibration equipment so that the specified vibration condition is transmitted to the outer container. Attach the accelerometer to the platform as close as possible to the test specimen.

9.2.2 Adjust the vibration test apparatus to produce the specified constant acceleration amplitude (zero-to-peak) over the specified range of frequencies. Starting at the lowest frequency, sweep the frequency of the vibration at a continuous logarithmic rate of 0.5 to 1.0 octaves per minute to the upper frequency limit and back to the lower limit. Repeat this complete cycle twice, recording all the resonant responses of the test specimen.

NOTE 4—If no test severity is specified, an acceleration amplitude of 0.5 *g*'s (zero to peak) over the frequency range from 3 to 100 Hz is recommended. Practice D 4169 may also be referred to for test level and duration recommendations.

NOTE 5—It should be noted that some existing vibration test equipment has limited frequency range capability and additional equipment may be needed to cover the entire recommended frequency range.

9.2.3 Dwell for the specified length of time at each resonant frequency determined in 9.2.2 (limited to a maximum of the four most severe resonances), or until damage to the container is noted, whichever occurs first. Adjust the frequency of vibration as necessary to maintain resonance.

NOTE 6—If no dwell time is specified, a dwell of 15 min is recommended. Practice D 4169 may also be referred to for test durations.

9.2.4 Repeat the procedures of 9.2.1, 9.2.2, and 9.2.3 with the container oriented in the other two mutually perpendicular axes, if handling and transportation in these orientations might occur.

9.2.5 Inspect the container and its contents and record any damage or deterioration resulting from the test.

9.3 *Method C—Palletized Load, Unitized Load, or Vertical Stack Resonance Test:*

9.3.1 Place the full-size unitized or palletized load(s) of test specimen(s) on the test machine platform to a height equal to that used in the mode of shipment. A single vertical column of containers may be used if vertical stacking alignment is used in shipping. Attach restraining devices to the platform to prevent the specimen(s) from horizontal movement off the platform, and to prevent toppling and excessive rocking. Adjust the restraining devices to permit free movement of the test specimen(s) of approximately 10 mm (0.4 in.) in any horizontal direction. Attach the accelerometer to the platform as close as possible to the test specimen(s), but protected so that it will not be contacted.

9.3.2 Adjust the vibration test apparatus to produce the specified constant acceleration amplitude (zero-to-peak) over the specified range of frequencies. Starting at the lowest frequency, sweep the frequency of vibration at a continuous logarithmic rate of 0.5 to 1.0 octaves per minute to the upper frequency limit, and then back to the lower limit. Repeat this complete cycle twice, and record all the resonance responses of the test load.

NOTE 7—Multiple-unit loads are likely to exhibit several resonant responses.

NOTE 8—If no severity is specified, an acceleration amplitude of 0.25 *g*'s (zero-to-peak) over the frequency range from 2 to 100 Hz is recommended. Practice D 4169 may be referred to for test level recommendations.

9.3.3 Dwell for the specified length of time at each resonant frequency determined in 9.3.2 (limited to a maximum of the four most severe resonances), or until damage is noted in the load, whichever occurs first. Adjust the frequency of the vibration as necessary to maintain resonance.

NOTE 9—If no dwell time is specified, a dwell time of 15 min is recommended. Practice D 4169 may be referred to for test durations.

9.3.4 Inspect the container(s) and contents and record any damage or deterioration resulting from the test.

10. Report

10.1 The report shall include the following:

10.1.1 Identification and description of the test specimens, including the container, the interior packaging, and the product (give size, weight, and any other pertinent details),

10.1.2 If Method C is used for unitized loads, describe the unitized load and the height of the stack, and the unitizing method employed,

10.1.3 Purpose of the test and the applicable performance specification, if any,

10.1.4 Sequence of test methods and the test intensities, frequencies, and durations used,

10.1.5 Verification of compliance with the test method or describe any deviations,

10.1.6 Number of replications of each test,

10.1.7 Atmospheric conditions the test specimens were subjected to, both prior and during the test,

10.1.8 Any other test the specimens were subjected to prior to this test,

10.1.9 Description of the apparatus and the instrumentation used,

10.1.10 Results of the test,

10.1.11 Descriptions and photographs of any damage or deterioration to the containers or their contents as a result of the test(s),

10.1.12 State whether the damage created in the laboratory testing replicates actual field damage to the container and contents,

10.1.13 All resonant responses and any observations that may assist in correct interpretation of results or lead to improvements in the design of the container, interior packaging, or product, and

10.1.14 Statement of whether or not the specimen(s) complied with the requirements of the applicable specification.

11. Precision and Bias

11.1 The precision and bias of these test methods cannot practicably be determined.

11.2 Testing results may vary significantly between Methods A1 and A2. This may not be noted in previous use of these test methods.

Standard Test Methods for Mechanical Handling of Unitized Loads and Large Shipping Cases and Crates[1]

This standard is issued under the fixed designation D 1083; the number immediately following the designation indicates the year of original adoption or, in the case of revision, the year of last revision. A number in parentheses indicates the year of last reapproval. A superscript epsilon (ε) indicates an editorial change since the last revision or reapproval.

1. Scope

1.1 These test methods are suitable for testing the integrity of large shipping cases and unitized loads as well as the ability of the contents to endure normal handling. The test methods described are applicable to common means of material handling including: forklift, pull pack, clamp truck, crane, and spade lift type handling systems as follows:

1.1.1 *Drop Test*—For measuring the ability of the case or unitized load to withstand rough handling and provide information useful in improving the design of the container (see 8.1.2 and 9.1).

1.1.2 *Spade Lift Test*—For lifting by spade lift attachment to determine the ability of the handling flap of the case or shipping unit to withstand repeated lifting and handling by this method (see 8.2).

1.1.3 *Clamp Lift Test*—For lifting by hydraulic clamp attachment, to determine the ability of the case or shipping unit, to withstand squeeze clamp handling consisting of repeated side compressions and lifting (see 8.3).

1.1.4 *Pull Pack Handling Test*—For testing the ability of the slip sheet tab to withstand repeated pulls by the gripper jaws without tearing (see 8.4).

1.1.5 *Grabhook Test*—For lifting by grabhooks to determine the ability of the case or unitized load to withstand the horizontal pressures of grabhooks (see 9.2).

1.1.6 *Sling Test*—For lifting by wire rope, cable, or woven fiber slings to determine the ability of the case or unitized load to withstand the compression of slings (see 9.3).

1.1.7 *Tip Test*—For determining the ability of tall or top heavy cases or crates to resist tipping over (see Section 10).

1.1.8 *Tipover Test*—For determining the ability of large shipping containers to resist the impacts associated with tipover hazards, and for determining the ability of the packaging and packing methods to provide protection to the contents, when the pack is tipped over (see Section 11).

1.2 *This standard does not purport to address the safety problems associated with its use. It is the responsibility of whoever uses this standard to consult and establish appropriate safety and health practices and determine the applicability of regulatory limitations prior to use.*

2. Referenced Documents

2.1 *ASTM Standards:*

D 775 Test Method for Drop Test for Loaded Boxes[2]

D 996 Terminology of Packaging and Distribution Environments[2]

D 3332 Test Methods for Mechanical-Shock Fragility of Products, Using Shock Machines[2]

D 4332 Practice for Conditioning Containers, Packages, or Packaging Components for Testing[2]

E 171 Specification for Standard Atmospheres for Conditioning and Testing Materials[2]

3. Terminology

3.1 *Definitions*—For definitions of terms used in these test methods, refer to Terminology D 996.

3.2 *Descriptions of Terms Specific to This Standard:*

3.2.1 *large shipping case or crate—for the purposes of these test methods,* a shipping container constructed of any material, and of such size and weight to require mechanical handling. A case or crate of this type may weigh from 100 lb (45 kg) up to many tons and measure proportionately. The case or crate may be secured to or carried by a base or pallet. Frame members may be provided for rigidity throughout the container.

3.2.2 *unitized load—for purposes of these test methods,* consists of a number of packages (two or more) wrapped or banded together as a shipping unit. These packages when unitized typically weigh more than 100 lb. The unitized method may be shrink wrapping, stretch wrapping, banding, strapping, taping, or gluing. A base consisting of a pallet or slip sheet may or may not be used.

3.2.3 *spade lift attachment*—a lift truck attachment used for "top handling" products packaged in interlocking flange cartons (I.F.C.) or folded cap or folded flap style cartons.

4. Significance and Use

4.1 The test methods described are designed for use in most cases with the actual equipment to be used in load handling.

4.2 These test methods may be used in evaluating cases, crates, or unitized loads as to suitability for mechanical handling by standard user specified load handling equipment.

4.3 The test methods given will allow the user to determine integrity and stability of the load as well as provide

[1] These test methods are under the jurisdiction of ASTM Committee D-10 on Packaging and are the direct responsibility of Subcommittee D10.22 on Handling and Transportation.

Current edition approved Sept. 30, 1988. Published November 1988. Originally published as D 1083 – 50 T. Last previous edition D 1083 – 85.

[2] *Annual Book of ASTM Standards*, Vol 15.09.

guidance to improve the design of the unit load where deficiencies are found.

4.4 Damage to products or packages observed during testing can be expected to correlate at least in a qualitative way to damage observed in actual distribution handling systems.

4.5 The results received from shock machine testing and free fall drop testing are different for certain products. Where this test is performed to satisfy a regulatory or contractual requirement, its use is subject to approval by the agency concerned.

5. Apparatus

5.1 *Drop Test Apparatus*—In making the drop test, the case or crate may be handled with any convenient equipment, such as a block and tackle, a hoist, or jacks with provision made for releasing the case or crate either by pulling supporting blocks from under it when tackle and jacks are used or by a release hook when a hoist is used. Other apparatus may be used for dropping a case or crate as follows:

5.1.1 Lift with heavy band iron or wire (to release, the band or wire is cut), or

5.1.2 Lift with a fork truck (while restraining the container, the fork truck is removed from under it), or

5.1.3 Use a tractor with block and tackle (blocks are pulled from under loaded container).

5.2 *Spade Lift Test Apparatus*—In making the spade lift test, use an actual lift truck with spade lift attachment to lift the shipping unit under the lifting flap. The lift truck and blade(s) must be capable of handling the size and quantity of products as per current practice.

5.3 *Clamp Lift Test Apparatus*—In making the clamp lift test, use an actual lift truck equipped with the appropriate load clamping device to lift the shipping unit. Provision must be made to measure the clamp force between the platens using a load cell(s) or other suitable device.

5.4 *Pull Pack Test Apparatus*—In making the pull pack test, use an actual lift truck equipped with an appropriate gripper jaw for pulling the load by the slip sheet tab onto the load plate. Alternatively, any suitable pulling device equipped with a gripper jaw may be used.

5.5 *Grabhook Test and Sling Test Apparatus*—In lifting the case or crate with grabhooks or slings, a hoist with grabhooks or slings attached is the simplest and easiest, but the test may be made with a tackle or by lifting the case or crate with jacks and setting it into the grabhooks or slings supported by some form of truss.

5.6 *Tip Test and Tipover Test Apparatus*—In making the tip or tipover test, the container may be handled with any convenient equipment, such as a forklift truck, a hoist, a block and tackle, or by hand. A smooth, level, concrete slab, pavement, or similarly unyielding surface with a mass at least 50 times that of the load and large enough to ensure that the test package falls entirely upon the surface shall be available upon which to perform these tests.

6. Test Specimen and Number of Tests

6.1 Test several cases or unitized loads of a given design, if possible, to obtain replication of results. If in the instance of a number of cases wrapped or banded together, it is not possible to test an entire unitized load, then sufficient cases should be assembled such that the height and one base dimension are nearly the same as the proposed unitized load. For the clamp, spade lift, or pull pack test, use an entire unit load. The same case or unitized load may often be used for all applicable tests if not tested to failure. Load the case or unitized load for the test with the actual contents for which it was designed, or if this is not possible, with a dummy load simulating such contents in mass, shape, and position in the case or unitized load.

6.2 Identify faces, edges, and corners of containers as shown in Test Method D 775, Fig. 1.

7. Conditioning

7.1 If desired, the test specimen may be conditioned to a representative temperature and humidity expected in actual distribution (see Specification E 171). When no conditioning requirements are given and container materials are moisture-sensitive, a standard conditioning atmosphere of 23 ± 2°C (73.4 ± 4°F) and 50 ± 2 % relative humidity is recommended. (See Practice D 4332.)

8. Unitized Load Tests

8.1 *Palletized Load Test Procedure:*

8.1.1 *Handling on Test Course:*

8.1.1.1 Use an actual lift truck equipped with forks to be used in actual practice. Seen Annex A1 for recommended standard course for lift truck handling test.

8.1.1.2 The operator is to engage, lift, and transport the load to each observation point in a manner typical of current practice until the course has been completed one cycle.

8.1.1.3 If the pallet allows fork entry from both directions, distribute the test cycles accordingly based on the probability of forking from each direction.

8.1.1.4 The minimum number of handlings required shall be specified by the user; however, it should be noted that the effects of temperature and humidity in the testing environment may vary greatly and, if so, may affect the strength of the shipping unit. For this reason, it may be useful to specify the minimum number of handlings based on test periods representing these extremes.

8.1.1.5 Examine the load during and after the test cycle to determine if failure has occurred.

8.1.2 *Raised Edge Free Fall Drop:*

8.1.2.1 With one edge of the unit load supported by the floor, raise the other end to prescribed heights and release it to fall flat on the impact surface. Where cases or crates are tall or top heavy, provision must be made to prevent the case or crate from tipping over after the drop is made.

8.1.2.2 *Measurements During Drop Tests*—Measure deflections and distortions of the unitized load and record before the test is started and after each drop. The differences in successive measurements permit a study of the distortions and deflections that have taken place. In addition to these measurements, a record should be made of any changes or breakage in the shipping unit and their location. A recommended simple form for recording the measurements and notations made during the drop test is shown in Fig. 1.

8.2 *Spade Lift Test Procedure:*

8.2.1 Use an actual lift truck equipped with the spade lift attachment to be used in actual practice. See Annex A1 for

SHIPPING UNIT DESCRIPTION... DATE.............

FLAT-EDGE DROP TEST: EDGE TESTED...................................

Height of Drop	Diagonal Measurement of Each Panel										Skid Deflection		Top Edge Deflection (Side Panel)		Remarks
	End *5*		Side *2*		End *6*		Side *4*		Top *1*		Side *2*	Side *4*	Side *2*	Side *4*	
	A–B	*C–D*	*A–B*	*C–D*	*A–B*	*C–D*	*A–B*	*C–D*	*A–B*	*C–D*					
0															

CORNER DROP TEST: CORNER TESTED...................................

Height of Drop	Diagonal Measurement of Each Panel										Skid Deflection		Top Edge Deflection (Side Panel)		Remarks
	End *5*		Side *2*		End *6*		Side *4*		Top *1*		Side *2*	Side *4*	Side *2*	Side *4*	
	A–B	*C–D*	*A–B*	*C–D*	*A–B*	*C–D*	*A–B*	*C–D*	*A–B*	*C–D*					
0															

FIG. 1 Suggested Form for Recording Drop Test Data

recommended standard course for lift truck handling test.

8.2.2 The operator is to engage, lift, and transport the load to each observation point in a manner typical of current practice.

8.2.3 If normal handling requirements specify two or more units high, the test should be conducted with an equivalent load on the test specimen. This test can be repeated for a predetermined number of handlings or until failure occurs. The minimum number of handlings required shall be specified by the user; however, it should be noted that the effects of temperature and humidity in the testing environment may vary greatly and, if so, will affect the strength of the handling flap. For this reason, it may be useful to specify the minimum number of handlings based on test periods representing these extremes.

8.2.4 Examine the shipping unit during and after the test cycle to determine if failure of the handling flap occurred.

8.3 *Clamp Handling Test Procedure:*

8.3.1 Test the ability of the unitized load or case to withstand repeated side compressions using a lift truck equipped with hydraulic carton clamp or other suitable load clamping device. If the cases or unitized loads are normally handled two-high, a second case or equivalent dummy load should be placed on the unit to be tested prior to clamping.

8.3.2 Provision shall be made for measuring the total force applied to the load using a load cell(s) or other suitable device at critical locations.

8.3.3 Measure the clamp force between the platen and record hydraulic pressure at all available clamp settings (that is, low, medium, and high).

8.3.4 Clamp the package or unitized load in a manner typical of current or expected practice starting with the lowest clamp force and increasing the force until load can be successfully transported. The test then consists of clamping the package or unitized load at either:

8.3.4.1 The minimum clamp force on the truck to allow repeated handlings in the particular distribution system, or

8.3.4.2 The next highest clamp setting on the truck to allow repeated handlings in the particular distribution system.

8.3.5 See Annex A1 for recommended standard course for lift truck handling tests.

8.3.5.1 The operator is to clamp, lift, and transport to each observation point until the course has been completed one cycle.

8.3.5.2 If the case or unitized load is normally clamped from both directions, distribute the test cycles accordingly based on the probability of clamping from each direction.

8.3.5.3 The minimum number of handlings and platen size required shall be specified by the user; however, it should be noted that the effects of temperature and humidity in the testing environment may vary greatly and, if so, may affect the strength of the shipping unit. For this reason, it may be useful to specify the minimum number of handlings based on test periods representing these extremes.

8.3.6 Examine the package and product after the test cycle to determine if failure occurred.

8.4 *Pull Pack Handling Test Procedure:*

8.4.1 *Handling on Test Course:*

8.4.1.1 Use an actual lift truck equipped with gripper jaw for pulling the load by the slip sheet tab onto the load plate or platens. See Annex A1 for recommended standard course for lift truck handling tests.

8.4.1.2 The operator is to engage, lift, and transport the load to each observation point in a manner typical of current practice.

8.4.1.3 The minimum number of handlings required shall be specified by the user; however, it should be noted that the effects of temperature and humidity in the testing environment may vary greatly and, if so, may affect the strength of the shipping unit. For this reason, it may be useful to specify the minimum number of handlings based on test periods representing these extremes.

8.4.1.4 Examine the load during and after each test cycle to determine if failure has occurred.

8.4.2 *Slip Sheet Tab Clamp and Pull:*

8.4.2.1 Use an actual lift truck equipped with gripper jaw for pulling the load by the slip sheet tab onto the load plate or platens.

8.4.2.2 The test sequence for handling the load on and off the lift truck plate or platens consists of the following:

8.4.2.3 Position the gripper jaw for clamping the slip sheet tab.

8.4.2.4 Clamp the slip sheet tab and pull the load onto the load plate or platens or pull for a representative distance along a smooth surface.

8.4.2.5 Release the gripper jaw and push the load off of the load plate.

8.4.2.6 This test can be repeated for a predetermined number of handlings or until failure occurs. The minimum number of handlings required shall be specified by the user; however, it should be noted that the effects of temperature and humidity in the testing environment may vary greatly and, if so, will affect the strength of the slip sheet tab. For this reason, it may be useful to specify the minimum number of handlings based on test periods representing these extremes.

8.4.2.7 Examine the slip sheet tab after the test cycle to determine if failure has occurred.

9. Large Case and Crate Tests

9.1 *Drop Test Procedure:*

9.1.1 *Method A—Rotational Edge Drop Test*—Raise one end of the case or crate and set it upon a timber or other support, placed at right angles to the length of the case or crate. The height of the support shall be sufficient to ensure that there will be no support for the base between the ends of the case or crate when dropping takes place, but should not be high enough to cause the case or crate to slide on the support when the drop end is raised for the drop. Raise the other end of the case or crate successively to prescribed heights and release to fall freely on a hard, solid surface. (Concrete is satisfactory.) Where cases or crates are tall or top heavy, provision must be made to prevent the case or crate from tipping over after the drop is made. If the stability of the load is in question, the tip test or the tipover test, or both, as described in Sections 10 and 11 should be performed. See Annex A2 for alternative drop test procedures.

9.1.2 *Method B—Corner Drop Test*—Place a 4 to 10-in. (100 to 254-mm) block flat under one corner of the end already supported in order to raise one corner higher than the other so that impacts on the diagonally opposite corner can be obtained. Raise the unsupported end of the case or crate so that the lower corner of that end reaches a specified height from the surface of the platform and allow to fall freely. Where cases or crates are tall or top heavy, provision must be made to prevent the case or crate from tipping over after the drop is made. If the stability of the load is in question, the tip test or the tipover test, or both, as described in Sections 10 and 11 of this standard should be performed. See Annex A2 for alternative drop test procedures.

9.1.3 *Method C—Raised Edge Free Fall Drop Test*—With one edge of the unit load supported by the floor, raise the other end to prescribed heights and release to fall flat on the impact surface. Where cases or crates are tall or top heavy, provision must be made to prevent the case or crate from tipping over after the drop is made. If the stability of the load is in question, the tip test or the tipover test, or both, as described in Sections 10 and 11 should be performed. See Annex A2 for alternative drop test procedures.

9.1.4 *Method D—Unsupported Free Fall Drop Test*—Raise the case or unitized load above the impact surface. Position load while in the air for flat bottom drop, edge, or corner drop as is desired. Release the load and allow to fall freely. The angle of drop is at the user's option but must be recorded. Where cases or crates are tall or top heavy, provision must be made to prevent the case or crate from tipping over after the drop is made. If the stability of the load is in question, the tip test or the tipover test or both as described in Sections 10 and 11 should be performed. See Annex A2 for alternative drop test procedures.

NOTE—When conducting a drop test using slings on a corrugated fiberboard case, it is necessary to install a nominal 2 by 6-in. wooden board (at least 1 in. longer than the dimension of the case) on top of the case and in line with each sling. This board will prevent each sling from collapsing the top of the corrugated fiberboard case when it is being lifted prior to the drop test.

9.1.5 *Shock Machine Method*—The free fall drop test may, in some instances, be performed using a programmable shock machine of appropriate dimensions using the apparatus and procedures described in Test Methods D 3322. See 4.5 for an equivalency statement for the shock machine method and free fall method of drop testing. See Annex A2 for alternative drop test procedures.

9.1.6 *Measurements During Drop Tests*—Measure deflections and distortions of the case, crate, or unitized load and record before the test is started and after each drop. The differences in successive measurements permit a study of the distortions and deflections that have taken place. In addition to these measurements, a record should be made of any changes or breakage in the shipping unit and their location. A recommended simple form for recording the measurements and notations made during the drop tests is shown in Fig. 1.

9.2 *Grabhook Test Procedures:*

9.2.1 Close the case or crate fully loaded with actual contents or a dummy load of the same total mass and distribution as for shipment.

9.2.2 Grabhooks may consist of one or two pair depending upon the case or crate dimensions, mass, and design. Points of contact for grabhooks may be stenciled on the case or crate. If no marks are shown, then contact may be made by either of the following methods:

9.2.2.1 One pair of grabhooks at any point (usually the center of balance), or

9.2.2.2 Two pair of grabhooks at any two points along case or crate length.

9.2.3 To determine whether the top bracing members are adequate to withstand the horizontal pressures of grabhooks, test the case or crate as follows: Place the grabhooks in contact with the sides a few inches from the top of the case or crate and lift the case or crate several minutes. Return the case to the ground for inspection. The angle of cable or chain between the grabhooks and the lifting hook of lifting device should be approximately 45° from the level of the container top.

9.2.4 Measure deflection of the case or crate while suspended a few inches in the air and again after releasing the hooks. Again make observations to determine if the deflections were such as to damage or dislodge the contents of the case or crate and if the case or crate returns to its original dimensions after the test.

9.3 *Sling Test Procedure*—Conduct sling tests in a manner similar to the grabhook test (see 9.2), with the exception that each sling shall encircle the sides and bottom of the case or crate at points of contact. The angle of slings between the top of the container and the lifting hook of lifting device shall be approximately 45° from the level of the container top. Sling tests shall consist of lifting test container with a single sling, and lifting test container with a double sling.

10. Tip Test Procedure

10.1 Use a case or crate fully loaded with the actual contents. If use of actual contents is not practical, a dummy load of the same total mass, size, and weight distribution may be substituted. The contents or dummy load shall be blocked, braced, and cushioned in place as for shipment.

10.2 Tip the case or crate to an angle of 22° from vertical in the most critical direction and release it.

10.3 The case or crate should right itself and not tip over in either the tipped direction or the opposite direction.

10.4 The test should be repeated in all potentially unstable directions.

10.5 If the case or crate tips over in any direction, the center of gravity must be lowered or the dimensions of the base must be increased until the test can be run successfully or the case or crate and product must be able to pass the tipover test procedure described in Section 11.

10.6 A special safety feature may be added for operator protection during the test. Loose chains, slings, or other similar restraints may be positioned around the load to prevent a complete tipover and impact that could result in personal injury. Such restraints should not be used as an alternative to conducting the test in an area large enough to ensure operator safety.

11. Tipover Test Procedure

11.1 Use a case or crate fully loaded with the actual contents. If use of actual contents is not practical, a dummy load of the same total mass, size, and weight distribution may be substituted. The contents or dummy load shall be blocked, braced, and cushioned in place as for shipment.

11.2 The case or crate shall be slowly tipped until it falls freely on its side on the prescribed surface. Unless otherwise specified, two of these tipovers shall be made, one on each side or 180° apart on a cylinder.

11.3 Examine the case or crate and contents after each tipover and record any signs of damage.

12. Report

12.1 The report shall include the following:

12.1.1 Dimensions of the case or unitized load under test, its complete structural specifications, kind of material, description and specifications for blocking and cushioning, if used, spacing, size, and kind of fasteners, if used, method of closing and unitizing, if any, dimensions and specifications of the pallet or slip sheet, if used, and net and gross weights,

12.1.2 Description of the contents of the case or unitized load under test,

12.1.3 Condition of the test specimen(s) after the test, a detailed description of any damage, and a summary of the criteria used to define damage.

12.1.4 Complete description of equipment used, including load handling attachment, forces measured, and operator of equipment,

12.1.5 Detailed record of tests including procedures used, angle of free fall drop test, if used, covering both container and contents, unitizing method, together with any observation that may assist in correctly interpreting the results or aid in improving the design of the container or the method of packing and unitizing. Also record the orientation of clamps or spade, if used, and the number of handlings for spade, clamp, or pull pack tests, if used, and

12.1.6 Statement to the effect that all tests were made in full compliance with these test methods.

13. Precision and Bias

13.1 No statement is made about the precision or the bias of these test methods since the results merely state whether there is conformance to the criteria for success specified in these procedures.

ANNEXES

(Mandatory Information)

A1. RECOMMENDATIONS FOR STANDARD TEST COURSE FOR LIFT OR CLAMP TRUCK HANDLING TESTS

A1.1 A standard course shall be established on a flat rigid surface representative of worst-case warehouse conditions.

A1.2 The *L* shaped course will include 10 to 11½-ft wide aisles with a 90° turn, appropriate acceleration/deceleration zones, and observation points (OP-1 through OP-5), per Fig. A1.1.

A1.3 User-defined test course obstacles may be used and are at the user's option. Lift truck velocity through this area with or without obstacles shall be at a uniform "walking" speed not to exceed 5 ft/s.

A1.4 The velocity of the lift truck in the turn shall be 3¼ ft/s, maximum.

A1.5 The unitized load under test should be examined after each cycle of handling (pickup/transport/setdown).

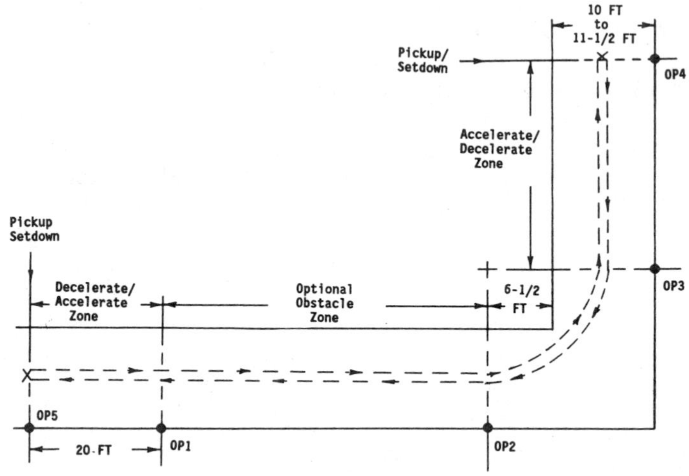

FIG. A1.1 Suggested Handling Course Layout

A2. DROP TEST ALTERNATIVE PROCEDURES

A2.1 Begin all drops at some predetermined height and increase by increments of 1 in. (25 mm) or more until the maximum desired test height is reached or failure occurs.

A2.2 Alternatively, the shipping unit may be given a predetermined number of drops from a fixed height.

A2.3 Edge drops may be made on alternate or adjacent edges.

A2.4 Corner drops may be made on alternate or successive corners.

A2.5 Freefall drops may be made in the same orientation or the dropping orientation varied for a specific number of drops.

The American Society for Testing and Materials takes no position respecting the validity of any patent rights asserted in connection with any item mentioned in this standard. Users of this standard are expressly advised that determination of the validity of any such patent rights, and the risk of infringement of such rights, are entirely their own responsibility.

This standard is subject to revision at any time by the responsible technical committee and must be reviewed every five years and if not revised, either reapproved or withdrawn. Your comments are invited either for revision of this standard or for additional standards and should be addressed to ASTM Headquarters. Your comments will receive careful consideration at a meeting of the responsible technical committee, which you may attend. If you feel that your comments have not received a fair hearing you should make your views known to the ASTM Committee on Standards, 1916 Race St., Philadelphia, PA 19103.

Standard Test Methods for Pallets and Related Structures Employed in Materials Handling and Shipping[1]

This standard is issued under the fixed designation D 1185; the number immediately following the designation indicates the year of original adoption or, in the case of revision, the year of last revision. A number in parentheses indicates the year of last reapproval. A superscript epsilon (ε) indicates an editorial change since the last revision or reapproval.

1. Scope

1.1 These test methods cover the determination of the strength and performance characteristics of pallets and related structures (such as skids, bases, platforms, and bins) employed in material and product storage, handling, and shipping. These test methods are provided for manufacturers, distributors, and users of pallets and related structures.

1.2 The test methods described apply to pallets and related structures constructed specifically for test purposes as well as to pallets selected at random from productions runs. These pallets are to be loaded in line with the applicable specifications.

1.3 *Limitations*—These test methods are not intended to lead to exact simulation of what happens to the structure in the field. The tests are designed to relate to practice with the added proviso that they are repeatable, do not depend on exceptional skill for validity, and are safe to conduct without any elaborate precautions (see 1.8).

1.3.1 These test methods are designed to provide stiffness, flexural strength, and durability data. The test findings are not expected to lead to unalterable conclusions. Field testing under actual use conditions may be necessary to verify the results obtained from laboratory tests.

1.4 These test methods relate to three major pallet strength and performance categories:

1.4.1 *Stiffness and Strength of Pallets*, as determined by flexural tests (see Sections 8 to 13).

1.4.1.1 Static tests are performed in preference to impact and dynamic tests because of the relative simplicity of performing static tests. Static tests may not always be directly applicable to pallet field conditions, in which case impact and dynamic load-capacity tests are recommended. In general, the impact and dynamic tests are more difficult to conduct.

1.4.2 *Physical Durability of Pallets*, as indicated by:

1.4.2.1 Separation resistance of pallet joints as determined by test in the direction perpendicular to the deck plane and parallel to the fastener axis (see Sections 14 to 18).

1.4.2.2 Shear resistance of pallet joints as determined by test in the direction parallel to the deck plane and perpendicular to the fastener axis (see Sections 19 to 23).

1.4.2.3 Lateral stability of pallet as determined by edge pallet-drop test (see Sections 24 to 29).

1.4.2.4 Rigidity of pallet as determined by corner pallet-drop test in accordance with Terminology D 996 (see Sections 30 to 35).

1.4.2.5 Flexural resistance of stringers and decks as determined by pallet-bending tests in both principal directions (see Sections 8 to 13).

1.4.2.6 Splitting resistance of deckboard ends as determined on a relative basis by cleavage test, described in Method D 143.

NOTE—The physical durability of pallets is also measured by the interaction of such physical pallet characteristics as the quality of the pallet materials, pallet-member placement, and pallet-handling environment. These durability factors do not require any special physical tests; hence, are not given consideration in these test methods.

1.4.3 *Special Pallet Properties*, as determined by tests on:

1.4.3.1 Incline-impact resistance of leading-edge deckboard (see Sections 36 to 42).

1.4.3.2 Fork-tip impact resistance of stringer or block (see Sections 43 to 48).

1.4.3.3 Impact resistance of pallet (see Sections 49 to 54).

1.4.3.4 Sliding-friction resistance of pallet (see Sections 55 to 60).

1.4.3.5 Sliding-wear resistance of pallet (see Sections 61 to 66).

1.4.3.6 Sluing resistance of pallet (see Sections 67 to 72).

1.4.3.7 Sling hoist-lift resistance of winged pallet (see Sections 73 to 78).

1.4.3.8 Vibration resistance of pallet (see Sections 79 to 84).

1.5 Only those tests required to determine the particular properties to be evaluated need to be performed.

1.6 The values stated in inch-pound units are to be regarded as the standard.

1.7 In case of dissatisfaction with the test results and rejection based on nonconformance with the requirements of these test methods, claims for a rehearing shall be made promptly and in writing to all involved.

1.8 *This standard does not purport to address all of the safety problems associated with its use. It is the responsibility of the user of this standard to establish appropriate safety and health practices and determine the applicability of regulatory limitations prior to use.* Specific precautions are given in Sections 9, 38, and in 75.2.

2. Referenced Documents

2.1 *ASTM Standards:*

[1] These test methods are under the jurisdiction of ASTM Committee D-10 and are the direct responsibility of Subcommittee D10.22 on Handling and Transportation.

Current edition approved March 29, 1985. Published October 1985. Originally published as D 1185 – 51 T. Last previous edition D 1185 – 73 (1981).

D 143 Methods of Testing Small Clear Specimens of Timber[2]
D 880 Method of Incline Impact Test for Shipping Containers[3]
D 996 Terminology of Packaging and Distribution Environments[3]
D 999 Methods for Vibration Testing of Shipping Containers[3]
D 1761 Test Methods for Mechanical Fasteners in Wood[2]
D 2016 Test Methods for Moisture Content of Wood[2]
D 2395 Test Methods for Specific Gravity of Wood and Wood-Base Materials[2]
E 171 Specification for Standard Atmospheres for Conditioning and Testing Materials[3]
F 680 Test Methods for Nails[4]

2.2 *ANSI Standards:*
MH1.1.2 Pallet Definitions and Terminology[5]
MH1.2.2 Pallet Sizes[5]
MH1.4.1 Procedures for Testing Pallets[5]

2.3 *Bulletins of VPI and SU Pallet and Container Research Laboratory:*
No. 122 Reconfirmation of MIBANT Test Criteria for Pallet Nails[6]
No. 126 Design of Pallet Deckboard-Stringer Joints[6]
No. 139 Auto-Stapled Pallets[6]

2.4 *Other Publication:*
National Design Specification for Stress-Grade Lumber and Its Fastenings (NDS)[7]

3. Terminology

3.1 For definitions of terms in these test methods, see Terminology D 996, and MH1.1.2.

4. Significance and Use

4.1 Flexural tests provide specific numerical data that may be used in stiffness and strength design equations and formulas.

4.2 Physical durability evaluation may be based on static, impact, and dynamic test data. They can be used to arrive at physical and economic performance evaluations.

4.2.1 The MIBANT test (Methods F 680 and VPI and SU Bulletin No. 122) provides information on the stiffness and toughness of the fasteners used for pallet assembly.

4.2.2 The joint tests provide information on the effectiveness of the fasteners in the material used for pallet assembly.

4.2.3 The edge pallet drop test provides information on the lateral stability of pallets.

4.2.4 The corner pallet drop test provides information on the rigidity of pallets.

4.2.5 The stringer and composite deck flexural tests provide information on the pallet load-carrying capacity.

4.3 Special tests may be necessary to assess the performance of specific materials and joining devices, design features, use applications, and other variables encountered.

5. Materials

5.1 These test methods are pertinent to pallets made of any materials that provide satisfactory performance under anticipated use conditions. In the case of hygroscopic materials, see Section 7.

6. Sampling

6.1 Sampling shall provide for the selection of representative specimens. The number of repetitions per type of test required can vary. At least five replications per joint test and three replications per pallet test shall be conducted.

7. Conditioning

7.1 Pallets and related structures may be assembled with materials that are susceptible to changes in moisture content. Prior to testing, it may be necessary to condition these assemblies to approximate moisture-content equilibrium of an appropriate environment such as that expected during pallet use. Pallets assembled and tested in green condition do not require any conditioning. Pallets assembled green and tested, for instance, at 12 % moisture content (simulating certain indoor use conditions) require at least six- to ten-week storage in an environment of 21°C (70°F) and 50 % relative humidity, depending on the wood species and thickness of the members used for pallet assembly.

7.2 It may be desirable to expose pallets and related structures to various temperatures.

7.3 Exposure may be constant or cyclic.

7.4 For specific standard conditioning atmospheres, see Specification E 171.

TEST METHODS COVERING PALLET STIFFNESS AND FLEXURAL STRENGTH

8. Apparatus

8.1 For laboratory testing, the following equipment is required for application of concentrated line loads:

8.1.1 *Calibrated Compression Testing Machine*, with a minimum load capacity of 45 000 N (10 000 lb) and a sufficiently large bed to provide support for the pallet. A 67 500-N (15 000-lb) capacity testing machine may be needed if a uniformly distributed flexible load is to be supported by the pallet.

8.1.2 *Plywood Line-Load Spreaders*—A suitable number of 100-mm (4-in.) wide and 13-mm (½-in.) thick, 5-ply plywood load spreaders, of sufficient length to support the pallet and transmit the line load along the whole length or width of the pallet (see Fig. 1) is required.

8.1.3 *Steel Pipes*, of at least 50-mm (2-in.) outside diameter and sufficient length to be placed below the supporting load spreaders and above the line-load spreaders, with the pipe to be of such stiffness as not to deflect more than 2 mm (0.08 in.) while the pallet is tested.

8.1.4 *Load-Distributing Plate or I beams*, hanging from the swivel or floating head of the testing machine.

8.2 In the case of application of a concentrated load at the center of the pallet, instead of the loading pipes, a spreader plate, with rounded edges of the plate bottom, of satisfactory

---

[2] *Annual Book of ASTM Standards*, Vol 04.09.
[3] *Annual Book of ASTM Standards*, Vol 15.09.
[4] *Annual Book of ASTM Standards*, Vol 15.08.
[5] Available from the American National Standards Institute, 1430 Broadway, New York, NY 10018.
[6] Available from Virginia Polytechnic Institute and State University, Blacksburg, VA 24061.
[7] Available from National Forest Products Association, 1619 Massachusetts Ave., N.W., Washington, DC 20036.

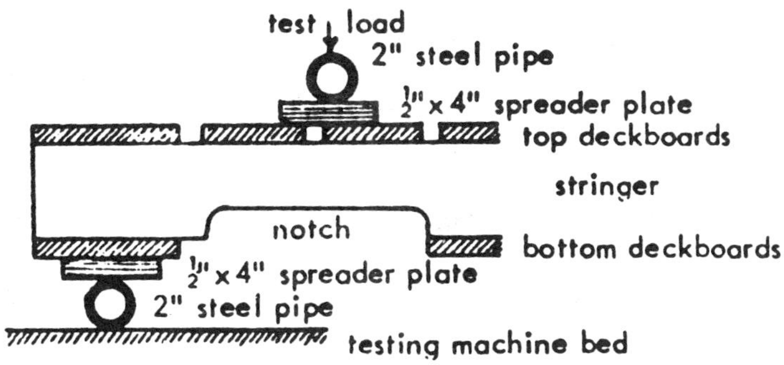

FIG. 1 Loading and Supporting Details for Pallet to Be Tested

strength and such size as to exceed the base of the concentrated load not more than 13 mm (½ in.). In the case of application of several concentrated loads, an adequate number of such spreader plates.

8.3 For applying a more or less uniformly distributed load, a flat, smooth, load-supporting surface of sufficient length to support the test apparatus and a suitable number of individual weights may take the place of a large-bed testing machine.

8.4 For applying a uniformly distributed load, a vacuum testing device or pneumatic pressure bag of sufficient size and strength to be placed between the pallet and the flat surface of a loading box or frame.

8.5 For deformation readings, micrometer dial gages or suitable measuring devices, such as calibrated sensors, with a capability of measuring 0.025 mm (0.001 in.).

9. Precautions

9.1 Possible injury to personnel and damage to the test equipment and instrumentation during and after load application by any unexpected release of potential strain energy accumulated during testing can be prevented by placing, for instance, restraining springs next to the pallet, which prevent the pallet from slipping sideways.

10. Load Application

10.1 The type of pallet, support conditions, and type of loading determine the flexural test conditions.

10.1.1 Pallet decks are loaded independently as single or multiple-span beams while the pallet (*1*) supports uniformly or partially distributed, concentrated or line loads, and (*2*) is placed on a level surface or stacked on a loaded pallet (see Fig. 2).

10.1.2 When a loaded pallet is placed in a rack that supports the pallet along its sides or ends, (*1*) the decks, which include stringerboards in the case of block pallets, or stringers, or both, act as a composite structure in flexure, and (*2*) the stringers act as simple beams (see Fig. 3).

10.1.3 Wing pallets supported along the pallet ends or sides are stressed in the same manner as flush pallets. Wing pallets supported by the deck ends produce high stresses in the overhangs.

10.1.4 Support of pallet by the fork tines may cause the pallet to act as a cantilevered composite structure.

11. Procedure

11.1 If not specifically required otherwise, test at least three, preferably five, new and unused pallets that meet the governing specifications.

11.2 *Pallet Support*—Place the pallet to be tested on supporting line-load spreaders centric on top of steel pipes located on top of the bed of the testing machine or surface serving in place of the testing machine bed.

11.3 *Load Application:*

11.3.1 Place line-load spreaders between the steel pipes and the top of the pallet crosswise to the span at the quarter points or any other desirable location or locations below the load-distributing I beams.

11.3.2 Place load-spreader plates between the surface serving as loading plate and the top of the pallet.

11.3.3 If individual weights are used, place them in such a way on the top of the pallet that they do not provide an arched body when the pallet deflects during loading.

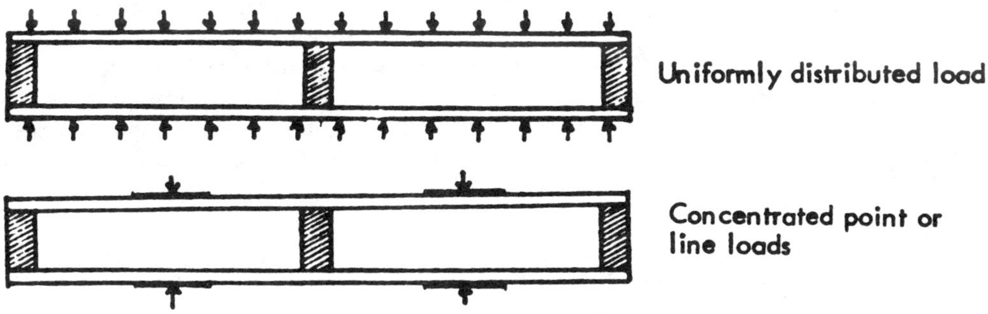

FIG. 2 Pallet With Uniformly Distributed and Concentrated Point or Line Loads

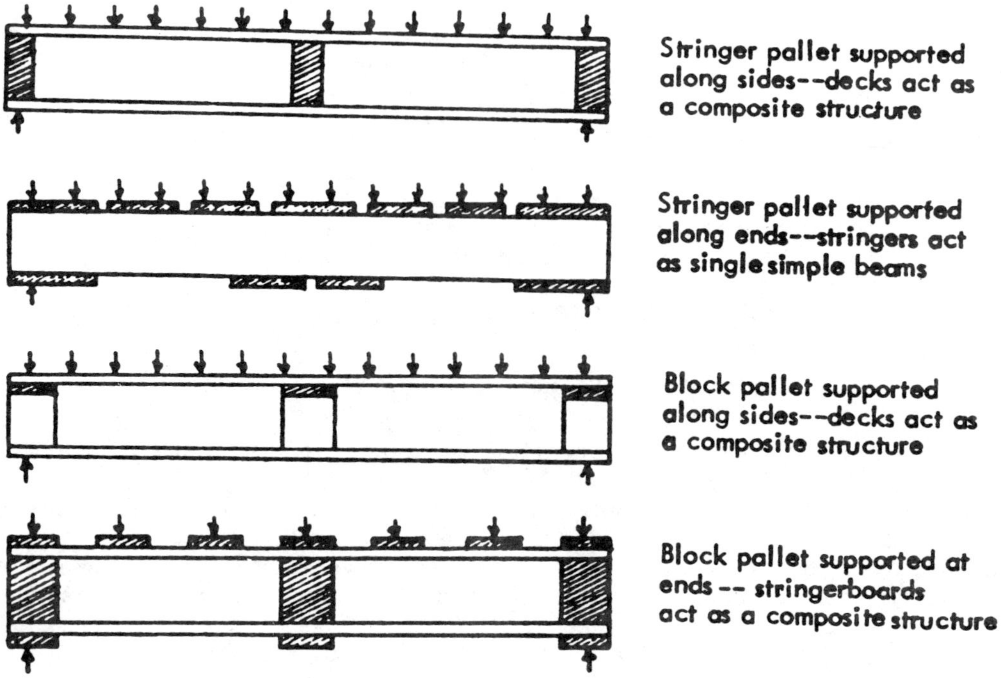

FIG. 3 Stringer and Block Pallets Supported Along Their Sides or Ends, Acting as Beams or Composite Structures

11.3.4 If a vacuum testing device or penumatic pressure bag is used, place it on top of the pallet within the loading box or frame.

11.4 Apply the test load at a uniform rate of motion of the cross-head or base of the testing machine of not more than 12.5 mm (0.50 in.)/min.

11.5 Observe the deflection of the pallet at its center and wherever else desirable at least at intervals approximating one fifth of the total test load. Maintain load-deformation-time records throughout the test. If application of a given load for a certain period, such as 24 h, is required, take deflection readings at the beginning of, at intervals during, and at the end of this period, to allow satisfactory plotting of a time-deformation curve for the complete period.

11.6 Observe and record any failures.

11.7 Determine properties of the materials, as are appropriate, such as moisture content and oven-dry specific gravity of the wood and wood-base products fastened (Test Methods D 2016 and D 2395) and the toughness of the nails used, that is, their MIBANT angles (Test Methods F 680, Section 7).

12. Report

12.1 The test report shall include the following information if applicable:

12.1.1 Date of test and date of report.

12.1.2 Test sponsor and test agency.

12.1.3 Identification of pallet or related structure: manufacturer, model design, size, style, type, and weight; materials and finish; pretest conditioning, dimensions, fasteners and their location; and other pertinent information and observations such as cracks and other defects.

12.1.4 Detailed drawings or photographs of typical test assemblies before, during, and after testing, if not otherwise fully described.

12.1.5 Species, oven-dry specific gravity, and moisture content at time of test, in percent of oven-dry weight of any wood members or components.

12.1.6 Relevant mechanical properties of the materials and fasteners used for the test assemblies.

12.1.7 Description of procedure used for the assembly.

12.1.8 Age of the assembly, in days, at time of test, if this information is of any significance.

12.1.9 Description of any conditioning and time between conditioning and testing, in days, if this information is of any significance.

12.1.10 Description of test method and loading procedure, especially if deviating from these test methods.

12.1.11 Number of replicate specimens tested.

12.1.12 All test data, including statistical analysis, to provide at least mean deflection and load values, in millimetres (inches) and newtons (pounds-force), respectively, and standard deviations.

12.1.13 Photographs, sketches, or word descriptions of failures observed.

12.1.14 Summary of findings.

12.1.15 Relevant test limitations and recommendations.

12.1.16 Listing of observers of tests and signatures of responsible persons.

13. Precision and Bias

13.1 No statement is made on the precision and bias of these test methods, since no data are available at this time based on the use of the test methods described.

TEST METHODS COVERING PALLET DURABILITY

Separation Resistance of Pallet Joints Perpendicular to Deckboard Plane

14. Apparatus

14.1 *Calibrated, Suitable Testing Machine*, with a minimum load capacity of 4500 N (1000 lbf), a swivel head fastened to the underside of the movable crosshead of the testing machine, and a loading yoke to apply one half of the total test load to each of the deckboard ends. Furthermore, a suitable gripping device to pull the fastener through the deckboard or from the stringer after completion of the deckboard-stringer separation test.

14.2 If data are to be obtained on the deckboard-stringer separation or on the fastener slip during the subsequent pull-through or withdrawal test, a micrometer dial gage or other suitable measurement device, such as a calibrated sensor, with a capability of measuring 0.025 mm (0.001 in.).

15. Load Application

15.1 It is feasible to apply an impact load to a pallet joint once or repetitively by dropping a given mass from a given height on the joint perpendicular to the deckboard plane, in order to determine the resistance to separation of the deckboard from the stringer. Usually, preference is given to applying a static load (as shown in Fig. 4) to the deckboard crossing the stringer in the direction perpendicular to that of the plane of the deckboard until failure occurs (Test Methods D 1761, 8.1.3; and, VPI and SU Bulletin No. 139, Fig. 2).

16. Procedure

16.1 Assemble at least five replicate joints. Each joint shall be assembled with a single fastener unless specified otherwise.

16.2 Use fasteners that are unused and that meet the governing specifications.

16.3 Drive the fastener perpendicular to the deckboard surface and flush or countersunk as specified. Drive the nail with a hand-hammer, an automatic nailing machine, or an appropriate pneumatic tool. Drive the staple with an appropriate pneumatic tool at the predetermined location. Insert any other fastener appropriately in line with the procedure to be used for pallet assembly.

16.4 If immediate tests are to be performed, test as quickly after fastener driving as is practical, at most within 15 min after joint assembly.

16.5 Place the specimen in the testing machine or grip the fastener with the gripping device (Test Method D 1761, 10.2.2).

16.6 Apply load by way of the swivel head, until failure occurs, at the constant rate of travel of 2.5 mm (0.10 in.) ±25 % per minute of movable crosshead or base of the testing machine (Test Method D 1761, 10.4).

16.7 In the case of a nailed joint, measure both the shank withdrawal and head pull-through resistance. In the case of a stapled joint, measure both the leg withdrawal and crown pull-through resistance.

16.8 Determine the properties of the materials as appropriate, such as the moisture content and the oven-dry specific gravity of the wood and wood-base products fastened (Test Methods D 2016 and D 2395).

17. Report

17.1 The report shall include the following information if applicable:

17.1.1 Date of test and date of report.

17.1.2 Test sponsor and test agency.

17.1.3 Relevant properties of the materials and fasteners used.

17.1.4 Description of procedure used for joint assembly.

17.1.5 Description of test specimens, including species of wood and its moisture content and oven-dry specific gravity; dimensions of the members; fastener type and size, and end and edge distances; and, for threaded fasteners, also wire diameter, thread-crest diameter, thread angle or number of helixes per inch of thread length.

17.1.6 Description of any conditioning and time between conditioning and testing, in days.

17.1.7 Detailed description of test method and loading procedure, if deviating from these test methods.

17.1.8 Number of replicate specimens tested.

17.1.9 Types of failures observed and all test data including statistical analysis to provide at least mean maximum load values and such load values per unit of fastener penetration into stringer as well as standard deviations.

17.1.10 Load-slip curves, if required.

17.1.11 Any other relevant observations, test limitations, and recommendations.

17.1.12 Listing of observers of tests and signatures of responsible persons.

18. Precision and Bias

18.1 The provisions of Section 13 are applicable.

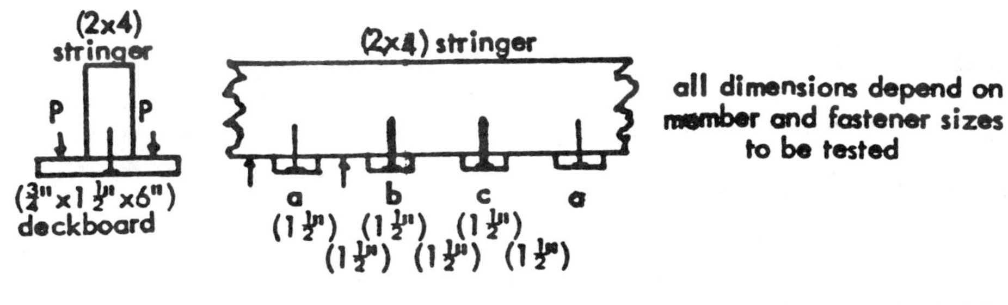

NOTE—For matched testing of different fasteners (a, b, c), cut all deckboard sections from the same piece along the grain and locate all fasteners along the same grain of the stringer.

FIG. 4 Test Set-Up for Application of Static Load for Determination of Deckboard-Stringer Separation Resistance, With Fastener Loaded Axially

Shear Resistance of Pallet Joints Parallel to Deckboard Plane

19. Apparatus

19.1 *Suitable Calibrated Testing Machine*, with a minimum load capacity of 4500 N (1000 lbf) and a swivel head fastened to the underside of the movable crosshead of the testing machine.

19.2 If data are to be obtained on the slip of the deckboard along the stringer, a micrometer dial gage or other suitable measurement device, such as a calibrated sensor, with a capability of measuring 0.025 mm (0.001 in.).

20. Load Application

20.1 It is feasible to apply an impact load to the pallet joint once or repetitively by dropping a given mass from a given height on the joint parallel to the deckboad plane and perpendicular to the stringer side, in order to determine the shear resistance of the joint (Test Method D 1761; also, MH1.4.1, App. B; also, VPI and SU Bulletin No. 126, Figs. 2a and 2b). Usually, preference is given to applying a static load (as shown in Fig. 5) to the deckboard crossing the stringer, thereby pushing the deckboard along the stringer in the direction parallel to that of the plane of the deckboard until failure occurs (Test Methods D 1761, Section 17; also, MH1.4.1, App. B; also, VPI and SU Bulletin No. 139, Fig. 2).

21. Procedure

21.1 Assemble at least five replicate joints for testing. Each joint shall be assembled with a single fastener, unless specified otherwise.

21.2 Use fasteners that are unused and that meet the governing specifications.

21.3 Drive the fastener perpendicular to the deckboard surface and flush or countersunk as specified. Drive the nail with a hand-hammer, an automatic nailing machine or an appropriate pneumatic tool. Drive the staple with an appropriate pneumatic tool at the predetermined location. Insert any other fastener appropriately in line with the procedure to be used for pallet assembly.

21.4 If immediate tests are to be performed, test as quickly after fastener driving as is practical, at most within 15 min after joint assembly.

21.5 Place the specimen in the testing machine (Test Method D 1761, 17.1).

21.6 Apply load by way of swivel head, until failure occurs, at the constant rate of travel of 2.5 ± 0.625 mm (0.10 ± 0.025 in.)/min of the movable crosshead or base of the testing machine (Test Method D 1761, 10.4).

21.7 In the case of a nailed or stapled joint, observe failure that occurs as a result of fastener shank or leg bending, withdrawal or shear, or combination thereof, or fastener head or crown pull-through.

21.8 Determine the properties of the materials as appropriate, such as the moisture content and the oven-dry specific gravity of the wood and wood-base products fastened (Test Methods D 2016 and D 2395) and the nail toughness, that is, the MIBANT angles for 25 samples from each lot of nails (Test Methods F 680).

22. Report

22.1 The report shall include the following information:

22.1.1 Date of test and date of report.

22.1.2 Test sponsor and test agency.

22.1.3 Relevant properties of materials and fasteners used.

22.1.4 Description of procedure used for joint assembly.

22.1.5 Description of test specimens, including species of wood and its moisture content and oven-dry specific gravity; dimensions of the members; fastener type and size, and end and edge distances.

22.1.6 Description of any conditioning and time between conditioning and testing, in days.

22.1.7 Detailed description of test method and loading procedure, if deviating from these test methods.

22.1.8 Number of replicate specimens tested.

22.1.9 Types of failures observed and all test data including statistical analysis to provide at least mean maximum load values as well as standard deviations.

22.1.10 Load-slip curves, if required.

22.1.11 Any other relevant observations, test limitations, and recommendations.

22.1.12 Listing of observers of tests and signatures of responsible persons.

23. Precision and Bias

23.1 The provisions of Section 13 are applicable.

Lateral Stability of Pallets

24. Significance and Use

24.1 The purpose of this test method is to determine the lateral stability of the pallet, that is, its resistance to impact forces in the pallet plane perpendicular to the pallet sides.

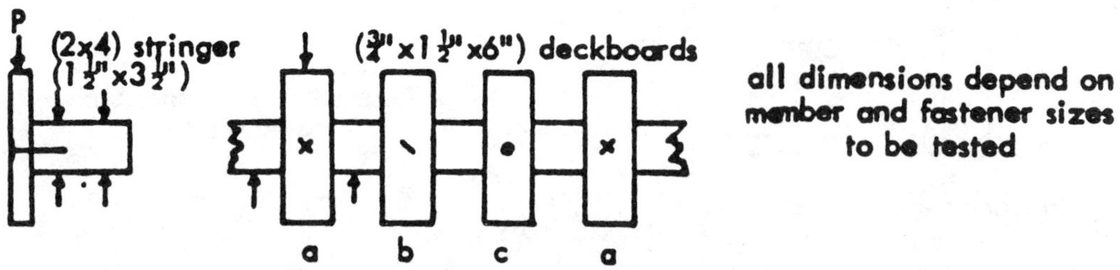

NOTE—For matched testing of different fasteners (a, b, c), cut all deckboard sections from same piece along the grain and locate all fasteners along the same grain of the stringer.

FIG. 5 Test Set-Up for Application of Static Load for Determination of Shear Resistance of Deckboard-Stringer Joint, With Load Applied Perpendicular to Fastener Axis

25. Apparatus

25.1 *Suitable Lifting Device,* such as a hoist or lift truck, allowing accurate positioning of the suspended pallet; a suspension and quick-release mechanism, such as a solenoid release; and a rigid, massive block of metal or concrete, or both, with a level, smooth, and hard surface.

26. Impact Application

26.1 Suspend the top edge of the pallet in such a manner that the pallet sides are horizontal and one of the cross-sectional pallet diagonals is vertical (as shown in Fig. 6). Drop the suspended pallet repetitively to fall freely onto the same side for a constant distance from the lowest pallet side to a level, smooth, and hard impact surface of a rigid, massive block of metal or concrete, or both, (4.3 of MH1.4.1).

27. Procedure

27.1 Unless specified otherwise, test at least three, preferably five, new and unused representative pallets that meet the governing specifications.

27.2 Obtain, if feasible, 25 samples of each of the lots of nails used for pallet assembly for the performance of tests on the physical and mechanical nail properties (27.8).

27.3 Weigh the unloaded pallet immediately prior to drop testing.

27.4 Suspend the pallet from the two corners of the top edge in such a manner that the pallet edges are horizontal and the cross-sectional diagonal from the suspension edge to the impact edge is vertical and the free-fall drop height from the impact edge to the impact area is 1016 ± 3 mm (40 ± 1/8 in.).

27.5 Drop pallet and restrain it after impacting in such a manner as to prevent a secondary fall. Repeat this procedure, always dropping the pallet onto the same edge from the specified height between lowest pallet edge and impact area.

27.6 Terminate the test after the sixth or twelfth drop, as specified, or when the pallet will have failed and will no longer function properly during its intended use. Failure has occurred when any pallet member will have been damaged to such an extent that it can no longer transmit the anticipated forces or when the joints between deck or deckboards and stringers are no longer connecting those members effectively or when the pallet is distorted to such an extent that the stringer face, instead of its edge, is impacted.

27.7 Observe and record any defects and damages immediately prior to and after each drop.

27.8 Determine the properties of the materials as appropriate, such as, the moisture content and the oven-dry specific gravity of the wood and wood-base products fastened (Test Methods D 2016 and D 2395) and the toughness of the nails, that is, the MIBANT angles for 25 samples from each lot of nails (Test Methods F 680).

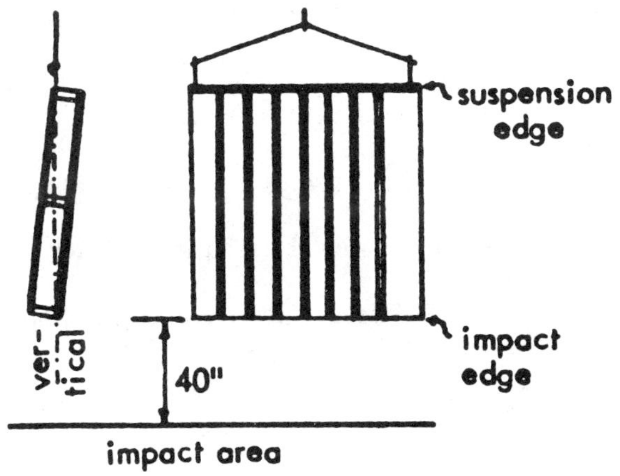

FIG. 6 Test Set-Up for Determination of Lateral Stability of Pallet

28. Report

28.1 The report shall include the following information if applicable:

28.1.1 Date of test and date of report.

28.1.2 Test sponsor and test agency.

28.1.3 Relevant properties of the materials and fasteners used.

28.1.4 Description of procedure used for assembly of pallet and its history if available.

28.1.5 Description of pallet, including size, style, type, design, and weight; species of wood and its moisture content and oven-dry specific gravity; member dimensions; fastener type, size, end and edge distances, and spacings.

28.1.6 Description of any conditioning and time between conditioning and testing, in days.

28.1.7 Detailed description of test method, if deviating from these test methods.

28.1.8 Number of replicate pallets tested and number of drops.

28.1.9 Number of drops to failure and types of failure observed immediately prior to and after each drop.

28.1.10 Any other relevant observations, test limitations, and recommendations.

28.1.11 Listing of observers of tests and signatures of responsible persons.

29. Precision and Bias

29.1 The provisions of Section 13 are applicable.

Rigidity of Pallet

30. Significance and Use

30.1 The purpose of this test method is to determine the rigidity of the pallet, that is, its resistance to impact and racking forces in the pallet plane.

31. Apparatus

31.1 *Suitable Lifting Device,* such as a hoist or lift truck, allowing accurate positioning of the suspended pallet; a suspension and quick-release mechanism, such as a solenoid release; and a rigid, massive block of metal or concrete, or both, with a level, smooth, and hard surface.

31.2 *Graduated Scale,* of sufficient length or a calibrated rod with attached scale, for determining the length of the pallet diagonals, with a capability of measuring 0.25 mm (0.01 in.).

32. Impact Application

32.1 Drop the vertically and diagonally suspended pallet (as shown in Fig. 7) repetitively to fall freely onto the same corner for a constant distance from the lowest pallet corner to a level, smooth, and hard impact surface of a rigid, massive block of metal or concrete, or both, (4.3 of MH1.4.1).

33. Procedure

33.1 Unless otherwise specified, test at least three, preferably five, new and unused, representative pallets that meet the governing specifications.

33.2 Obtain, if feasible, 25 samples of each of the lots of nails used for pallet assembly for the performance of tests on the physical and mechanical nail properties (27.8).

33.3 Weigh the unloaded pallet immediately prior to drop testing, in order to allow adjustment of the test data.

33.4 Suspend the pallet at one of its four corners in such a manner that the diagonal across each pallet face from the suspension corner to the impact corner is vertical and the free-fall drop height from the impact corner to the impact area is 1016 ± 3 mm (40 ± ⅛ in.).

33.5 Drop pallet and restrain it after impact in such a manner as to prevent a secondary fall. Repeat this procedure, always dropping the pallet onto the same corner from the specified height between lowest pallet corner and impact area.

33.6 Prior to the first drop and immediately after each subsequent drop, measure and record the length of each of the two top-deck and two bottom-deck pallet diagonals. Measure between the reference points, marks or pins, located 25 ± 2 mm (1 ± 1/16 in.) from both edges leading to each of the four pallet corners. Place a second mark or pin 25 ± 2 mm (1 ± 1/16 in.) from the first mark or pin along the diagonal at the pallet impact corner, since damage to the impact corner might extend to and beyond the first mark.

33.7 Terminate the test after the sixth or twelfth drop, as specified, or when deformation information becomes meaningless.

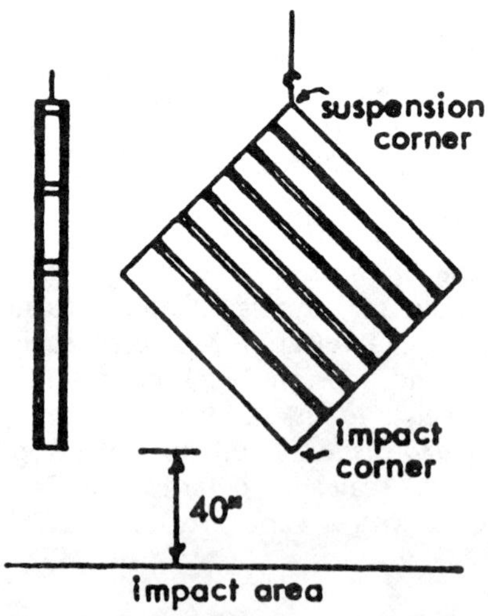

FIG. 7 Test Set-Up for Determination of Rigidity of Pallet

33.8 Compute and average the changes in length of the four diagonals with respect to the pre-drop dimensions after each drop. Plot the average length changes against the number of drops to provide a visual record of the test values. Compute, average, and tabulate the length changes as a percentage of the pre-drop lengths of the four pallet diagonals. List the average percentage length changes after the sixth and twelfth drops. Compute and list the cumulative average length changes and these changes divided by the pallet weight, in order to obtain overall ratings of the pallet rigidity.

33.9 Observe and record any defects and damages immediately prior to and after each drop.

33.10 Determine the properties of the materials as appropriate, such as, the moisture content and the oven-dry specific gravity of the wood and wood-base products fastened (Test Methods D 2016 and D 2395) and the toughness of the nails, that is, the MIBANT angles for 25 samples from each lot of nails (Test Methods F 680).

34. Report

34.1 The report shall include the following information if applicable:

34.1.1 Date of test and date of report.

34.1.2 Test sponsor and test agency.

34.1.3 Relevant properties of the materials and fasteners used.

34.1.4 Description of procedure used for assembly of pallet and its history, if available.

34.1.5 Description of pallet, including size, style, type, design, and weight; species of wood and its moisture content and oven-dry specific gravity; member dimensions; fastener type, size, end and edge distances, and spacings.

34.1.6 Description of any conditioning and time between conditioning and testing, in days.

34.1.7 Detailed description of test method, if deviating from these test methods.

34.1.8 Number of replicate pallets tested and number of drops.

34.1.9 Individual and average length changes of pallet diagonals, average percentage length changes, cumulative average length changes, and average length changes divided by the pallet weight as well as statistical analysis to provide mean test values obtained after the sixth and twelfth drops as well as standard deviations.

34.1.10 Plot of average percentage length changes against number of drops.

34.1.11 Types of failures observed immediately prior to and after each drop.

34.1.12 Any other relevant observations, test limitations, and recommendations.

34.1.13 Listing of observers of tests and signatures of responsible persons.

35. Precision and Bias

35.1 The provisions of Section 13 are applicable.

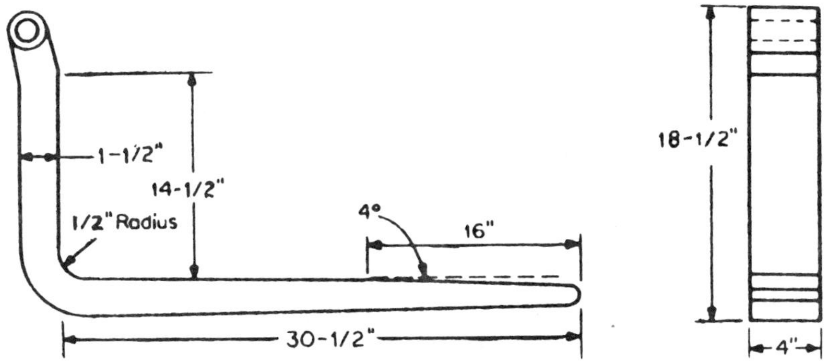

FIG. 8 Standard Fork Used for Determination of Incline-Impact Resistance of Leading-Edge Deckboard of Pallet

TEST METHODS COVERING SPECIAL PALLET PROPERTIES

Incline-Impact Resistance of Leading-Edge Deckboard

36. Significance and Use

36.1 The purpose of this test method is to determine the resistance of the top leading-edge deckboard of the pallet to impact by two lift-truck forks and this deckboard's separation resistance from the stringers or blocks.

37. Apparatus

37.1 *Incline Impact Tester* (see Method D 880), modified by firmly fastening to the rigid backstop of the tester two standard forks of 38 by 100 by 910-mm (1½ by 4 by 36-in.) size, with a 13-mm (½-in.) inside radius of the fork heel and a 4° taper for a distance of 400 mm (16 in.) from the tine tips (as shown in Fig. 8), in such a manner that they are spaced 700 mm (28 in.) on centers and that the flat, nontapered, upper section of the forks is angled 4° downward from the direction of the dolly-travel plane. Such forks are used with fork-lift trucks of 13500-N (3000-lb) or greater capacity.

37.2 *Weight Box,* of a width equal to that of the pallet, however, 400 mm (16 in.) shorter than the pallet length and a suitable number of weights.

38. Precautions

38.1 Possible premature triggering of the dolly-release mechanism shall be avoided by the use of appropriate safety devices. Caution shall be exercised and safety requirements shall be adhered to in order to prevent impacting any personnel by the dolly, pallet, weight box, and weights, as well as pallet debris.

39. Impact Application

39.1 Place the loaded pallet on a dolly on an inclined track in such a way that the bottom edge of the top deckboard, after release of the dolly at a predetermined

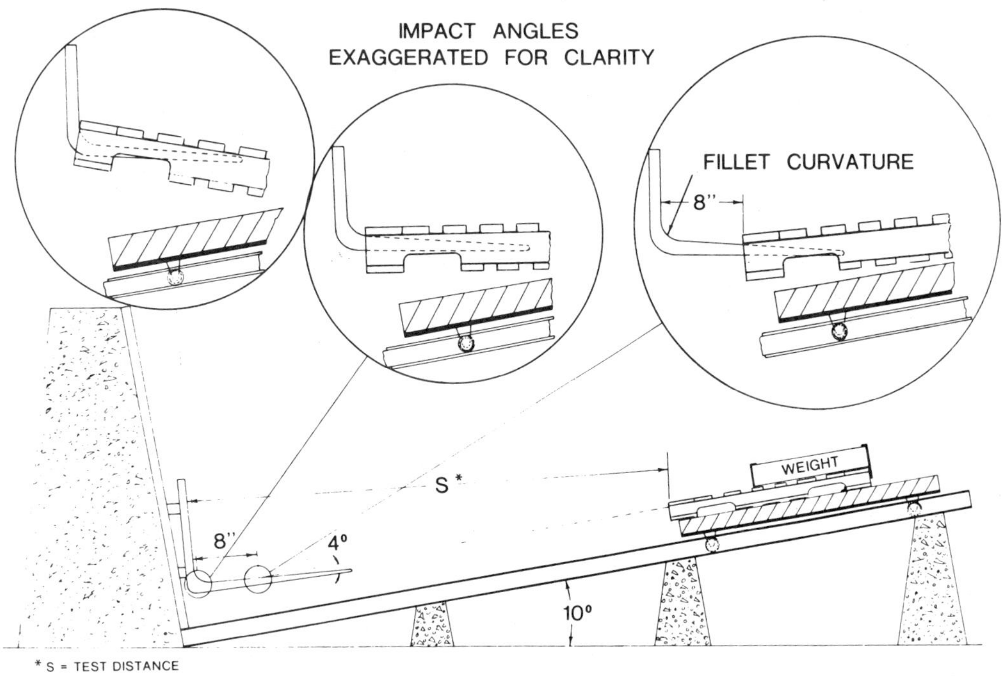

FIG. 9 Test Set-Up for Determination of Incline-Impact Resistance of Leading-Edge Deckboard of Pallet

location and its downward travel on the track, rides (as shown in Fig. 9) on the top of the fork tines for a distance of 200 mm (8 in.) from the fork heels prior to impacting them (5.6 of MH1.4.1).

40. Procedure

40.1 Place the pallet on the dolly of the tester in such a way that the forward-leading edge of the top and bottom forward-leading-edge deckboard is parallel to and overhangs the forward-leading-edge of the dolly by 50 mm (2 in.). Place the weight box in such a way that the trailing upper box edge is flush with the trailing upper pallet edge. Fasten the weight box to the pallet in such a way that the kinetic inertia of the box is transferred at impact directly to the pallet stringers or blocks. This can be accomplished by angling the trailing box edge over the upper trailing pallet edge in such a way that the steel angle of the box is in contact with the stringers or blocks. Alternatively, the box can be fastened directly to the stringers or blocks. A number of suitable weights, as specified, shall be distributed uniformly over the box area and fastened to the box in such a way that they move with the box during impact.

40.2 Adjust the fork height in such a manner that the bottom leading edge of the forward-leading-edge deckboard rides on the top of the tines for the distance of 200 mm (8 in.) from the fork heels prior to impacting them.

40.3 Move the loaded dolly of the tester up the incline to result in a 356 ± 3-mm (14 ± ⅛-in.) clearance between dolly front and front of the fork back. If necessary, adjust this distance of travel for pallet failure to occur during not less than five and not more than ten impacts.

40.4 Release the loaded dolly to travel freely until the bottom leading edge of the forward-leading-edge deckboard makes contact with the top of the tines, for this deckboard to ride the top of the tines prior to impacting the fork heels. Repeat this impact cycle or progressively increase the distance of free travel by such increments, as specified, until pallet failure occurs.

40.5 Observe and record any defects and damages immediately prior to testing and after each impact.

40.6 If the pallet or pallet top deck bends and rides beyond the tines up the forks during impact, place and rigidly fasten to the pallet top two 25 by 100-mm (nominal 1 by 4-in.) stiffener boards of full pallet length as close as feasible near the pallet sides without causing interference with the forks during impact.

40.7 Pallet failure shall be considered to have occurred when the forks have destroyed the leading-edge deckboard in such a way that the pallet will no longer function with the particular type and size of handling devices intended to be used or have torn the leading edge of the top leading-edge deckboard or deck from the stringers or blocks or when the pallet will no longer support the load it is designed to carry.

41. Report

41.1 The report shall include the following information if applicable:

41.1.1 Date of test and date of report.

41.1.2 Test sponsor and test agency.

41.1.3 Relevant properties of materials and fasteners used.

41.1.4 Description of procedure used for assembly of pallet and its history.

41.1.5 Description of pallet, including size, style, type, design, and weight; species of wood and its moisture content and oven-dry specific gravity; member dimensions; fastener type, size, end and edge distances, and spacings.

41.1.6 Number of replicate pallets tested.

41.1.7 Detailed description of test method if deviating from these test methods, as well as use of any pallet deck stiffeners.

41.1.8 Description of any conditioning and time between conditioning and testing, in days.

41.1.9 Weight of weight box and weights in box.

41.1.10 Distance of travel during first and any subsequent impact cycles.

41.1.11 Energy applied during first and any subsequent impacts as well as total energy required to cause failure.

41.1.12 Types of defects and damages observed immediately prior to testing and after each impact.

41.1.13 Cause of pallet failure and any other relevant observations, test limitations, and recommendations.

41.1.14 Listing of observers of tests and signatures of responsible persons.

42. Precision and Bias

42.1 The provisions of Section 13 are applicable.

Fork-Tip Impact Resistance of Stringer or Block

43. Significance and Use

43.1 The purpose of this test method is to determine the resistance of the pallet stringers or blocks to impact by the tine tips of the forks of lift trucks and other mechanical handling devices.

44. Apparatus

44.1 The provisions of Section 37 are applicable.

45. Impact Application

45.1 Place the loaded pallet on a dolly on an inclined impact track in such a way that the face of the outer stringer or the faces of the blocks; or the ends of the stringers or blocks; or the outer corner of an outer stringer or block are impacted by the tine tips or tip after release of the dolly at a predetermined location and its downward travel on the track.

46. Procedure

46.1 The provisions of Section 38 are applicable. Use a new pallet for testing in the direction of the second pallet axis or the pallet diagonal.

46.2 Adjust the fork height in such a manner that the mid-height of the outer stringer face or block faces or of the stringer or block ends or of the stringer or block outer corner impact the tine tips or tip.

46.3 The provisions of 40.3 are applicable.

46.4 Release the loaded dolly to travel freely until the face(s), ends or corner impact the tine tips or tip. Repeat the impact cycle or progressively increase the free-travel distance by such increments, as specified, until stringer or block failure occurs.

46.5 The provisions of 40.5 are applicable.

46.6 Stringer or block failures shall be considered to have occurred when a tine tip splits a stringer or block or the pallet will no longer support the load it is designed to carry as a result of stringer or block failure.

46.7 *Precautions*—The provision of Section 38 are applicable.

47. Report

47.1 The provisions of Section 41 are applicable.

48. Precision and Bias

48.1 The provisions of Section 13 are applicable.

Impact Resistance of Pallet

49. Significance and Use

49.1 The purpose of this test method is to determine the resistance of the loaded pallet to sudden velocity changes.

50. Apparatus

50.1 *Incline Impact Tester* (Method D 880), provided with a 10 by 15-mm (4 by 6-in.), solid, high-density (NDS, Table 13, Group 1, such as hard maple) timber of at least the length and the width of the pallet, with this timber firmly fastened to the rigid backstop of the tester.

50.2 *Weight Box*, as in 37.2.

51. Impact Application

51.1 Place the loaded pallet on a dolly on an inclined track in such a way that the leading edge of the bottom leading-edge deckboard or, in the case of a skid, that is, a pallet without a bottom deckboard, the bottom 6 mm (¼ in.) of the face of the outer stringer or the faces of the blocks or the ends of the stringers or blocks are impacted by a hardwood hazard after release of the dolly at a predetermined location and its downward travel on the track.

52. Procedure

52.1 The provisions of 40.1 are applicable. Use a new pallet for testing in the direction of the second pallet axis.

52.2 Adjust the height of the hardwood-timber hazard in such a manner that the leading edge of the bottom leading-edge deckboard or, in the case of a skid, the bottom 6 mm (¼ in.) of the outer-stringer face or block faces or of the stringer or block ends will be impacted by this hazard.

52.3 The provisions of 40.3 are applicable.

52.4 Pallet failure shall be considered to have occurred when the pallet will no longer maintain the load it is designated to carry.

52.5 *Precautions*—The provisions of Section 38 are applicable.

53. Report

53.1 The provisions of Section 41 are applicable.

54. Precision and Bias

54.1 The provisions of Section 13 are applicable.

Sliding Friction Resistance of Pallet

55. Significance and Use

55.1 The purpose of this test method is to determine the resistance of the loaded pallet to sliding on a flat horizontal surface.

56. Apparatus

56.1 *Sheet of Galvanized Steel*, of twice the maximum size of the pallet tested, lift truck or other suitable pulling device, rope or cable, and scale with a minimum capacity of 4500 N (1000 lb) and a capability of measuring 4.5 N (1 lb).

57. Force Application

57.1 Place the loaded pallet on a flat sheet of galvanized steel on a horizontal, flat, smooth, concrete floor. Pull the pallet with a rope or a cable attached to a suitable scale between lift truck or other suitable pulling device and pallet (C.2 of MH1.4.1).

58. Procedure

58.1 Place the loaded pallet on one end of a sheet of galvanized steel located on a horizontal, flat, smooth, concrete floor. Attach a suitable rope or cable to the pallet and a suitable scale to the other rope or cable end.

58.2 Pull on the scale with a lift truck or any other suitable vehicle in such a way as to allow measuring the horizontal force required to pull the loaded pallet gradually along the sheet of steel on the floor. Observe the force required for the loaded pallet to begin to move. Subsequently, pull loaded pallet in the direction perpendicular to the first direction and observe the horizontal force required for the loaded pallet to begin to move.

58.3 Calculate the coefficient of friction, for both test directions, as follows:

$$C_f = L/W$$

where:

L = horizontal force required to start to move loaded pallet, and

W = weight of loaded pallet.

59. Report

59.1 The report shall include the following information if applicable:

59.1.1 The provisions of 41.1.1 to 41.1.8 and 41.1.14 are applicable.

59.1.2 Weight of loaded pallet.

59.1.3 Force required for pallet to start to move.

59.1.4 Coefficient of friction.

60. Precision and Bias

60.1 The provisions of Section 13 are applicable.

Sliding-Wear Resistance of Pallet

61. Significance and Use

61.1 The purpose of this test is to determine the resistance to damage of the loaded pallet sliding across a flat horizontal surface.

62. Apparatus

62.1 *Fork-Lift Truck or Other Suitable Pushing Device,* moving at a rate of approximately 7.6 m (25 ft)/min.

63. Force Application

63.1 Place the loaded pallet on a horizontal, flat, smooth, concrete floor and push the pallet with a fork-lift truck or other suitable pushing device for a distance of 10.7 m (35 ft) at a constant rate in approximately 85 s (C.1 of MH1.4.1).

64. Procedure

64.1 Place the loaded pallet on a horizontal, flat, smooth, concrete floor.

64.2 Position fork-lift truck or other suitable pushing device in such a way that the front of the fork backs is vertical and is in contact with the leading edge of the leading-edge deckboard, and that the fork tines extend between pallet stringers or blocks without touching the pallet.

64.3 Operate lift truck in such a way as to push the loaded pallet for a distance of 10.7 m (35 ft) at a constant rate in approximately 85 s. Subsequently, push loaded pallet in the direction perpendicular to the first direction.

64.4 Observe damage, if any, after each of the two tests.

65. Report

65.1 The report shall include the following information if applicable:

65.1.1 The provisions of 41.1.1 to 41.1.8 and 41.1.14 are applicable.

65.1.2 Weight of loaded pallet.

65.1.3 Damage observed, if any, after test.

66. Precision and Bias

66.1 The provisions of Section 13 are applicable.

Sluing Resistance of Pallet

67. Significance and Use

67.1 The purpose of this test is to determine the resistance to damage of the pallet rotated on a flat horizontal surface.

68. Apparatus

68.1 *Forklift Truck.*

69. Force Application

69.1 Place the loaded pallet on a horizontal, flat, smooth, concrete floor and push the pallet with the fork tine of a fork-lift truck in contact with the outer face of the outer stringer or block near its end in such a manner that the pallet is slued 180° (C.6 of MH1.4.1).

70. Procedure

70.1 Place the loaded pallet on a horizontal, flat, smooth, concrete floor.

70.2 Position forklift truck in such a way that the fork tip is in contact with the outer face of the outer stringer or outer block near its end.

70.3 Operate lift truck in such a way as to slue the loaded pallet 180°. Repeat such sluing several times, as specified, by contacting the stringer or block in the same manner or until the pallet will no longer function during its intended use.

70.4 Observe damage, if any, during the test and after its completion.

71. Report

71.1 The report shall include the following information if applicable:

71.1.1 The provisions of 41.1.1 to 41.1.8 and 41.1.14 are applicable.

71.1.2 Weight of the loaded pallet.

71.1.3 Damage observed, if any, during and after the test.

72. Precision and Bias

72.1 The provisions of Section 13 are applicable.

Sling Hoist-Lift Resistance of Winged Pallet

73. Significance and Use

73.1 The purpose of this test method is to determine the stiffness and resistance to damage of the sling-supported loaded winged pallet.

74. Apparatus

74.1 *Hoist Lift,* with two 19-mm (¾-in.) nylon-rope or other adequate slings without spreaders and suitable scale with capability of measuring 3 mm (⅛ in.).

75. Force Application

75.1 Lift the loaded pallet off the ground for not less than 2 min with two slings, without spreaders, placed under the wings near the stringers or blocks. Bring the slings to a common point located above the center of balance at a distance from the ground which is equal to twice the height of the loaded pallet (C.3 of MH1.4.1).

75.2 **Caution**—Lifting a load as simulated by this test method (without a net) must be considered unsafe practice; therefore, is not recommended.

76. Procedure

76.1 Place the two slings under the wings of the loaded pallet near the outer stringers or blocks.

76.2 Lift loaded pallet with hoist off the ground and hold it suspended for not less than 2 min.

76.3 Measure maximum deflection of sides and ends of lifted pallet and observe any inadequacies or damage during the 2-min period.

77. Report

77.1 The report shall include the following information if applicable:

77.1.1 The provisions of 41.1.1 to 41.1.8 and 41.1.14 are applicable.

77.1.2 Weight and total height of pallet including its load.

77.1.3 Maximum deflection of sides and ends of suspended pallet.

77.1.4 Inadequacies and damage, if any, of suspended pallet.

78. Precision and Bias

78.1 The provisions of Section 13 are applicable.

Vibration Resistance of Pallet

79. Significance and Use

79.1 Determine the vibration resistance of the loaded pallet to vertical and lateral sinusoidal vibratory motions to which the loaded pallet may be exposed during its handling and transportation.

80. Apparatus

80.1 *Vibration Tester*, provided with vibrating platform of suitable size, load-carrying capacity, and frequency and amplitude adjustors, supported by a mechanism that keeps the platform surface essentially horizontal during any vertical or lateral sinusoidal vibrations.

81. Force Application

81.1 Place the loaded pallet on the vibration platform and vibrate the pallet at a frequency and an amplitude as well as for a duration as specified (Section 7 of MH1.4.1; also, Methods D 999).

82. Procedure

82.1 Weigh and place the loaded pallet on, without fastening it to, the vibrating platform. Secure to the top deck of the pallet the specified load or a simulated load of approximately the same density as that specified. Use side rails or gates to prevent any excessive lateral movement of the loaded pallet on the platform during the test vibrations.

82.2 Test the pallet in two positions: First place and test the pallet in such a way that the horizontal component of the vibratory motion is parallel to the end-to-end pallet axis and, second, place and test the pallet turned perpendicular so that the horizontal component of the vibratory motion is perpendicular to the end-to-end pallet axis.

82.3 Apply the specified frequency and amplitude of vibrations during the specified test duration, except that the test shall be terminated when the pallet will no longer support the load, it is designed to carry, as a result of premature failure or the pallet will no longer function properly with the particular type and size of handling device to be used.

82.4 Observe and record any inadequacies and damages, if any, during and after each test phase.

83. Report

83.1 The report shall include the following information if applicable:

83.1.1 The provisions of 41.1.1 to 41.1.8 and 41.1.14 are applicable.

83.1.2 Description of load and weight of loaded pallet.

83.1.3 Description of securing load to pallet.

83.1.4 Frequency, amplitude, type, and duration of vibratory motions.

83.1.5 Inadequacies and damages, if any, of pallet and its load during and after each test phase.

84. Precision and Bias

84.1 The provisions of Section 13 are applicable.

Standard Test Method for Dynamic Shock Cushioning Characteristics of Packaging Material[1]

This standard is issued under the fixed designation D 1596; the number immediately following the designation indicates the year of original adoption or, in the case of revision, the year of last revision. A number in parentheses indicates the year of last reapproval. A superscript epsilon (ε) indicates an editorial change since the last revision or reapproval.

1. Scope

1.1 This test method covers a procedure for obtaining dynamic shock cushioning characteristics of packaging materials through acceleration-time data achieved from dropping a falling guided platen assembly onto a motionless sample. This test method does not address any effects or contributions of exterior packaging assemblies.

1.2 The data acquired may be used for a single point or for use in developing a dynamic cushion curve for the specific material being tested. Curves are used either to predict performance of materials under use conditions or for comparison among different materials at specific input conditions. Caution should be used when attempting to compare data from different methods or when using such data for predicting in-package performance. Depending upon the particular materials of concern, correlation of such data (from among differing procedures or for predicting in-package performance) may be highly variable.

NOTE 1—Alternative and related method for possible consideration is Test Method D 4168.

1.3 The values stated in inch-pound units are to be regarded as the standard. The SI units given in parentheses are for information only.

1.4 *This standard does not purport to address all of the safety problems, if any, associated with its use. It is the responsibility of the user of this standard to establish appropriate safety and health practices and determine the applicability of regulatory limitations prior to use.*

2. Referenced Documents

2.1 *ASTM Standards:*

D 966 Terminology of Packaging and Distribution Environments[2]

D 4168 Test Method for Transmitted Shock Characteristics of Foam-in-Place Cushioning Materials[3]

D 4332 Practice for Conditioning Containers Package or Package Components for Testing[3]

E 105 Practice for Probability Sampling of Materials[4]

E 122 Practice for Choice of Sample Size to Estimate a Measure of Quality for a Lot or Process[4]

3. Terminology

3.1 *Definitions:*

3.1.1 *cushioning material*—a material used to isolate or reduce the effect of externally applied shock or vibration forces, or both.

3.1.2 For additional definitions applicable to this test method, see Terminology D 996.

3.2 *Description of Terms Specific to This Standard:*

3.2.1 *acceleration*—the rate of change of velocity of a body with respect to time, measured in in./s^2 (m/s^2).

3.2.2 *displacement*—the magnitude of movement of a body, point, or surface from a fixed reference point, measured in inches (metres).

3.2.3 *dynamic cushion curve*—a graphic representation of dynamic shock cushioning or transmitted shock (in *G*'s) over a variety of static loading conditions (psi or kg/square m) for a specific cushioning material thickness (or structure) at a specific equivalent free fall drop height.

3.2.3.1 Such representations can encompass the average response readings in *G*'s of a number of drops, the average of drops, two to five for each test phase, or represent a single, specific drop number in a drop sequence (that is, first or third drop data).

3.2.4 *equivalent free-fall drop height*—the calculated height of free fall in vacuum required for the dropping platen to attain a measured or given impact velocity.

3.2.5 *equivalent free-fall impact velocity*—the calculated impact velocity of the dropping platen if it were to free fall in a vacuum from a specific test drop height.

3.2.6 *platen drop height*—the actual drop height of the test machine platen required to obtain an equivalent free fall impact velocity.

3.2.7 *reaction mass*—a mass, consisting of the impact surface and any other rigidly attached mass that reacts in an opposing manner to the forces produced during the impact of the dropping platen on the impact surface.

3.2.8 *static loading*—the applied mass in pounds (kilograms) divided by the area, measured in square inches (metres) to which the mass is applied (lbs/in.^{2} or kg/m^2). Sometimes referred to as static stress loading.

3.2.9 *velocity*—the rate of change of position of a body in a specified direction with respect to time, measured in inches per second (metre per second).

3.3 *Symbols:*

3.3.1 *g*—symbol for the acceleration due to the effects of the earth's gravitational pull. While somewhat variable, it is usually considered a constant of value 386 in./s^2 or (9.8 m/s^2).

3.3.2 *G*—symbol for the dimensionless ratio between an

---

[1] This test method is under the jurisdiction of ASTM Committee D-10 on Packaging and is the direct responsibility of Subcommittee D10.13 on Interior Packaging.

Current edition approved June 3, 1991. Published July 1991. Originally published as D 1596 – 59 T. Last previous edition D 1596 – 78a (1983)$^{\epsilon 2}$.

[2] Discontinued—See *1980 Annual Book of ASTM Standards*, Vol 06.01.

[3] *Annual Book of ASTM Standards*, Vol 15.09.

[4] *Annual Book of ASTM Standards*, Vol 14.02.

acceleration in length per time squared units and the acceleration of gravity in the same units.

4. Summary of Test Method

4.1 An apparatus having a guided dropping platen capable of having variable mass, inputs a dynamic force into a test specimen placed on a rigid impact surface at a predetermined impact velocity that equates to a free fall drop height. An accelerometer rigidly mounted to the dropping platen and connected to a data acquisition system records the shock experienced (acceleration-time history) during the impact on the test specimen. By changing the variables, such as impact velocity, static loading (dropping platen mass), and the number of test impacts for any given test, dynamic shock cushioning characteristics (cushion curves) of the packaging material can be developed.

5. Significance and Use

5.1 Dynamic cushioning test data obtained by this test method are applicable to the cushioning material and not necessarily the same as obtained in a package. In addition to the influence of the package, the data can also be affected by the specimen area, thickness, loading rate, and other factors.

6. Apparatus

6.1 *Testing Machine:*

6.1.1 Any guided vertical drop testing system that will produce test conditions conforming to the requirements specified in this section is acceptable. The system shall consist of a rigid flat faced dropping platen, a rigid flat impact surface whose face is parallel to the dropping platen face.

NOTE 2—Lack of rigidity can cause undesirable vibrations in the apparatus that are recorded in the acceleration-time curve. This condition has also been a suspected cause for discontinuities in dynamic data where the mass of the dropping platen is varied at constant height. The existence of flexing in the apparatus often can be verified with aid of high-speed video and subsequently corrected.

6.1.2 The dropping platen should have provisions for firmly mounting additional mass to adjust its total mass to a desired value. Its mass may be determined by consideration of the static loading of the items the dynamic test is simulating. Various testing systems will have different ranges of testing capability, both for drop height and static loadings.

6.1.3 All dropping platens are influenced by guide system friction and air resistance. The significance of these effects varies with the type of apparatus and the mass on the dropping platen. For this reason, the equivalent free fall impact velocity of the dropping platen is equated to a free fall drop height rather than an actual platen drop height. (*Example*—Using the formula in 11.2, a 30 in. (0.7 m) free fall is equated to a 152 in./s (3.9 m/s) impact velocity of the dropping platen.)

6.2 *Reaction Mass:*

6.2.1 The testing machine shall be attached to a reaction mass. The reaction mass shall be sufficiently heavy and rigid so that not more than 2 % of the impact acceleration is lost to the reaction mass while conducting dynamic tests. The rigid impact surface should be in intimate contact with the reaction mass so that the two bodies move as one (surface preparation or grouting may be required). This performance may be verified by using shock sensors, one located on the dropping platen and one on the impact surface or on the reaction mass immediately next to the impact surface to measure the acceleration levels. The ratio of the measured impact acceleration of the reaction mass divided by the measured acceleration of the dropping platen shall be equal to or less than 2 %.

6.2.2 As an alternative to measuring the acceleration level of the reaction mass for each test condition, the reaction mass is acceptable if it is 50 times the maximum mass of the dropping platen. Neither the depth nor the width of such a mass shall be less than half the length.

6.3 *Instrumentation and Shock Sensors:*

6.3.1 Instrumentation is required to measure the impact velocity to an accuracy of ±2 % of the true value.

6.3.2 Accelerometers, signal conditions, and data storage apparatus are required to monitor acceleration versus time histories. The instrumentation systems shall have the following minimum properties:

6.3.2.1 Frequency response range from 2 Hz or less to at least 1000 Hz.

6.3.2.2 Accuracy reading to be within ±5 % of the actual value.

6.3.2.3 Cross axis sensitivity less than 5 % of full scale.

7. Sampling

7.1 The choice of sampling plans for materials depends on the purpose of the testing. Practice E 105 is recommended.

7.2 The number of test specimens for each condition of test (for example, static loading) depends on the desired degree of precision and the availability of materials. Practice E 122 and many statistical tests provide excellent guidance on the choice of sample size. It is recommended that at least three replicate test specimens be used for each initial test condition. Then, depending on the accuracy and degree of certainty required, this sample size may be increased or decreased.

7.3 Randomization of test specimens from the sample of material and randomization of the order of testing are recommended. This may be accomplished by the use of random number tables, lottery, or other accepted procedures of randomization.

8. Test Specimens

8.1 Test specimens shall be right square prisms or other configuration as desired with the maximum length and width dimensions less than the corresponding drop platen dimensions. It is recommended that the minimum length and width dimensions be 4 by 4 in. (101.6 by 101.6 mm). Because pneumatic effects and buckling properties of cushioning materials may be influenced by size and shape of the specimen, 8 by 8 in. (203.2 by 203.2 mm) specimens are recommended whenever possible. When comparing data for different cushions, identically shaped specimens should be used.

NOTE 3—Not all test apparatus are able to provide static loadings across the usable range of all possible materials to be tested. For this reason it is important to prominently note the sizes of specimens tested in the test report.

9. Conditioning

9.1 Materials, such as cellulosic materials, that undergo changes in physical properties as the temperature and the relative humidity to which they are exposed are varied need to be preconditioned in accordance with Practice D 4332. For polymeric cushions, condition test specimens prior to test for a sufficient length of time to essentially achieve and maintain equilibrium in accordance with any requirements. In the absence of other requirements, use standard conditioning atmosphere of 23 ± 2°C (73.4 ± 3.6°F) and 50 ± 2 % relative humidity.

10. Procedure

10.1 *Thickness*—Load the the specimen by a rigid plate or machine platen to 0.025 psi (17.55 kg/m^2) of top surface area. (*Example*—Specimen size of 8 by 8 in. (203.2 by 203.2 mm) = 64 in.^{2} (41290 mm^2); 64 in.^{2} × 0.025 psi (41290 mm^2 ÷ (1 × 10^6) × 17.55 kg/m^2 = a load of 1.6 lb (0.725 kg).) After a 30 s interval, and while the specimen is still under load, measure the thickness to an accuracy of 1/32 in. or 1 mm, by averaging the thickness measurement taken on the four corners of the specimen. Record this value as the specimen thickness. For odd shapes report where measurements were taken.

10.2 *Area and Mass*—Measure the top surface area of the specimen with apparatus yielding values accurate to 1/32 in. or 1 mm. Measure the mass of the specimen with apparatus yielding values accurate to 30 g.

10.3 *Dynamic Test*—Center the test specimen on the impact surface face and prepare the dropping platen to strike the cushion on its top surface area. Then impact the specimen with a series of five drops at a predetermined static loading and impact velocity in the dynamic tester, allowing a minimum of 1 min between each drop. Take a complete acceleration-time record for each drop and measure the impact velocity of the platen just before impact to ensure it is representative of the impact velocity equated to the desired free drop height. To obtain dynamic data of a general nature for a given cushion, it is necessary to repeat the five test drops on a new specimen varying some condition of test such as static loading, impact velocity, or cushion thickness. As an option, upon the completion of the five drops measure the final thickness of the specimen in accordance with the procedure outlined in 10.1 to determine dynamic set.

11. Calculation

11.1 Calculate the density of a test specimen as follows:

$$\text{Inch-pound units } D = (3.81 \times M)/(L_1 \times L_2 \times T)$$
$$\text{(Metric) } D = ((1 \times 10^6) \times M)/(L_1 \times L_2 \times T)$$

where:

D = density, lb/ft^3 (kg/m^3),
M = mass of specimen, grams,
L_1 = length of specimen, in. (mm),
L_2 = width of specimen, in. (mm), and
T = original thickness of specimen, in. (mm).

11.2 Equate an impact velocity to a free fall drop height or vice versa as follows:

$$h = V_i^2/2\,g \text{ (solving for free fall drop height)}$$
$$V_i = \sqrt{2\,gh} \text{ (solving for impact velocity)}$$

where:

h = free fall drop height, in. (m),
V_i = measured impact velocity, in./s (m/s), and
g = acceleration due to gravity, 386 in./s^2 (9.8 m/s^2).

11.3 Calculate the dynamic set as follows:

$$\text{Dynamic set, \%} = [(T - F)/T] \times 100$$

where:

T = original thickness of specimen, in. (mm), and
F = thickness of specimen after test, in. (mm).

12. Report

12.1 Report the following information:
12.2 A description of cushioning material tested,
12.2.1 Name of manufacturer,
12.2.2 Generic name of material,
12.2.3 Date of cushion manufacture, if known,
12.2.4 Date and method of cushion fabrication, if known,
12.3 Number of specimens tested and date of test,
12.3.1 The size and shape of the specimens, and
12.3.2 The original thickness of the specimen.
12.3.2.1 *Optional*—The final thickness in inches (millimetres) and the resultant dynamic set in percent.
12.3.3 The density of each specimen in pounds per cubic foot (kg/m^3) as specified by the manufacturer or if unknown, as calculated by the method shown in Section 11.
12.4 Conditioning parameters.
12.5 Identification of apparatus and instrumentation used, including dates of last instrument calibrations, manufacturer's names and model numbers. Details of any modifications thereto, if known, shall be included.
12.5.1 State method of verification of proper reaction mass; measurement or mass ratio.
12.6 The dropping platen mass, in pounds (kilograms), and the static loading on the specimen, in lb/in.^{2} (kg/m^2), for each test condition,
12.6.1 The impact velocity and equivalent free fall drop height for each test sequence ($h = V_i^2/2\,g$),
12.6.2 A representative sample of the acceleration-time curve for each phase of the testing, along with a reading of the acceleration amplitude for each drop, and
12.6.3 *Optional*—A best fit dynamic shock cushion curve can be plotted with a minimum of five points using a french curve, drawing spline, or by use of a mathematical curve fitting analysis.
12.7 Descriptions of any deviations from the specified test method.

13. Precision and Bias

13.1 *Precision*—The precision of this test method has not been determined and is under investigation by the subcommittee. It is believed that different testing apparatus and instrumentation used to test a consistent material should produce averages that agree within 10 % of each other.

13.2 *Bias*—No measure of bias is possible for this test method because an accepted reference or referee value is not available.

14. Keywords

14.1 cushion curves; cushioning materials; dynamic shock characteristics; guided platen method

Standard Practice for Methods of Closing, Sealing, and Reinforcing Fiberboard Shipping Containers[1]

This standard is issued under the fixed designation D 1974; the number immediately following the designation indicates the year of original adoption or, in the case of revision, the year of last revision. A number in parentheses indicates the year of last reapproval. A superscript epsilon (ε) indicates an editorial change since the last revision or reapproval.

1. Scope

1.1 This practice describes several methods for closing, sealing, and reinforcing fiberboard (solid and corrugated, excluding triple wall) shipping boxes used for shipment. One or several methods listed in this practice may be referenced in regulations, specifications, or contracts. (See Appendix for cross reference).

1.2 This practice is intended to be workable in both inch-pound (U.S. customary) units and SI metric units. Unit conversions are intended to be functional but are not necessarily exact; this should not significantly affect the performance of the closing, sealing, or reinforcing methods.

NOTE 1—All references to this standard should be made by closure number.

1.3 *This standard does not purport to address all of the safety problems, if any, associated with its use. It is the responsibility of the user of this standard to establish appropriate safety and health practices and determine the applicability of regulatory limitations prior to use.*

2. Referenced Documents

2.1 *ASTM Standards:*

D 996 Terminology of Packaging and Distribution Environments[2]

D 3654 Test Method for Holding Power of Pressure-Sensitive Tapes[2]

D 3759 Test Method for Tensile Strength and Elongation of Pressure-Sensitive Tapes[2]

D 3950 Specification for Strapping, Nonmetallic (and Joining Methods)[2]

D 3953 Specification for Flat Steel Strapping and Connectors[2]

D 4169 Practice for Performance Testing of Shipping Containers and Systems[2]

D 4675 Guide for Selection and Use of Flat Strapping Materials[2]

D 5118 Practice for Fabrication of Fiberboard Shipping Boxes[2]

F 592 Definitions of Terms Relating to Collated and Cohered Fasteners and Their Application Tools[3]

2.2 *U.S. Government Documents:*[4]

CID A-A-884 Tape, Pressure Sensitive Adhesive, Box Closure

CID A-A-1492 Tape, Gummed, Paper, Plain

CID A-A-1671 Tape, Gummed, (Paper, Reinforced, Asphalt Laminated)

CID A-A-1672 Tape, Gummed, (Paper, Reinforced, Non-Asphalt Laminated)

CID A-A-1683 Tape, Pressure Sensitive Adhesive, (Packaging, Paper)

CID A-A-1685 Tape, Pressure Sensitive Adhesive, (High Tensile Strength, Black, Weather Resistant, Glass Filament Reinforced)

CID A-A-1687 Tape, Pressure Sensitive Adhesive, (Medium Tensile Strength, Glass Filament Reinforced)

CID A-A-1830 Tape, Pressure Sensitive Adhesive, Box Closure

MMM-A-250 Adhesive, Water Resistant (For Closure of Fiberboard Boxes)

PPP-T-45 Tape, Gummed Paper, Reinforced or Plain

PPP-T-60 Tape, Packaging, Waterproof

PPP-T-76 Tape, Pressure Sensitive Adhesive Paper

PPP-T-97 Tape, Pressure Sensitive Adhesive, Filament Reinforced

2.3 *Other Standards:*

International Fiberboard Case Code[5]

NOTE 2—FEFCO[5] established a code of box styles to facilitate international trade. The box codes consist of four-digit numbers; for example, a regular slotted container was assigned the code 0201.

3. Terminology

3.1 Packaging terms used in this practice shall be as found in Terminology D 996.

3.2 *Definitions:*

3.2.1 *crown*—staple end opposite staple point, connecting both staple legs and providing bearing area (F 592).

3.2.2 *leg*—staple part connecting staple crown with staple point, driven through and into or through materials being fastened (F 592).

3.2.3 *leg thickness*—maximum dimension of staple-leg cross section measured parallel to staple-crown axis (F 592).

3.2.4 *leg width*—maximum dimension of staple-leg cross section measured perpendicular to staple-crown axis (F 592).

3.2.5 *stitch*—staple cut and formed from wire immedi-

[1] This practice is under the jurisdiction of ASTM Committee D-10 on Packaging and is the direct responsibility of Subcommittee D10.14 on Closure and Reinforcement.

Current edition approved Jan. 25, 1991. Published June 1991.

[2] *Annual Book of ASTM Standards*, Vol 15.09.

[3] *Annual Book of ASTM Standards*, Vol 15.08.

[4] Standardization Documents, Order Desk, Building 4, Section D, 700 Robbins Avenue, Philadelphia, PA 19111-5094, ATTN: NPODS.

[5] International Fiberboard Case Code, European Federation of Manufacturers of Corrugated Board (FEFCO), 37 rue d'Amsterdam, 75008, Paris, France, 874-63-76.

ately prior to driving by same machine (F 592).

3.3 *Descriptions of Terms Specific to This Standard:*

3.3.1 *closure*—the means of securing the flaps or covers of a box so that the box will not accidentally open during normal shipment, handling, and storage.

3.3.2 *reinforcing*—the means (when required) of improving the stability and integrity of a fiberboard box or closure so as to increase the performance level of the box during normal handling, shipping, or storage.

3.3.3 *sealing*—the means (when required) of making a box reasonably impervious to the entrance of extraneous material (such as dust) or to prevent the leakage of the box contents during distribution.

3.3.4 *setup*—the forming of a knocked-down box (or box section) into a box (section) ready for filling, including the end, cover, and section closures made before the box is filled.

3.3.5 *staple*—preformed *U*-shaped wire fastener usually with two same-size pointed or pointless legs connected by crown located opposite staple-point ends; designed to be driven by strike, pneumatic, electric, manual, or spring tools through or through and into layers of penetrable material and to hold two or more pieces together.

3.3.6 *wide crown staple*—a staple whose nominal width is 1¼ in. (approximately 32 mm) or greater.

3.3.7 *wide crown stitch*—a stitch whose nominal width is 1¼ in. (approximately 32 mm) or greater.

4. Significance and Use

4.1 The means of closing the flaps of corrugated shipping containers is one of the important decisions for a shipper. It may affect the overall material cost, the labor costs, the packaging equipment costs, and the performance of the box during distribution.

4.2 This practice provides a list of many available procedures for closing, sealing, and reinforcing boxes. Most of these describe a particular packaging material and a particular use of it on a box. Some methods involve package performance testing; any closure or reinforcement which passes prescribed tests (usually on the proposed box and contents) is considered to comply with that section of this practice. Performance-based methods are encouraged because they allow for considerable flexibility in the choice of packaging materials and methods yet provide assurance of a given level of performance.

4.3 A diverse range of methods are described in this practice but relative performance levels are not given. Laboratory package testing such as Practice D 4169 and field experience are recommended to rate alternative methods on performance for the intended use.

4.4 The descriptions of materials and methods in this practice provide direction for the procurement of materials and for their common uses on boxes. These descriptions may be sufficient for some specification purposes, but others will require more thorough material and application requirements than those given here.

4.5 The list of methods in this practice is not exhaustive; deviation or variation from the listed methods is possible with the approval of the appropriate parties referencing this practice. Depending on the characteristics of the contents, the specific shipping container, the acceptable package costs, the distribution hazards, and the desired degree of assurance of successful shipment, more stringent methods than those listed may be needed or less stringent methods than those listed may be adequate. The methods described are typical uses of the closure, reinforcing, or sealing materials; many of these materials can be used for other box styles or other methods on boxes but all of these are not included in this practice. Materials for box closure, reinforcement, or sealing are commercially available which have higher or lower performance characteristics than those listed in this practice; these may often be used in a lesser or greater quantity than the described applications of the listed materials.

4.6 This practice assumes the following:

4.6.1 The shipping container and inner packing are properly designed for the contents and the distribution hazards (see Practices D 4169 and D 5118),

4.6.2 Normal, reasonable care is taken in shipment, including storage, handling, loading, and bracing, and

4.6.3 Normal, reasonable care is taken in the application of the closure, reinforcing, or sealing materials to the shipping container.

5. Materials and Equipment

5.1 *Adhesives:*

5.1.1 *Hot Melt Adhesives*—Hot melt adhesives are solid at room temperature, liquified by heat, applied molten, and solid upon cooling to bond two surfaces. They are based on thermoplastic polymers modified with resins or waxes, or both, and are usually applied in the range of approximately 300 to 400°F (approximately 150 to 200°C). The adhesive should maintain its functionality in environmental conditions expected in distribution and storage.

5.1.2 *Water-Borne Adhesives*—Several types of adhesives are available in which water is the vehicle or solvent. These are typically applied at room temperature and the surfaces are compressed together until the adhesive develops sufficient strength to maintain substrate contact. The adhesive should maintain its functionality in environmental conditions expected in distribution and storage. The federal specification applicable to adhesives is MMM-A-250.

5.1.3 Semi-automatic and fully automatic box closure equipment is available for the application of adhesives. Hot melt adhesives are also applied by hand held "guns." Adhesives work best on clean, dry surfaces. It is important to have the box flaps held in place until the hot melt adhesive solidifies or the water borne adhesive develops sufficient bond.

5.2 *Pressure-Sensitive Tape:*

5.2.1 *Pressure-Sensitive, Film-Backed Tape*—This tape consists of a film backing, usually a biaxially oriented polylpropylene or polyester (some polyvinyl chloride films are also used), with a high tack pressure-sensitive adhesive. No solvents, heat, or moistening are needed. The tape should maintain its functionality in the environmental conditions expected in distribution and storage. The applicable federal documents are PPP-T-60 and CID's A-A-884 and A-A-1830.

5.2.2 *Pressure-Sensitive Filament Tape*—This type of tape typically consists of fiberglass filaments embedded in a pressure-sensitive adhesive with a protective film backing. Several grades of tape are available. The tape should maintain its functionality in environmental conditions expected in distribution and storage. The applicable federal documents are PPP-T-97 and CID's A-A-1685, and A-A-1687.

5.2.3 *Pressure-Sensitive Paper Tape*—This type of tape has a backing of kraft paper and a pressure-sensitive adhesive. The tape should maintain its functionality in environmental conditions expected in distribution and storage. The applicable federal documents are PPP-T-76 and CID's A-A-1492 and A-A-1683.

5.2.4 A wide variety of tape dispensing and applying equipment is available. Hand-held dispensers are common. Pressure-sensitive tapes also capable of being applied by equipment at high speeds. Tapes work best on clean, dry surfaces. It is important to give an adequate rubdown to pressure-sensitive tapes to develop sufficient bond to hold the flaps closed.

5.3 *Gummed (Remoistenable) Tape:*

5.3.1 *Reinforced Gummed Tapes*—This type of tape has a water activated adhesive and contains glass reinforcing fibers embedded between two sheets of paper, using asphaltic or nonasphaltic binders. Many constructions are available. The tape should maintain its functionality in environmental conditions expected in distribution and storage. The applicable federal documents are PPP-T-45, and CID's A-A-1671 and A-A-1672.

5.3.2 *Gummed Paper Tape*—This type of tape has a backing of kraft paper and a water activated adhesive. The tape should maintain its functionality in environmental conditions expected in distribution and storage. The applicable federal documents are PPP-T-45 and CID's A-A-1492 and A-A-1683.

5.3.3 A wide variety of tape dispensing and applying equipment is available. Table top dispensers are common. Gummed tapes are capable of being applied by equipment at high speeds. Tapes work best on clean, dry surfaces. It is important to hold water activated tapes in place until they develop sufficient bond to hold the flaps closed.

5.4 *Staples and Stitches:*

5.4.1 *Staples:*

5.4.1.1 Staples shall be of good commercial quality steel and have a minimum tensile strength of 80 000 psi (550 mPa). They shall have a rust-resistant coating of copper, zinc, or other material. Staples are produced in the following nominal sizes:

Crown Width	Leg Width	Leg Thickness
0.50 in. (12.5 mm)	0.100 in. (2.5 mm)	0.017 in. (0.43 mm)
1.25 in. (32 mm)	0.072 in. (1.9 mm)	0.033 in. (0.84 mm)

5.4.1.2 Staples are collated and cohered into strips or rolls. The strips are designed for loading into straight magazines. The roll, also called coils, allow the loading of a large number of staples into a tool.

5.4.1.3 Staples are driven by staplers. Two types of staplers are in common use.

(*a*) One type of stapler uses moving anvils to clinch the staple legs around the fiberboard. These tools drive the staple legs through the flaps, simultaneously inserting anvils through the flap surface to meet and clinch the staple legs so they grab the inner flaps and secure them to the outer flaps. The anvils then retract from the flaps.

(*b*) The other type of stapler uses an anvil to clinch the staple legs. The anvil may be supported by a pedestal or by an arm attached to the stapler.

(*c*) Each type of stapler is available as a manual device and as a powered device. The manually operated staplers are usually portable although some are mounted on frames and activated by a foot pedal. The latter are commonly called "bottomers" used to close the bottom flaps of a box before filling. The powered tools may be electric or pneumatic. Staples may be driven by portable staplers or by stationary heads mounted in semi- or fully-automatic equipment for moderate production requirements.

5.4.2 *Stitches:*

5.4.2.1 Stitches shall be of good commercial quality steel. Common nominal crown widths are 0.50 in. (12.5 mm) and 1.25 in. (32 mm). Nominal leg width is 0.100 in. (2.5 mm). Nominal leg thicknesses are 0.014 in. (0.36 mm) and 0.017 in. (0.43 mm). The minimum tensile strength is 100 000 psi (690 mPa) for wire of thickness (0.014 in. (0.36 mm). The minimum tensile strength is 80 000 psi (550 mPa) for wire of thickness 0.017 in. (0.43 mm). They shall have a rust resistant coating of copper, zinc, or other material.

5.4.2.2 Stitches are driven by electric or pneumatic stitchers. These tools cut a section of wire from a spool of wire, form a stitch (with crown and legs) and drive the legs through the box flaps. The legs are clinched on a clinch block, thereby securing the inner and outer flaps. Stitchers are stationary machines.

5.5 *Strapping*—See Guide D 4675 for a discussion of strappings and their application equipment.

6. Methods of Box Closure

6.1 *Performance-Based Closures* (for all box styles)—When a Method 1 box closure is referenced, any reasonable means of closing the box is allowed which is able to successfully complete the prescribed performance test on the proposed package with contents. Care must be taken with the choice of acceptance criteria for a closure. Often the criterion is that after the tests, the box remain closed in a condition suited for normal sale or use; some limited box or closure damage may be acceptable.

6.1.1 *Closure Method 1A*—The closure must be able to complete the test sequence of Practice D 4169, Distribution Cycle 3 or 4, Assurance Level 1.

6.1.2 *Closure Method 1B*—The closure must be able to complete the test sequence of Practice D 4169, Distribution Cycle 3 or 4, Assurance Level II.

6.1.3 *Closure Method 1C*—The closure must be able to complete the test sequence of Practice D 4169, Distribution Cycle 3 or 4, Assurance Level III.

6.2 *Closure Methods for Regular Slotted Containers,* Box Code 0201, Center Special Slotted Container, 0204, and similar box styles

NOTE 2—See 2.3 (and Note 1) for information in box codes

6.2.1 *Adhesive Closures of RSC's and Similar Boxes:*

6.2.1.1 *Closure Method 2A1*—The box shall be closed with a hot-melt adhesive in strips a minimum of 1 in. (25 mm) long and ¼ in. (6 mm) wide after compression. There shall be a minimum of four strips, two on each side of the center line; the separation between strips shall be a maximum of 2 in. (50 mm) with the innermost strips being ½ in. (13 mm) from each side of the center line.

6.2.1.2 *Closure Method 2A2*—The box shall be closed

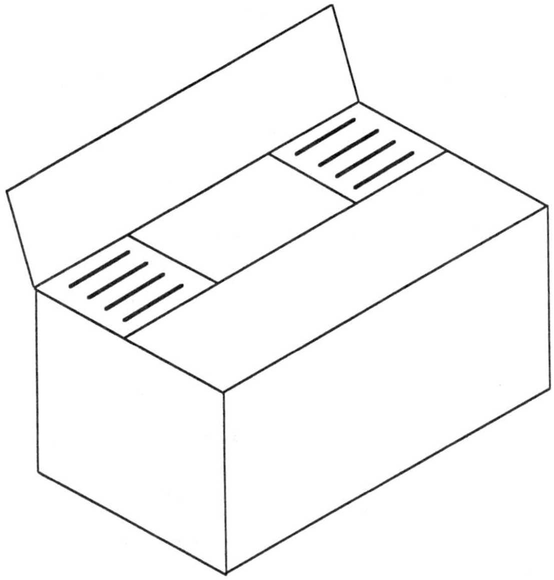

FIG. 1 Example of Closure Method 2A2

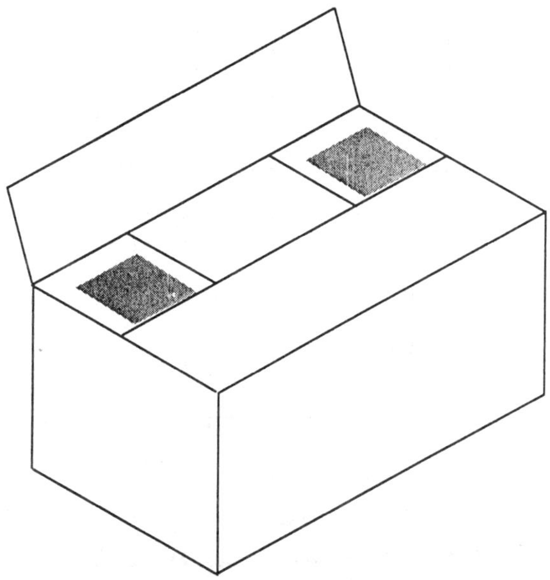

FIG. 2 Example of Closure Methods 2A3 and 2C5

with a hot-melt adhesive which will maintain bond at temperatures for a range from −20°F (−29°C) to 165°F (74°C). Adhesive shall be applied in at least eight full-length strips on each inner flap, each strip having a minimum width of 3⁄16 in. (5 mm) after compression. Strips shall not be more than 1½ in. (38 mm) apart with one strip not more than ½ in. (13 mm) from each side of the center seam. See Fig. 1.

6.2.1.3 *Closure Method 2A3*—The box shall be closed with a waterborne adhesive by securely bonding at least 50 % of the contact area of the flaps. The bonded areas shall extend to within ½ in. (13 mm) of the center seam. See Fig. 2.

6.2.1.4 *Closure Method 2A4*—The box shall be closed with the application of adhesive conforming with MMM-A-250 by securely bonding at least 75 % of the contact area of the flaps. The adhesive shall be applied within ¼ in. (6 mm) of the edges and scorelines of the inner flaps.

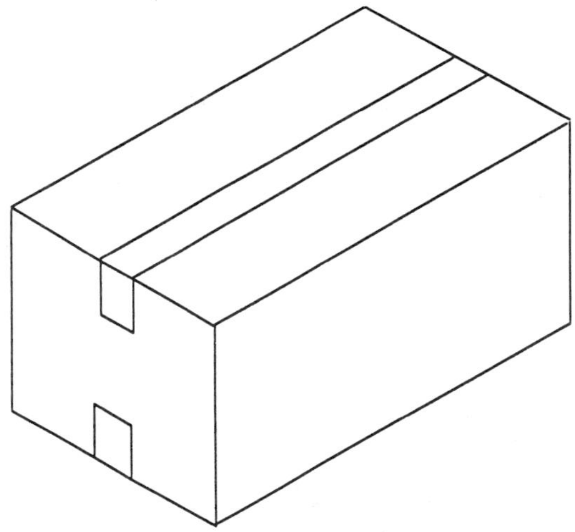

FIG. 3 Example of Closure Methods 2B1, 2B2, 2B3, 2B4, 2B5, 2B6, 2C1, 2C2, 2C3, and 2C4

6.2.1.5 *Closure Method 2A5*—The box shall be closed with a hot melt or a waterborne adhesive securely bonding at least 25 % of the contact area of the flaps. The bonded areas shall extend to within ½ in. (13 mm) of the center seam.

6.2.2 *Pressure-Sensitive Tape Closure of RSC's and Similar Boxes:*

6.2.2.1 *Closure Method 2B1*—The box shall be closed with 2 in. (48 mm) wide pressure-sensitive film backed tape applied over the center seams of the box, continuing at least 2 in. (50 mm) onto the box ends. The tape shall have both a nominal machine direction and cross direction tensile strength of 20 lb/in. (350 N/100 mm) in accordance with Test Method D 3759 and an adhesive with a holding power to fiberboard, Test Method D 3654, Procedure A, of at least 1000 min. See Fig. 3.

6.2.2.2 *Closure Method 2B2*—The box shall be closed with 2 in. (48 mm) wide pressure-sensitive film backed tape applied over the center seams of the box, continuing at least 2 in. (50 mm) onto the box ends. The tape shall have a 2-mil (0.05-mm) biaxially oriented polypropylene backing with both an average machine direction and cross direction tensile strength of at least 33 lb/in. (575 N/100 mm) in accordance with Test Method D 3759. The adhesive shall have a holding power to fiberboard, Test Method D 3654, Procedure A, of at least 2000 min. See Fig. 3.

6.2.2.3 *Closure Method 2B3*—The box shall be closed with 2 in. (48 mm) wide pressure-sensitive film backed tape applied over the center seams of the box, continuing at least 2 in. (50 mm) onto the box ends. The colored tape shall conform with PPP-T 60, Type III, Class 1. See Fig. 3.

6.2.2.4 *Closure Method 2B4*—The box shall be closed with 2 in. (48 mm) wide pressure-sensitive film backed tape applied over the center seams of the box, continuing at least 2 in. (50 mm) onto the box ends. The transparent tape shall conform with PPP-T-60, Type III, Class 2. See Fig. 3.

6.2.2.5 *Closure Method 2B5*—The box shall be closed with 2 in. (48 mm) wide pressure-sensitive film backed tape applied over the center seams of the box, continuing at least 2 in. (50 mm) onto the box ends. The tape shall have a 2-mil biaxially oriented polyester backing and conform with CID

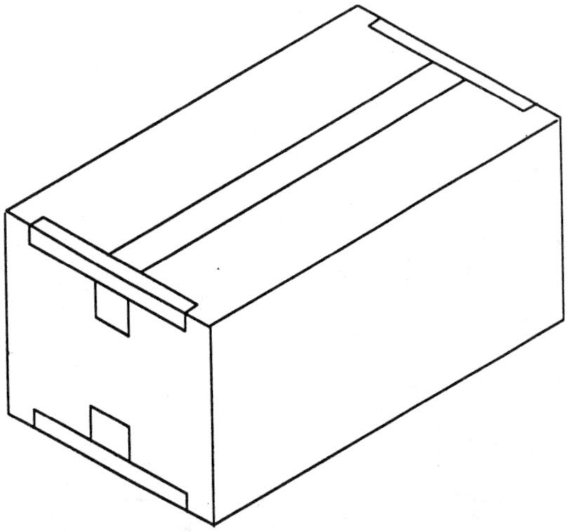

FIG. 4 Example of Closure Method 2B7

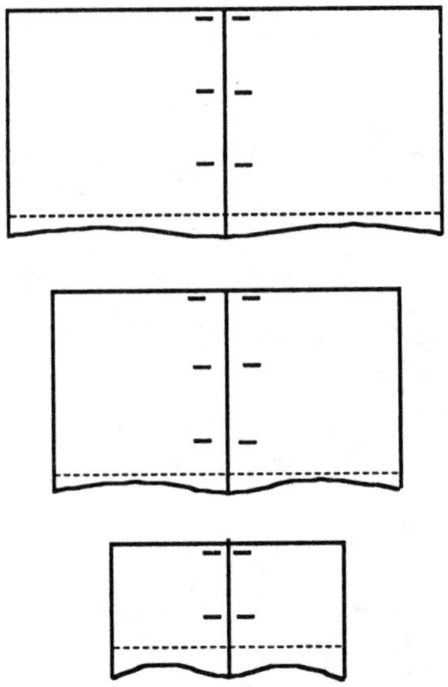

FIG. 5 Example of Closure Method 2D1

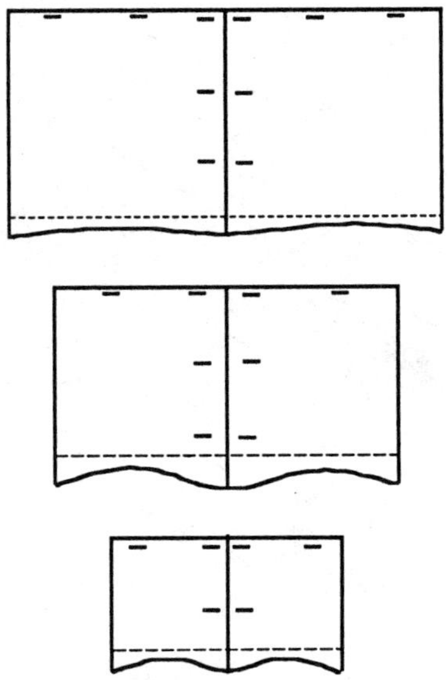

FIG. 6 Example of Closure Method 2D2

A-A-884 (tan) or CID A-A-1830 (transparent). See Fig. 3.

6.2.2.6 *Closure Method 2B6*—The box shall be closed with 2 in. (48 mm) wide paper tape applied over the center seams of the box, continuing at least 2 in (50 mm) onto the box ends. The tape shall comply with CID A-A-1683. See Fig. 3.

6.2.2.7 *Closure Method 2B7*—The box shall be closed with 2 in. (48 mm) wide paper tape applied over the center seams of the box, continuing at least 2 in. (50 mm) onto the box ends. Four additional strips of tape shall be centered over the seam formed by the edges of the outer flaps and the end panels; the tape shall extend within ¾ in. (19 mm) of the box corners. The tape shall conform with PPP-T-76. See Fig. 4.

6.2.3 *Gummed Tape Closure of RSC's and Similar Boxes:*

6.2.3.1 *Closure Method 2C1*—The box shall be closed with 3 in. (75 mm) wide reinforced gummed tape applied over the center seams of the box, continuing at least 2½ in. (65 mm) onto the box ends. The tape shall have glass reinforcements in both the machine and cross machine direction (or at angles to provide some cross direction strength). See Fig. 3.

6.2.3.2 *Closure Method 2C2*—The box shall be closed with 3 in. (75 mm) wide reinforced gummed tape applied over the center seams of the box, continuing at least 2½ in. (65 mm) onto the box ends. The tape shall comply with PPP-T-45, Type II. See Fig. 3.

6.2.3.3 *Closure Method 2C3*—The box shall be closed with 3 in. (75 mm) wide reinforced gummed tape applied over the center seams of the box, continuing at least 2½ in. (65 mm) onto the box ends. The tape shall comply with CID A-A-1671 (asphaltic) or A-A-1672 (nonasphaltic). See Fig. 3.

6.2.3.4 *Closure Method 2C4*—The box shall be closed with 2 in. (48 mm) wide paper tape applied over the center seams of the box, continuing at least 2 in. (50 mm) onto the box ends. The tape shall comply with CID A-A-1492 Grade B. See Fig. 3.

6.2.3.5 *Closure Method 2C5*—The box shall be closed with 2 in. (48 mm) wide paper tape applied over the center seams of the box, continuing at least 2 in. (50 mm) onto the box ends. Four additional strips of tape shall be centered over the seam formed by the edges of the outer flaps and the end panels; the tape shall extend within ¾ in. (19 mm) of the box corners. The tape shall comply with CID A-A-1492 Grade B. See Fig. 4.

6.2.4 *Staple and Stitch Closure and Setup of RSC's and Similar Boxes—General*—Staples are used both for box setup and final closure. Stitcher operation requires that the stitcher's head and clinch block be on opposite sides of the flaps being joined. Hence, stitches are used only for the setup where access of the clinching block is not restricted. The following methods apply to the use of staples for both setup and final closure and for box setup with stitches.

6.2.4.1 *Closure Method 2D1*—The box shall be closed with ½-in. crown staples or stitches applied to each side of the center seam in an area where outer flaps overlay inner flaps; fasteners should not be more than 2½ in. (65 mm) apart. See Fig. 5. See Table 1 for quantities required for various box widths.

TABLE 1 Number of ½-in. Crown Staples or Stitches for Regular Slotted Containers

Inside Width of Box, in.[A]	Number of Fasteners At Each End of Box, by Method		
	2D1, Commercial, Fig. 5	2D2, Rule 41, Fig. 6	2D3, Government, Fig. 7
6	2	2	6
7	2	4	6
9	4	6	6
10	4	6	8
11	4	6	8
12	4	8	10
14	6	10	12
16	6	10	14
17	6	12	14
18	6	12	16
19	8	14	16
20	8	14	18
22	8	16	20
24	10	16	22

[A] One inch = 25.4 mm.

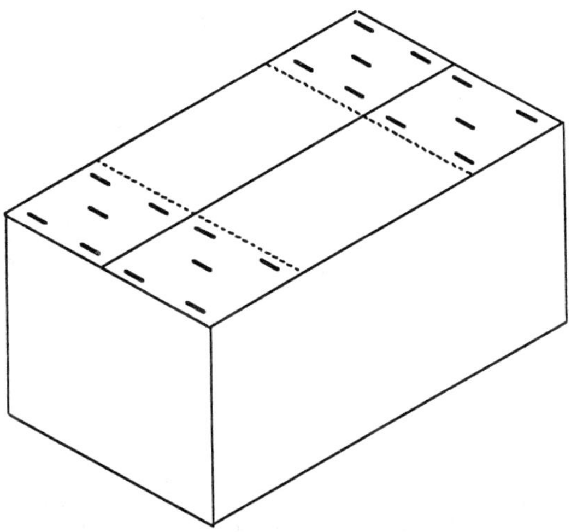

FIG. 7 Example of Closure Method 2D3

6.2.4.2 *Closure Method 2D2*—The box shall be closed with ½-in. crown staples or stitches applied to each side of the center seam in the area where outer flaps overlay inner flaps and to the sides of the outside flaps: fasteners shall not be more than 2½ in. (65 mm) apart. See Fig. 6. See Table 1 for quantities required for various widths.

6.2.4.3 *Closure Method 2D3*—The box shall be closed with ½-in. crown staples or stitches distributed uniformly over the areas where outer flaps overlay inner flaps. Table 1 presents the number of staples/stitches for each end for stated box widths. See Fig. 7.

6.2.4.4 *Closure Method 2D4*—The box shall be closed with wide crown staples or stitches applied along the center seam not more than 5 in. (125 mm) apart in the area where outer flaps overlay inner flaps. See Fig. 8. See Table 2 for quantities required for various box widths.

6.2.4.5 *Closure Method 2D5*—The box shall be closed with wide crown staples or stitches applied across the center seam in the area where the outer flaps overlay inner flaps and to the sides of the outside flaps: fasteners shall not be more than 5 in. (125 mm) apart. See Fig. 9.

6.2.4.6 *Closure Method 2D6*—The box shall be closed with wide crown staples or stitches distributed uniformly over the areas where outer flaps overlay inner flaps. Table 2 presents the number of staples or stitches for each end of stated box widths. See Fig. 10.

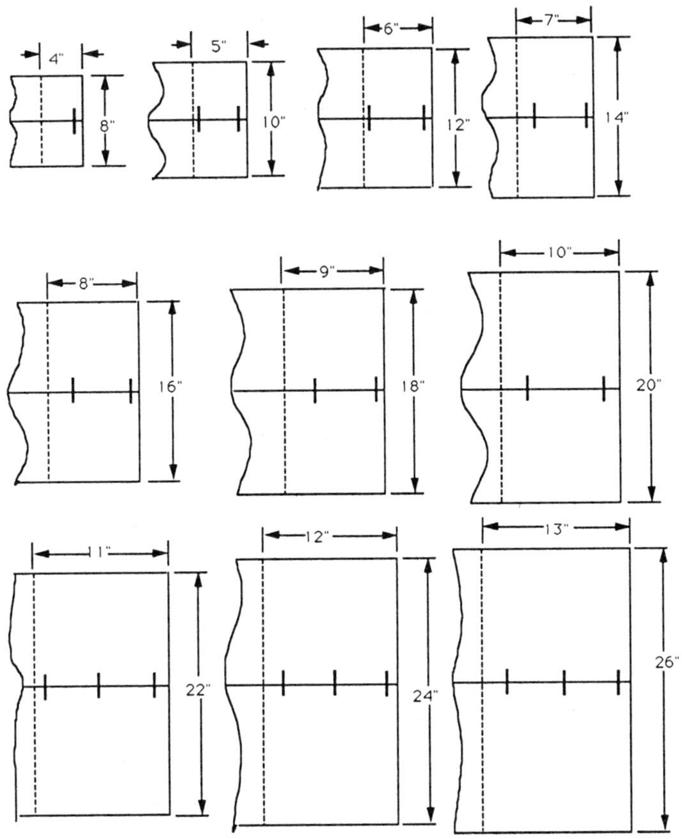

FIG. 8 Example of Closure Method 2D4

TABLE 2 Number of Wide Crown Staples or Stitches for Regular Slotted Containers

Inside Width of Box, in.[A]	Number of Fasteners At Each End of Box, by Method		
	2D4, Commercial, Fig. 8	2D5, Rule 41, Fig. 9	2D6, Government, Fig. 10
8	1	1	4
10	2	2	5
12	2	2	6
14	2	2	8
16	2	4	9
18	2	4	10
20	2	4	11
22	3	5	12
24	3	5	14
26	3	5	15

[A] One inch = 25.4 mm.

6.2.5 *Strapping Closure of RSC's and Similar Boxes:*

6.2.5.1 *Closure Method 2E*—The box shall be closed as described in Guide D 4675, Section S1

6.3 *Closure Methods for Overlap Slotted Container, Box Code 0202, Center Special Overlap Slotted Container, Code 0205, and Other Partially Overlapping Boxes:*

6.3.1 *Adhesive Closure of Partial Overlap Boxes:*

6.3.1.1 *Closure Method 3A1*—The box shall be closed with hot melt or waterborne adhesive securely bonding at least 50 % of the two overlapped flaps as shown in Fig. 11.

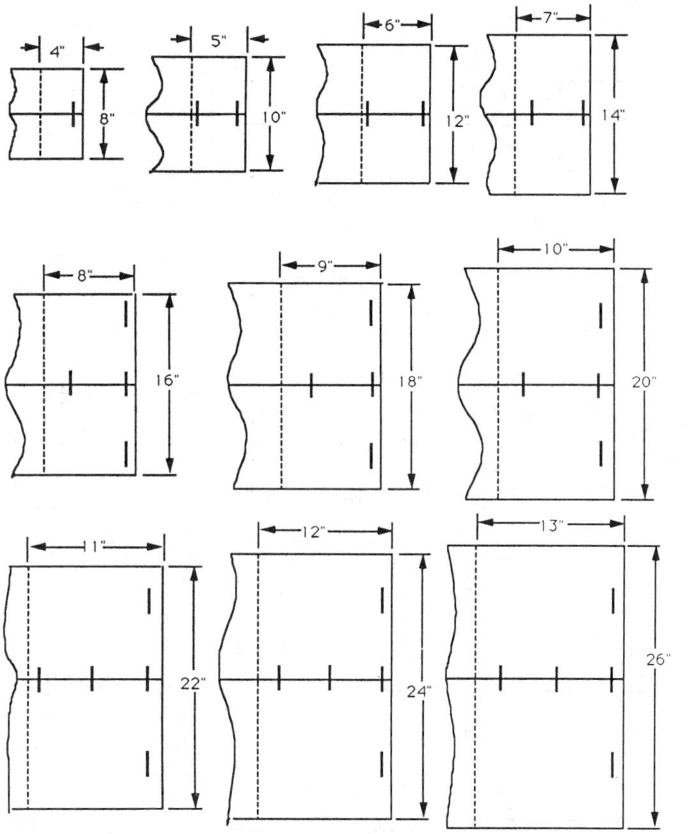

FIG. 9 Example of Closure Method 2D5

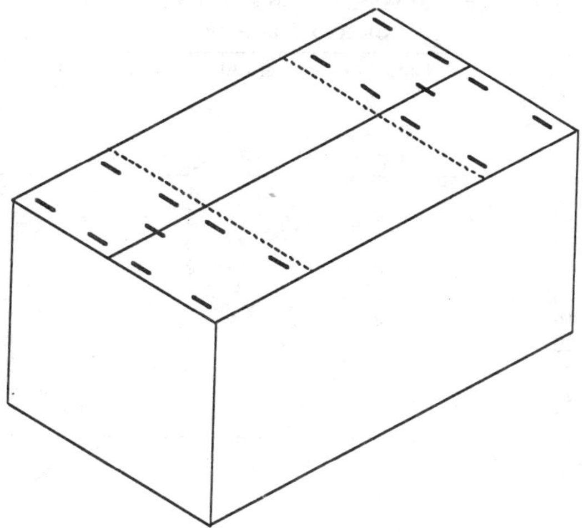

FIG. 10 Example of Closure Method 2D6

6.3.1.2 *Closure Method 3A2*—The box shall be closed with the application of adhesive conforming with MMM-A-250 by securely bonding at least 50 % of the area of contact of the flaps.

6.3.2 *Pressure-Sensitive Tape Closure of Partial Overlap Boxes:*

6.3.2.1 *Closure Method 3B1*—The box shall be closed with pressure-sensitive tape applied so that one half of the width is on the outer flap and one half is applied to the under flap. The tape shall be 2 in. (or 48 mm) wide and conform

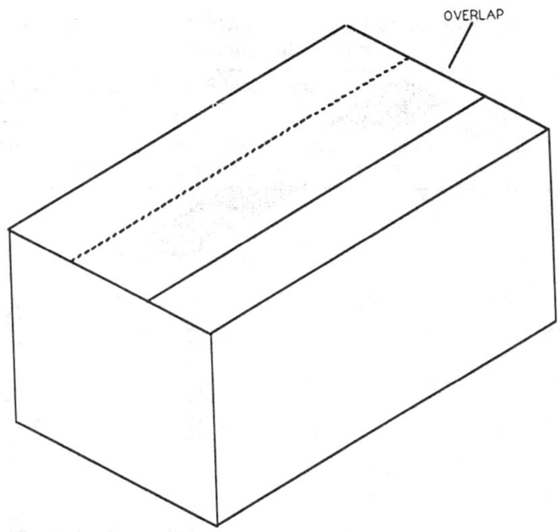

FIG. 11 Example of Closure Method 3A1 and 3A2

TABLE 3 Number of ½-in. Crown Staples or Stitches for Boxes With Partially or Fully Overlapping Flaps

Number of Fasteners Along Length of Flap Overlap		Number of Fasteners Along Each Side of the Overlapping Flap	
Methods 3D1, 3D2, 4D1, and 4D2		Methods 3D2 and 4D2	
Length	Fasteners Along Length of Overlap	Width	Fasteners Along Each Edge[A]
below 6	1	below 4½	0
6 up to 9	2	up to 7	1
up to 12	3	up to 9½	2
up to 15	4	up to 12	3
up to 18	5	up to 14½	4
up to 21	6	up to 17	5
up to 24	7	up to 19½	6
up to 27	8	up to 22	7
up to 30	9	up to 24½	8

[A] This quantity excludes the fastener in the line of fasteners running the length of the overlap of the flaps.

with PPP-T-60 Type III. Tape shall extend down onto the end panels not less than 2 in. (50 mm).

6.3.3 *Gummed Tape Closure of Partial Overlap Boxes:*

6.3.3.1 *Closure Method 3C1*—The box shall be closed with gummed tape applied so that one half of the width is on the outer flap and one half is applied to the under flap. The tape shall be 3 in. (75 mm) wide and conform with PPP-T 45, Type II.

6.3.4 *Staple and Stitch Closure and Setup of Partial Overlap Boxes—General*—Staples are used both for box setup and final closure. Stitcher operation requires that the stitcher's head and clinch block be on opposite sides of the flaps being joined. Hence, stitches are used only for the setup where access of the clinching block is not restricted. The following methods apply to the use of staples for both setup and final closure and for box setup with stitches.

6.3.4.1 *Closure Method 3D1*—The box shall be closed with ½-in. crown staples or stitches applied along the length of the overlap; fasteners should not be more than 2½ in. (63 mm) apart. See the length column of Table 3 for quantity required for various box lengths. See Fig. 12.

6.3.4.2 *Closure Method 3D2*—The box shall be closed with ½-in. crown staples or stitches applied along the length

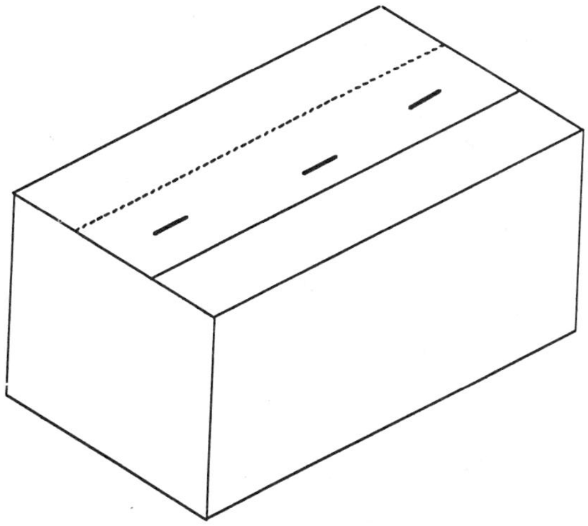

FIG. 12 Example of Closure Method 3D1

TABLE 4 Number of Wide Crown Staples or Stitches for Boxes With Partially or Fully Overlapping Flaps

Number of Fasteners Along Length of Flap Overlap		Number of Fasteners Along Each Side of the Overlapping Flap	
Methods 3D3, 3D4, 4D3, and 4D4		Methods 3D4 and 4D4	
Length	Fasteners Along Length of Overlap	Width	Fasteners Along Each Edge[A]
up to 10	2	up to 6	0
16	3	11	1
22	4	16	2
28	5	21	3
34	6	26	4
40	7	31	5
46	8	36	6
52	9	41	7
58	10	46	8

[A] This quantity excludes the fastener in the line of fasteners running the length of the overlap of the flaps.

of the overlap and the sides of the outside flaps not more than 2½ in. (63 mm) apart to join the inner and outer flaps. See Table 3 for quantities required for various box lengths and widths. See Fig. 13.

6.3.4.3 *Closure Method 3D3*—The box shall be closed with wide crown staples or stitches applied along the length of the overlap; fasteners shall not be more than 5 in. (125 mm) apart. See Fig. 12. See the length column of Table 4 for quantity required for various box lengths.

6.3.4.4 *Closure Method 3D4*—The box shall be closed with wide crown staples or stitches applied along the length of the overlap and the sides of the outside flaps not more than 5 in. (125 mm) apart to join the inner and outer flaps. See Table 4 for quantities required for various box lengths and widths. See Fig. 13.

6.3.4.5 *Closure Method 3D5*—The box shall be closed with staples or stitches distributed over the areas where outer flaps overlay inner flaps. The number of staples or stitches is given in Table 5. When the inner flaps gap by more than 3 in. (75 mm), additional staples shall be driven in a row parallel to and approximately ½ to 1 in. (12 to 25 mm) from the long edge of the outer flaps for the full length of the gap. See Fig. 14.

6.3.5 *Strapping Closure of Overlap Boxes:*

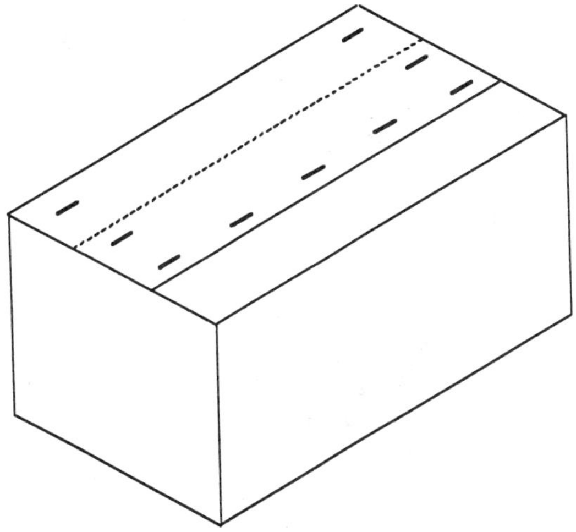

FIG. 13 Example of Closure Methods 3D2, 3D3, and 3D4

TABLE 5 Number of Staples or Stitches for Partial and Full Overlap Boxes[A,B]

Overlapping Flap Width (Equal to Box Width for Full Overlap Boxes), in.[C]	Number of Fasteners at Each End of Box	
	Wide Crown	½ in. Crown
6	. . .	6
7	. . .	6
8	4	6
9	4	6
10	5	8
11	5	8
12	6	10
14	8	12
16	9	14
17	9	14
18	10	16
19	10	16
20	11	18
22	12	20
24	14	22

[A] Closure Method 3D5, Partial Overlap—Fig. 12.
[B] Closure Method 4D5, Full Overlap—Fig. 19.
[C] 1 in. = 25.4 mm.

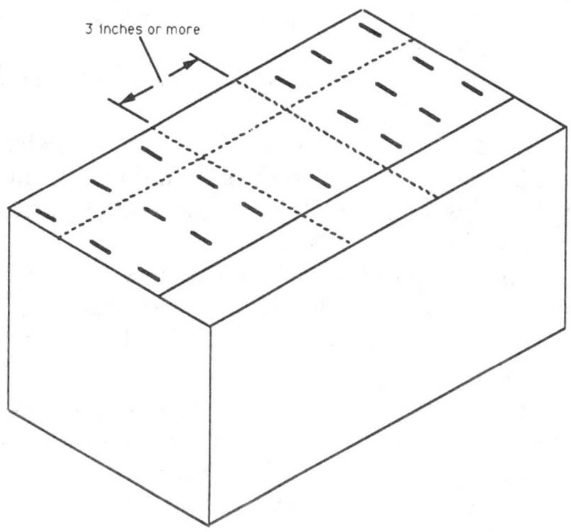

FIG. 14 Example of Closure Method 3D5

6.3.5.1 *Closure Method 3E*—The box shall be closed as described in Guide D 4675, Section S1

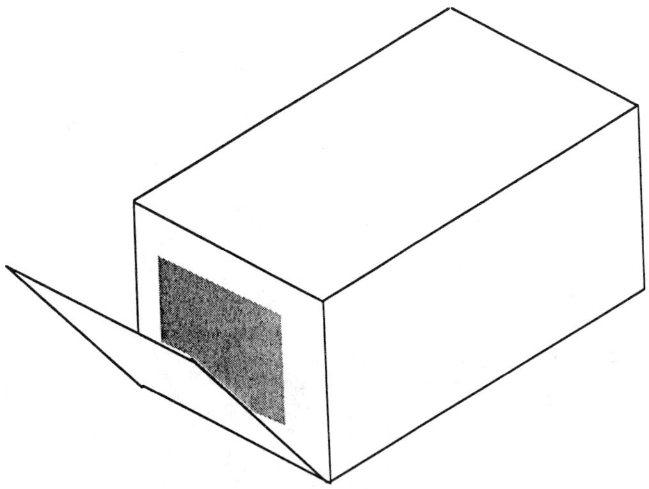

FIG. 15 Example of Closure Methods 4A1 and 4A2

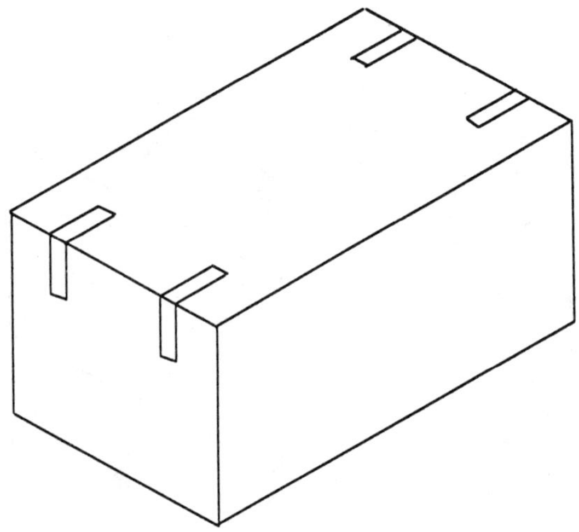

FIG. 17 Example of Closure Methods 4B3, 4B4, 4B5, and 4C2

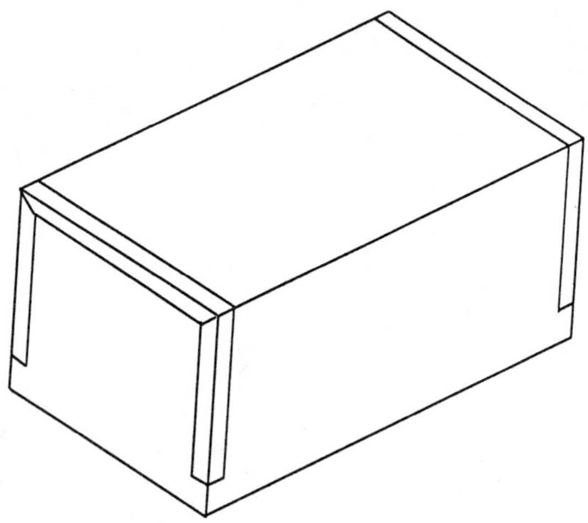

FIG. 16 Example of Closure Methods 4B1, 4B2, and 4C1

6.4 *Closure Methods for Full Overlap Box, Box Code 0203, Center Special Full Overlap Slotted Container, Code 0206, Five Panel Folder, Code 0410, and Other Boxes with Fully Overlapping Flaps:*

6.4.1 *Adhesive Closures of Boxes with Fully Overlapping Flaps:*

6.4.1.1 *Closure Method 4A1*—The box shall be closed with the application of hot melt or waterborne adhesive by securely bonding at least 50 % of the contact area of the flaps. See Fig. 15.

6.4.1.2 *Closure Method 4A2*—The box shall be closed with the application of adhesive conforming with MMM-A-250 by securely bonding at least 50 % of the contact area of contact of all flaps. See Fig. 15.

6.4.1.3 *Closure Method 4A3*—The box shall be closed with the application of a hot melt or waterborne adhesive by securely bonding at least 25 % of the contact area of the flaps. The bonded areas shall extend to within ½ in. (13 mm) of the flap edges.

6.4.2 *Pressure-Sensitive Tape Closure of Boxes with Fully Overlapping Flaps:*

6.4.2.1 *Closure Method 4B1*—The box shall be closed with 3 in. (72 mm) wide pressure-sensitive paper tape applied to all seams. The tape shall conform with PPP-T-76. See Fig. 16.

6.4.2.2 *Closure Method 4B2*—The box shall be closed with 2 in. (48 mm) wide pressure-sensitive tape applied to all seams. The tape shall conform with PPP-T-60. See Fig. 16.

6.4.2.3 *Closure Method 4B3*—The box shall be closed with pressure-sensitive filament tape in ½ in. (12 mm) by at least 4 in. (100 mm) long strips placed across the box seams. Strips shall extend onto each box panel at least 2 in. (50 mm) with at least one strip for each 15 in. (380 mm) of flap length. Tape shall comply with CID A-A-1687 or PPP-T-97, Type II, Class B. See Fig. 17.

6.4.2.4 *Closure Method 4B4*—The box shall be closed with pressure-sensitive filament tape in ¾ in. (18 mm) by at least 6 in. (150 mm) long strips placed across the box seams. Strips shall extend onto each box panel at least 3 in. (75 mm) with at least one strip for each 15 in. (380 mm) of flap length. Tape shall comply with CID A-A-1687 or PPP-T-97, Type II, Class B. See Fig. 17.

6.4.2.5 *Closure Method 4B5*—The box shall be closed with pressure-sensitive filament tape in ¾ in. (18 mm) by at least 6 in. (150 mm) long strips placed across the box seams. Strips shall extend onto each box panel at least 3 in. (75 mm) with at least one strip for each 15 in. (380 mm) of flap length. Tape shall comply with CID A-A-1685 or PPP-T-97, Type IV. See Fig. 17.

6.4.3 *Gummed Tape Closure of Boxes with Fully Overlapping Flaps:*

6.4.3.1 *Closure Method 4C1*—The box shall be closed with 3 in. (75 mm) wide reinforced paper tape applied to all seams. The tape shall conform with PPP-T-45, Type II. See Fig. 16.

6.4.3.2 *Closure Method 4C2*—The box shall be closed with reinforced gummed tape in 3 in. (75 mm) wide by 10 in. (250 mm) long strips placed across the box seams. Strips should extend onto each box panel at least 5 in. (125 mm) with at least one strip for each 20 in. (500 mm) of flap length. Tape must comply with CID A-A-1671 (asphaltic) or CID A-A-1672 (nonasphaltic). See Fig. 17.

6.4.4 *Staple and Stitch Closure and Setup of Boxes with Fully Overlapping Flaps—General*—Staples are used both for box setup and final closure. Stitcher operation requires

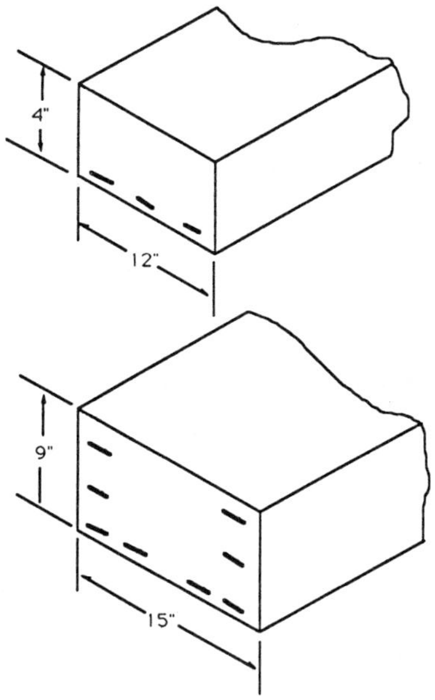

FIG. 18 Example of Closure Methods 4D1 and 4D2

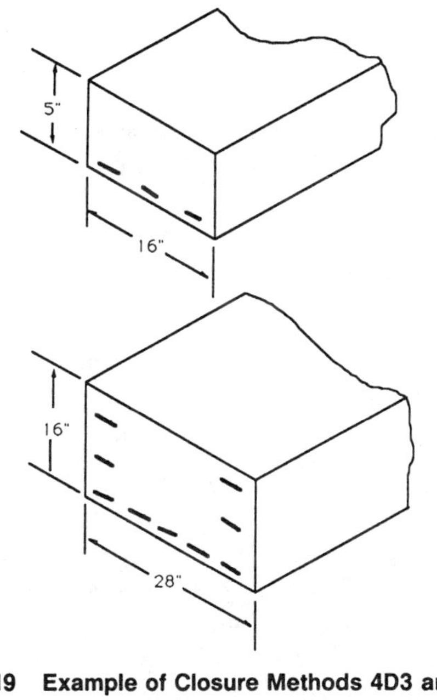

FIG. 19 Example of Closure Methods 4D3 and 4D4

that the stitcher's head and clinch block be on opposite sides of the flaps being joined. Hence, stitches are used only for the setup where access of the clinching block is not restricted. The following methods apply to the use of staples for both setup and final closure and for box setup with stitches.

6.4.4.1 *Closure Method 4D1*—The box shall be closed with ½-in. crown staple or stitches applied along the length of the overlap; fasteners should not be more than 2½ in. (63 mm) apart. See Fig. 18. See the length column of Table 3 for quantity required for various box lengths.

6.4.4.2 *Closure Method 4D2*—The box shall be closed with ½-in. crown staples or stitches applied along the length of the overlap and the sides of the outside flaps not more than 2½ in. (63 mm) apart. See Fig. 18. See Table 3 for quantities required for various box lengths and widths.

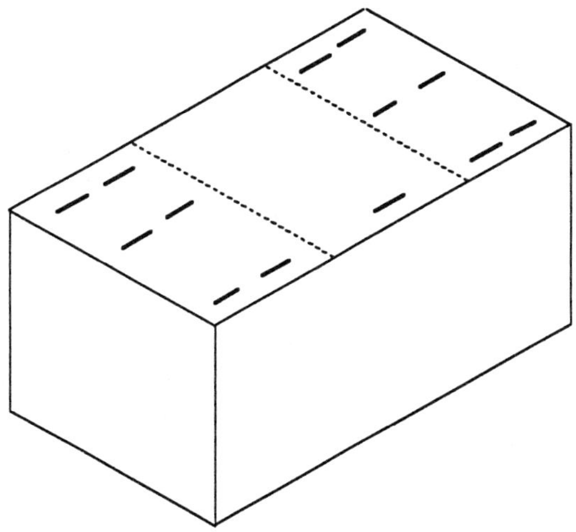

FIG. 20 Example of Closure Method 4D5

6.4.4.3 *Closure Method 4D3*—The box shall be closed with wide crown staples or stitches securely placed along the length of the overlap not more than 5 in. (125 mm) apart to join the inner and outer flaps. See Fig. 19. See the length column of Table 4 for quantity required for various box lengths.

6.4.4.4 *Closure Method 4D4*—The box shall be closed with wide crown staples or stitches securely placed along the length of the overlap and the sides of the outside flaps not more than 5 in. (125 mm) apart to join the inner and outer flaps. See Fig. 19. See Table 4 for quantities required for various box lengths and widths.

6.4.4.5 *Closure Method 4D5*—The box shall be closed with staples or stitches distributed over the areas where outer flaps overlay inner flaps. The number of staples or stitches is given in Table 5. When the inner flaps gap by more than 3 in. (75 mm), additional staples shall be driven in a row parallel to and approximately ½ to 1 in. (12 to 25 mm) from the long edge of the outer flaps for the full length of the gap. See Fig. 20.

6.4.5 *Strapping Closure of Boxes with Fully Overlapping Flaps:*

6.4.5.1 *Closure Method 4E*—The box shall be closed as described in Guide D 4675, Section S1.

6.5 *Closure Methods for Full Telescope Boxes, Box Codes 0300, 0301, 0302, 0303, 0304, 0307, 0308, 0309, 0320, 0321, and Other Boxes Which Fully Telescope* (for half-slotted containers, the top and bottom must be closed by an appropriate method in 6.2 for Method 2 closures):

6.5.1 *Pressure-Sensitive Tape Closure of Full Telescope Boxes:*

6.5.1.1 *Closure Method 5A1*—The box shall be closed with 3 in. (72 mm) wide pressure-sensitive paper tape applied to all seams. The tape shall conform with PPP-T-76. See Fig. 21.

6.5.1.2 *Closure Method 5A2*—The box shall be closed with 2 in. (48 mm) wide pressure-sensitive tape applied to all seams. The tape shall conform with PPP-T-60. See Fig. 21.

6.5.1.3 *Closure Method 5A3*—The box shall be closed with pressure-sensitive filament tape in ½ in. (12 mm) by at least 4 in. (100 mm) long strips placed across the box seams. Strips shall extend onto each box panel at least 2 in. (50 mm)

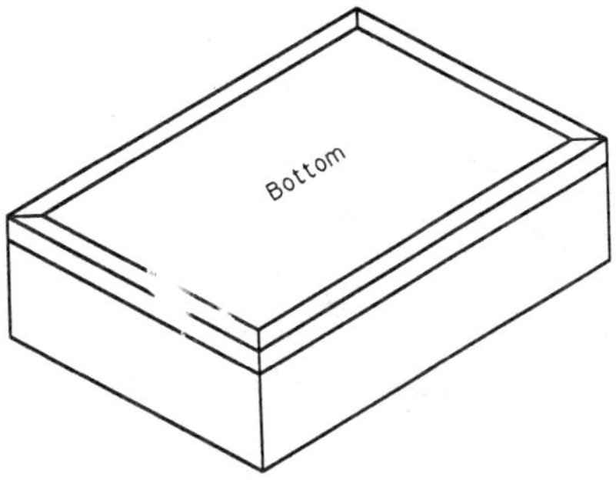

FIG. 21 Example of Closure Methods 5A1, 5A2, and 5B1

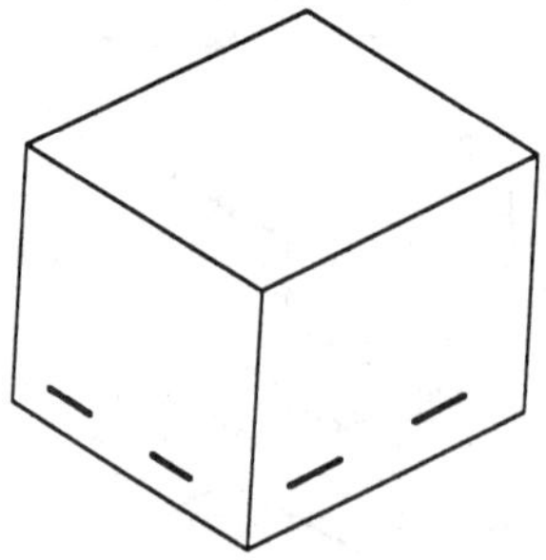

FIG. 23 Example of Closure Methods 5C1 and 5C2

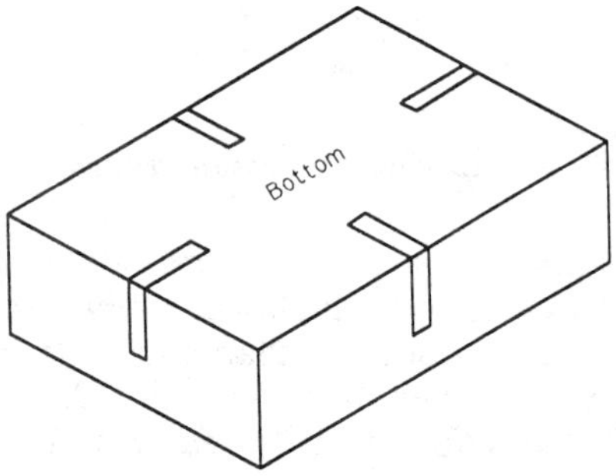

FIG. 22 Example of Closure Methods 5A3, 5A4, 5A5, and 5B2

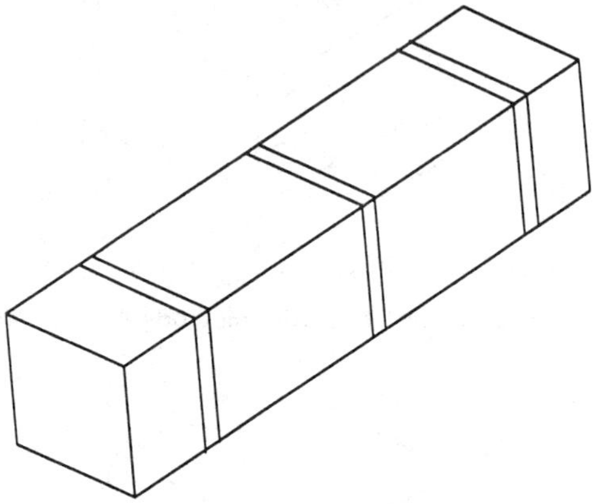

FIG. 24 Example of Closure Method 6D2 and Reinforcement Methods 2A and 2B

with at least one strip for each 15 in. (380 mm) of flap length for box sides and ends. Tape shall comply with CID A-A-1687 or PPP-T-97, Type II, Class B. See Fig. 22.

6.5.1.4 *Closure Method 5A4*—The box shall be closed with pressure-sensitive filament tape in ¾ in. (18 mm) by at least 6 in. (150 mm) long strips placed across the box seams. Strips shall extend onto each box panel at least 3 in. (75 mm) with at least one strip for each 15 in. (380 mm) of flap length for box sides and ends. Tape shall comply with CID A-A-1687 or PPP-T-97, Type II, Class B. See Fig. 22.

6.5.1.5 *Closure Method 5A5*—The box shall be closed with pressure-sensitive filament tape in ¾ in. (18 mm) by at least 6 in. (150 mm) long strips placed across the box seams. Strips shall extend onto each box panel at least 3 in. (75 mm) with at least one strip for each 15 in. (380 mm) of flap length for box sides and ends. Tape shall comply with CID A-A-1685 or PPP-T-97, Type IV. See Fig. 22.

6.5.2 *Gummed Tape Closure of Full Telescope Boxes:*

6.5.2.1 *Closure Method 5B1*—The box shall be closed with 3 in. (75 mm) wide reinforced paper tape applied to all seams. The tape shall conform with PPP-T-45, Type II. See Fig. 21.

6.5.2.2 *Closure Method 5B2*—The box shall be closed with reinforced gummed tape in 3 in. (75 mm) wide by 10 in. (250 mm) long strips placed across the box seams. Strips shall extend onto each box panel at least 5 in. (125 mm) with at least one strip for each 20 in. (500 mm) of flap length for box sides and ends. Tape shall comply with CID A-A-1671 (asphaltic) or CID A-A-1672 (nonasphaltic). See Fig. 22.

6.5.3 *Staple Closure of Full Telescope Boxes:*

6.5.3.1 *Closure Method 5C1*—The box shall be closed with ½-in. crown staples spaced 2½ in. (63 mm) apart around the perimeter of the box cover. Staple crowns shall be parallel to the edge of the top section. See Fig. 23 for an example.

6.5.3.2 *Closure Method 5C2*—The box shall be closed with wide crown staples spaced 5 in. (125 mm) apart around the perimeter of the box cover. Staple crowns shall be parallel to the edge of the top section. See Fig. 23 for an example.

6.5.4 *Strapping Closure of Full Telescope Boxes:*

6.5.4.1 *Closure Method 5D1*—The box shall be closed as described in Guide D 4675, Section S1.

6.5.4.2 *Closure Method 5D2*—The box shall be closed with steel or plastic strapping conforming with Specifications D 3953 or D 3950 placed around the package. Three parallel bands are required, one in the center and one within 6 in. (150 mm) of ends of box. See Fig. 24.

6.5.4.3 *Closure Method 5D3*—The box shall be closed with steel or plastic strapping conforming with Specifications D 3953 or D 3950 placed around the package. A minimum of two crossing bands shall be used. Boxes exceeding 30 in. (760 mm) in length shall have additional bands placed around the box. See Fig. 25.

6.6 *Closure Methods for Boxes with Partially Telescoping Covers, Box Codes 0306, 0310, 0311, 0312, 0313* (for half-slotted containers, the top and bottom must be closed by an appropriate method in 6.2 for Method 2 closures):

6.6.1 *Pressure-Sensitive Tape Closures for Boxes with Covers:*

6.6.1.1 *Closure Method 6A1*—The box shall be closed with pressure-sensitive filament tape in ½-in. (12-mm) strips extending 2 in. (50 mm) onto the top surface of the box, down the side or end, and onto the bottom surface of the box

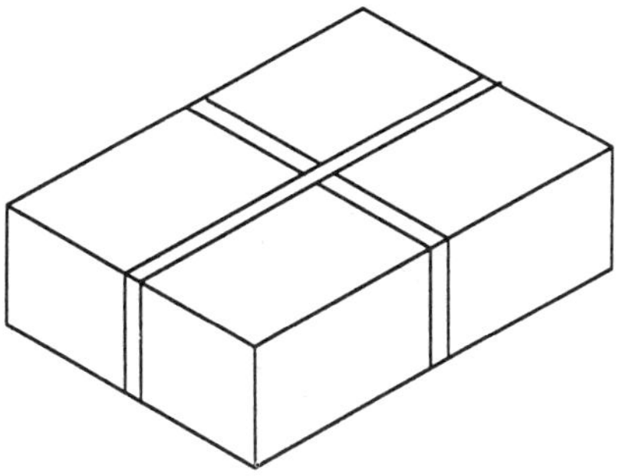

FIG. 25 Example of Closure Method 6D3

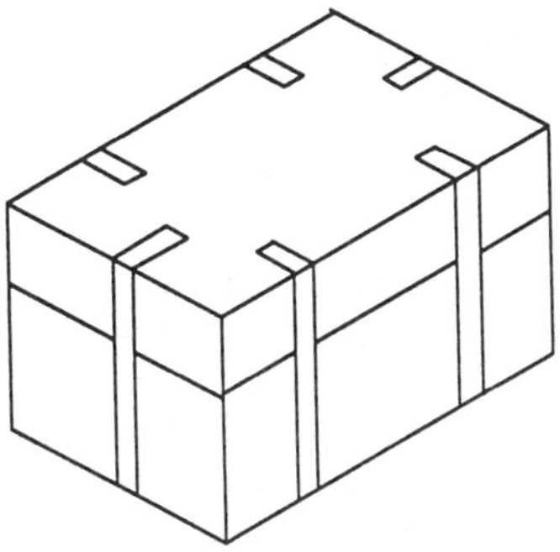

FIG. 26 Example of Closure Methods 6B1, 6A2, and 6A1

2 in. (50 mm). One strip shall be used for each 15 in. (380 mm) of length for box sides and ends. Boxes weighing over 50 lb (23 kg) shall use ¾ in. (18 mm) wide tape. Tape must comply with CID A-A-1687 or PPP-T-97, Type II, Class B. See Fig. 26.

6.6.1.2 *Closure Method 6A2*—The box shall be closed with pressure-sensitive filament tape in ¾-in. (18-mm) strips extending 3 in. (75 mm) onto the top surface of the box, down the side or end, and onto the bottom surface of the box 3 in. (75 mm). One strip shall be used for each 15 in. (380 mm) of length for box sides and ends. Boxes weighing over 70 lb (32 kg) shall use 1 in. (24 mm) wide tape. Tape shall comply with CID A-A-1685 or PPP-T-97, Type IV. See Fig. 26.

6.6.2 *Gummed Tape Closure of Boxes with Covers:*

6.6.2.1 *Closure Method 6B1*—The box shall be closed with reinforced gummed tape in 3 in. (75 mm) wide strips extending 5 in. (250 mm) onto the top surface of the box, down the side or end, and onto the bottom surface of the box 3 in. (75 mm). One strip shall be used for each 15 in. (380 mm) of length for box sides and ends. Tape shall comply with CIC A-A-1671 (asphaltic) or CID A-A-1672 (nonasphaltic). See Fig. 26.

6.6.3 *Staple Closure of Boxes with Covers:*

6.6.3.1 *Closure Method 6C1*—The box shall be closed with ½-in. crown staples spaced 2½ in. (63 mm) apart around the perimeter of the box cover. Staple crowns shall be parallel to the edge of the top section. See Fig. 27.

6.6.3.2 *Closure Method 6C2*—The box shall be closed with wide crown staples spaced 5 in. (125 mm) apart around the perimeter of the box cover. Staple crowns shall be parallel to the end of the top section. See Fig. 27.

6.6.4 *Strapping Closures for Boxes with Covers:*

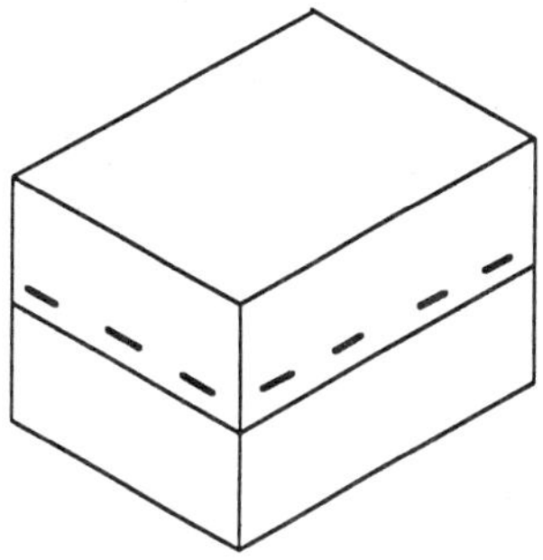

FIG. 27 Example of Closure Methods 6C1 and 6C2

6.6.4.1 *Closure Method 6D1*—The box shall be closed as described in Guide D 4675, Section S1.

6.6.4.2 *Closure Method 6D2*—The box shall be closed with steel or plastic strapping conforming with Specifications D 3953 or D 3950 placed around the package. Three parallel bands are required, one in the center and one within 6 in. (150 mm) of ends of box. See Fig. 24.

6.6.4.3 *Closure Method 6D3*—The box shall be closed with steel or plastic strapping conforming with Specifications D 3953 or D 3950 placed around the package. A minimum of two crossing bands shall be used. Boxes exceeding 30 in. (760 mm) in length shall have additional bands placed around the box. See Fig. 25.

7. Methods of Box Sealing

7.1 Boxes shall have a closure based on Section 6. In some instances, a sealing method also conforms with a closure method. When strapping is used as a closure, the sealing material shall be placed on the box prior to the application of strapping.

7.1.1 *Sealing Method A*—The box shall be sealed with 2 in. (48 mm) wide pressure-sensitive film backed tape applied over all seams and corners of the box. The tape shall have both a nominal machine direction and cross direction tensile strength of at least 20 lb/in. (350 N/100 mm) in accordance with Test Method D 3759 and an adhesive holding power to fiberboard, Test Method D 3654, Procedure A, of at least 1000 min. See Fig. 28.

7.1.2 *Sealing Method B*—The box shall be sealed with 2 in. (48 mm) wide pressure-sensitive tape meeting the requirements of PPP-T-60 or PPP-T-76 applied over all seams, corners, and joints of the box. Tape applied to a manufacturer's joint shall cover the joint but not extend beyond the corners. The tape applied to the seams shall be centered over the seams and shall extend over the corners and edges of the box a minimum of 2½ in. (65 mm) onto adjacent box panels. See Fig. 29.

7.1.3 *Sealing Method C*—The box shall be sealed with 3 in. (72 mm) wide pressure-sensitive tape meeting the requirements of PPP-T-60 or PPP-T-76 applied over all seams, corners, and joints of the box. Tape applied to a manufacturer's joint shall cover the joint but not extend beyond the corners. The tape applied to the seams shall be centered over the seams and shall extend over the corners and edges of the box a minimum of 2½ in. (65 mm) onto adjacent box panels. See Fig. 29.

7.1.4 *Sealing Method D*—The box shall be sealed with 2

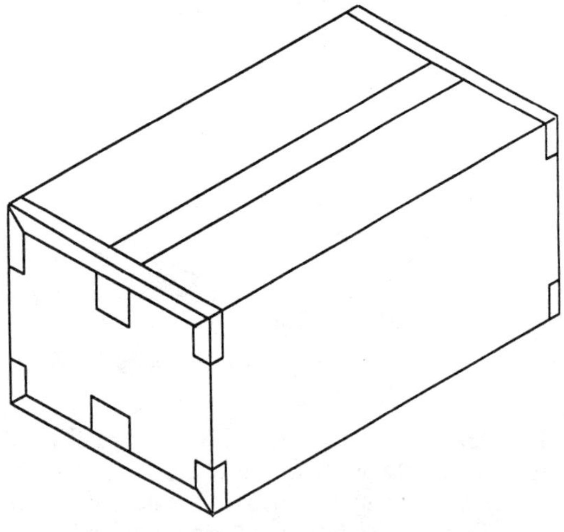

FIG. 28 Example of Sealing Method A

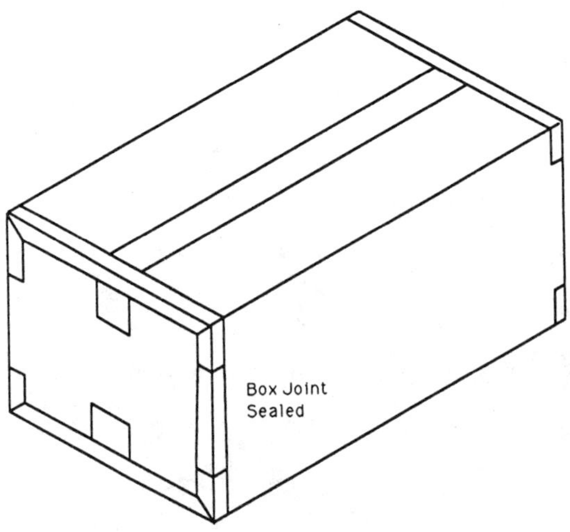

FIG. 29 Example of Sealing Methods B, C, and D

in. (48 mm) wide gummed tape meeting the requirements of PPP-T-45, Type III applied over all seams, corners, and joints of the box. Tape applied to a manufacturer's joint shall cover the joint but not extend beyond the corners. The tape applied to the seams shall be centered over the seams and shall extend over the corners and edges of the box a minimum of 2½ in. (65 mm) onto adjacent box panels. See Fig. 29.

8. Methods of Box Reinforcement

8.1 Boxes shall have a closure based on Section 6 and, if specified, a sealing method based on Section 7. In some instances, a reinforcement method may also conform with a closure method. When strapping is used as a reinforcement, the sealing material shall be placed on the box prior to the application of strapping.

8.1.1 *Performance-Based Reinforcement*—When a Method 1 box reinforcement is referenced, any reasonable means of reinforcing the box is allowed which is able to successfully complete the prescribed performance test on the proposed package with contents. If the box and closure successfully completes the prescribed performance without a separate reinforcement method, none is required. Care must be taken with the choice of acceptance criteria for a reinforcement. Often it could be that the box remain closed in a condition suited for normal sale or use; some box, closure, or reinforcement damage often may be acceptable.

8.1.1.1 *Reinforcement Method 1A*—The box, closure, and reinforcement must be able to complete the test sequence of Practice D 4169, Distribution Cycle 3 or 4, Assurance Level I.

8.1.1.2 *Reinforcement Method 1B*—The box, closure, and reinforcement must be able to complete the test sequence of Practice D 4169, Distribution Cycle 3 or 4, Assurance Level II.

8.1.1.3 *Reinforcement Method 1C*—The box, closure, and reinforcement must be able to complete the test sequence of Practice D 4169, Distribution Cycle 3 or 4, Assurance Level III.

8.1.2 *Tape Reinforcements:*

8.1.2.1 *Reinforcement Method 2A*—The box shall be reinforced with pressure-sensitive filament tape in ½ in. (12 mm) wide bands placed around the girth (smallest circumference) of the box. Bands shall overlap onto themselves at least 2 in. (50 mm) with at least one band for each 15 in. (380 mm) of box length. Boxes over 70 lb (32 kg) shall use ¾ in. (18 mm) wide tape. Tape shall comply with CID A-A-1687 or PPP-T-97, Type II, Class B. See Fig. 24.

8.1.2.2 *Reinforcement Method 2B*—The box shall be reinforced with pressure-sensitive filament tape in ¾ in. (18 mm) wide bands placed around the girth (smallest circumference) of the box. Bands shall overlap onto themselves at least 3 in. (75 mm) with at least one band for each 15 in. (380 mm) of box length. Boxes over 70 lb (32 kg) shall use 1 in. (24 mm) wide tape. Tape shall comply with CID A-A-1685 or PPP-T-97, Type IV. See Fig. 24.

8.1.2.3 *Reinforcement Method 2C*—The box shall be reinforced with pressure-sensitive filament tape in ½ in. (12 mm) wide strips placed over a box seam and extend at least 2 in. (50 mm) onto adjacent box panels. At least one strip shall be applied for each 15 in. (380 mm) of box length. Boxes over 70 lb (32 kg) shall use ¾ in. (18 mm) wide tape. Tape shall comply with CID A-A-1687 or PPP-T-97, type II, Class B.

8.1.2.4 *Reinforcement Method 2D*—The box shall be reinforced with pressure-sensitive filament tape in ¾ in. (18 mm) wide strips placed over a box seam and extend at least 3 in. (75 mm) onto adjacent box panels. At least one strip shall be applied for each 15 in. (380 mm) of box length. Boxes over 70 lb (32 kg) shall use 1 in. (24 mm) wide tape. Tape shall comply with CID A-A-1685 or PPP-T-97, Type IV.

8.1.2.5 *Reinforcement Method 2E*—The box shall be reinforced with pressure-sensitive filament tape in ½ in. (12 mm) wide by at least 4 in. (100 mm) long strips placed over end edges of a box and extend at least 2 in. (50 mm) onto adjacent box panels. At least one strip shall be applied to each end flap. Boxes over 70 lb (32 kg) shall use ¾ in. (18 mm) wide tape. Tape shall comply with CID A-A-1687 or PPP-T-97, Type II, Class B.

8.1.2.6 *Reinforcement Method 2F*—The box shall be reinforced with pressure-sensitive filament tape in ¾ in. (18 mm) wide by at least 6 in. (150 mm) long strips placed over end edges of a box and extend at least 3 in. (75 mm) onto adjacent box panels. At least one strip shall be applied to each end flap. Boxes over 70 lb (32 kg) shall use 1 in. (or 24 mm) wide tape. Tape shall comply with CID A-A-1685 or PPP-T-97, Type IV.

8.1.3 *Strapping Reinforcements:*

8.1.3.1 *Reinforcement Method 3A*—The box shall be reinforced as described in Guide D 4675, Section S1.

9. Keywords

box; close; closure; package; reinforcement; seal

APPENDIX

(Nonmandatory Information)

X1. CROSS REFERENCE

X1.1 Some of the methods described in this guide are based on previously published standards while others are based on useful, but not previously standardized, current industry practice. Tables provide a cross reference of box closure methods to two other standards. This indicates at least a similarity to some of the referenced documents, but it does not assure equivalence.

TABLE X1.1 Closure Methods

ASTM Procedure	UFC Rule 41	Fed Std 224	ASTM Procedure	UFC Rule 41	Fed Std 224
1A	...	...	4A1	7a, 8e	...
1B	...	...	4A2	...	121
1C	...	...	4A3	...	...
1D	...	...	4B1	7c, 8e	106, 125
2A1	...	...	4B2	...	106, 125
2A2	7a	101	4B3	...	125
2A3	7a	...	4B4	...	125
2A4	7a	101	4B5	...	125
2A5	...	...	4C1	8e	125
2B1	...	...	4C2	...	...
2B2	...	...	4D1	7b	...
2B3	7e	102, 121	4D2	7b	...
2B4	7e	102, 121	4D3	7b	...
2B5	7e	...	4D4	7b	...
2B6	...	...	4D5	...	101
2B7	7c	121	4E	...	...
2C1	...	...	5A1	8a	106, 125
2C2	7d	121	5A2	...	106, 125
2C3	7d	...	5A3	...	125
2C4	...	...	5A4	8a	125
2C5	7c	...	5A5	8a	125
2D1	7b	...	5B1	8a	125
2D2	7b	...	5B2	8a	...
2D3	7b	...	5C1	...	...
2D4	7b	...	5C2	...	...
2D5	...	101	5D1	...	...
2D6	...	101	5D2	8a	...
2E	...	...	5D3	8a	...
3A1	...	...	6A1	...	...
3A2	7a	...	6A2	...	105
3B1	...	103	6B1	...	...
3C1	7d	...	6B2	...	...
3D1	7b	...	6C1	...	107
3D2	7b	...	6C2	...	...
3D3	7b	...	6D1	...	...
3D4	7b	...	6D2	...	...
3D5	...	103	6D3	...	...
3E	...	...			

TABLE X1.2 Sealing Methods

ASTM Procedure	Fed Std 224
A	...
B	150, 160
C	151
D	...

TABLE X1.3 Reinforcing Methods

ASTM Procedure	Fed Std 224
1A	...
1B	...
1C	...
2A	...
2B	200
2C	200
2D	200
2E	200
3A	...

Designation: D 2176 – 89

Standard Test Method for Folding Endurance of Paper by the M.I.T. Tester[1]

This standard is issued under the fixed designation D 2176; the number immediately following the designation indicates the year of original adoption or, in the case of revision, the year of last revision. A number in parentheses indicates the year of last reapproval. A superscript epsilon (ε) indicates an editorial change since the last revision or reapproval.

1. Scope

1.1 This test method describes the use of the M.I.T.-type folding apparatus for determining folding endurance of paper. The M.I.T. tester can be adjusted for papers of any thickness; however, if the outer fibrous layers of paper thicker than about 0.25 mm (0.01 in.) rupture during the first few folds, the test loses its significance. The procedure for the Schopper-type apparatus is given in Test Method D 643. This test method is the technical equivalent of TAPPI T 511.

1.2 *This standard does not purport to address the safety problems associated with its use. It is the responsibility of the user of this standard to establish appropriate safety and health practices and determine the applicability of regulatory limitations prior to use.*

2. Referenced Documents

2.1 *ASTM Standards:*

D 585 Method for Sampling and Accepting a Single Lot of Paper, Paperboard, Fiberboard, or Related Product[2]

D 643 Test Method for Folding Endurance of Paper by the Schopper Tester[3]

D 685 Method of Conditioning Paper and Paper Products for Testing[2]

3. Terminology

3.1 *Definitions:*

3.1.1 *fold number*—the number of double folds.

3.1.2 *folding endurance*—the $\log_{10}$ of the folding number.

4. Apparatus

4.1 *Folding Tester,* consisting of:

4.1.1 A spring-loading clamping jaw constrained to move without rotation in a direction perpendicular to the axis of rotation of the folding head specified below and having its clamping surfaces in the plane of this axis. The load shall be applied through a spring attached to the jaw assembly, which is easily adjustable to provide any desired tension on the specimen from 0.5 to 1.5 kg. The deflection of the spring when loaded shall be at least 17 mm (0.67 in.)/kg.

4.1.2 An oscillating folding head supporting two smooth, cylindrical folding surfaces parallel to, and symmetrically placed with respect to, the axis of rotation. The portion of the axis of rotation shall be midway between the common tangent planes of the two folding surfaces. The folding head shall be provided with a clamping jaw with its nearest edge 9.5 mm (0.375 in.) beyond the axis of rotation. The rotary oscillating movement of the head shall be such as to fold the paper through an angle of 135 ± 2° both to the right and to the left of the position of the unfolded specimen. Each of the two folding surfaces shall have a radius of curvature of 0.38 ± 0.015 mm (0.015 ± 0.001 in.) and a width of not less than 19 mm (0.75 in.). The distance separating the folding surfaces shall be greater than the uncompressed thickness of the paper being tested but by no more than 0.25 mm (0.01 in.). Various size folding heads are required for testing different thicknesses of paper. Heads available will accommodate thicknesses from 0 to 0.01, 0.01 to 0.02, 0.02 to 0.03, 0.03 to 0.04, and 0.04 to 0.05 in.

4.1.3 *Motor*—A power-driven device for imparting a rotary oscillating motion of 175 ± 25 periods/min to the folding clamp.

4.1.4 *Counter,* for registering the number of double folds required to break the specimen.

4.2 *Strip Cutter,* to cut 15-mm wide parallel strips to within 0.02 mm with clean edges.

4.3 *Exhaust Fan,* centrifugal, mounted so that its inlet, at least 5 cm in diameter, is next to the oscillating head and the specimen being tested, thus drawing the conditioning room air rapidly over both specimen and head. It capacity shall be great enough to prevent the temperature of the head from rising more than 0.5°C.

5. Test Specimen

5.1 The sample to be tested shall be obtained in accordance with Method D 585. From each sample shall be obtained, so as to be representative of the test unit, at least ten specimens in each principal direction of the paper cut accurately to a width of 15 ± 0.02 mm and a length of 13, preferably 15 cm. Strips shall be selected that are free of wrinkles or blemishes not inherent in the paper, and care shall be taken that the area where the flexing takes place does not contain any portion of a watermark. The long edges of the specimen shall be clean cut and parallel.

6. Adjustment and Calibration

6.1 Make sure that the folding edges are free from rust or dirt and that the counter operates properly.

6.2 Measure the plunger friction by determining the additional load required to move the plunger perceptibly when displaced under a load of 1.0 kg or the load tension used in testing. The friction should not be greater than 25 g.

---

[1] This test method is under the jurisdiction of ASTM Committee D-6 on Paper and Paper Products and is the direct responsibility of Subcommittee D06.92 on Test Methods.

Current edition approved June 30, 1989. Published August 1989. Originally published as D 2176 – 63 T. Last previous edition D 2176 – 69 (1982)$^{\epsilon 2}$. Formerly part of Methods D 643.

[2] *Annual Book of ASTM Standards,* Vol 15.09.

[3] Discontinued; see *1980 Annual Book of ASTM Standards,* Part 20.

6.3 Measure the change in tension due to eccentricity of rotation of folding edges as follows: Place a test strip of strong paper of the proper thickness, cut in the machine direction, in the tester as for making a folding test, and apply a tension of 1 kg or that to be used for the testing. Rotate the folding head slowly throughout the entire folding cycle and measure the maximum change in displacement of the plunger to an accuracy of 0.1 mm (0.004 in.). This should not be greater than that produced by a weight of 35 g. The curvature of the folding edges can be measured by making casts, magnifying them in profile and comparing them to true circles. A folding tester in steady use should be adjusted and calibrated at intervals of not more than 1 month. If not in steady use, then immediately before a standard test.

7. Conditioning

7.1 Condition and test the paper in an atmosphere in accordance with Method D 685. Handle a specimen by the ends and do not touch it with the hands in the region in which it is to be folded.

NOTE 1—As folding endurance is very sensitive to the moisture content of the specimen, it is important to observe strictly the requirements for preconditioning from the dry side, for conditioning, and for conditions during testing.

8. Procedure

8.1 Turn the oscillating folding head so that the opening is vertical. Place a weight on the top of the plunger equivalent to the tension desired on the specimen, normally 1 kg, tap the plunger sideways to eliminate friction, check and set the load indicator, and remove the weight. Depress and lock the plunger in position corresponding to the load. Without touching the part of the strip to be folded, clamp the specimen firmly and squarely in the jaws, with the surface of the specimen lying wholly within one plane, that is, flat, and with the sides not touching the oscillating jaw mounting-plate.

8.2 Unscrew the plunger lock to apply the specified tension to the test strip. If the reading of the load indicator then changes, reclamp the specimen under the proper tension. Whenever possible use a tension of 1 kg, but if this does not give good results (that is, between 10 and 1000 double folds), use a greater or lesser tension.

NOTE 2—The number of the folds will vary by about the cube of the applied tension.

8.3 Set the counter to zero and fold the strip at a uniform rate of 175 ± 25 double folds/min until it breaks.

8.4 Record the number of double folds made before fracture.

9. Report

9.1 *Folding Endurance*—Convert raw data to $\log_{10}$. Calculate mean of logs and report as "$\log_{10}$ M.I.T. Folding Endurance."

9.2 *Fold Number*—Report the number of double folds.

9.3 State clearly if a tension other than 1 kg is used.

10. Precision and Bias

10.1 *Repeatability:*

Folding Endurance	0.12 $\log_{10}$
Fold Number	5.8 %

10.2 *Reproducibility:*

Folding Endurance	0.39 $\log_{10}$
Fold Number	18.7 %

10.3 *Bias*—The bias for this test method is unknown.

Standard Test Method for
Creep Properties of Package Cushioning Materials[1]

This standard is issued under the fixed designation D 2221; the number immediately following the designation indicates the year of original adoption or, in the case of revision, the year of last revision. A number in parentheses indicates the year of last reapproval. A superscript epsilon (ε) indicates an editorial change since the last revision or reapproval.

[ε1] NOTE—Section 12 was added editorially in September 1984.

1. Scope

1.1 This test method covers the determination of creep properties of package cushioning materials. It is applicable to materials available in bulk, sheet, or molded form used for the cushioning of articles during storage, handling, and shipment.

1.2 *This standard does not purport to address the safety problems associated with its use. It is the responsibility of the user of this standard to establish appropriate safety and health practices and determine the applicability of regulatory limitations prior to use.*

2. Referenced Documents

2.1 *ASTM Standards:*

E 105 Practice for Probability Sampling of Materials[2]

E 122 Practice for Choice of Sample Size to Estimate a Measure of Quality of a Lot or Process[2]

3. Summary of Test Method

3.1 The test apparatus consists of a suitable testing device having a base plate and a guided movable platen which can be loaded with weights. The loaded movable platen is placed on a cushion to simulate static compressive loading of cushioning material in actual packaging. By measuring the change in thickness of the loaded cushion with time, creep properties of the cushioning material can be obtained.

4. Significance and Use

4.1 This test determines the extent and nature of cushion thickness change under a given static load. Creep data obtained by this method are applicable to the cushion under the conditions of the particular test and are not necessarily the same as obtained in a complete pack in actual packaging environments. In addition to the influence of the complete pack, the data may also be affected by specimen area and thickness, by varying ambient conditions of temperature, humidity, and vibration, and by impact conditions.

5. Descriptions of Terms Specific to This Standard

5.1 *cushioning material*—a material, as distinguished from a built-up structure, used as a shock or vibration isolator. The majority of materials classified under the term cushioning materials exhibit a high degree of compressibility and recovery and are either interlocking or bonded fibers, cellular plastics, or elastomeric substances, in bulk, sheet, or molded form.

5.2 *creep*—the deformation of a material occurring with time and due to an externally applied constant stress. For cushioning materials specifically, it may be defined as the change in thickness of a cushion under static compressive load over a period of time.

5.3 *permanent set*—the permanent change in thickness of an unloaded cushion as a result of an applied compressive load for any given time interval and any given unloaded recovery time period.

6. Apparatus

6.1 The apparatus shall consist of a modified Kerstner static-load box,[3] or similar testing apparatus that will yield the same results (see Fig. 1 for an isometric drawing of the modified Kerstner apparatus), as follows:

6.1.1 *Base Plate (Outer Box),* constructed of nominal 1-in. white pine (¾ in. (19.0 mm) actual thickness) and fabricated so that the grain of adjacent panels is aligned in opposite directions to reduce swelling which may occur at high temperatures and humidity, or both. A machined aluminum plate, 6½ by 8 by ¼ in. (165.1 by 203.2 by 6.4 mm), is placed on the inside bottom surface and used as the base plate.

6.1.2 *Movable Guided Platen,* that can be loaded with weights (lead shot or lead weights molded to the dimensions of the inner box), and constructed and fabricated in the same manner as described in 6.1.1. A machined aluminum plate is placed between the top of the test specimen and the movable platen and serves as a reference for measuring the height of the specimen, as well as to apply the required 0.025-psi (0.17-kPa) loading to the specimen during the test. Dimensions of this plate will vary depending on the cross-sectional area of the samples under test (Table 1).

6.1.3 *Means of Measurement,* between the base plates and the movable platen. Two vertical lines, one scribed at the center of the edges of the top plate and the other at the center

[1] This test method is under the jurisdiction of ASTM Committee D-10 on Packaging and is the direct responsibility of Subcommittee D10.13 on Interior Packaging.

Current edition approved June 18, 1968. Published August 1968. Originally published as D 2221 – 63 T. Last previous edition D 2221 – 68 (1979).

[2] *Annual Book of ASTM Standards,* Vol 14.02.

[3] This apparatus is described in O. S. Kerstner, "General Principles of Package Design, Part I, Cushioning, NAI-57-187, MPR 705572," Northrop Aircraft, Inc., Hawthorne, CA, Feb. 8, 1957.

TABLE 1 Variable Dimensions of Sample Plate

Sample Size, in. (mm)	Sample Area, in.^{2} (cm^2)	Aluminum Plate Dimensions, in. (mm)		
		Length	Width	Thickness
2 by 2 (51 by 51)	4 (26)	4.25 (108.0)	4.25 (108.0)	0.056 (1.42)
3 by 3 (76 by 76)	9 (58)	4.25 (108.0)	4.25 (108.0)	0.125 (3.18)
4 by 4 (102 by 102)	16 (103)	5.25 (133.4)	5.25 (133.4)	0.145 (3.68)
5 by 5 (127 by 127)	25 (161)	5.25 (133.4)	5.25 (133.4)	0.227 (5.77)
6 by 6 (152 by 152)	36 (232)	6.25 (158.8)	6.25 (158.8)	0.230 (5.84)
2 (51) dia	3.14 (20.2)	4.25 (108.0)	4.25 (108.0)	0.043 (1.09)
3 (76) dia	7.07 (49.6)	4.25 (108.0)	4.25 (108.0)	0.100 (2.54)
4 (102) dia	12.57 (81.1)	5.25 (133.4)	5.25 (133.4)	0.118 (3.00)
5 (127) dia	19.64 (126.7)	5.25 (133.4)	5.25 (133.4)	0.178 (4.52)
6 (152) dia	28.27 (183.3)	6.25 (158.8)	6.25 (158.8)	0.182 (4.62)

of the upper surface of the base plate (Fig. 1) serve as location references when the thickness of the specimen (distance between the aluminum plates) is determined at various time intervals. The measuring apparatus used to determine specimen thickness under test shall be a steel rule which is graduated to 0.01 in. (0.3 mm).

7. Test Specimens

7.1 Test specimens shall be right square prisms or right cylinders with the lateral dimensions at least the same as the original thickness, and with minimum dimensions of not less than 2 by 2 by 1 in. (51 by 51 by 25 mm) thick. The preferred size is 6 by 6 by 4 in. (152 by 152 by 102 mm) thick. If the cushioning material, as supplied, is less than 1 in. thick, the required thickness may be obtained by using two or more layers of the material. Specimens with larger areas are recommended whenever possible and may be dictated by the apparatus used to measure creep. Fiber length, pore size, or the nature of a material may also be determining factors as regards specimen size.

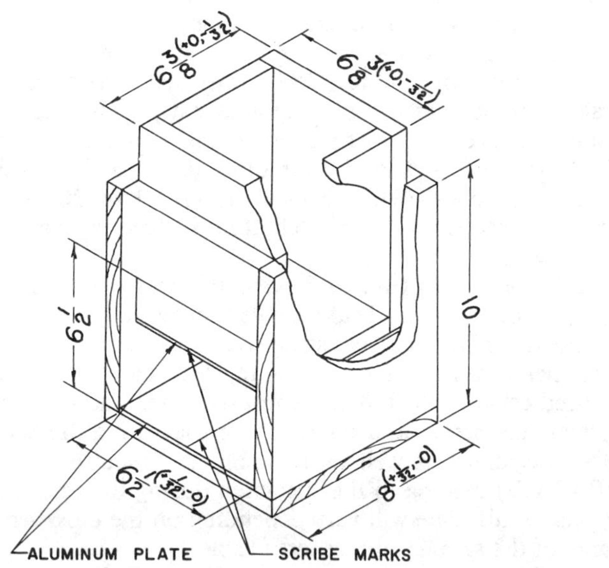

INSIDE DIMENSIONS OF OUTER BOX

$6\frac{1}{2}\ (+\frac{1}{32}, -0) \times 6\frac{1}{2}\ (+\frac{1}{32}, -0) \times 10''$

in.	1/32	6⅜	6½	8	10
mm	0.8	161.9	165.1	203.2	254

FIG. 1 Modified Kerstner Creep Apparatus (Metric Equivalents)

7.2 The number of specimens tested as a sample may vary widely, depending on the intended use of the data. It is recommended that at least four specimens be used for the initial sample of a material. Then, depending on the accuracy and degree of certainty required, this sample size may be increased or decreased. To assure better representation of the sample, individual specimens should be selected by systematic randomization. This can be done by assigning a consecutive number to each of the specimens of the sample, and then selecting the specimens which have numbers that correspond to a series drawn by lottery. Sampling procedures for selecting specimens are discussed in Practice E 105. Procedures for determining the number of specimens required for each sample are given in Practice E 122.

8. Conditioning

8.1 Precondition all specimens at 35 max % relative humidity and at 25°C (77°F) max for 24 h, and then condition at 50 ± 5 % relative humidity and 23 ± 2°C (73.4 ± 3.6°F). Length of conditioning shall be a minimum of 16 h, and until the difference between two successive weights of the specimen determined at 1-h intervals is less than 1 % of the average specimen weight.

8.2 Conduct the test at 23 ± 2°C (73.4 ± 3.6 °F) and 50 ± 5 % relative humidity. If creep is determined at other temperatures and humidities, or both (according to the option of 9.7), conduct only the procedures described in 9.4 to 9.6 at the optional temperatures and humidities, or both.

9. Procedure

9.1 *Original Thickness (T_o)*—Load evenly and gently by a machined aluminum platen (described in Section 6) the conditioned specimen, as furnished or cut, to 0.025 psi (0.17 kPa) of top surface area. After a 30-s interval, and while the specimen is still under 0.025-psi load; measure the thickness to the nearest 0.01 in. (0.3 mm) at the geometric center of the top surface of the specimen, or average the thickness measurements taken on the four corners of the specimen. Record this value as the original thickness (T_o).

9.2 *Length, Width, and Weight*—Determine the measurements with an apparatus yielding values accurate to 0.01 in. for length and width and 0.01 g for weight.

9.3 *Preworking (Optional)*—For cushioning applications where a high degree of compressibility and recovery of the cushion is required, a preworking of the creep test specimen prior to loading is recommended. A suggested preworking procedure may consist of compressing the specimen 65 % of its thickness twice at a rate not to exceed 1 cps. Rest the specimen for a minimum period of 16 h. Following the rest period, record as the preworked thickness (T_p) the thickness of the specimen, as determined in 9.2.

9.4 *Loading of the Specimen*—Using the apparatus described in Section 6, center the aluminum plate on the specimen and apply the desired load (aluminum plate, inner box, and the molded weights) evenly and gently to the entire upper surface of each specimen with the sccibe marks in vertical alignment. Start to measure or determine the thickness of the specimen while under load, 60 ± 5 s after the load has been applied. Determine the thickness by averaging the vertical perpendicular distance between the aluminum plates at the scribe marks on both the front and back of the test apparatus. Record this thickness as the initial thickness under load (T_i).

9.5 *Creep Determination*—Measure the distance between the aluminum plates of the loaded specimen at any desired time interval, but at least at 6 min, 1 h, 24 h, 96 h (4 days) and 168 h (7 days) after the application of the load. Determine the distance between the aluminum plates in accordance with 9.4. More frequent and continued readings are recommended to establish a percent creep-versus-time curve. Record the distance measured at each specified time interval as the deflection thickness (T_d).

9.6 *Recovery Determination*—At the end of the creep loading test time, remove the test load from the specimen. At three intervals, 30 s, 30 min, and 24 h after removal of the test load, make thickness determinations of the specimen as specified in 9.1. Between these determinations, the test load shall not remain on the specimen. Record as the thickness after recovery period ($T_{r(30\ s)}$, $T_{r(30\ min)}$, $T_{r(24\ r)}$).

NOTE 1—Since creep tests especially are quite sensitive to shock and vibration, select the location of the testing apparatus for a minimum of disturbance. When the possible locations are not free from vibrations, design the test equipment and mounting so that the specimen is isolated from shock and vibration.

9.7 *Creep Determinations at Other Temperatures and Humidities (Optional)*—In order to establish data that will simulate actual conditions of experience, it may be necessary to perform creep tests as described, but at varied temperatures and humidities, or both, which can be expected in service. List extremes of such temperatures and humidities, or both, as from −54 to 71°C (−65 to +160°F), and from 35 to 95 % relative humidity at 23°C (73.4°F). Use temperatures and humidities, or both, other than these extremes as desired, and record. Note that instruments or equipment may not operate at the same efficiency at these temperatures.

10. Calculation

10.1 Calculate the density as follows:

$$D = (3.81 \times W)/(l \times w \times T_o)$$

where:
D = density, lb/ft^3,
W = weight of specimen, g,
l = length of specimen, in.,
w = width of specimen, in., and
T_o = original thickness of specimen, in.

10.2 Calculate the static stress for the given loading as follows:

$$\text{Static stress, psi} = F/(l \times w)$$

where:
F = load applied, lb,
l = length of specimen, in., and
w = width of specimen, in.

10.3 Calculate the strain for the given static stress as follows:

$$\text{Strain, \%} = [(T_o - T_i)/T_o] \times 100$$

where:
T_i = initial thickness under load, in., and
T_o = original thickness, in.

NOTE 2—Where preworking is utilized, substitute T_p, preworked thickness, for T_o above.

10.4 Calculate creep based on original thickness as follows:

$$\text{Creep, at any given time interval, \%} = [(T_o - T_d)/T_o] \times 100$$

where:
T_o = original thickness, in., and
T_d = deflection thickness, at the given time interval in., measured between aluminum plates.

NOTE 3—Where preworking is utilized, substitute T_p, preworked thickness, for T_o above.

10.5 Calculate creep based on initial thickness under test load as follows:

$$\text{Creep, at any given time interval, \%} = [(T_i - T_d)/T_i] \times 100$$

where:
T_i = initial thickness under load, in., and
T_d = deflected thickness at the given time interval, in., measured between aluminum plates.

10.6 Calculate permanent set after 24 h as follows:

$$\text{Permanent set, \%} = [(T_o - T_{r(24\ h)})/T_o] \times 100$$

where:
T_o = original thickness, in., and
$T_{r(24\ h)}$ = thickness after recovery period of 24 h, in.

NOTE 4—Where preworking is utilized, substitute T_p, preworked thickness, for T_o above.

11. Report

11.1 Report the following:

11.1.1 Number of specimens tested, origin and description of material, and dates tested,

11.1.2 Original thickness of the specimen, in inches, as determined in 9.2, and the area, in square inches, as calculated from measurements determined in 9.2,

11.1.3 Density of each specimen in pounds per cubic foot, as calculated in 10.1,

11.1.4 Preworking procedure utilized for those materials which are preworked (optional),

11.1.5 Preworked thickness of the preworked specimen, in inches, as determined in 9.3 (optional),

11.1.6 Static stress, in pounds per square inch, as calculated in 10.2,

11.1.7 Strain, in percent, as calculated in 10.3,

11.1.8 Initial thickness under load, in inches, as determined in 9.4,

11.1.9 Deflection thicknesses, in inches, at 6 min, 1 h, 24 h, 96 h (4 days), and 168 h (7 days), and at any additional time intervals chosen, as determined in 9.5,

11.1.10 Creep, in percent, based on the original or preworked thickness, as calculated in 10.4,

11.1.11 Creep, in percent, based on the initial thickness under load, as calculated in 10.5 (optional),

11.1.12 Plot of creep determined versus the corresponding time interval,

11.1.13 Thickness after recovery period, in inches, as determined in 9.6,

11.1.14 Permanent set, in percent, based on the originalor preworked thickness, as calculated in 10.6,

11.1.15 Detailed description of any deviations from the procedure as specified herein, and

11.1.16 Compilation of data from 11.1.1 to 11.1.15 for any other conditions, as described in 9.7 (optional).

12. Precision and Bias

12.1 Precision and bias measurements are currently under study by ASTM Subcommittee D10.13.

Standard Test Method for Environmental Stress-Crack Resistance of Blow-Molded Polyethylene Containers[1]

This standard is issued under the fixed designation D 2561; the number immediately following the designation indicates the year of original adoption or, in the case of revision, the year of last revision. A number in parentheses indicates the year of last reapproval. A superscript epsilon (ε) indicates an editorial change since the last revision or reapproval.

$^{\epsilon 1}$ NOTE—Editorial changes were made throughout in January 1989.

1. Scope

1.1 Under certain conditions of stress, and in the presence of environments such as soaps, wetting agents, oils or detergents, blow-molded polyethylene containers may exhibit mechanical failure by cracking at stresses appreciably below those that would cause cracking in the absence of these environments.

1.2 This test method measures the environmental stress crack resistance of containers, which is the summation of the influence of container design, resin, blow-molding conditions, post treatment, or other factors that can affect this property. Three procedures are provided as follows:

1.2.1 *Procedure A, Stress-Crack Resistance of Containers to Commercial Liquids*—This procedure is particularly useful for determining the effect of container design on stress-crack resistance or the stress-crack resistance of a proposed commercial package containing a proprietary liquid product.

1.2.2 *Procedure B, Stress-Crack Resistance of a Specific Container to Polyoxyethylated Nonylphenol, a Stress-Cracking Agent*—The conditions of test described in this procedure are designed for testing containers made from Type III polyethylene Specification D 1248. Therefore, this procedure is recommended for containers made from Type III polyethylene only. This procedure is particularly useful for determining the effect of resin on the stress-crack resistance of the container.

1.2.3 *Procedure C, Controlled Elevated Pressure Stress-Crack Resistance of a Specific Container to Polyoxyethylated Nonylphenol, a Stress-Cracking Agent*—The internal pressure is controlled at a constant elevated level.

1.3 These procedures are not designed to test the propensity for environmental stress cracking in the neck of containers, such as when the neck is subjected to a controlled strain by inserting a plug.

1.4 The values stated in SI units are to be regarded as the standard.

1.5 *This standard may involve hazardous materials, operations, and equipment. This standard does not purport to address all of the safety problems associated with its use. It is the responsibility of the user of this standard to establish appropriate safety and health practices and determine the applicability of regulatory limitations prior to use.* Specific precautionary statements are given in Section 8 and Notes 1 and 9.

2. Referenced Documents

2.1 *ASTM Standards:*

D 374 Test Methods for Thickness of Solid Electrical Insulation[2]

D 618 Methods of Conditioning Plastics and Electrical Insulating Materials for Testing[3]

D 1248 Specification for Polyethylene Plastics Molding and Extrusion Materials[3]

E 145 Specification for Gravity-Convection and Forced-Ventilation Ovens[4]

3. Terminology

3.1 *Description of Term Specific to This Standard:*

3.1.1 *failure*—during this test, the formation of any imperfection, such as a crack, which results in a loss of pressurizing gas or stress-cracking agent. A container has failed when:

3.1.1.1 It has lost pressure through any aperture other than heat seal areas; or, in Procedure C, when there is a detectable flow of supply air into the bottle,

3.1.1.2 There is visible to an observer with normal eyesight any crack completely through the container wall, or

3.1.1.3 There is evidence of the contained liquid on the outside of the container through any aperture other than one at the heat-seal area, or the contained liquid volume has been reduced.

4. Summary of Test Method

4.1 Procedure A consists of exposing any filled, sealed, blow-molded container to the action of a potential stress-cracking agent within the container, at an elevated temperature. The time to failure is observed.

4.2 Procedure B consists of exposing a partly filled and sealed blow-molded standard container to the action of polyoxyethylated nonylphenol,[5] a stress-cracking agent, within the container, as well as to the action of this agent as

---

[1] This test method is under the jurisdiction of ASTM Committee D-20 on Plastics and is the direct responsibility of Subcommittee D20.12 on Olefin Plastics.

Current edition approved. Aug. 14, 1970. Published October 1970. Originally published as D 2561 – 66 T. Last previous edition D 2561 – 67 T.

[2] *Annual Book of ASTM Standards*, Vols 07.02, 10.01, 10.02, and 15.09.

[3] *Annual Book of ASTM Standards*, Vol 08.01.

[4] *Annual Book of ASTM Standards*, Vols 10.01 and 14.02.

[5] Igepal CO-630 (Antarox CO-630) obtained from GAF Corp., Dyestuff and Chemical Div., 140 West 51 Street, New York, NY 10020.

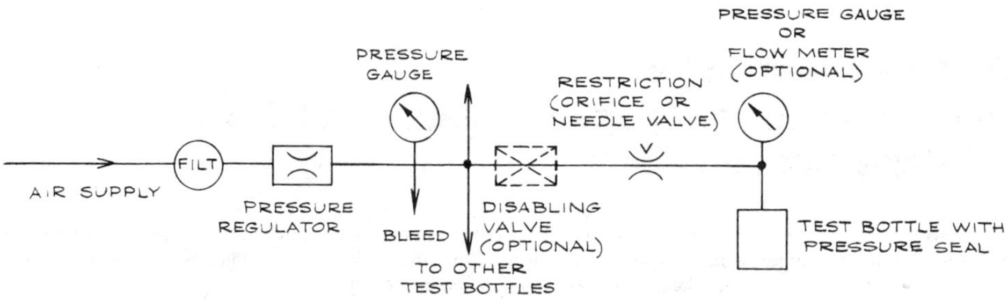

FIG. 1 Apparatus for Procedure C

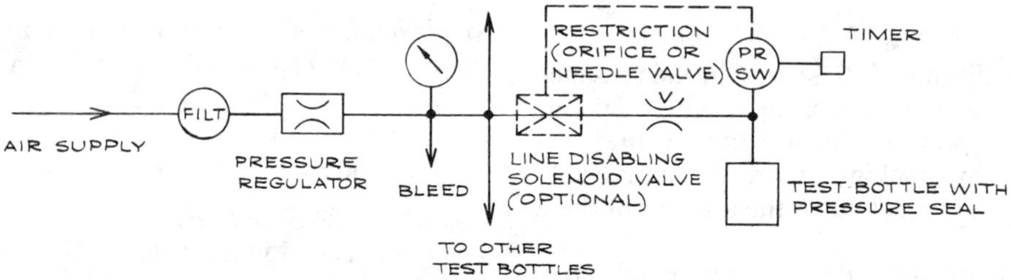

FIG. 2 Apparatus for Procedure C, Including Refinements in Failure Detection

an external environment, at an elevated temperature. The time to failure is observed.

4.3 Procedure C consists of exposing a partly filled blow-molded standard container to the action of polyoxyethylated nonylphenol, a stress-cracking agent, within the container, as well as to a constant elevated pressure internally applied and at an elevated temperature. The time-to-failure can be determined in a tactual-visual manner, or instrumentally.

5. Significance and Use

5.1 When properly used, these procedures may serve to isolate such factors as material, blow-molding conditions, post-treatment, etc., on the stress-crack resistance of the container.

5.2 Environmental stress cracking of blow-molded containers is governed by many factors. Since variance of any of these factors can change the environmental stress-crack resistance of the container, the test results are representative only of a given test performed under defined conditions in the laboratory. The reproducibility of results between laboratories on containers made on more than one machine from more than one mold has not been established.

5.3 Results can be used for estimating the shelf life of blow-molded containers in terms of their resistance to environmental stress cracking provided this is done against a rigorous background of practical field experience and reproducible test data.

6. Apparatus

6.1 *For Procedures A, B, and C:*

6.1.1 *Circulating-Air Oven,* consistent with ovens prescribed in Specification E 145, except for size, capable of maintaining a temperature of 60 ± 1°C (140 ± 1.8°F) and an air-flow rate of 8.5 to 17 m^3/min (300 to 600 ft^3/min).

NOTE 1: **Caution**—A high-temperature safety switch is highly recommended on this oven. Some test liquids can cause extreme pressure build-up upon heating. Under these conditions bottles can rupture with explosive force. This condition can cause injury to the operator as well as damage to the ovens. The override cut-off switch should be set to turn off the oven heat if the test temperature is exceeded by as much as 10°C (18°F).

6.1.2 *Balance,* accurate to within ±1.0 g (for weighing containers and contents) or a volumetric filling apparatus accurate to ±1 mL.

6.2 *For Procedures A and B Only:*

6.2.1 *Heat-Seal Laminate*[6] for sealing the containers.

6.2.2 *Heat-Sealing Unit.*[7]

6.2.3 *Torque Meter.*[8]

6.2.4 *Glass Beakers,* large enough to hold the contents of one test container.

6.3 *For Procedures A and C Only:*

6.3.1 *Polyethylene Bags,* approximately 0.038-mm (1.5-mils) thick, large enough to enclose completely a test container. The bag should fit loosely around the container and be long enough so that the bag opening can be closed above the container closure.

6.4 *For Procedure C Only:*

6.4.1 The essential parts of this apparatus are schematically shown in Fig. 1. Additional refinements in fail detection can be added as shown in Fig. 2. The *necessary* equipment is as follows:

6.4.1.1 *Clear Air Supply* of sufficient pressure to operate regulator and maintain regulated pressure to manifold.

6.4.1.2 *Air Filter,* to remove oil, water, dust, and other contaminants.

---

[6] One such laminate is a polyethylene-foil laminate meeting specification MIL-B-131C, Class 2. This is available from the Robertson Co., Jamaica, Long Island, NY.

[7] Selector Hand Sealing Iron (165 W) made by Selector of Shelton, CT, or Super Sealer made by Clamco Div., Cleveland Detroit Corp., 5400 Brookpark Rd., Cleveland, OH, is suitable.

[8] Model 25 Owens-Illinois Spring Torque Tester, Owens-Illinois Glass Co., Toledo, OH, is suitable.

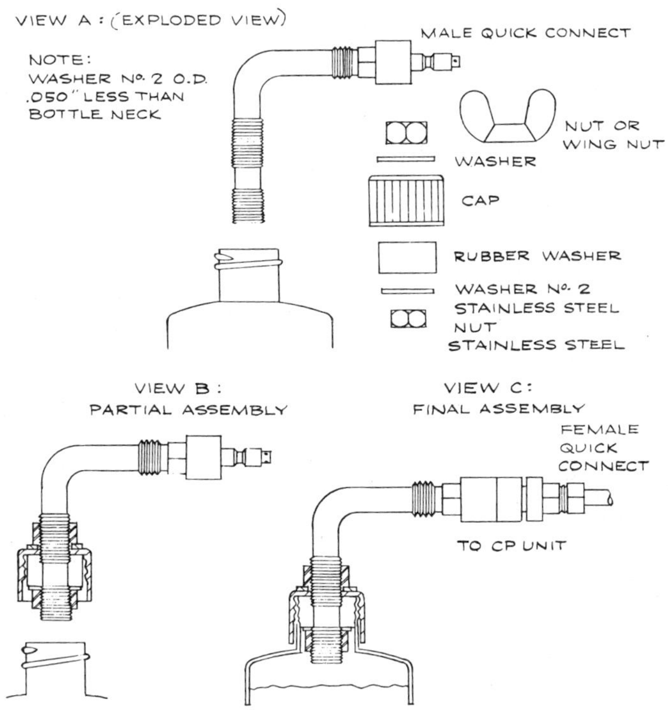

FIG. 3 Bottle Pressure Seal and Tube

6.4.1.3 *Pressure Regulator,* to reduce line pressure to 34.5 ± 1.72 kPa (5.0 ± 0.25 psig).

6.4.1.4 *Pressure Gages,* calibrated to indicate a pressure of 34.5 kPa (5.0 psig) with a precision of 0.34 kPa (0.05 psig).

NOTE 2—A mercury manometer is of benefit in calibrating the pressure gages, and monitoring precise pressure measurements.

6.4.1.5 *Air Valves.*

6.4.1.6 *Restricting Line Orifice or Needle Valve*—This restriction retards the flow of air to the bottle so that supply pressure remains constant after bottle failure, enabling a number of bottles to be pressurized from a single regulated supply. Pressure drop on the bottle side of this restriction is one indication of bottle failure. The orifice size or restriction used will depend upon the sensitivity of the pressure switch or gage. Orifices that pass 300 cm^3/min at 6.9 kPa (1 psi) differential pressure have been found satisfactory. Needle valves, which may be adjusted to flow rates as low as 5.0 cm^3/min, may be useful in cases where greater sensitivity to small failures is desired.

6.4.1.7 *Bottle Cap Assemblies*—Each bottle must be securely sealed and attached to the test fixture. Assemblies essentially like those shown in Fig. 3 have been found satisfactory.

7. Reagents

7.1 *For Procedure A:*

7.1.1 Any reagent or proprietary liquid that is potentially an environmental stress-cracking agent.

7.2 *For Procedure B:*

7.2.1 *Polyoxyethylated Nonylphenol,* a stress-cracking agent.[5]

NOTE 3—Polyoxyethylated nonylphenol is hygroscopic and the undiluted agent should be kept tightly stoppered.

7.2.2 *Polyoxyethylated Nonylphenol Solution*—Prepare a 10 % solution, by volume, of the stress-cracking agent in distilled water in sufficient volume to fill a minimum of fifteen 473-mL (16-oz) containers to ⅓ of overflow capacity (178 mL).

NOTE 4—It has been found to be helpful due to the viscosity of the stress-cracking agent, to prepare the solution at an elevated temperature. A temperature of 50°C (120°F) has been found suitable.

7.2.3 *Dye Indicator Solution*—Add 0.1 % by weight of a wetting agent[9] to distilled water. Dissolve 0.001 % by weight of Gentian Violet in the water.

NOTE 5—Since only about 0.1 mL (2 drops) of this solution is added to each bottle, only a small volume is needed.

7.3 *For Procedure C:*

7.3.1 *Polyoxyethylated Nonylphenol,* a stress-cracking agent.[5] See Note 3.

7.3.2 *Polyoxyethylated Nonylphenol Solution*—Prepare a 33⅓ % solution by volume, of the stress-cracking agent in distilled water in sufficient volume to fill a minimum of fifteen 473-mL (16-oz) containers to ¼ of the overflow capacity (133 mL). See Note 4.

8. Safety Precautions

8.1 Proper precautions should be taken to prevent overheating of the containers during testing since some products which may be tested by Procedure A may cause an extreme pressure build-up in the container and could cause the

[9] Aerosol OT Solution from Fisher Scientific Corp., 203 Fisher Building, Pittsburgh, PA 15219, has been found suitable.

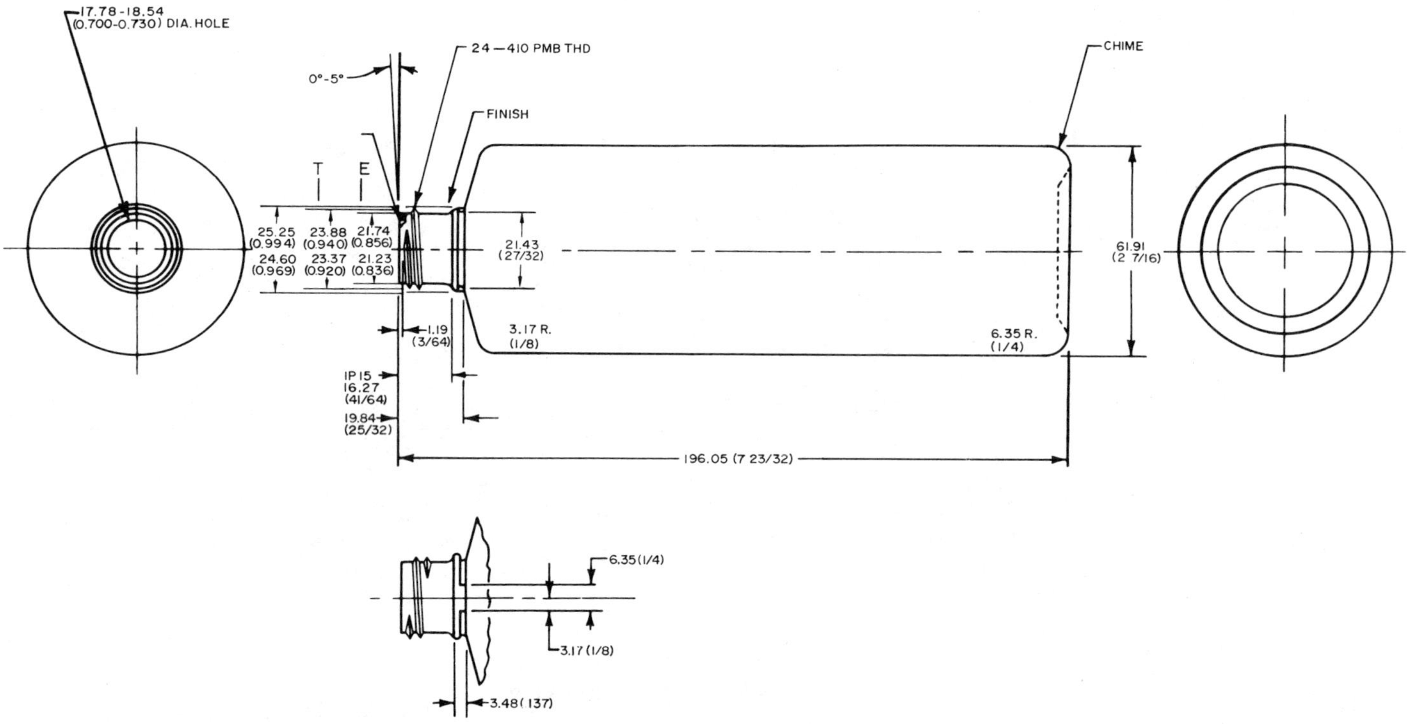

NOTE—Dimensions are in millimetres with inches in parentheses.

FIG. 4 Standard 473-mL (16-oz) Blown Container

container to rupture explosively. Proper safety measures against over-heating are described in Note 1 under 6.1.1.

8.2 Sometimes a container will fail by means of a small pinhole. Since the container is under pressure during the test, liquid may be forced out of the opening spraying the inside of the oven and the operator, if an inspection is being made. Precautions to prevent this from happening are described in 11.1.4, Note 9.

8.3 Care should be taken in handling the stress cracking agent since there is some possibility of its causing dermatitis.

8.4 Proper precautions should be taken in handling compressed air equipment when following Procedure C.

9. Test Specimen

9.1 *For Procedure A:*

9.1.1 A minimum of 15 blow-molded containers, representative of the lot to be tested, and fitted with a screw closure affording a leakproof seal, shall be selected.

9.2 *For Procedures B and C:*

9.2.1 A standard blow-molded container shall be used for this test. It is a 473-mL (16-oz) cylindrical bottle weighing approximately 20 g, as shown in Fig. 4. A minimum of 15 containers shall be selected as in 9.1.1. The minimum wall thickness of the container shall be not less than 0.305 mm (12 mils). The pinch-off area of the container shall not extend into the chime radius.

NOTE 6—Test Methods D 374, modified to use a ball tip micrometer, may be used to measure the thickness of the container.

10. Conditioning

10.1 *Conditioning*—Condition the test specimens at 23 ± 2°C (73.4 ± 3.6°F) and 50 ± 5 % relative humidity for not less than 40 h prior to test in accordance with Procedure A of Methods D 618 for those tests where conditioning is required. In cases of disagreement, the tolerances shall be ±1°C (±1.8°F) and ±2 % relative humidity.

10.2 *Test Conditions*—Conduct all tests at 60°C, unless instructed otherwise.

11. Procedure

11.1 *Procedure A:*

11.1.1 Obtain a minimum of 15 containers of the type and size chosen for test. Fill each to nominal capacity with the chosen test liquid, usually a proprietary liquid product or an aqueous solution of polyoxyethylated nonylphenol.

NOTE 7—Partial filling, that is, ⅓ of nominal capacity, has been found to increase the severity of the test with many test liquids. Thus, the partial fill may be used to accelerate the test. The use of an elevated controlled pressure as in Procedure C may also accelerate the test.

11.1.2 Heat-seal the containers with a suitable triple laminate, polyethylene side to bottle. Apply a polyethylene or suitably lined closure with sufficient torque to ensure a double seal.

NOTE 8—The pressure applied during sealing should be a minimum to ensure no deformation of the container. The container should not be handled in such a manner as to deform the walls during sealing. Any deformation of the container during sealing may result in a volume change which will affect the final test pressure. An application torque of 1.7 N·m (15 in·lbf) has been found sufficient for the standard container described in Fig. 4.

11.1.3 After sealing, invert the containers to coat the inside walls with the agent.

11.1.4 Place each container in a plastic bag and close the bag opening above the closure by folding or by means of a rubber band, string, tape, etc. Do not heat-seal the bag.

NOTE 9: **Caution**—A plastic bag is used to protect the other containers on test from the possibility of one container failing and spraying the other containers. It also protects the operator during inspection of the containers.

11.1.5 Place the containers in a vertical position with the finish up, in beakers, in the oven at the test temperature of 60 ± 1°C (140 ± 1.8°F). Check the temperature periodically for constancy.

11.1.6 Inspect the containers for environmental stress-crack failure hourly for the first 8 h and thereafter at least once each 24 h. Remove containers that fail and record for each, failure exposure time, position of failure with relation to mold number or parting line, and type of failure.

NOTE 10—It is not necessary to remove the plastic bag to inspect the containers. Failures are easily detectable with the bag in place.

NOTE 11—During each inspection for failures, the bottles remaining on test may be moved in a random manner to new positions in the oven to eliminate any effect due to a static oven temperature profile, if one does exist.

11.1.7 Continue exposure of nonfailures until all fail, or to a maximum of 360 h, and record the number of containers still under test at that time.

11.2 *Procedure B:*

11.2.1 Fill a minimum of 15 containers to ⅓ of overflow capacity (178 mL) with the stress-cracking solution described in 7.2.2.

11.2.2 Put approximately 0.1 mL (2 drops) of the dye solution (7.2.3) in each container.

11.2.3 Heat-seal and invert the filled containers as in Procedure A, 11.1.2 and 11.1.3.

11.2.4 Place the containers in the oven at the test temperature of 60 ± 1°C (140 ± 1.8°F) in a vertical position with the finish up, in beakers containing sufficient undiluted stress-cracking agent (7.2.1) to cover the chime area of the container. Check the temperature periodically for constancy.

11.2.5 Inspect containers for failure to a maximum exposure time of 360 h as detailed in Procedure A, 11.1.6 and 11.1.7.

11.3 *Procedure C:*

11.3.1 Fill a minimum of 15 containers to ¼ of overflow capacity (133 mL) with the stress-cracking solution described in 7.3.2. Avoid spilling the solution on the outside of the container since this might lead to premature failure.

11.3.2 Place the cap assembly (Fig. 3) on each bottle and force the rubber stopper into the bottle far enough to start the cap. Avoid forcing the stopper in too far, or bending the copper tubing.

11.3.3 Invert the container while holding a finger over the opening to coat the inside walls completely with the agent.

11.3.4 Connect the cap assembly to the bottle tester, noting the test position of each bottle. Enclose each bottle assembly in a plastic bag as described in 11.1.4. See also Note 9.

11.3.5 Close the oven door to maintain the proper temperature while the remaining bottles are assembled.

11.3.6 Set the pressure regulator at minimum pressure, that is, unscrew the handle until there is no pressure against the diaphragm.

11.3.7 Adjust the line pressure to 69 kPa (10 psi) and then adjust bottle pressure to 34.5 ± 1.72 kPa (5.0 ± 0.25 psi).

NOTE 12—Since an error exceeding 1.72 kPa (0.25 psi) can be critical, it is essential that the pressure gage be accurate and that it is read properly. Another critical variable is temperature. The temperature of the air near the bottles should be checked and maintained at 60 ± 1°C (140 ± 1.8°F).

11.3.8 Record the starting time. After 5 min, check to make sure no bottle cap assemblies are leaking. The containers must be connected to the source of pressure during the test in order to assure that the pressure remains 34.5 ± 1.72 kPa (5.0 ± 0.25 psig).

NOTE 13—Leaking cap assemblies may be checked by immersing the bottles on test in water to a level above the seal. See Fig. 3.

11.3.9 Inspect containers for failure to a maximum exposure time of 360 h as detailed in Procedure A, 11.1.6 and 11.1.7.

NOTE 14—Automatic timing equipment is useful to record the failure time of the containers.

12. Calculation

12.1 Calculate the percentage of the containers that have failed at any given time by the equations in 12.1.1 or in 12.1.2, depending upon the number of containers tested and upon the frequency of inspection.

12.1.1 For 30 or more containers tested, or where inspection for failures are made only once every 24 h after the first 8 h, or both:

$$\text{Failures, \%} = (n/N) \times 100$$

where:

n = cumulative number of containers that have failed at the given time, and

N = number of containers tested.

12.1.2 For less than 30 containers tested *and* where inspections for failures are made more frequently than regular 24-h intervals after the first 8 h, determine the percentage of the containers that have failed at any given time by the equation in 12.1.1, except that the divisor is $N + 1$ instead of N.[10]

12.2 F_{50} *Failure Time:*

12.2.1 Plot the data on log probability graph paper[11] with hours on the log scale and percentage failure on the probability scale. Draw the best fitting straight line for the plot. The hours indicated at the intersection of the data line with the 50 % failure level probability line shall be reported as the F_{50} failure time.

[10] E. J. Gumbel, "Statistical Theory of Extreme Values and Some Practical Applications," National Bureau of Standards Applied Mathematics Series 33, February 12, 1954, pp. 13–15.

[11] Probability paper is available from Keuffel & Esser Co., No. 359-22G (probability × logarithmic).

12.2.2 At least one half of the containers must have failed before an F_{50} value can be reported.

12.3 If extremes of the distribution need to be studied, additional testing will be necessary.

13. Report

13.1 The report shall include the following:

13.1.1 Procedure that was used (Procedure A, B, or C) and test temperature, if different from 60°C,

13.1.2 Complete identification and description of the containers tested, including base resin, blow-molding conditions, coloring system, weight, any unusual material distribution, description of geometry, and any other available information,

13.1.3 For Procedure A, description of the test liquid used and the percentage fill of the containers,

13.1.4 Number of bottles tested,

13.1.5 F_i, the time of the first observed failure,

13.1.6 F_{50}, the estimated time at which 50 % of the containers failed as determined from the plot described in Section 12, Calculation.

13.1.7 F_{100}, the observed time in hours at which 100 % of the containers failed, or if all containers did not fail in 360 h, report F_{100} as >360 h,

13.1.8 Locations and types of failure, and

13.1.9 Any unusual occurrences noted during testing.

14. Precision and Bias

14.1 *Procedure B:*

14.1.1 Reproducibility was determined for Procedure B for containers made from a Type III polyethylene (Specification D 1248). The containers were made from a single laboratory from a single mold and tested in seven different laboratories. The reproducibility that was achieved between laboratories for F_{50} values was $\times/\div$ 1.81 when expressed as two-sigma limits.

14.1.2 The reproducibility that can be achieved in a single laboratory between groups of containers, from the same mold, and the same raw material, has not been determined.

14.2 *Procedure C:*

14.2.1 Reproducibility was determined for Procedure C for containers made from Type III polyethylene (Specification D 1248). The containers were made in a single laboratory from a single mold and tested in five different laboratories. The reproducibility that was achieved between laboratories for F_{50} values was $\times/\div$ 1.58 when expressed as two-sigma limits. The repeatability that was achieved in a single laboratory from the same mold and the same raw material for F_{50} values was $\times/\div$ 1.58 when expressed as two-sigma limits.

NOTE 15—The within- (intra) and between- (inter) laboratory two-sigma limits of $\times/\div$ 1.58 for F_{50} means that if a single laboratory reported a single F_{50} value of 10 h then one expects that repeated evaluations from the same population of bottles at this laboratory would be between 15.8 and 6.3 h, 95 % of the time. If one laboratory reports an F_{50} of 10 h for the average of many specimens, then another laboratory's F_{50} average of many specimens would be between 15.8 and 6.3 h, 95 % of the time. The number 15.8 comes from multiplying 1.58 by 10 and 6.3 comes from dividing 10 by 1.58.

NOTE 16—The precision may be improved by adding automatic recording devices to monitor a specified bottle failure. Such failures may be indicated by a pressure drop or by gas flow to the bottle. See Figs. 1 and 2.

Designation: D 2808 – 69 (Reapproved 1990)

Standard Test Method for Compressive Strength of Corrugated Fiberboard (Short Column Test)[1]

This standard is issued under the fixed designation D 2808; the number immediately following the designation indicates the year of original adoption or, in the case of revision, the year of last revision. A number in parentheses indicates the year of last reapproval. A superscript epsilon (ε) indicates an editorial change since the last revision or reapproval.

1. Scope

1.1 This test method covers the determination of the edgewise compressive strength, parallel to the flutes, of single-wall, double-wall, or triple-wall corrugated fiberboard.

1.2 *This standard does not purport to address the safety problems associated with its use. It is the responsibility of the user of this standard to establish appropriate safety and health practices and determine the applicability of regulatory limitations prior to use.*

2. Referenced Documents

2.1 *ASTM Standards:*

D 585 Method for Sampling and Accepting a Single Lot of Paper, Paperboard, Fiberboard, or Related Product[2]

D 641 Method for Conditioning Paperboard, Fiberboard, and Paperboard Containers for Testing[3]

3. Significance and Use

3.1 This test method measures the edgewise compressive strength of a short column of combined corrugated fiberboard. Research has shown that the edgewise compressive strength of specimens with flutes vertical in combination with the flexural stiffness of the combined board relate to the top-to-bottom compressive strength of corrugated fiberboard shipping containers **(1)**.[4] This test method may be used for comparing the edgewise compressive strength of different lots of similar combined boards, or for comparing different material combinations **(2,3)**.

4. Apparatus

4.1 *Compression Testing Machine*, having the following:

4.1.1 An upper and lower platen, one spring supported and the other driven. The surfaces of the platens shall be smooth, flat, and shall remain parallel to each other within 1 part in 2000 throughout the test. The platens shall have not more than 0.002-in. (0.050-mm) lateral relative movement.

4.1.2 A rate of force increase of 25 ± 5 lb/s (11 ± 2 kgf/s) when the platens are in contact and driven at 1.25 ± 0.75 in. (31.75 ± 19.05 mm)/min **(4)**.

4.1.3 A capacity of at least 500 lb (227 kg).

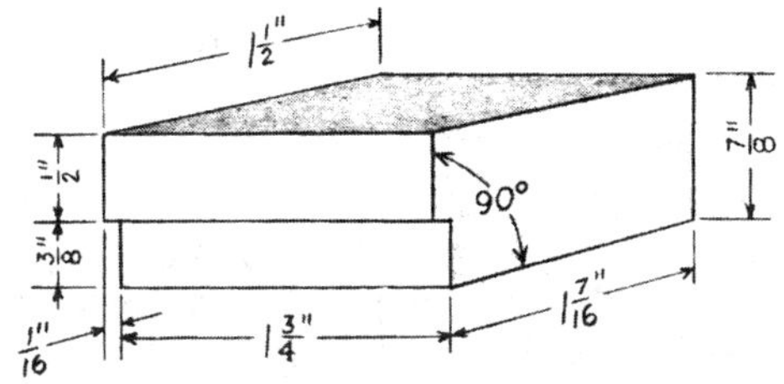

FIG. 1 Metal Guide Block

4.1.4 A means for measuring and indicating the applied load within 1 lb (0.5 kg).

4.1.5 An indicating mechanism that can be checked accurately with dead-weight loads, load cell, or proving ring. The accuracy required is 0.5 % or 0.5 lb (0.25 kg), whichever is greater.

4.2 Two metal guide blocks, like the one shown in Fig. 1, are required to align the specimen vertically in the testing machine.

5. Test Specimens (5,6)

5.1 From each test unit of the sample obtained in accordance with Test Method D 585, accurately cut with a sharp, no-set, (hollow-ground or taper-ground is desirable) saw blade ten representative specimens, each 1.25 ± 0.0625 in. (31.75 ± 1.59 mm) high by 2.00 ± 0.03125 in. (50.8 ± 0.79 mm) wide with the long edges parallel to each other and perpendicular to the axes of the flutes (Fig. 2). Assume that the saw blade is at a 90° angle to the table supporting the specimen.

5.2 Dip each loading edge (long edge) in molten paraffin (melting point 125°F, (52°C) approximately) to a depth of ¼ in. (6.0 mm) and hold there until the absorbed paraffin, as determined visually, begins to migrate above the ¼-in. (6.0-mm) dipped zone. Normally, a 3-s dip in molten paraffin at a temperature of 156 to 166°F (69 to 74°C) is satisfactory. If excessively rapid migration is encountered, reduce the temperature of the molten paraffin. Immediately after dipping, momentarily blot the loading edges of the specimen on paper toweling, preheated on a hot plate maintained at 170 to 180°F (77 to 82°C).

NOTE 1—The following alternative procedure for impregnating the loading edge of a specimen with paraffin is permissible: Place the loading edge on a paraffin-saturated pad, such as a paper toweling, heated on a hot plate maintained at 170 to 180°F (77 to 82°C) until the paraffin impregnates the specimen to the desired ¼-in. (6.0-mm) depth. Generally this method is slower than the dipping method, and therefore permits better control of the depth of paraffin penetration for specimens

[1] This test method is under the jurisdiction of ASTM Committee D-10 on Packaging and is the direct responsibility of Subcommittee D10.17 on Paper and Paperboard Products.

Effective July 18, 1969.

[2] *Annual Book of ASTM Standards*, Vol 15.09.

[3] Discontinued, see *1977 Annual Book of ASTM Standards*, Part 20.

[4] The **boldface** numbers in parentheses refer to the list of references at the end of this test method.

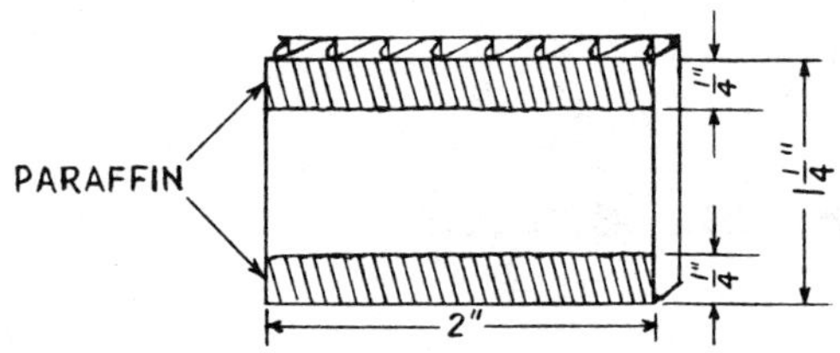

FIG. 2 Edgewise Test Specimen

in which paraffin migration is rapid.

6. Conditioning

6.1 Precondition and condition the prepared specimens in accordance with Test Method D 641.

7. Procedure

7.1 Perform all tests in the conditioning atmosphere.

7.2 Determine the rate of platen movement to use for each compression machine as follows:

7.2.1 Bring the platens together (no specimen) and adjust the speed of the driven platen so the force applied to the spring-supported platen increases at 25 ± 5 lb/s (11 ± 2 kgf/s).

7.2.2 Record the platen movement rate thus determined. On most machines this rate will be 0.5 to 2.0 in. (12.7 to 50.8 mm)/min.

7.3 Measure the width (nominal 2 in. (50.8 mm) dimension) of each specimen to the nearest 0.03125 (1/32) in. (0.79 mm).

7.4 Center the specimen on the bottom platen. Place a guide block on each side of the specimen centrally located relative to it so that the flutes are held perpendicular to the platen. The guide blocks shall be placed largest face up, with the offset ends adjacent and in contact with the specimen above the paraffined areas.

7.5 Apply a compressive force to the specimen. Use the platen movement rate determined in 6.2. When the load on the specimen is between 5 and 15 lb (2 and 7 kg) remove both guide blocks and without altering the platen movement rate, continue to apply pressure until the specimen fails.

7.6 Record the maximum load.

NOTE 2—Valid tests shall have the failure of the specimen occur in the region between the paraffin reinforcement zones and shall not fail by bending.

8. Report

8.1 For each test specimen, report:

8.1.1 Maximum load to cause failure (pounds or kilograms), and

8.1.2 Whether or not the specimen exhibited a valid failure.

8.2 For each test unit, report:

8.2.1 Average maximum load from valid tests (pounds or kilograms),

8.2.2 Average maximum load per unit width for valid tests, calculated from average maximum load and specimen width (pounds per inch or kilograms per centimeter),

8.2.3 Maximum and minimum values, or standard deviation, from valid tests (pounds or kilograms),

8.2.4 Number of valid test specimens,

8.2.5 A description of material tested, and

8.2.6 A statement that the test was conducted in compliance with this procedure, or description of any deviations.

9. Precision (7)

9.1 *Repeatability*—Two test results (average of ten specimens) obtained by the same operator under essentially the same conditions may be expected 95 % of the time to agree with each other to within 5 % of the average of the two test results.

9.2 *Reproducibility*—Two test results (average of ten specimens) obtained in different laboratories for the same sample of material may be expected 95 % of the time to agree within 12 % of the average of the two test results.

REFERENCES

(1) McKee, R. C., Gander, J. W., and Wachuta, J. R., "Compression Strength Formula for Corrugated Boxes," *Paperboard Packaging*, PPACA, Vol 48, August, 1963, p. 149.

(2) Maltenfort, G. G., "Compression Strength of Corrugated." *Paperboard Packaging*, PPACA, Vol 48, August, 1963, p. 160.

(3) Moody, R. C., "Edgewise Compressive Strength of Corrugated Fiberboard as Determined by Local Instability," *U. S. Forest Service Research Paper FPL 46*, December, 1965.

(4) Moody, R. C., and Koning, J. W., Jr., "Effect of Loading Rate on the Edgewise Compressive Strength of Corrugated Fiberboard," *U. S. Forest Service Research Note FPL-0121*, April, 1966.

(5) Koning, J. W., Jr., "Comparison of Two Specimen Shapes for Short Column Test of Corrugated Fiberboard," *U. S. Forest Service Research Note FPL-0109*, October, 1965.

(6) Koning, J. W., Jr., "A Short Column Crush Test of Corrugated Fiberboard," *Tappi*, TAPPA, Vol 47, No. 3, March 1964, p. 134.

(7) Moody, R. C., "An Interlaboratory Study of the Short Column Testing of Corrugated Fiberboard." A study conducted in cooperation with ASTM Committee D-6, Subcommittee IV, October, 1966.

Designation: D 2860 – 90

Standard Test Method for Adhesion of Pressure-Sensitive Tape to Fiberboard at 90° Angle and Constant Stress[1]

This standard is issued under the fixed designation D 2860; the number immediately following the designation indicates the year of original adoption or, in the case of revision, the year of last revision. A number in parentheses indicates the year of last reapproval. A superscript epsilon (ε) indicates an editorial change since the last revision or reapproval.

1. Scope

1.1 This test method covers the determination of the property of adhesion of pressure-sensitive tape to a fiberboard surface. The user of this test method has the option of choosing the paper surface for the test. The test surface may be the standard linerboard, another linerboard (or paper) of individual interest, or a specific fiberboard.

1.2 *This standard does not purport to address the safety problems associated with its use. It is the responsibility of the user of this standard to establish appropriate safety and health practices and determine the applicability of regulatory limitations prior to use.*

NOTE 1—A complete metric companion to Test Method D 2860 has been developed—D 2860M; therefore, no metric equivalents are presented in this method. Test Method D 2860M uses a 24 by 24-mm tape area rather than the 1 by 1-in. (25.4 by 25.4-mm) area used in this method. This difference in test parameters will generally produce lower time to failure for Test Method D 2860M than Test Method D 2860 but not necessarily by a constant factor.

2. Referenced Documents

2.1 *ASTM Standards:*

D 996 Terminology of Packaging and Distribution Environments[2]

D 2904 Practice for Interlaboratory Testing of a Textile Test Method that Produces Normally Distributed Data[3]

D 2906 Practice for Statements on Precision and Bias for Textiles[3]

D 3715 Practice for Quality Assurance of Pressure-Sensitive Tapes[2]

D 4332 Practice for Conditioning Containers, Packages, or Packaging Components for Testing[2]

3. Terminology

3.1 *Definitions*—Terminology found in Terminology D 996 shall apply.

4. Summary of Test Method

4.1 *Procedure A*—The pressure-sensitive tape is mounted on a rigid panel. The adhesive side is exposed. A strip of linerboard, which represents the fiberboard, is placed against the tape adhesive with controlled pressure. The area of contact is specified. The panel is mounted in a horizontal position. A specified mass is fastened to one end of the linerboard. The mass acts to peel the linerboard away from the adhesive at an angle of 90°. The time required to peel the linerboard from the tape is measured.

4.2 *Procedure B*—In this procedure, the fiberboard or the linerboard representing it is adhered to a rigid panel and the tape peeled from it. In all other respects, the procedure is the same as Procedure A.

5. Significance and Use

5.1 This test method is designed to show the adhesion of a pressure-sensitive tape to fiberboard when the tape is under a constant stress. Test data from this test method for some types of packaging tape relates to tape performance on fiberboard box closure.

5.2 Procedure A should be used when comparing tapes to an established requirement. The requirement may be established using either the standard linerboard or another linerboard representing a specific fiberboard.

5.3 Procedure B should be used to rank tapes as to their ability to adhere to fiberboard.

6. Apparatus and Materials

6.1 *Tape,* pressure-sensitive, double-coated, 2-in. width. The tape shall exhibit sufficient adhesion to the panel and test specimen or fiberboard to avoid the failures cited in 10.1.9 and 10.2.14.

6.2 *Panel*—Any steel piece at least 2 by 5 in. and no less than 0.043 in. thick, having one surface (2 by 5 in.) that does not deviate from a plane by more than 0.002 in.

6.3 *Stop Watch.*

6.4 *Razor Blade,* single-edge.

6.5 *Mask*—Any paper no more than 0.002 in. thick formed to a rectangle approximately 5 by 2.5 in. with a rectangular hole exactly 1.0 ± 0.01 by 1.25 in. centered in the piece. The 1-in. hole dimension shall be in the 5-in. direction (see Figs. 1 and 2).

6.6 *Cutter,* for preparing fiberboard or linerboard and tape specimens.

6.7 *Paperboard Piece,* approximately 1 by 1 in., for reinforcement.

6.8 *Paper Punch,* single-hole.

6.9 *Fiberboard or Linerboard Surface,*[4] against which the test is made.

[1] This test method is under the jurisdiction of ASTM Committee D-10 on Packaging, and is the direct responsibility of Subcommittee D10.14 on Closure and Reinforcement.

Current edition approved June 29, 1990. Published August 1990. Originally published as D 2860 – 70. Last previous edition D 2860 – 87.

[2] *Annual Book of ASTM Standards,* Vol 15.09.

[3] *Annual Book of ASTM Standards,* Vol 07.01.

[4] The linerboard accepted as the standard test surface, Standard Reference Material 1810A, is available from the Office of Standard Reference Materials, National Institute of Standards and Technology, Washington, DC 20234.

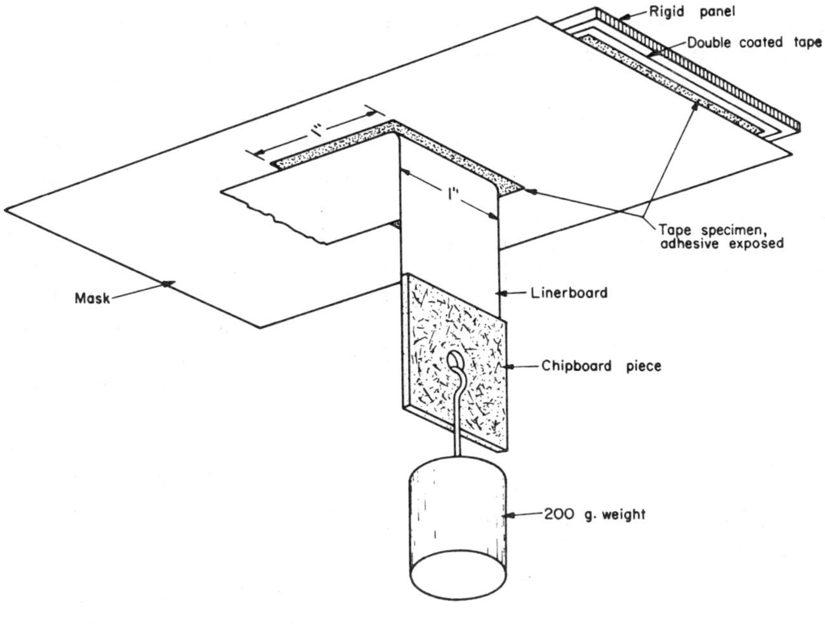

NOTE—Fix panel securely in a horizontal plane.

FIG. 1 Test Assembly for Procedure A

6.9.1 For Procedure A, cut a rectangle exactly 1.0 ± 0.01 in. wide by at least 5 in. in length, from the linerboard representing the fiberboard on which the tape will be used. The long dimension shall lie at a right angle to the machine direction of the linerboard. Reinforce one end of the strip with a chipboard piece adhered with double-coated tape. Punch a hole at the center of the reinforcement.

6.9.2 For Procedure B, cut a rectangle, preferably 5 by 2 in., but no less than 2 by 2 in., from the fiberboard on which the tape will be used (or linerboard representing it). The long test dimension shall lie at a right angle to the machine direction of the linerboard.

6.9.3 For either procedure, prepare the linerboard or fiberboard in advance of testing and mark it to show which side to use. When comparing one roll of tape to another or one lot of tape to another, the linerboard or fiberboard should be from a single lot.

6.10 *Steel Roller,* rubber-covered, 3.25 ± 0.1 in. in diameter and 1.75 ± 0.05 in. in width, covered with rubber approximately 0.25 in. in thickness having a Shore A durometer hardness of 80 ± 5. The surface of the rubber shall form a true cylinder devoid of any convex or concave deviations. The mass of the roller shall be 4.5 ± 0.1 lb. The mass of the handle shall not add to the mass of the roller during use.

NOTE 2—A simple check to determine if the rubber surface is cylindrical is to wrap the roller in a very thin paper (onionskin) and drag it across a flat glass plate on which is placed a carbon paper, face up. The carbon rubs off onto the thin paper wrapper to reveal high spots or hollows on the rubber surface.

6.11 *Test Stand,* to support the panel securely in a horizontal plane during the test period.

6.12 *Mass,* 200 ± 0.2 g, with a hook.

6.13 *Timing System,* for measuring test time in minutes.

6.14 *Paper,* clean, unprinted, and without a coating of any kind, for use in adhering linerboard or fiberboard to the panel (Procedure B).

7. Sampling

7.1 Sampling shall be in accordance with Practice D 3715.

8. Test Specimens

8.1 A roll or rolls of tape for testing must be at least 1.5 in. in width.

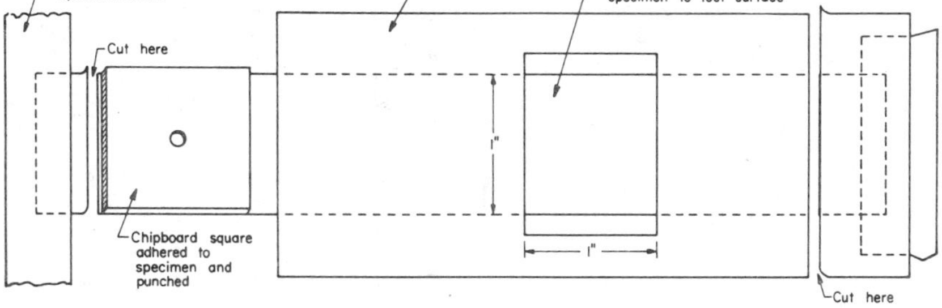

FIG. 2 Preparation of Specimen for Procedure B

8.2 Unwind and discard at least three, but no more than six, outer wraps from the sample roll before taking specimens for testing.

8.3 Test five specimens from each roll of tape.

9. Conditioning

9.1 Condition sample rolls of tape and linerboard or fiberboard in the standard conditioning atmosphere as described in Practice D 4332 for a period of not less than 24 h. Test at these conditions.

10. Procedure

10.1 *Procedure A:*

10.1.1 Adhere the double-coated tape to the panel with firm finger pressure and remove the liner.

10.1.2 Lightly sand the backing of the test tape on the roll for approximately 10 in. with 300-grit sandpaper to remove the easy unwind treatment. Start timing with the stop watch. Immediately unwind 12 to 18 in. of tape from a freely rotating roll at a rate of 20 to 30 in.

10.1.3 Center the sanded area of this strip, adhesive side up, on the panel, aligning the long edges parallel with the panel long edges. Trim it to the panel dimensions. Do not allow any object to contact the adhesive.

10.1.4 Place a mask at the center of the specimen so that the 1-in. dimension of the hole in the mask lies parallel to the long dimension of the specimen edges (see Fig. 1). Rub against the mask to assure that the tape specimen is well adhered to the double-coated tape on the panel. Avoid touching the adhesive.

10.1.5 At 90 ± 5 s, position the linerboard strip (end opposite the hole) to span the mask and center it lengthwise on the specimen adhesive. Assure that the correct side of the linerboard is placed against the specimen adhesive. Do not apply any finger pressure to the specimen. Continue timing.

10.1.6 Roll the linerboard against the specimen lengthwise once each way at a rate of 12 in./min using the rubber-covered roller.

10.1.7 Immediately set the panel, specimen side down, into the test stand, assuring that it will remain in a horizontal plane throughout the test.

10.1.8 At 180 ± 5 s (3 min from unwinding the tape from the roll), add the 200-g mass so that no shock load is imparted to the linerboard, and immediately start the time. Allow the test to continue until the linerboard has peeled free from the specimen adhesive.

10.1.9 Void all tests where the tape (test specimen) has pulled away from the panel (either the specimen from the double-coated tape or the double-coated tape from the panel). When this occurs, repeat the test assuring that the panel, the double-coated tape, and the test specimen are well bonded together.

10.2 *Procedure B:*

10.2.1 Adhere the double-coated tape to the panel with firm finger pressure and remove the liner.

10.2.2 Superimpose the fiberboard (or linerboard) on the double-coated tape, exposing the correct side and assuring that the machine direction is at a right angle to the long dimension of the panel. Handle the fiberboard only at the edges and corners so that an area at least 1.5 in. square at its center is absolutely free of contamination.

10.2.3 Place a clean sheet of paper (7.14) on top of the fiberboard. Rub against this to assure that the fiberboard is well bonded to the double-coated tape on the panel. Do not rub the fiberboard surface with the paper or fingers.

10.2.4 Start timing with the stop watch. Immediately unwind 12 to 18 in. of tape from a freely rotating roll at a rate of 20 to 30 in.

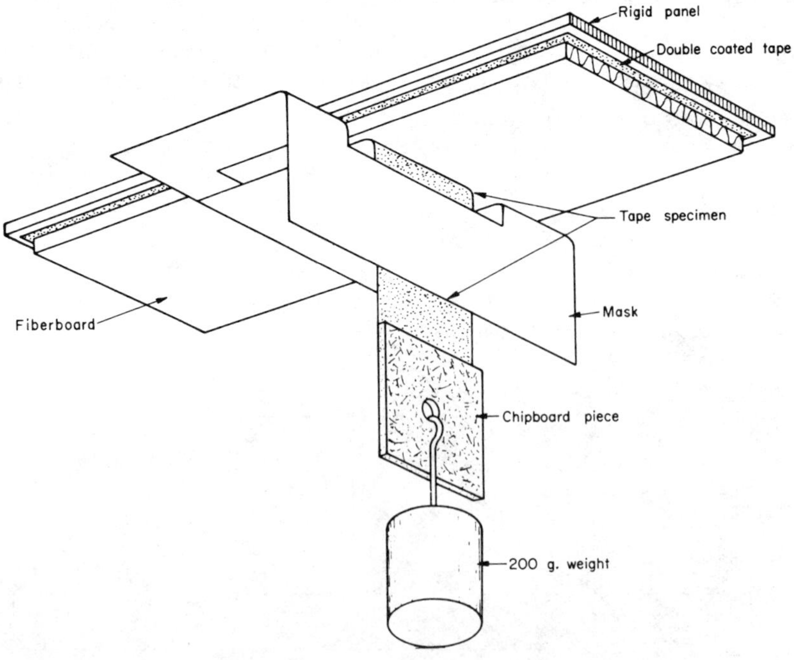

NOTE—Fix panel secure a horizontal plane.

FIG. 3 Test Assembly for Procedure B

10.2.5 From the center of this strip, cut a specimen 1.0 ± 0.01 in. wide and at least 7 in. in length. Handle this strip by its ends only. Do not allow adhesive to contact any object.

10.2.6 Lay the specimen, adhesive side up, on the work bench and tape the ends down to hold the specimen flat.

10.2.7 Apply a mask at one end of the specimen so that the 1-in. dimension of the hole lies parallel to the long edges of the specimen (see Fig. 2).

10.2.8 Apply the chipboard square to the exposed adhesive at the other end of the specimen (see Fig. 2).

10.2.9 Cut away the taped ends of the specimen. Punch a hole for the hook of the mass in the chipboard.

10.2.10 At 90 ± 5 s, position the specimen lengthwise on the fiberboard, adhesive down, so that the 1 by 1-in. adhesive area is approximately centered on the fiberboard. Do not apply any finger pressure to the specimen. Continue timing.

10.2.11 Roll the specimen lengthwise once each way at a rate of 12 in./min, using the rubber-covered roller.

10.2.12 Immediately set the panel, specimen side down, into the test stand, assuring that it will remain in a horizontal plane throughout the test.

10.2.13 At 180 ± 5 s (a total of 3 min from unwinding the tape from the roll), add the 200-g mass so that no shock load is imparted to the specimen and start the timer. Allow the test to continue until the specimen has peeled free from the fiberboard (see Fig. 3).

10.2.14 Void all tests where the fiberboard has pulled away from the panel (either the fiberboard from the double-coated tape or the double-coated tape from the panel). When this occurs, repeat the test, assuring that the panel, double-coated tape, and the fiberboard are well bonded together.

11. Calculation

11.1 To determine the test result for each roll of tape, convert each of the five specimen test results to its common logarithm. Obtain the arithmetic mean of all logarithms and then convert the arithmetic mean logarithm back to time by obtaining its antilogarithm. This final step gives the test result for the roll of tape under consideration. No single specimen value shall be considered as representative of the roll under test.

12. Report

12.1 The report shall include the following:

12.1.1 A statement that this test method was used, indicating which procedure (A or B), and any deviations from the method as written,

12.1.2 Complete identification of each roll of tape tested including type, source, manufacturer's code number, and form,

12.1.3 Linerboard or fiberboard identification including the type, source, manufacturer's code number, and conditioning history prior to the test,

12.1.4 Number of specimens tested from each roll of tape,

12.1.5 Anomalous behavior during the test, that is, adhesive transfer or splitting paper fiber pulled free from the linerboard or fiberboard, delamination of the linerboard or fiberboard, and

12.1.6 Mean time value calculated in 11.1 for each of the rolls of tape under test.

TABLE 1 Critical Difference, Percent of Grand Average for the Conditions Noted[A]

Number of Observations in Each Average	Single-Operator Precision	Within-Laboratory Precision	Between-Laboratory Precision
1	27.7	28.6	30.3
5	14.4	16.1	16.4
10	11.7	13.8	14.1

[A] Before comparing two averages of observed data, transform them using Eq 1.

TABLE 2 Width of 95 % Confidence Limits, Percent of the Grand Average for the Conditions Noted[A]

Number of Observations in Each Average	Single-Operator Precision	Within-Laboratory Precision	Between-Laboratory Precision
1	±19.6	±20.2	±21.4
5	±10.2	±11.4	±11.6
10	±8.3	±9.7	±10.0

[A] Before evaluating the precision of a single average expressed in observed units of measure; transform the average using Eq 1, calculate in transformed units the average plus and minus the tabulated values, and convert the resulting values into observed units of measure using Eq 1. The values of the upper and lower confidence limits in observed units of measure will not be symmetrical about the average.

13. Precision and Bias

13.1 *Summary*—The difference between two single observations should not exceed 27.7 % of the average of the two observations in transformed units in 95 out of 100 cases when both observations are taken by the same well-trained operator using the same piece of test equipment and specimens randomly drawn from the same sample of material. Larger differences may occur under all other circumstances. The true value of adhesion to fiberboard at 90° angle under constant stress can only be defined in terms of a specific test method. Within this limitation, Test Method D 2860 has no known bias. Sections 13.2 through 13.5 explain the basis for this summary and for evaluations made under other conditions. The information applies either to comparisons of specimens of the same material or to specimens of different materials.

NOTE 3—This precision statement applies only to Procedure A.

13.2 *Interlaboratory Test Data*[5]—An interlaboratory study was made in 1980 in which randomly drawn samples of two materials were tested in each of four laboratories. Two operators in each laboratory, each tested three specimens from each of three rolls of each material. The observed data did not conform to the assumptions underlying the analysis of variance. Therefore, before analysis, the data were transformed using the following equation:

$$C = \log_{10} D$$

where:

C = transformed datum, and

D = observed datum.

13.2.1 The components of variance for adhesion to fiberboard at 90° angle under constant stress expressed as coefficient of variation (Note 4) were calculated to be:

[5] Supporting data are available from ASTM Headquarters. Request RR:D-10-1002.

Single-operator component	3.0 % of the average in transformed units
Within-laboratory component	2.6 % of the average in transformed units
Between-laboratory component	1.1 % of the average in transformed units
Replication component	9.5 % of the average in transformed units

13.3 *Critical Differences*—For the components of variance reported in 12.2.1, two averages of observed values should be considered significantly different at the 95 % probability level if the difference equals or exceeds the differences shown in Table 1.

NOTE 4—The calculations for coefficient of variation and other statistics found in subsequent sections of this statement are described in Practice D 2906 and Appendix A3 of Practice D 2904.

NOTE 5—The tabulated values of the critical differences and confidence limits should be considered to be a general statement particularly with respect to between-laboratory precision. Before a meaningful statement can be made about two specific laboratories, the amount of the statistical bias between them, if any, must be established with each comparison being based on recent data obtained on specimens randomly drawn from one sample of the material to be evaluated.

13.4 *Confidence Limits*—For the components of variance reported in 12.2, the single averages of observed values have the 95 % confidence limits (Note 5) shown in Table 2.

13.5 *Bias*—No justifiable statement can be made on the bias of Test Method D 2860 for testing adhesion to fiberboard at 90° angle and constant stress since the true value cannot be established by an accepted referee method.

14. Keywords

14.1 adhesion at 90° angle and constant stress; pressure sensitive tape

Standard Test Method for Adhesion of Pressure-Sensitive Tape to Fiberboard at 90° Angle and Constant Stress [Metric][1]

This standard is issued under the fixed designation D 2860M; the number immediately following the designation indicates the year of original adoption or, in the case of revision, the year of last revision. A number in parentheses indicates the year of last reapproval. A superscript epsilon (ε) indicates an editorial change since the last revision or reapproval.

1. Scope

1.1 This test method covers the determination of the property of adhesion of pressure-sensitive tape to a fiberboard surface. The user of this test method has the option of choosing the paper surface for the test. The test surface may be the standard linerboard, another linerboard (or paper) of individual interest, or a specific fiberboard.

1.2 *This standard does not purport to address the safety problems associated with its use. It is the responsibility of the user of this standard to establish appropriate safety and health practices and determine the applicability of regulatory limitations prior to use.*

NOTE 1—This test method is the metric counterpart of Test Method D 2860. It uses a 24 mm × 24 mm tape area rather than the 1 in. × 1 in. (25.4 mm × 25.4 mm) area used in Test Method D 2860. This difference in test parameters will generally produce lower time to failure for Test Method D 2860M than Test Method D 2860 but not necessarily by a constant factor.

2. Referenced Documents

2.1 *ASTM Standards:*

D 996 Terminology of Packaging and Distribution Environments[2]

D 2904 Practice for Interlaboratory Testing of a Textile Test Method that Produces Normally Distributed Data[3]

D 2906 Practice for Statements on Precision and Bias for Textiles[3]

D 3715 Practice for Quality Assurance of Pressure-Sensitive Tapes[2]

D 4332 Practice for Conditioning Containers, Packages, or Packaging Components for Testing[2]

3. Terminology

3.1 *Definitions*—Terminology found in Terminology D 996 shall apply.

4. Summary of Test Method

4.1 *Procedure A*—The pressure-sensitive tape is mounted on a rigid panel. The adhesive side is exposed. A strip of linerboard, which represents the fiberboard, is placed against the tape adhesive with controlled pressure. The area of contact is specified. The panel is mounted in a horizontal position. A specified mass is fastened to one end of the linerboard. The mass acts to peel the linerboard away from the adhesive at an angle of 90°. The time required to peel the linerboard from the tape is measured.

4.2 *Procedure B*—In this procedure, the fiberboard or the linerboard representing it is adhered to a rigid panel and the tape peeled from it. In all other respects, the procedure is the same as Procedure A.

5. Significance and Use

5.1 This test method is designed to show the adhesion of a pressure-sensitive tape to fiberboard when the tape is under a constant stress. Test data from this test method for some types of packaging tapes relate to tapes performance on fiberboard box closure.

5.2 Procedure A should be used when comparing tapes to an established requirement. The requirement may be established using either the standard linerboard or another linerboard representing a specific fiberboard.

5.3 Procedure B should be used to rank tapes as to their ability to adhere to fiberboard.

6. Apparatus and Materials

6.1 *Tape,* pressure-sensitive, double-coated, at least 48 mm width. The tape shall exhibit sufficient adhesion to the panel and test specimen or fiberboard to avoid the failures cited in 10.1.9 and 10.2.14.

6.2 *Panel*—Any steel piece at least 50 by 125 mm and no less than 1.1 mm thick, having one surface that does not deviate from a plane by more than 0.050 ± 0.002 mm.

6.3 *Stop Watch.*

6.4 *Razor Blade,* single-edge.

6.5 *Mask*—any paper no more than 0.050 mm thick formed to a rectangle approximately 125 by 60 mm with a rectangular hole exactly 24 ± 0.25 by 32 mm centered in the piece. The 24-mm hole dimension shall be in the 125-mm direction (see Figs. 1 and 2).

6.6 *Cutter,* for preparing fiberboard or linerboard and tape specimens.

6.7 *Paperboard Piece,* approximately 25 by 25 mm for reinforcement.

6.8 *Paper Punch,* single-hole.

6.9 *Fiberboard or Linerboard Surface,*[4] against which the test is made.

6.9.1 For Procedure A, cut a rectangle exactly 24 ± 0.25

[1] This test method is under the jurisdiction of ASTM Committee D-10 on Packaging and is the direct responsibility of Subcommittee D10.14 on Closure and Reinforcement.

Current edition approved June 29, 1990. Published August 1990. Originally published as D 2860M – 82. Last previous edition D 2860M – 87.

[2] *Annual Book of ASTM Standards,* Vol 15.09.

[3] *Annual Book of ASTM Standards,* Vol 07.01.

[4] The linerboard accepted as the standard test surface, Standard Reference Material 1810A, is available from the office of Standard Reference Materials, National Institute of Standards and Technology, Washington, DC 20234.

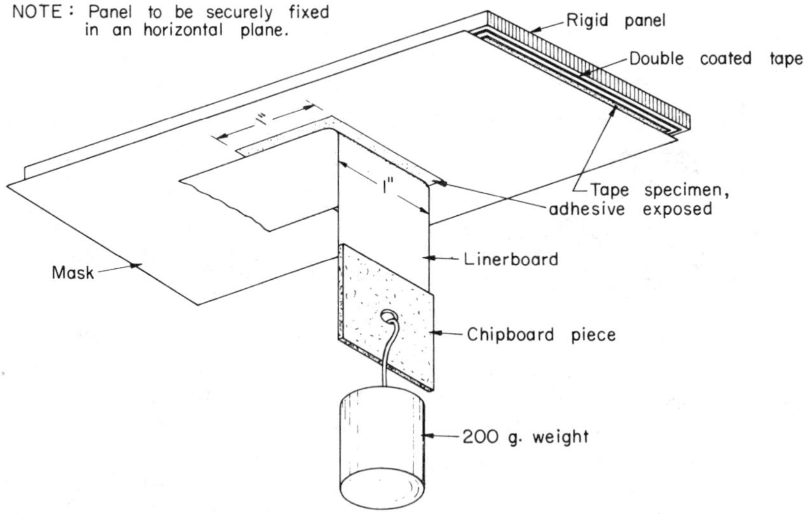

NOTE—Fix panel securely in a horizontal plane.

FIG. 1 Test Assembly for Procedure A

mm wide by at least 125 mm in length, from the linerboard representing the fiberboard on which the tape will be used. The long dimension shall lie at a right angle to the machine direction of the linerboard. Reinforce one end of the strip with a chipboard piece adhered with double-coated tape. Punch a hole at the center of the reinforcement.

6.9.2 For Procedure B, cut a rectangle, preferably 125 by 50 mm, but no less than 50 by 50 mm, from the fiberboard on which the tape will be used (or linerboard representing it). The long test dimension shall lie at a right angle to the machine direction of the linerboard.

6.9.3 For either procedure, prepare the linerboard or fiberboard in advance of testing and mark it to show which side to use. When comparing one roll of tape to another or one lot of tape to another, the linerboard or fiberboard should be from a single lot.

6.10 *Steel Roller,* rubber-covered, 82.5 ± 2.5 mm in diameter and 45 ± 1.5 mm in width, covered with rubber approximately 6 mm in thickness having a Shore A durometer hardness of 80 ± 5. The surface of the rubber shall form a true cylinder devoid of any convex or concave deviations. The mass of the roller shall be 2040 ± 45 g. The mass of the handle shall not add to the mass of the roller during use.

NOTE 2—A simple check to determine if the rubber surface is cylindrical is to wrap the roller in a very thin paper (onionskin) and drag it across a flat glass plate on which is placed a carbon paper, face up. The carbon rubs off onto the thin paper wrapper to reveal high spots or hollows on the rubber surface.

6.11 *Test Stand,* to support the panel securely in a horizontal plane during the test period.

6.12 *Mass,* 200 ± 0.2 g, with a hook.

6.13 *Timing System,* for measuring test time in minutes.

6.14 *Paper,* clean, unprinted, and without a coating of any kind, for use in adhering linerboard or fiberboard to the panel (Procedure B).

7. Sampling

7.1 Sampling shall be in accordance with Practice D 3715.

8. Test Specimens

8.1 A roll or rolls of tape for testing must be at least 36 mm in width.

8.2 Unwind and discard at least three, but no more than six, outer wraps from the sample roll before taking specimens for testing.

8.3 Test five specimens from each roll of tape.

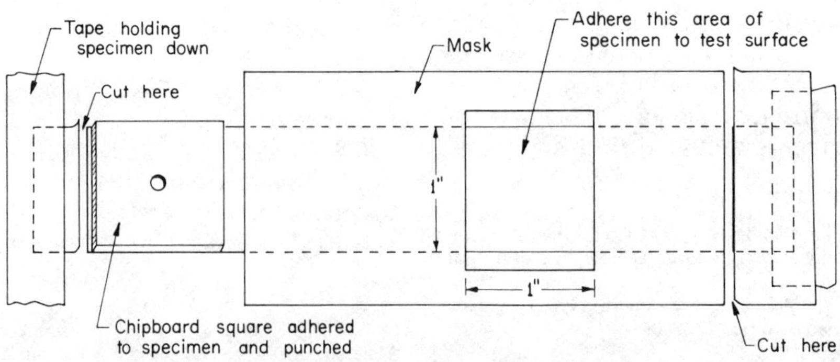

FIG. 2 Preparation of Specimen for Procedure B

9. Conditioning

9.1 Condition sample rolls of tape and linerboard or fiberboard in the standard conditioning atmosphere as described in Practice D 4332 for a period of not less than 24 h. Test at these conditions.

10. Procedure

10.1 *Procedure A:*

10.1.1 Adhere the double-coated tape to the panel with firm finger pressure and remove the liner.

10.1.2 Lightly sand the backing of the test tape on the roll for approximately 250 mm with 300-grit sandpaper to remove the easy unwind treatment. Start timing with the stop watch. Immediately unwind approximately 300 to 450 mm of tape from a freely rotating roll at a rate of approximately 500 to 750 mm/s.

10.1.3 Center the sanded area of this strip, adhesive side up, on the panel, aligning the long edges parallel with the panel long edges. Trim it to the panel dimensions. Do not allow any object to contact the adhesive.

10.1.4 Place a mask at the center of the specimen so that the 24-mm dimension of the hole in the mask lies parallel to the long dimension of the specimen edges (see Fig. 1). Rub against the mask to assure that the tape specimen is well adhered to the double-coated tape on the panel. Avoid touching the adhesive.

10.1.5 At 90 ± 5 s, position the linerboard strip (end opposite the hole) to span the mask and center it lengthwise on the specimen adhesive. Assure that the correct side of the linerboard is placed against the specimen adhesive. Do not apply any finger pressure to the specimen. Continue timing.

10.1.6 Roll the linerboard against the specimen lengthwise once each way at a rate of 5.0 ± 0.2 mm/s using the rubber-covered roller.

10.1.7 Immediately set the panel, specimen side down, into the test stand, assuring that it will remain in a horizontal plane throughout the test.

10.1.8 At 180 ± 5 s (3 min from unwinding the tape from the roll), add the 200-g mass so that no shock load is imparted to the linerboard, and immediately start the time. Allow the test to continue until the linerboard has peeled free from the specimen adhesive.

10.1.9 Void all tests where the tape (test specimen) has pulled away from the panel (either the specimen from the double-coated tape or the double-coated tape from the panel). When this occurs, repeat the test assuring that the panel, the double-coated tape, and the test specimen are well bonded together.

10.2 *Procedure B:*

10.2.1 Adhere the double-coated tape to the panel with firm finger pressure and remove the liner.

10.2.2 Superimpose the fiberboard (or linerboard) on the double-coated tape, exposing the correct side and assuring that the machine direction is at a right angle to the long dimension of the panel. Handle the fiberboard only at the edges and corners so that an area at least 35 mm square at its center is absolutely free from contamination.

10.2.3 Place a clean sheet of paper (6.14) on top of the fiberboard. Rub against this to assure that the fiberboard is well bonded to the double-coated tape on the panel. Do not rub the fiberboard surface with the paper or fingers.

10.2.4 Start timing with the stop watch. Immediately unwind approximately 300 to 450 mm of tape from a freely rotating roll at a rate of approximately 500 to 750 mm/s.

10.2.5 From the center of this strip, cut a specimen 24 ± 0.25 mm wide and at least 180 mm in length. Handle this strip by its ends only. Do not allow adhesive to contact any object.

10.2.6 Lay the specimen, adhesive side up, on the work bench and tape the ends down to hold the specimen flat.

10.2.7 Apply a mask at one end of the specimen so that the 24-mm dimension of the hole lies parallel to the long edges of the specimen (see Fig. 2).

10.2.8 Apply the chipboard square to the exposed adhesive at the other end of the specimen (see Fig. 2).

10.2.9 Cut away the taped ends of the specimen. Punch a hole for the hook of the mass in the chipboard.

10.2.10 At 90 ± 5 s, position the specimen lengthwise on

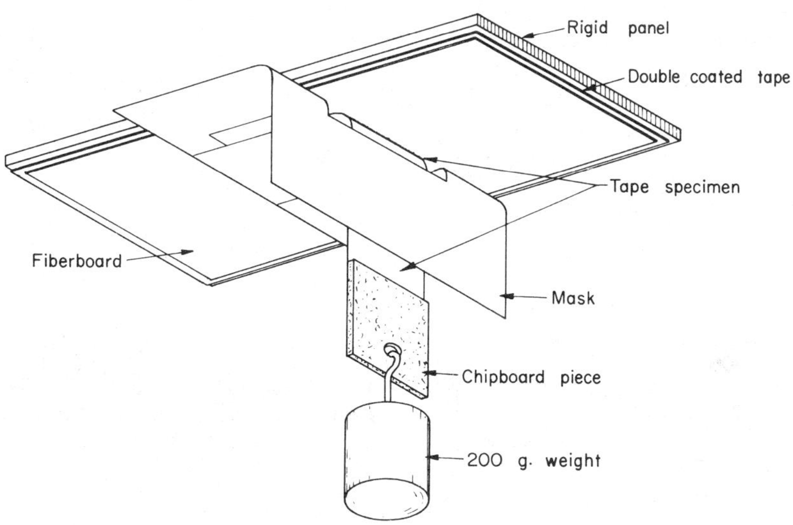

NOTE—Fix panel securely in a horizontal plane.

FIG. 3 Test Assembly for Procedure B

the fiberboard, adhesive down, so that the 24 by 24-mm adhesive area is approximately centered on the fiberboard. Do not apply any finger pressure to the specimen. Continue timing.

10.2.11 Roll the specimen lengthwise once each way at a rate of 5.0 ± 0.2 mm/s using the rubber-covered roller.

10.2.12 Immediately set the panel, specimen side down, onto the test stand, assuring that it will remain in a horizontal plane throughout the test.

10.2.13 At 180 ± 5 s (a total of 3 min from unwinding the tape from the roll), add the 200-g mass so that no shock load is imparted to the specimen and start the timer. Allow the test to continue until the specimen has peeled free from the fiberboard (see Fig. 3).

10.2.14 Void all tests where the fiberboard has pulled away from the panel (either the fiberboard from the double-coated tape or the double-coated tape from the panel). When this occurs, repeat the test, assuring that the panel, double-coated tape, and the fiberboard are well bonded together.

11. Calculation

11.1 To determine the test result for each roll of tape, convert each of the five specimen test results to its common logarithm. Obtain the arithmetic mean of all logarithms and then convert the arithmetic mean logarithm back to time by obtaining its antilogarithm. This final step gives the test result for the roll of tape under consideration. No single specimen value shall be considered as representative of the roll under test.

12. Report

12.1 The report shall include the following:

12.1.1 A statement that this test method was used, indicating which procedure (A or B), and any deviations from the method as written,

12.1.2 Complete identification of each roll of tape tested including type, source, manufacturer's code number, and form,

12.1.3 Linerboard or fiberboard identification including the type, source, manufacturer's code number, and conditioning history prior to test,

12.1.4 Number of specimens tested from each roll of tape,

12.1.5 Anomalous behavior during the test, that is, adhesive transfer or splitting paper fiber pulled free from the linerboard or fiberboard, delamination of the linerboard or fiberboard, and

12.1.6 Mean time value calculated in 11.1 for each of the rolls of tape under test.

13. Precision and Bias

13.1 *Summary*—The difference between two single observations should not exceed 27.7 % of the average of the two observations in transformed units in 95 out of 100 cases when both observations are taken by the same well-trained operator using the same piece of test equipment and specimens randomly drawn from the same sample of material. Larger differences may occur under all other circumstances. The true value of adhesion to fiberboard at 90° angle under constant stress can only be defined in terms of a specific test method. Within this limitation, Test Method D 2860M has no known bias. Sections 13.2 through 13.5 explain the basis for this summary and for evaluations made under other conditions. The information applies either to comparisons of specimens of the same material or to specimens of different materials.

TABLE 1 Critical Difference, Percent of Grand Average for the Conditions Noted[A]

Number of Observations in Each Average	Single-Operator Precision	Within-Laboratory Precision	Between-Laboratory Precision
1	27.7	28.6	30.2
5	14.4	16.1	16.4
10	11.7	13.8	14.1

[A] Before comparing two averages of observed data, transform them using Eq 1.

TABLE 2 Width of 95 % Confidence Limits, Percent of the Grand Average for the Conditions Noted[A]

Number of Observations in Each Average	Single-Operator Precision	Within-Laboratory Precision	Between-Laboratory Precision
1	±19.6	±20.2	±21.4
5	±10.2	±11.4	±11.6
10	±8.3	±9.7	±10.0

[A] Before evaluating the precision of a single average expressed in observed units of measure; transform the average using Eq 1, calculate in transformed units the average plus and minus the tabulated values, and convert the resulting values into observed units of measure using Eq 1. The values of the upper and lower confidence limits in observed units of measure will not be symmetrical about the average.

NOTE 3—This precision statement applies only to Procedure A.

13.2 *Interlaboratory Test Data*[5]—An interlaboratory study was made in 1980 in which randomly drawn samples of two materials were tested in each of four laboratories. Two operators in each laboratory, each tested three specimens from each of three rolls of each material. The observed data did not conform to the assumptions underlying the analysis of variance. Therefore, before analysis, the data were transformed using the following equation:

$$C = \log_{10} D \quad (1)$$

where:
C = transformed datum, and
D = observed datum.

13.2.1 The components of variance for adhesion to fiberboard at 90° angle under constant stress expressed as coefficient of variation (Note 4) were calculated to be:

Single-operator component	3.0 % of the average in transformed units
Within-laboratory component	2.6 % of the average in transformed units
Between-laboratory component	1.1 % of the average in transformed units
Replication component	9.5 % of the average in transformed units

13.3 *Critical Differences*—For the components of variance reported in 13.2.1, two averages of observed values should be considered significantly different at the 95 % probability level if the difference equals or exceeds the differences shown in Table 1.

[5] Supporting data are available from ASTM Headquarters. Request RR:D-10-1002.

NOTE 4—The calculations for coefficient of variation and other statistics found in subsequent sections of this statement are described in Practice D 2906 and Appendix A3 of Practice D 2904.

NOTE 5—The tabulated values of the critical differences and confidence limits should be considered to be a general statement particularly with respect to between-laboratory precision. Before a meaningful statement can be made about two specific laboratories, the amount of the statistical bias between them, if any, must be established with each comparison being based on recent data obtained on specimens randomly drawn from one sample of the material to be evaluated.

13.4 *Confidence Limits*—For the components of variance reported in 12.2, the single averages of observed values have 95 % confidence limits (Note 5) shown in Table 2.

13.5 *Bias*—No justifiable statement can be made on the bias of Test Method D 2860M for testing adhesion to fiberboard at 90° angle and constant stress since the true value cannot be established by an accepted referee method.

14. Keywords

14.1 adhesion at 90° angle and constant stress; pressure sensitive tape

Standard Specification for Dimensions and Tolerances for Plastic Bottles[1]

This standard is issued under the fixed designation D 2911; the number immediately following the designation indicates the year of original adoption or, in the case of revision, the year of last revision. A number in parentheses indicates the year of last reapproval. A superscript epsilon (ε) indicates an editorial change since the last revision or reapproval.

$^{\epsilon 1}$ NOTE—Section 1.3 was added editorially October 1988.

1. Scope

1.1 This specification covers the thread configuration and dimensions for finishes for plastic bottles with screw-type closures, having a maximum capacity of 18.9 L (5 gal) and a maximum bottle dimension of 305 mm (12 in.). Included are tolerances for bottle capacity and body dimensions.

1.2 The values stated in SI units are to be regarded as the standard.

1.3 The following precautionary caveat pertains only to the test method portion, section 8 of this specification. *This standard may involve hazardous materials, operations, and equipment. This standard does not purport to address all of the safety problems associated with its use. It is the responsibility of the user of this standard to establish appropriate safety and health practices and determine the applicability of regulatory limitations prior to use.*

2. Referenced Documents

2.1 *ASTM Standards:*

D 618 Methods of Conditioning Plastics and Electrical Insulating Materials for Testing[2]

D 1898 Practice for Sampling of Plastics[3]

3. Terminology

3.1 *Definitions:*

3.1.1 *bottle finish*—the configuration of the neck or opening of the bottle which serves to engage specific parts of the closure in order to securely attach it to the bottle.

3.1.2 *bottle height*—the maximum dimension of a bottle in a plane perpendicular to the bottle base when the bottle is setting in its normal upright position.

3.1.3 *bottle thickness*—the smaller dimension in a plane parallel to the bottle base.

3.1.4 *bottle width*—the larger dimension in a plane parallel to the bottle base.

4. Dimensions and Recommended Variations

4.1 *Recommended Variations in Overflow Capacity*—In overflow capacity recommended variations of plastic bottles up to 18.9 L (5 gal) in size shall be in accordance with Table 1.

4.2 *Recommended Variations in Body Dimensions*—The recommended variations in the body dimensions of plastic bottles covered by this specification shall be in accordance with Table 2.

4.3 *Finish Dimensions and Design*—The bottle finish dimensions and design for the nine finishes covered by this specification shall be in accordance with Figs. 1 through 8 and Tables 3 through 10.

5. Sampling

5.1 A batch or lot shall be considered as a unit of manufacture.

5.2 Unless otherwise agreed upon between manufacturer and purchaser, the material shall be sampled in accordance with the procedure described in Section 9 of Practice D 1898. Adequate statistical sampling prior to packaging shall be considered an acceptable alternative.

6. Conditioning

6.1 *Conditioning*—Condition the test specimens at 23 ± 2°C (73.4 ± 3.6°F) and 50 ± 5% relative humidity for not less than 40 h prior to test in accordance with Procedure A of Methods D 618 for those tests where conditioning is required. In cases of disagreement, the tolerances shall be ±1°C (±1.8°F) and ±2 % relative humidity.

6.2 *Test Conditions*—Conduct tests in the Standard Laboratory Atmosphere of 23 ± 2°C (73.4 ± 3.6°F) and 50 ± 5 % relative humidity, unless otherwise specified in the test methods or in this specification. In cases of disagreement, the tolerances shall be ±1°C (±1.8°F) and ±2 % relative humidity.

7. Apparatus

7.1 *For Determining Bottle Capacity:*

7.1.1 *Balance*, having an accuracy of ±0.1 % or better at rated capacity (full scale).

7.1.2 *Pipet, Graduated Cylinder, and Beaker*, as required.

7.1.3 Conditioned water at 23 ± 2.0°C (73.4 ± 3.6°F) containing a wetting agent sufficient to flatten the meniscus and eliminate air bubbles.

NOTE 1—Any liquid detergent may be used as the wetting agent.

7.1.4 *Stop Watch.*

7.2 *For Determining Finish and Body Dimensions:*

7.2.1 *Micrometers, Vernier Height Gages, or Vernier Calipers* with an accuracy of ±0.025 mm (±0.001 in.).

7.2.2 *Internal Micrometers or Telescoping Gages* with an accuracy of ±0.025 mm (±0.001 in.).

---

[1] This specification is under the jurisdiction of Committee D-20 on Plastics and is the direct responsibility of Subcommittee D 20.12 on Olefin Plastics.

Current edition approved Oct. 29, 1982. Published July 1983. Originally published as D 2911 – 70. Last previous edition D 2911 – 72 (1978).

[2] *Annual Book of ASTM Standards*, Vol 08.01.

[3] *Annual Book of ASTM Standards*, Vol 08.02.

TABLE 1 Bottle Overflow Capacity Tolerances for Plastic Bottles

Fluid Ounces		Millilitres	
Bottle Overflow Capacity	Tolerance (±)	Bottle Overflow Capacity	Tolerance (±)
less than 0.75	0.05	less than 22	1.5
0.75 and less than 1.2	0.07	22 and less than 35	2.0
1.2 and less than 1.6	0.08	35 and less than 47	2.5
1.6 and less than 2.1	0.10	47 and less than 62	3.0
2.1 and less than 2.8	0.12	62 and less than 83	3.5
2.8 and less than 3.9	0.14	83 and less than 115	4
3.9 and less than 5.4	0.17	115 and less than 159	5
5.4 and less than 7.4	0.20	159 and less than 218	6
7.4 and less than 9.8	0.24	218 and less than 289	7
9.8 and less than 13	0.30	289 and less than 384	9
13 and less than 18	0.37	384 and less than 531	11
18 and less than 26	0.44	531 and less than 767	13
26 and less than 37	0.51	767 and less than 1092	15
37 and less than 51	0.68	1092 and less than 1505	20
51 and less than 72	0.81	1505 and less than 2125	24
72 and less than 98	1.01	2125 and less than 2892	30
98 and less than 119	1.30	2892 and less than 3512	38
119 and less than 139	1.50	3512 and less than 4103	44
139 and less than 160	1.80	4103 and less than 4723	53
160 and less than 180	2.00	4723 and less than 5313	59
180 and less than 210	2.20	5313 and less than 6199	65
210 to 5 gallons	1 % of capacity	6199 to 18.89 litres	1 % of capacity

TABLE 2 Body Dimensional Tolerance

Range of Specific Dimensions		Height Dimensions		Width and Depth Dimensions	
in.	mm	in.	mm	in.	mm
0 up to but not including 1	0 to 25.40	0.030	0.76	0.030	0.76
1 up to but not including 2	25.40 to 50.80	0.030	0.76	0.050	1.27
2 up to but not including 4	50.80 to 101.60	0.050	1.27	0.060	1.52
4 up to but not including 6	101.60 to 152.40	0.050	1.27	0.080	2.03
6 up to but not including 8	152.40 to 203.20	0.060	1.52	0.090	2.29
8 up to but not including 10	203.20 to 254.00	0.060	1.52	0.110	2.79
10 up to but not including 12	254.00 to 304.80	0.080	2.03	0.120	3.05
12 up to but not including 15	304.80 to 381.00	0.090	2.29	0.150	3.81
15 up to but not including 18	381.00 to 457.20	0.110	2.79	0.150	3.81

7.2.3 *Commercial Scale*, good quality calibrated in 1-mm or 1/32-in. increments.

8. Test Methods

8.1 *Bottle Capacity:*

8.1.1 Weigh the empty bottle and record the weight in grams.

8.1.2 Fill the bottle to overflow capacity with conditioned water containing the wetting agent, adjusting the meniscus by use of a pipet, if necessary, until the meniscus is tangent to the top of the finish surface. No more than 2 min shall be allowed for filling the bottle and no additional water shall be added after the bottle is filled to overflow and free of air bubbles.

8.1.3 If the time exceeds 2 min, discard the sample and start over.

NOTE 2—The capacity of the container is somewhat time-dependent in that the force of the water in the container will cause the container sides to bulge with time and additional water will be needed to adjust to overflow capacity.

8.1.4 Weigh the filled bottle and record the weight in grams.

8.1.5 Calculate the bottle volume as follows:

$$B_v \text{ (mL)} = (B_f - B_e)/0.997$$

where:

B_v = volume of bottle, mL,
B_f = weight of filled bottle, g,
B_e = weight of empty bottle, g, and
0.997 = weight of water (g/mL) at 23 ± 2.0°C (73.4 ± 3.6°F).

$$B_v \text{ (fluid oz)} = (B_f - B_e)/29.50$$

where 29.50 = weight of water (g/fluid oz) at 23 ± 2.0°C (73.4 ± 3.6°F).

8.1.6 Determine the volume in millilitres (fluid ounces) for each test specimen.

8.1.7 Report the arithmetic average volume of the specimens tested as the bottle capacity of the batch or lot from which the samples were chosen.

8.2 *Body Dimensions:*

8.2.1 Using a suitable micrometer, a vernier height gage, or a scale, measure the height of the bottle.

NOTE 3—If the top of the finish is parallel to the base of the container and the height must be measured from the top of the finish, and a commercial scale is used, one method is to place the bottle on a flat surface and a straightedge across the top of the bottle finish. Then measure the perpendicular distance between the straightedge and the surface upon which the bottle is setting. Four measurements should be made, each 90° apart and the maximum used as the bottle height.

8.2.2 Using a suitable micrometer or vernier caliper, measure the width and thickness of the bottle.

8.2.2.1 In the case of a rectangular container, use the midpoints of the sides as the measuring points.

TABLE 3 SP-400 Finish for Plastic Bottles

NOTE—Top dimension in each column shown in inches. Bottom dimension in each column shown in millimetres.

mm	T		E		H		S		I	Helix Angle, β	Cutter Diameter	Threads per Inch
	max	min	max	min	max	min	max	min	min			
18	0.704	0.688	0.620	0.604	0.386	0.356	0.052	0.022	0.325	3°30′	0.375	8
	17.88	17.47	15.75	15.34	9.80	9.04	1.32	0.56	8.25		9.52	
20	0.783	0.767	0.699	0.683	0.386	0.356	0.052	0.022	0.404	3°7′	0.375	8
	19.89	19.48	17.75	17.35	9.80	9.04	1.32	0.56	10.26		9.52	
22	0.862	0.846	0.778	0.762	0.386	0.356	0.052	0.022	0.483	2°49′	0.375	8
	21.89	21.49	19.76	19.35	9.80	9.04	1.32	0.56	12.27		9.52	
24	0.940	0.924	0.856	0.840	0.415	0.385	0.061	0.031	0.516	2°34′	0.375	8
	23.88	23.47	21.74	21.34	10.54	9.78	1.55	0.79	13.11		12.70	
28	1.088	1.068	0.994	0.974	0.415	0.385	0.061	0.031	0.614	2°57′	0.500	6
	27.63	27.13	25.25	24.74	10.54	9.78	1.55	0.79	15.59		12.70	
30	1.127	1.107	1.033	1.013	0.418	0.388	0.061	0.031	0.653	2°51′	0.500	6
	28.62	28.12	26.24	25.73	10.62	9.85	1.55	0.79	16.59		12.70	
33	1.265	1.241	1.171	1.147	0.418	0.388	0.061	0.031	0.791	2°31′	0.500	6
	32.13	31.52	29.74	29.13	10.62	9.85	1.55	0.79	20.09		12.70	
35	1.364	1.340	1.270	1.246	0.418	0.388	0.061	0.031	0.875	2°21′	0.500	6
	34.64	34.04	32.26	31.65	10.62	9.85	1.55	0.79	22.22		12.70	
38	1.476	1.452	1.382	1.358	0.418	0.388	0.061	0.031	0.987	2°9′	0.500	6
	37.49	36.88	35.10	34.49	10.62	9.85	1.55	0.79	25.07		12.70	
40	1.580	1.550	1.486	1.456	0.418	0.388	0.061	0.031	1.091	2°0′	0.500	6
	40.13	39.37	37.74	36.98	10.62	9.85	1.55	0.79	27.71		12.70	
43	1.654	1.624	1.560	1.530	0.418	0.388	0.061	0.031	1.165	1°55′	0.500	6
	42.01	41.25	39.62	38.86	10.62	9.85	1.55	0.79	29.59		12.70	
45	1.740	1.710	1.646	1.616	0.418	0.388	0.061	0.031	1.251	1°49′	0.500	6
	44.20	43.43	41.81	41.05	10.62	9.85	1.55	0.79	31.77		12.70	
48	1.870	1.840	1.776	1.746	0.418	0.388	0.061	0.031	1.381	1°41′	0.500	6
	47.50	46.74	45.11	44.35	10.62	9.85	1.55	0.79	35.08		12.70	
51	1.968	1.933	1.874	1.839	0.423	0.393	0.061	0.031	1.479	1°36′	0.500	6
	49.99	49.10	47.60	46.71	10.74	9.98	1.55	0.79	37.57		12.70	
53	2.067	2.032	1.973	1.938	0.423	0.393	0.061	0.031	1.578	1°31′	0.500	6
	52.50	51.61	50.11	49.22	10.74	9.98	1.55	0.79	40.08		12.70	
58	2.224	2.189	2.130	2.095	0.423	0.393	0.061	0.031	1.735	1°25′	0.500	6
	56.49	55.60	54.10	53.21	10.74	9.98	1.55	0.79	44.07		12.70	
60	2.342	2.307	2.248	2.213	0.423	0.393	0.061	0.031	1.853	1°20′	0.500	6
	59.49	58.60	57.10	56.21	10.74	9.98	1.55	0.79	47.07		12.70	
63	2.461	2.426	2.367	2.332	0.423	0.393	0.061	0.031	1.972	1°16′	0.500	6
	62.51	61.62	60.12	59.23	10.74	9.98	1.55	0.79	50.09		12.70	
66	2.579	2.544	2.485	2.450	0.423	0.393	0.061	0.031	2.090	1°13′	0.500	6
	65.51	64.62	63.12	62.23	10.74	9.98	1.55	0.79	53.09		12.70	
70	2.736	2.701	2.642	2.607	0.423	0.393	0.061	0.031	2.247	1°8′	0.500	6
	69.49	68.60	67.11	66.22	10.74	9.98	1.55	0.79	57.07		12.70	
75	2.913	2.878	2.819	2.784	0.423	0.393	0.061	0.031	2.424	1°4′	0.500	6
	73.99	73.10	71.60	70.71	10.74	9.98	1.55	0.79	61.57		12.70	
77	3.035	3.000	2.941	2.906	0.502	0.472	0.075	0.045	2.546	1°1′	0.500	6
	77.09	76.20	74.70	73.81	12.75	11.99	1.90	1.14	64.67		12.70	
83	3.268	3.233	3.148	3.113	0.502	0.472	0.075	0.045	2.753	1°9′	0.500	5
	83.01	82.12	79.96	79.07	12.75	11.99	1.90	1.14	69.93		12.70	
89	3.511	3.476	3.391	3.356	0.550	0.520	0.075	0.045	2.918	1°4′	0.500	5
	89.18	88.29	86.13	85.24	13.79	13.21	1.90	1.14	74.12		12.70	
100	3.937	3.902	3.817	3.782	0.612	0.582	0.075	0.045	3.344	0°57′	0.500	5
	100.00	99.11	96.95	96.06	15.54	14.78	1.90	1.14	84.94		12.70	
110	4.331	4.296	4.211	4.176	0.612	0.582	0.075	0.045	3.737	0°51′	0.500	5
	110.01	109.12	106.96	106.07	15.54	14.78	1.90	1.14	94.92		12.70	
120	4.724	4.689	4.604	4.569	0.700	0.670	0.075	0.045	4.131	0°47′	0.500	5
	119.99	119.10	116.94	116.05	17.78	17.02	1.90	1.14	104.93		12.70	

8.2.2.2 For a cylindrical container, make one measurement on the parting line and another measurement 90° from the parting line. Use the average of these two measurements as the container width.

NOTE 4—If, because of the shape of the container, there is some question about where the measurements should be taken, the agreement should be reached between the purchaser and the seller.

8.3 *Finish Dimensions*—Using suitable micrometers, vernier calipers, or telescoping gages, measure the finish dimensions of the bottle.

9. Retest and Rejection

9.1 If the results of any test do not conform to the requirements of this specification, retesting to determine conformity may be performed as agreed upon between the purchaser and the seller.

TABLE 4 SP-410 Finish for Plastic Bottles

NOTE—Top dimension in each column shown in inches. Bottom dimension in each column shown in millimetres.

mm	T[A]		E[A, B]		H[C]		L[D]	S		I[E,F]	W	Helix Angle, β	Cutter Diameter	Threads[G] per Inch
	max	min	max	min	max	min	min	max	min	min	max			
18	0.704	0.688	0.620	0.604	0.538	0.508	0.361	0.052	0.022	0.325	0.084	3°30′	0.375	8
	17.88	17.47	15.75	15.34	13.66	12.90	9.17	1.32	0.56	8.25	2.13		9.52	
20	0.783	0.767	0.699	0.683	0.569	0.539	0.361	0.052	0.022	0.404	0.084	3°7′	0.375	8
	19.89	19.48	17.75	17.35	14.45	13.69	9.17	1.32	0.56	10.26	2.13		9.52	
22	0.862	0.846	0.778	0.762	0.600	0.570	0.376	0.052	0.022	0.483	0.084	2°49′	0.375	8
	21.89	21.49	19.76	19.35	15.24	14.48	9.55	1.32	0.56	12.27	2.13		9.52	
24	0.940	0.924	0.856	0.840	0.661	0.631	0.437	0.061	0.031	0.516	0.084	2°34′	0.375	8
	23.88	23.47	21.74	21.34	16.79	16.03	11.10	1.55	0.79	13.11	2.13		9.52	
28	1.088	1.068	0.994	0.974	0.723	0.693	0.463	0.061	0.031	0.614	0.094	2°57′	0.500	6
	27.63	27.13	25.25	24.74	18.36	17.60	11.76	1.55	0.79	15.59	2.39		12.70	

[A] *T* and *E* dimensions are the average of two measurements across the major and minor axis. The limits of ovality will be determined by the container supplier and container customer, as necessary.

[B] Consideration must be given to the sealing surface width for the sealing system being used.

[C] Dimension *H* is measured from the top of the finish to the point where diameter *T*, extended parallel to the centerline, intersects the shoulder.

[D] Contour of bead, undercut, or shoulder is optional. If bead is used, bead diameter and *L* min must be maintained.

[E] Unless otherwise specified, *I* min applies to the full length of the opening.

[F] Concentricity of *I* min with respect to diameters *T* and *E* is not included. *I* min is specified for filler tube only.

[G] A minimum of 1½ full turns of thread shall be maintained.

TABLE 5 SP-415 Finish for Plastic Bottles

NOTE—Top dimension in each column shown in inches. Bottom dimension in each column shown in millimetres.

mm	T[A]		E[A, B]		H[C]		L[D]	S		I[E,F]	W	Helix Angle, β	Cutter Diameter	Threads[G] per Inch
	max	min	max	min	max	min	min	max	min	min	max			
13	0.514	0.502	0.454	0.442	0.467	0.437	0.306	0.052	0.022	0.218	0.045	3°11′	0.375	12
	13.06	12.75	11.53	11.23	11.86	11.10	7.77	1.32	0.56	5.54	1.14		9.52	
15	0.581	0.569	0.521	0.509	0.572	0.542	0.348	0.052	0.022	0.258	0.045	2°48′	0.375	12
	14.76	14.45	13.23	12.93	14.53	13.77	8.84	1.32	0.56	6.55	1.14		9.52	
18	0.704	0.688	0.620	0.604	0.632	0.602	0.429	0.052	0.022	0.325	0.084	3°30′	0.375	8
	17.88	17.47	15.75	15.34	16.05	15.29	10.90	1.32	0.56	8.25	2.13		9.52	
20	0.783	0.767	0.699	0.683	0.757	0.727	0.456	0.052	0.022	0.404	0.084	3°7′	0.375	8
	19.89	19.48	17.75	17.35	19.23	18.47	11.58	1.32	0.56	10.26	2.13		9.52	
22	0.862	0.846	0.778	0.762	0.852	0.822	0.546	0.052	0.022	0.483	0.084	2°49′	0.375	8
	21.89	21.49	19.76	19.35	21.64	20.88	13.87	1.32	0.56	12.27	2.13		9.52	
24	0.940	0.924	0.856	0.840	0.972	0.942	0.561	0.061	0.031	0.516	0.084	2°34′	0.375	8
	23.88	23.47	21.74	21.34	24.69	23.93	14.25	1.55	0.79	13.11	2.13		9.52	
28	1.088	1.068	0.994	0.974	1.097	1.067	0.655	0.061	0.031	0.614	0.094	2°57′	0.500	6
	27.63	27.13	25.25	24.74	27.86	27.10	16.64	1.55	0.79	15.59	2.39		12.70	
33	1.265	1.241	1.171	1.147	1.289	1.259	0.772	0.061	0.031	0.791	0.094	2°31′	0.500	6
	32.13	31.52	29.74	29.13	32.74	31.98	19.61	1.55	0.79	20.09	2.39		12.70	

[A] *T* and *E* dimensions are the average of two measurements across the major and minor axis. The limits of ovality will be determined by the container supplier and container customer, as necessary.

[B] Consideration must be given to the sealing surface width for the sealing system being used.

[C] Dimension *H* is measured from the top of the finish to the point where diameter *T*, extended parallel to the centerline, intersects the shoulder.

[D] Contour of bead, undercut, or shoulder is optional. If bead is used, bead diameter and *L* min must be maintained.

[E] Unless otherwise specified, *I* min applies to the full length of the opening.

[F] Concentricity of *I* min with respect to diameters *T* and *E* is not included. *I* min is specified for filler tube only.

[G] A minimum of 2 full turns of thread shall be maintained.

TABLE 6 SP-425 Finish for Plastic Bottles

NOTE—Top dimension in each column shown in inches. Bottom dimension in each column shown in millimetres.

mm	T^A		$E^{A,B}$		H^C		S^D		$I^{E,F}$	Helix Angle β	Cutter Diameter	Threads[G] per Inch
	max	min	max	min	max	min	max	min	min			
13	0.514	0.502	0.454	0.442	0.325	0.295	0.052	0.022	0.218	3°11′	0.375	12
	13.06	12.75	11.53	11.23	8.25	7.49	1.32	0.56	5.54		9.52	
15	0.581	0.569	0.521	0.509	0.325	0.295	0.052	0.022	0.258	2°48′	0.375	12
	14.76	14.25	13.23	12.93	8.25	7.49	1.32	0.56	6.55		9.52	

[A] *T* and *E* dimensions are the average of two measurements across the major and minor axis. The limits of ovality will be determined by the container supplier and container customer, as necessary.

[B] Consideration must be given to the sealing surface width for the sealing system being used.

[C] Dimension *H* is measured from the top of the finish to the point where diameter *T*, extended parallel to the centerline, intersects the shoulder.

[D] Contour of bead, undercut, or shoulder is optional.

[E] Unless otherwise specified, *I* min applies to the full length of the opening.

[F] Concentricity of *I* min with respect to diameters *T* and *E* is not included in this standard. *I* min is specified for filler tube only.

[G] A minimum of 2 full turns of thread shall be maintained on the bead finish. A minimum of 1½ full turns of thread shall be maintained on the beadless finish.

TABLE 7 SP-100 Finish for Plastic Bottles

NOTE—Top dimension in each column shown in inches. Bottom dimension in each column shown in millimetres.

mm	T^A		$E^{A,B}$		H^C		LIP Outside Diameter[D]		$I^{E,F,G}$	Helix Angle β	Cutter Diameter	Threads[H] per Inch
	max	min	max	min	max	min	max	min	min			
22	0.862	0.846	0.758	0.742	0.559	0.529	0.748	0.732	05.12	2°49′	0.375	8
	21.89	21.49	19.25	18.85	14.20	13.44	19.00	18.59	13.00		9.52	
24	0.940	0.924	0.836	0.820	0.559	0.829	0.826	0.810	0.590	2°34′	0.375	8
	23.88	23.47	21.23	20.83	14.20	13.44	20.98	20.57	14.99		9.52	
26	1.009	0.989	0.905	0.885	0.559	0.529	0.895	0.875	0.655	2°24′	0.375	8
	25.63	25.12	22.99	22.48	14.20	13.44	22.73	22.22	16.64		9.52	
28	1.088	1.068	0.984	0.964	0.559	0.529	0.974	0.954	0.735	2°13′	0.375	8
	27.63	27.13	24.99	24.49	14.20	13.44	24.74	24.23	18.67		9.52	
38	1.476	1.452	1.372	1.348	0.604	0.574	1.362	1.338	1.098	1°37′	0.375	8
	37.49	36.88	34.85	34.24	15.34	14.58	34.59	33.98	20.89		9.52	

[A] *T* and *E* dimensions are the average of two measurements across the major and minor axis. The limits of ovality will be determined by the container supplier and container customer as necessary.

[B] Consideration must be given to the sealing surface width for the sealing system being used.

[C] Dimension *H* is measured from the top of the finish to the point where diameter *T*, extended parallel to the centerline, intersects the shoulder.

[D] Contour of undercut or shoulder is optional.

[E] Unless otherwise specified, *I* min applies to the full length of the opening.

[F] Concentricity of *I* min with respect to diameters *T* and *E* is not included. *I* min is specified for filler tube only.

[G] When valve style closures are used with this finish, special consideration must be given to a specific controlled inside diameter. In addition, dimensions indicated with an asterisk (*) on Fig. 5 may be varied to ensure adequate material for finishing the inside diameter.

[H] A minimum of 1⅛ turns of thread shall be maintained.

TABLE 8 SP-103 Finish for Plastic Bottles

NOTE—Top dimension in each column shown in inches. Bottom dimension in each column shown in millimetres.

mm	T^A		$E^{A,B}$		H^C		LIP Outside Diameter[D]		$I^{E,F,G}$	Helix Angle β	Cutter Diameter	Threads[H] per Inch
	max	min	max	min	max	min	max	min	min			
26	1.009	0.989	0.905	0.885	0.638	0.608	0.895	0.875	0.670	2°24′	0.375	8
	25.63	25.12	22.99	22.48	16.20	15.44	22.73	22.22	17.62		9.52	

[A] *T* and *E* dimensions are the average of two measurements across the major and minor axis. The limits of ovality will be determined by the container supplier and container customer as necessary.

[B] Consideration must be given to the sealing surface width for the sealing system being used.

[C] Dimension *H* is measured from the top of the finish to the point where dimension *T*, extended parallel to the centerline, intersects the shoulder.

[D] Contour of undercut or shoulder is optional.

[E] Hole diameter *I* to be measured through full length of finish unless otherwise specified.

[F] Concentricity of *I* min with respect to diameters *T* and *E* is not included. *I* min is specified for filler tube only.

[G] When valve style closures are used with this finish, special consideration must be given to a specific controlled inside diameter. In addition, dimensions indicated with asterisk (*) on Fig. 5 may be varied to ensure adequate material for finishing the inside diameter.

[H] A minimum of 1⅛ turns of thread shall be maintained.

TABLE 9 SP-200 Finish

NOTE—Top dimension in each column shown in inches. Bottom dimension in each column shown in millimetres.

mm	T^A		$E^{A,B}$		H^C		LIP Outside Diameter[D]		$I^{E,F,G}$	Helix Angle β	Cutter Diameter	Threads[H] per Inch
	max	min	max	min	max	min	max	min	min			
24	0.940 23.88	0.924 23.47	0.832 21.13	0.816 20.73	0.809 20.55	0.779 19.79	0.822 20.88	0.806 20.47	0.540 13.72	3°27′	0.375 9.52	6
28	1.118 28.40	1.098 27.89	1.010 25.65	0.990 25.15	0.809 20.55	0.779 19.79	1.000 25.40	0.980 24.89	0.710 18.03	2°53′	0.375 9.52	6

Alternate *H* dimensions are recognized and will be designated as SP200A:

Alt. *H* 24 mm, 0.789/0.759 20.04/19.28

Alt. *H* 28 mm, 0.823/0.793 20.90/20.14

[A] *T* and *E* dimensions are the average of two measurements across the major and minor axis. The limits of ovality will be determined by the container supplier and container customer, as necessary.

[B] Consideration must be given to the sealing surface width for the sealing system being used.

[C] Dimension *H* is measured from the top of the finish to the point where diameter *T*, extended parallel to the centerline, intersects the shoulder.

[D] Contour of undercut or shoulder is optional.

[E] Unless otherwise specified, *I* min applies to the full length of the opening.

[F] Concentricity of *I* min with respect to diameters *T* and *E* is not included. *I* min is specified for filler tube only.

[G] When valve style closures are used with this finish, special consideration must be given to a specific controlled inside diameter. In addition, dimensions indicated with asterisk (*) on Fig. 6 may be varied to ensure adequate material for finishing the inside diameter.

[H] A minimum of 1½ turns of thread shall be maintained.

TABLE 10 SP-444 Finish for Plastic Bottles

Finish Identity	T^A				E^A				H^B				S				$I^{C,D}$		Helix Angle β	Cutter Diameter		Threads per Inch	Pitch		LIP Outside Diameter			
	max in.	min in.	max mm	min mm	max in.	min in.	max mm	min mm	max in.	min in.	max mm	min mm	max in.	min in.	max mm	min mm	min in.	min mm		in.	mm		in.	mm	max in.	min in.	max mm	min mm
24	0.940	0.924	23.88	23.47	0.846	0.830	21.49	21.08	0.587	0.557	14.91	14.15	0.190	0.160	4.83	4.06	0.516	13.11	3°25′	0.375	9.52	6	0.167	4.24	0.836	0.820	21.23	20.83
28	1.088	1.068	27.63	27.13	0.994	0.974	25.25	24.74	0.602	0.572	15.29	14.53	0.190	0.160	4.83	4.06	0.614	15.59	2°57′	0.500	12.70	6	0.167	4.24	0.984	0.964	24.99	24.48
33	1.265	1.241	32.13	31.52	1.171	1.147	29.74	29.13	0.620	0.590	15.75	14.99	0.190	0.160	4.83	4.06	0.791	20.09	2°31′	0.500	12.70	6	0.167	4.24	1.161	1.137	29.49	28.88
38	1.476	1.452	37.49	36.88	1.382	1.358	35.10	34.49	0.671	0.641	17.04	16.28	0.190	0.160	4.83	4.06	0.987	25.07	2°9′	0.500	12.70	6	0.167	4.24	1.372	1.348	34.85	34.24
43	1.654	1.624	42.01	41.25	1.560	1.530	39.62	38.86	0.678	0.648	17.22	16.46	0.190	0.160	4.83	4.06	1.165	29.59	1°55′	0.500	12.70	6	0.167	4.24	1.550	1.520	39.37	38.61
45	1.740	1.710	44.20	43.43	1.646	1.616	41.81	41.05	0.678	0.648	17.22	16.46	0.190	0.160	4.83	4.06	1.251	31.77	1°49′	0.500	12.70	6	0.167	4.24	1.636	1.606	41.55	40.79
48	1.870	1.840	47.50	46.74	1.776	1.746	45.11	44.35	0.678	0.648	17.22	16.46	0.190	0.160	4.83	4.06	1.381	35.08	1°41′	0.500	12.70	6	0.167	4.24	1.766	1.736	44.86	44.09
53	2.067	2.032	52.50	51.61	1.973	1.938	50.11	49.22	0.678	0.648	17.22	16.46	0.190	0.160	4.83	4.06	1.578	40.08	1°31′	0.500	12.70	6	0.167	4.24	1.963	1.928	49.86	48.97
56	2.145	2.110	54.48	53.59	2.051	2.016	52.10	51.21	0.678	0.648	17.22	16.46	0.190	0.160	4.83	4.06	1.656	42.06	1°28′	0.500	12.70	6	0.167	4.24	2.041	2.006	51.84	50.95
58	2.224	2.189	56.49	55.60	2.130	2.095	54.10	53.21	0.678	0.648	17.22	16.46	0.190	0.160	4.83	4.06	1.735	44.07	1°25′	0.500	12.70	6	0.167	4.24	2.120	2.085	53.85	52.96
63	2.461	2.426	62.51	61.62	2.367	2.332	60.12	59.23	0.678	0.648	17.22	16.46	0.190	0.160	4.83	4.06	1.972	50.09	1°16′	0.500	12.70	6	0.167	4.24	2.357	2.322	59.87	58.98
70	2.770	2.735	70.36	69.47	2.645	2.610	67.18	66.29	0.678	0.648	17.22	16.46	0.109	0.079	2.77	2.01	2.247	57.07	1°42′	0.500	12.70	4	0.250	6.35	2.635	2.600	66.93	66.04
73	2.864	2.829	72.75	71.86	2.770	2.735	70.36	69.47	0.678	0.648	17.22	16.46	0.190	0.160	4.83	4.06	2.375	60.32	1°5′	0.500	12.70	6	0.167	4.24	2.760	2.725	70.10	69.21
75	2.913	2.878	73.99	73.10	2.819	2.784	71.60	70.71	0.678	0.648	17.22	16.46	0.190	0.160	4.83	4.06	2.424	61.57	1°4′	0.500	12.70	6	0.167	4.24	2.809	2.774	71.35	70.46
83	3.268	3.233	83.01	82.12	3.148	3.113	79.96	79.07	0.678	0.648	17.22	16.46	0.155	0.125	3.94	3.18	2.753	69.93	1°9′	0.500	12.70	5	0.200	5.08	3.138	3.103	79.70	78.82

[A] *T* and *E* dimensions are the average of two measurements across the major and minor axis. The limits of ovality will be determined by the container supplier and container customer, as necessary.

[B] Dimension *H* is measured from top of the finish to the point where diameter, *I*, extended parallel to the centerline, intersects the shoulder or bead.

[C] Unless otherwise specified, *I* min applies to the full length of the opening.

[D] Concentricity of *I* min with respect to diameters *T* and *E* is not included. *I* min is specified for filler tube only.

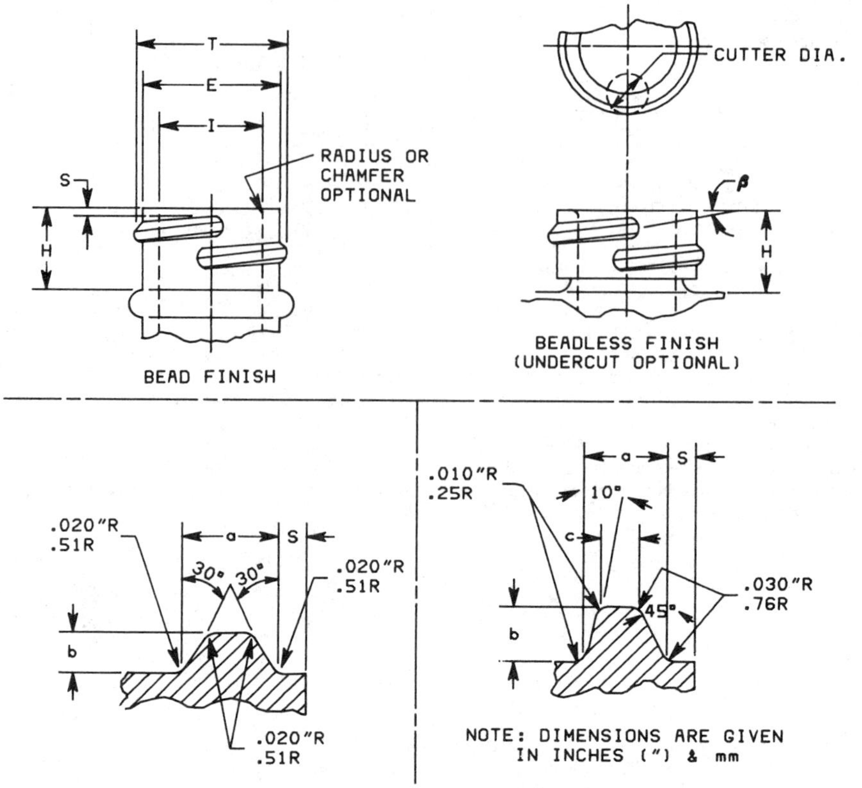

"L" Style				"M" Style			
All-Purpose Thread (Plastic or Metal Closures)				Modified Buttress Thread (Plastic Closures)			
Threads/in.	*a*	*b*	*c*	Threads/in.	*a*	*b*	*c*
5	in. 0.120	0.060	0.051	5	in. 0.120	0.060	0.049
	mm 3.05	1.52	1.30		mm 3.05	1.52	1.24
6	in. 0.094	0.047	0.040	6	in. 0.094	0.047	0.039
	mm 2.39	1.19	1.02		mm 2.39	1.19	0.99
8	in. 0.084	0.042	0.036	8	in. 0.084	0.042	0.035
	mm 2.13	1.07	0.91		mm 2.13	1.07	0.89

Example Thread Nomenclature
"L" Style: L28SP400
"M" Style: M28SP400

NOTE 1—*T* and *E* dimensions are the average of two measurements across the major and minor axis. The limits of ovality will be determined by the container supplier and container customer, as necessary.

NOTE 2—Dimension *H* is measured from the top of the finish to the point where diameter *T*, extended parallel to the centerline, intersects the bead or shoulder.

NOTE 3—Contour of bead, undercut or shoulder is optional.

NOTE 4—Unless otherwise specified, *I* min applies to the full length of the opening.

NOTE 5—Concentricity of *I* min with respect to diameters *T* and *E* is not included. *I* min is specified for filler tube only.

NOTE 6—A minimum of 1 full turn of thread shall be maintained.

NOTE 7—Corresponding dimensions and finish details are shown in Table 3.

NOTE 8—Consideration must be given to the sealing surface width for the sealing system being used.

NOTE 9—Many child resistant closures, etc.

FIG. 1 SP 400 Finish Thread Cross Sections

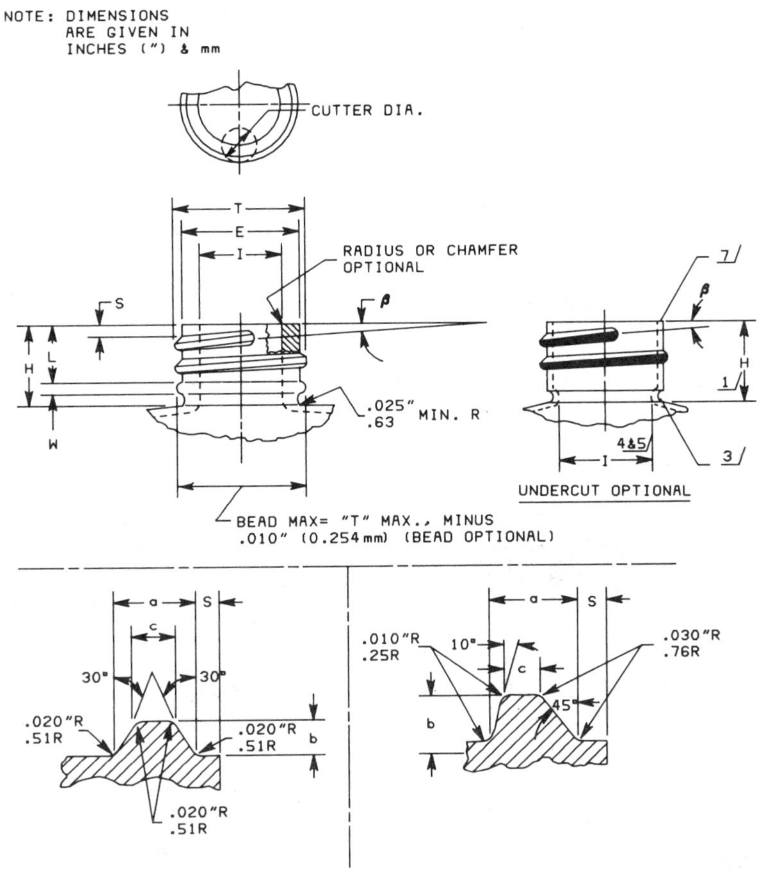

"L" Style				"M" Style			
All-Purpose Thread (Plastic or Metal Closures)				Modified Buttress Thread (Plastic Closures)			
Threads/in.	*a*	*b*	*c*	Threads/in.	*a*	*b*	*c*
6	in. 0.094	0.047	0.040	6	in. 0.094	0.047	0.039
	mm 2.239	1.19	1.02		mm 2.239	1.19	0.99
8	in. 0.084	0.042	0.036	8	in. 0.084	0.042	0.035
	mm 2.13	1.07	0.91		mm 2.13	1.07	0.89

Example Thread Nomenclature:
"L" Style: L22SP410
"M" Style: M22SP410

NOTE 1—Construction of neck from *B* to *D* must be held within the shaded area shown.

NOTE 2—A minimum of 1⅛ turns of thread shall be maintained.

NOTE 3—Unless otherwise specified, *I* min applies to the full length of the opening.

NOTE 4—Concentricity of *I* min with respect to diameters *T* and *E* is not included. *I* min is specified for filler tube only.

NOTE 5—*T* and *E* dimensions are the average of two measurements across the major and minor axis. The limits of ovality will be determined by the container supplier and container customer, as necessary.

NOTE 6—Consideration must be given to the sealing surface width for the sealing system being used.

NOTE 7—When valve style closures are used with this finish, special consideration must be given to a specific controlled inside diameter. In addition, dimensions indicated with asterisk (*) may be varied to ensure adequate material for finishing the inside diameter.

NOTE 8—Corresponding dimensions and details shown in Table 4.

FIG. 2 SP 410 Finish Thread Cross Sections

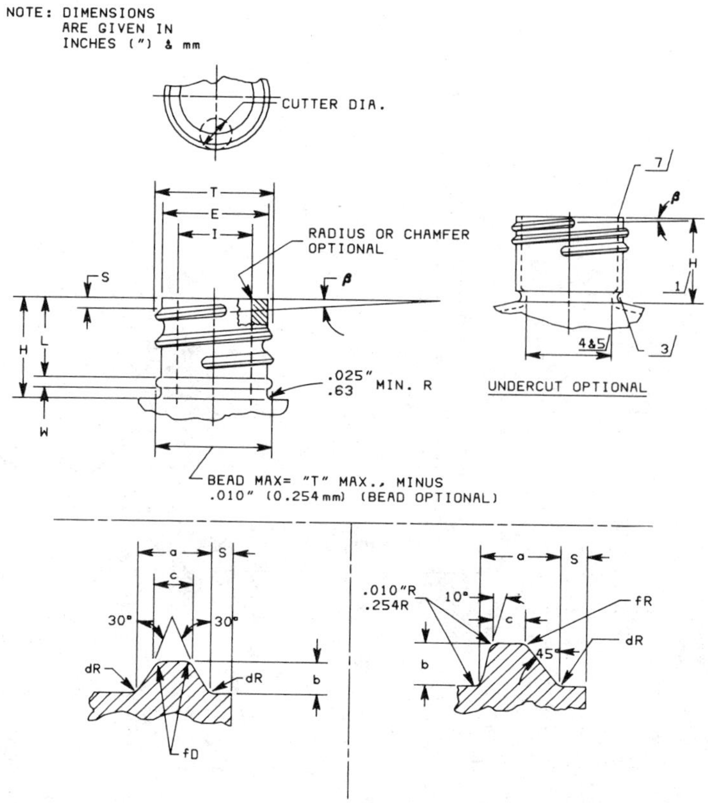

"L" Style — All-Purpose Thread (Plastic or Metal Closures)

Threads/in.		a	b	c	d	f
6	in.	0.094	0.047	0.040	0.020	0.020
	mm	2.39	1.19	1.02	0.51	0.51
8	in.	0.084	0.042	0.036	0.020	0.020
	mm	2.13	1.07	0.91	0.51	0.51
12	in.	0.045	0.030	0.011	0.015	0.005
	mm	1.14	0.76	0.28	0.38	0.13

"M" Style — Modified Buttress Thread (Plastic Closures)

Threads/in.		a	b	c	d	f
6	in.	0.094	0.047	0.039	0.030	0.030
	mm	2.39	1.19	0.99	0.76	0.76
8	in.	0.084	0.042	0.035	0.030	0.030
	mm	2.13	1.07	0.89	0.76	0.76
12	in.	0.051	0.030	0.016	0.020	0.008
	mm	1.29	0.76	0.41	0.51	0.22

Example Thread Nomenclature:
"L" Style: L22SP415
"M" Style: M22SP415

NOTE 1—Construction of neck from *B* to *D* must be held within the shaded area shown.

NOTE 2—A minimum of 1⅛ turns of thread shall be maintained.

NOTE 3—Unless otherwise specified, *I* min applies to the full length of the opening.

NOTE 4—Concentricity of *I* min with respect to diameters *T* and *E* is not included. *I* min is specified for filler tube only.

NOTE 5—*T* and *E* dimensions are the average of two measurements across the major and minor axis. The limits of ovality will be determined by the container supplier and container customer, as necessary.

NOTE 6—Consideration must be given to the sealing surface width for the sealing system being used.

NOTE 7—When valve style closures are used with this finish, special consideration must be given to a specific controlled inside diameter. In addition, dimensions indicated with asterisk (*) may be varied to ensure adequate material for finishing the inside diameter.

NOTE 8—Corresponding dimensions and finish details are shown in Table 5.

FIG. 3 SP 415 Finish Thread Cross Sections

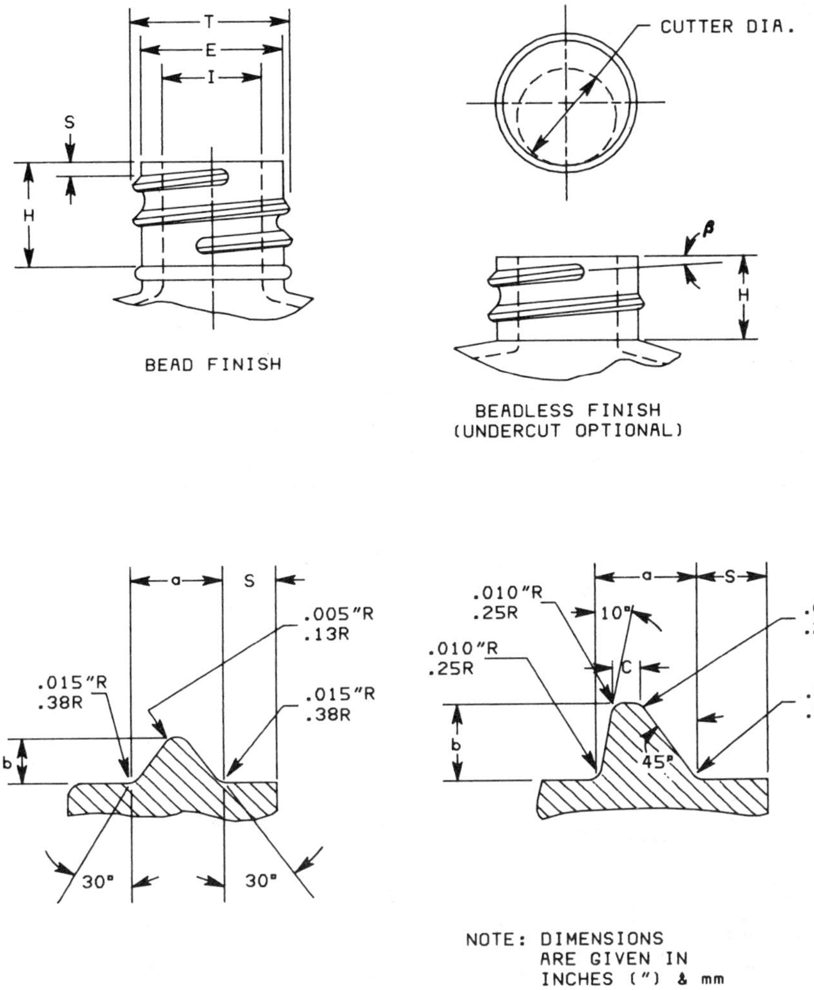

NOTE—Contour of bead, undercut, or shoulder is optional.

"L" Style				"M" Style				
All-Purpose Thread (Plastic or Metal Closures)				Modified Buttress Thread (Plastic Closures)				
Threads/in.		*a*	*b*	Threads/in.		*a*	*b*	*c*
12	in.	0.045	0.030	12	in.	0.051	0.030	0.016
	mm	1.14	0.76		mm	1.29	0.76	0.41
12	in.	0.045	0.030	12	in.	0.051	0.030	0.016
	mm	1.14	0.76		mm	1.29	0.76	0.41

Example Thread Nomenclature
"L" Style: L15SP425 or "M" Style M15SP425

NOTE—Corresponding dimensions and finish details are shown in Table 6.

FIG. 4 SP 425 Finish Thread Cross Sections

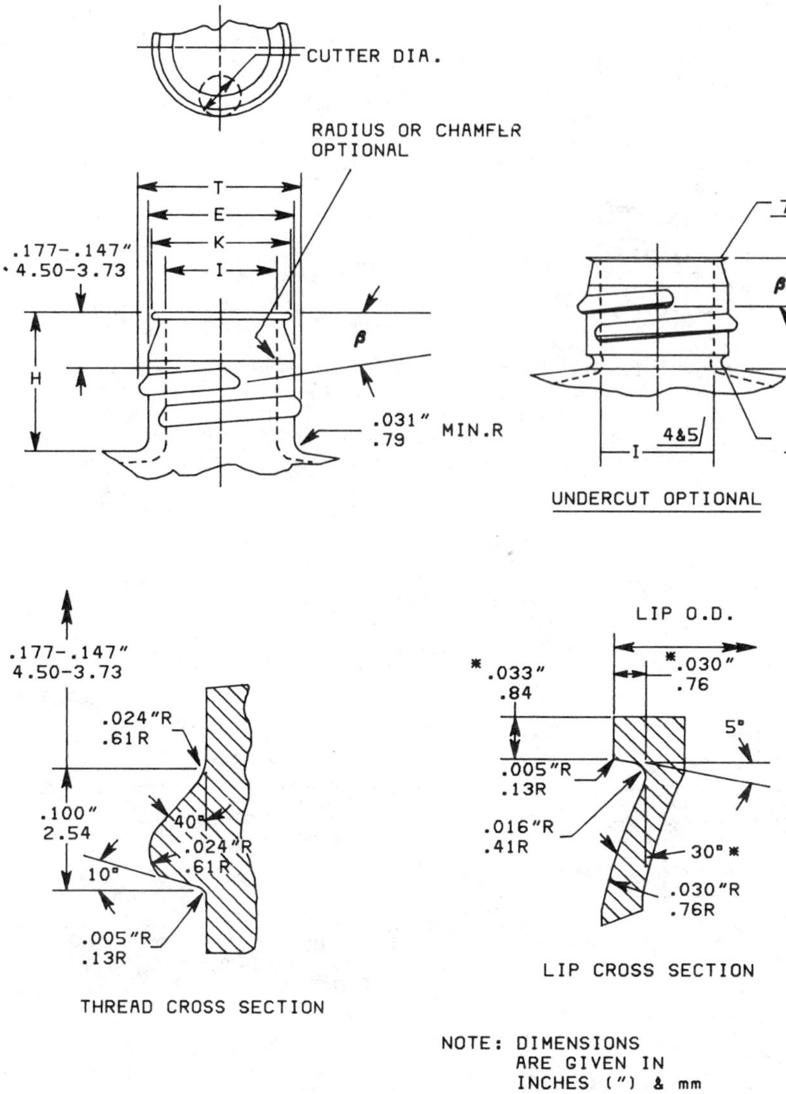

*(See Footnote, e, Table 7 and 8)

NOTE 1—Construction of neck from *B* to *D* must be held within the shaded area shown.

NOTE 2—A minimum of 1⅛ turns of thread shall be maintained.

NOTE 3—Unless otherwise specified, *I* min applies to the full length of the opening.

NOTE 4—Concentricity of *I* min with respect to diameters *T* and *E* is not included. *I* min is specified for filler tube only.

NOTE 5—*T* and *E* dimensions are the average of two measurements across the major and minor axis. The limits of ovality will be determined by the container supplier and container customer, as necessary.

NOTE 6—Consideration must be given to the sealing surface width for the scaling system being used.

NOTE 7—When valve style closures are used with this finish, special consideration must be given to a specific controlled inside diameter. In addition, dimensions indicated with asterisk (*) may be varied to ensure adequate material for finishing the inside diameter.

NOTE 8—Top dimension is shown in inches: bottom dimension in millimeters.

NOTE 9—Corresponding dimensions and finish details are shown in Tables 7 and 8.

FIG. 5 SP-103 Finish and SP-100 Finish, Thread and Lip Cross Section

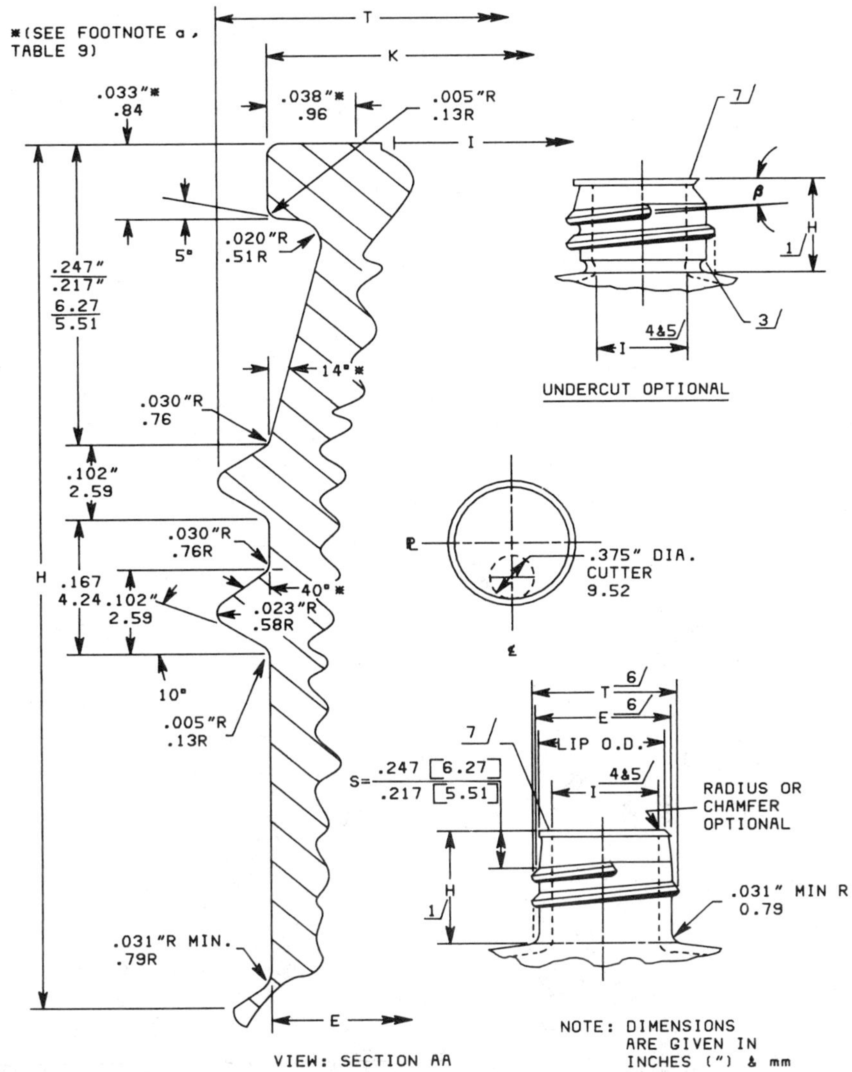

View: Section AA

NOTE 1—Construction of neck from *B* to *D* must be held within the shaded area shown.

NOTE 2—A minimum of 1⅛ turns of thread shall be maintained.

NOTE 3—Unless otherwise specified, *I* min applies to the full length of the opening.

NOTE 4—Concentricity of *I* min with respect to diameters *T* and *E* is not included. *I* min is specified for filler tube only.

NOTE 5—*T* and *E* dimensions are the average of two measurements across the major and minor axis. The limits of ovality will be determined by the container supplier and container customer, as necessary.

NOTE 6—Consideration must be given to the sealing surface width for the sealing system being used.

NOTE 7—When valve style closures are used with this finish, special consideration must be given to a specific controlled inside diameter. In addition, dimensions indicated with asterisk (*) may be varied to ensure adequate material for finishing the inside diameter.

NOTE 8—Top dimension is shown in inches; bottom dimension in millimeters.

FIG. 6 SP-200 Finish, Thread and Lip Section

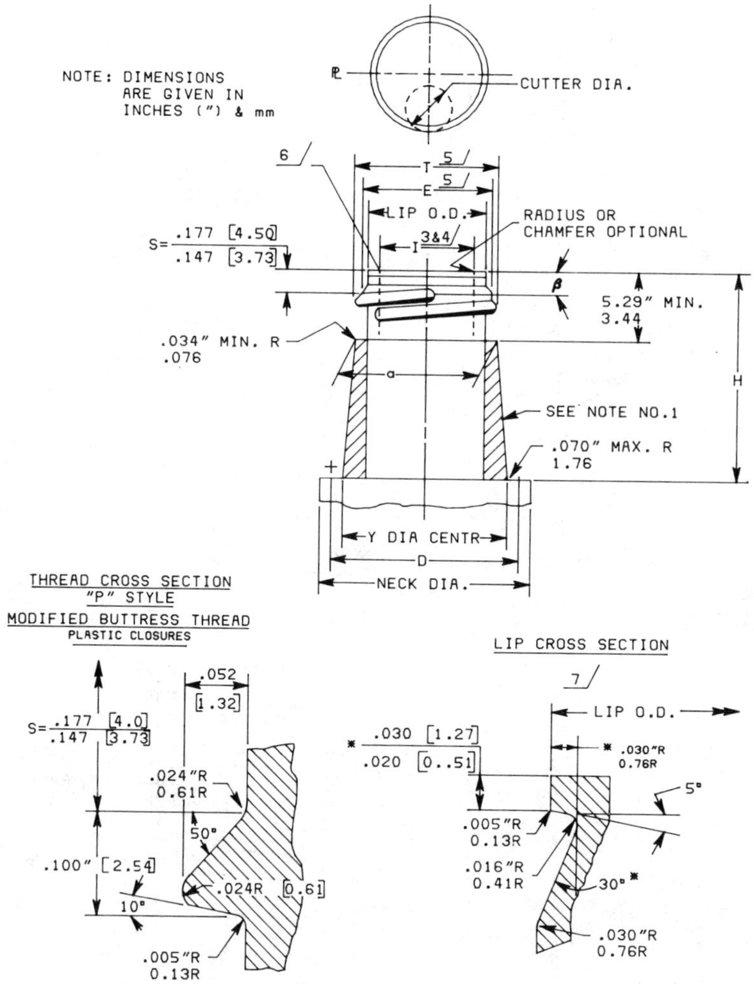

mm	T^A		E^A		LIP Outside Diameter		M		B	D	Y	$I^{B,C}$	Helix Angle	Cutter Diameter	Threads per Inch	Pitch
	max	min	max	min	max	min	max	min	max	max	construction	min				
28	1.088 27.63	1.068 27.13	0.984 24.99	0.964 24.49	0.974 24.74	0.954 24.23	1.700 43.10	1.670 42.42	1.062 26.97	1.320 33.53	1.188 30.17	0.735 18.67	2°13′	0.375 9.52	8	0.125 3.18

[A] *T* and *E* dimensions are the average of two measurements across the major and minor axis. The limits of ovality will be determined by the container supplier and container customer, as necessary.

[B] Unless otherwise specified, *I* min applies to the full length of the opening.

[C] Concentricity of *I* min with respect to diameters *T* and *E* is not included. *I* min is specified for filler tube only.

NOTE 1—Constriction of neck from *B* to *D* must be held within the shaded area shown.

NOTE 2—A minimum of 1⅛ turns of thread shall be maintained.

NOTE 3—Consideration must be given to the sealing surface width for the sealing system being used.

NOTE 4—When valve style closures are used with this finish, special consideration must be given to a specific controlled inside diameter. In addition, dimensions indicated with asterisk (*) may be varied to ensure adequate material for finishing the inside diameter.

FIG. 7 SP-110 Finish

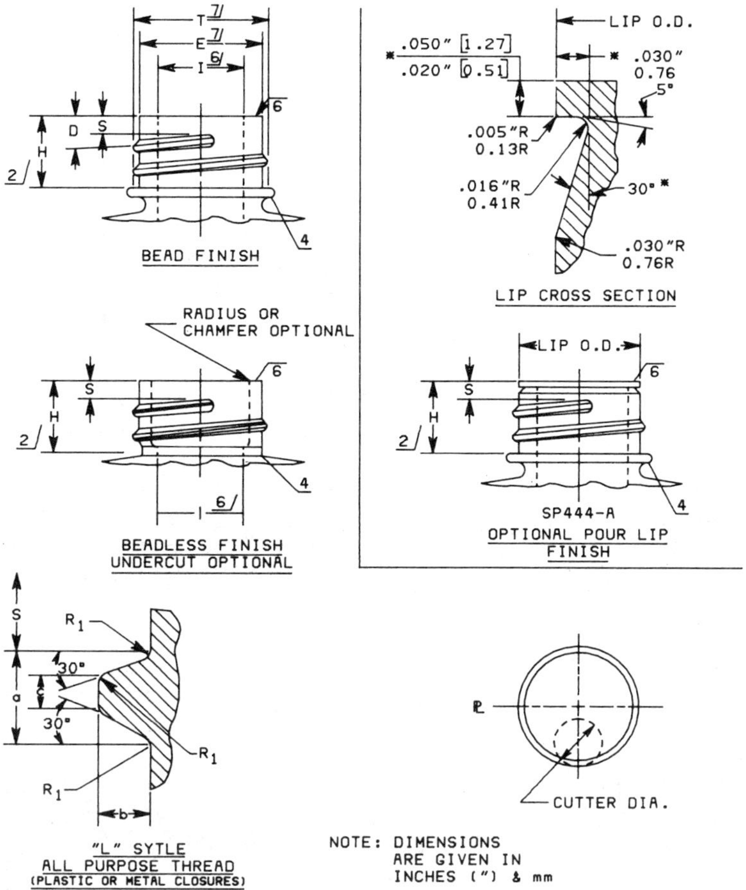

INCHES

T. P. I./IN	a	b	c	R_1
4	.125	.062	.053	.020
5	.120	.060	.051	.020
6	.094	.047	.040	.020

METRIC

PITCH	a	b	c	R_1
6.35	3.18	1.57	1.35	0.51
5.08	3.05	1.52	1.30	0.51
4.24	2.39	1.19	1.02	0.51

NOTE 1—This finish drawing was established by PBI to provide plastic bottle finishes comparable to 445, 450, and 480 glass finishes.

NOTE 2—Dimension *H* is measured from top of the finish to the point where diameter *T*, extended parallel to the centerline, intersects the shoulder or bead.

NOTE 3—A minimum of 1-1/8 turns of thread shall be maintained.

NOTE 4—Contour of bead, undercut, or shoulder is optional.

NOTE 5—Unless otherwise specified. *I* min applies to the full length of the opening.

NOTE 6—Concentricity of *I* min with respect to diameters and *T* and *E* is not included. *I* min is specified for filler tube only.

NOTE 7—*T* and *E* dimensions are the average of two measurements across the major and minor axis. The limits of ovality will be determined by the container supplier and container customer, as necessary.

NOTE 8—Consideration must be given to the sealing surface width for the sealing system being used.

NOTE 9—When valve style closures are used with this finish, special consideration must be given to a specific controlled inside diameter in addition dimensions indicated with asterisk (*) may be varied to ensure adequate material for finishing the inside diameter.

FIG. 8 SP-444 Finish

APPENDIX

(Nonmandatory Information)

XI. ESTABLISHING PLASTIC BOTTLE FINISHES

X1.1 Objective

X1.1.1 The following procedures are recommended as a guide for establishing plastic bottle finishes. The objective is to provide a basis of common understanding for all container suppliers that may be helpful to them in developing finish dimensions that fall outside the present range of finish dimensions developed and when converting glass finishes to plastic.

X1.2 Determination of Finish Dimensions

X1.2.1 *H Dimensions*—To determine *H*, use maximum *H* of closure minus compressed liner and add an amount for desirable clearance (preferably 0.015 in. or 0.38 mm). This will be minimum *H*. For maximum *H* dimensions, add 0.030 in. or 0.76 mm to minimum *H*. *H* dimension is measured from the top of the finish to the point where diameter *T* extended parallel to centerline intersects the bead or shoulder.

X1.2.2 *T and E Dimensions*—To determine *T* and *E* dimensions, maintain maximum SP finish dimensions as shown in the tables of Specification D 2911,[4] use the SP finish tolerances with corresponding neck size (millimetres).

[4] *Annual Book of ASTM Standards*, Vol 08.02.

X1.2.3 *T and E Dimensions* are the average of two measurements across the major and minor axis. The limits of ovality will be determined by the container supplier and container customer.

X1.2.4 *I Minimum*—Minimum dimension must be specified. It should be recognized that this is for filler tube clearance only.

X1.2.5 *S Dimension*—All conventional continuous thread finishes use the SP-400 *S* Dimension. When a pour-out finish is involved, the SP-100 finish should be used as a guide.

X1.2.6 *L Dimension*—Refer to SP-410 and SP-415 finishes where a bead is used within an *H* dimension.

X1.2.7 *Use of Internal Fitments*—Special consideration should be given whenever controlled internal diameters are required because of use with internal fitments. The controlled internal diameter must provide the required interference to satisfy the functional need. Specified *T* and *E* dimensions should be maintained after insertion of the fitment. It should be noted that the fitment is not a part of container *S* or *H* dimension.

X1.2.8 *Thread Profile*—The SP-400 *L* style or *M* style profile should be utilized when continuous threads are required.

X1.2.9 A minimum of 1 full turn of thread should be maintained.

Standard Test Method for
Leaks in Heat-Sealed Flexible Packages[1]

This standard is issued under the fixed designation D 3078; the number immediately following the designation indicates the year of original adoption or, in the case of revision, the year of last revision. A number in parentheses indicates the year of last reapproval. A superscript epsilon (ε) indicates an editorial change since the last revision or reapproval.

1. Scope

1.1 This test method covers the determination of *gross* leaks in heat-sealed flexible packages. It may be used to evaluate the performance of heat-sealed packages.

1.2 Small leaks may not be detected. Viscoelastic effects on the products or entrapped air become significant and prevent passage through small leaks. Positive pressure inside the pouch after the vacuum is drawn may force the product to plug small leaks. The size of the leak that can be detected is dependent upon the products contained and the nature of the packaging material.

1.3 *This standard does not purport to address the safety problems associated with its use. It is the responsibility of whoever uses this standard to consult and establish appropriate safety and health practices and determine the applicability of regulatory limitations prior to use.*

2. Definition

2.1 *leak*—any opening in a flexible package that contrary to intention either lets contents escape or permits substances to enter.

3. Apparatus

3.1 *Vacuum Chamber*—Any suitable transparent container capable of withstanding approximately one atmosphere pressure differential, fitted with a flat-vacuum-tight cover. A vacuum gage, an inlet tube from a source of vacuum, and an outlet tube to the atmosphere shall be sealed into the cover. The inlet and outlet tubes shall be equipped with hand valves. Attached to the underside of the cover shall be a transparent plate that will closely approximate the inside diameter of the container and be such a distance from the top of the container that when it is two thirds filled with water, the attached plate will be positioned 1 in. (25 mm) under the water.

4. Sampling

4.1 The number of specimens used in the test sample may be varied according to the nature of the product, its cost, its size, and whether the specimens are taken from a production line in a normal packaging operation, or are few in number, or are to be used only for purposes of comparative evaluation of procedures or materials.

5. Test Specimen

5.1 A specimen shall be one flexible package and its intended contents, or simulated contents, packed and sealed for shipment.

6. Conditioning

6.1 No special conditioning of the specimen is required but it shall be at equilibrium with normal room conditions.

7. Procedure

7.1 Submerge the specimen in water contained in the vessel within the vacuum chamber. The uppermost surface of the specimen shall be covered by not less than 1 in. (25 mm) of water.

NOTE—Two or more small packages may be tested at the same time, provided that they are placed in such a manner that all parts of every package under test can be observed for leakage during the test.

7.2 Set the cover on the jar, close the outlet valve, and turn on the vacuum so that the gage rises slowly (from 30 s to 1 mm) to one of the following gage readings: 5, 10, 15, or 27 in. (686 mm) Hg (1.68×10^4, 3.37×10^4, 5.05×10^4 and 9.1×10^4 Pa) (atmospheric pressure is 29.9 in. or 1×10^5 Pa).

7.3 During the rise in vacuum, observe the submerged specimen for leakage in the form of a steady progression of bubbles from the flexible container. Isolated bubbles caused by entrapped air are not considered as leaks.

7.4 Hold the vacuum for a minimum of 30 s.

7.5 Release the vacuum, remove the lid, and examine the specimen for the presence of water inside the specimen.

7.6 An alternative method of detecting seal leaks, when paper is the outer ply of a package, is to cut off an edge of the intact package and fill it with an aqueous dye solution. A leak will cause visible exterior straining in 30 s.

8. Interpretation of Results

8.1 If there are bubbles definitely attributable to leaks in a specimen during the rise of vacuum, or held at full vacuum, it fails the test.

8.2 If water is inside a specimen, it fails the test.

8.3 If there are no bubbles observed, attributable to leaks, and if no water is inside a specimen, it passes the test.

9. Report

9.1 The report shall include the following:

9.1.1 A statement that the test was conducted in compliance with this test method or a description of the deviations from this test method,

9.1.2 Identification of the specimen and the specific material tested,

[1] This test method is under the jurisdiction of ASTM Committee F-2 on Flexible Barrier Materials and is the direct responsibility of Subcommittee F02.30 on Test Methods.

Current edition approved March 30, 1984. Published May 1984. Originally published as D 3078 – 72. Last previous edition D 3078 – 72 (1977).

9.1.3 A statement whether or not leakage occurred, and if possible, a report of the location of each leak,

9.1.4 A statement of the number of specimens included in the test and the number of failures, if any,

9.1.5 When the test is performed to check compliance with requirements, a statement that the sample did or did not meet the requirement, and identification of the source for the requirement, and

9.1.6 When the test is conducted to evaluate or compare products, materials or methods, a statement of any observations that may lead to improvements.

10. Precision and Bias

10.1 The precision and bias of this test method have not been determined due to a lack of samples having the necessary distribution of leaking heat seals.

Standard Test Method for Water Vapor Transmission of Flexible Heat-Sealed Packages for Dry Products[1]

This standard is issued under the fixed designation D 3079; the number immediately following the designation indicates the year of original adoption or, in the case of revision, the year of last revision. A number in parentheses indicates the year of last reapproval. A superscript epsilon (ϵ) indicates an editorial change since the last revision or reapproval.

$^{\epsilon 1}$ NOTE—Paragraph 10.2 was added editorially in March 1984.

1. Scope

1.1 This test method covers the determination of the amount of water vapor transmission for flexible heat-sealed packages under specified conditions of exposure.

NOTE—Adequate heat-seal efficiency should be determined prior to this test.

1.2 *This standard does not purport to address the safety problems associated with its use. It is the responsibility of whoever uses this standard to consult and establish appropriate safety and health practices and determine the applicability of regulatory limitations prior to use.*

2. Summary of Test Method

2.1 Test samples are filled with desiccant and after sealing are exposed to 100 ± 2°F (37.8 ± 1.1°C) and 90 ± 2 % relative humidity for at least 1 month. Measurements of mass are taken at intervals to determine the average rate of mass change. Controls are used to monitor mass gain of package materials.

3. Significance and Use

3.1 This test determines the water vapor barrier properties of the package. With proper precautions and background experience, reproducible results can be obtained to aid in the selection of proper package materials required to provide the product shelf-life desired. This test may be used to establish a performance specification.

4. Definition

4.1 *water vapor transmission of a desiccant-filled package*—the increase in mass after exposure to a specified humid environment (90 ± 2 % relative humidity, 100 ± 2°F (37.8 ± 1.1°C)). The mass increase may be divided by a time unit to give a rate (for example, grams per 30 days). It is assumed that this mass increase is due to water vapor transmission (through the package) from the test environment to the desiccant.

5. Apparatus

5.1 *Weighing Balance*—Analytical balance accurate to 1 mg.

5.2 *Test Chamber*—A test room or cabinet provided with conditioned air that is continuously circulated around the specimens under test. The conditions in the chamber shall be such that no condensation occurs on the specimens.

6. Materials

6.1 *Desiccant*—Anhydrous calcium chloride in the form of small lumps is recommended for this test. This should pass a No. 8 (2.36-mm) sieve, and be free of fines that will pass a No. 30 (600-μm) sieve. The desiccant should be dried for at least 24 h at 200°C prior to use.

7. Test Specimens

7.1 Test specimens shall be representative of the packages being tested, and shall be closed and sealed in the normal manner.

7.2 At least ten test specimens should be prepared for each packaging variable being tested.

7.3 Each test specimen should be half full of desiccant up to 100 g of desiccant. If the half full capacity is greater than 100 g of desiccant, the package shall be less than half full, but each shall contain approximately the same quantity of desiccant.

7.4 If more than one type of package is tested for comparison purposes, all types should have the same (as close as possible) surface area and shape, and should contain the same quantity of desiccant.

7.5 Five identical control packages shall be prepared, containing no desiccant.

8. Procedure

8.1 Place the selected quantity of the desiccant inside the test specimen with no other contents, but with a suitable liner, if the package normally requires a liner, so that the package can be handled with the regular automatic packaging machinery. Close and seal the specimens and controls in exactly the same manner that packages with normal contents are closed and sealed. This method may be applied to hand-sealed packages, but take great care to ensure a complete seal. Note the use of the hand seal in the report.

8.2 Prewarm the sealed packages slowly to a temperature of approximately 100°F (37.8°C).

8.3 Place the warmed specimens and controls inside the test room or cabinet in a position where free access of the

[1] This test method is under the jurisdiction of ASTM Committee F-2 on Flexible Barrier Materials and is the direct responsibility of Subcommittee F02.30 on Test Methods.

Current edition approved July 28, 1972. Published September 1972.

conditioned circulating air is provided on all surfaces of the packages.

8.4 Make successive weighings of the packages and controls at suitable intervals and plot on arithmetical graph paper the mass gained against time after subtracting the average mass gain of the controls. Accuracy of the test is adversely affected by too frequent weighings. For highly permeable packages, a minimum weighing frequently of 3 days is recommended. For packages having a low rate of permeability, a weighing frequency of bi-weekly to monthly is recommended. All weighings shall preferably be made without removing the specimen from the test chamber. In instances where this is not possible, remove the specimens and controls from the test chamber and allow them to equilibrate for 1 h in the weighing area. Weigh in the same sequence each time. Specimens should not be out of the test chamber for more than 3 h at a time. Continue until a constant rate of gain is attained as indicated by at least three successive points in a straight line. The slope of this portion of the curve will furnish a measure of water vapor transmission.

9. Report

9.1 The report shall include the following:

9.1.1 Identification of the package including data on films, heat seal equipment, liners, etc.,

9.1.2 Desiccant used and quantity,

9.1.3 Temperature and relative humidity at which the test was conducted. If the tolerance of ±2 % relative humidity or ±2°F (1.1°C) temperature is exceeded, the variation shall be specified.

9.1.4 Number of specimens tested and number of controls,

9.1.5 Weighing frequency,

9.1.6 Water vapor transmission reported as grams of water per 30 days for the period of constant rate of gain,

9.1.7 Data on any previous tests to which the specimens might have been subjected, and

9.1.8 A statement to the effect that all tests were made in accordance with this method.

10. Precision and Bias

10.1 Duplicate determinations on uniform, undamaged specimens should agree within 10 % within a given laboratory.

10.2 More complete precision and bias for this test method are being developed.

Standard Test Method for Thermal Insulation Quality of Packages[1]

This standard is issued under the fixed designation D 3103; the number immediately following the designation indicates the year of original adoption or, in the case of revision, the year of last revision. A number in parentheses indicates the year of last reapproval. A superscript epsilon (ε) indicates an editorial change since the last revision or reapproval.

1. Scope

1.1 This test method covers the determination of the thermal insulation quality of a package and its enclosed packaging from temperature differentials between the packaged item and the outside environment. It is suitable for testing packages without and with various internal refrigerants, with or without interior packaging. Representative test conditions covered are indicated by Figs. 1 and 2. Depending upon the type of insulation material used, a water-vaporproof barrier might be used just inside the exterior packages of Fig. 1.

1.2 *This standard does not purport to address all of the safety problems associated with its use. It is the responsibility of whoever uses this standard to consult and establish appropriate safety and health practices and determine the applicability of regulatory limitations prior to use.* Specific precautionary statements are given in 5.2.3.1 and 9.2.2.1.

2. Referenced Document

2.1 *ASTM Standard:*

D 641 Method of Conditioning Paperboard, Fiberboard, and Paperboard Containers for Testing[2]

3. Significance and Use

3.1 Certain items, such as biological materials, pharmaceuticals, industrial adhesives, gyroscopes, blood, and fancy meats, must be shipped inside temperature-controlled packages. Factors affecting the rate of heat transfer of the package include the moisture content of the different package components and the thickness, continuity, density, position, and uniformity of the insulation.

3.2 Because of the variety of factors affecting the performance of a thermally insulated package, testing should be conducted with the actual package whenever possible. When simulated packages are used, special care must be exercised so that the simulated payload and coolant of the model will be as close as possible to the actual materials in temperature and other relevant physical properties.

4. Definitions

4.1 *draft-free atmosphere*—A relatively stationary atmosphere wherein the test specimens are remote from air currents.

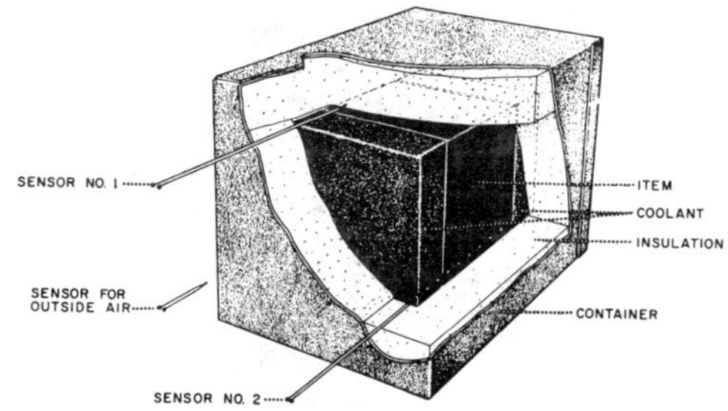

a—Containers Having Usable Inside Volume of 1 ft^3 (0.03 m^3) or Less.

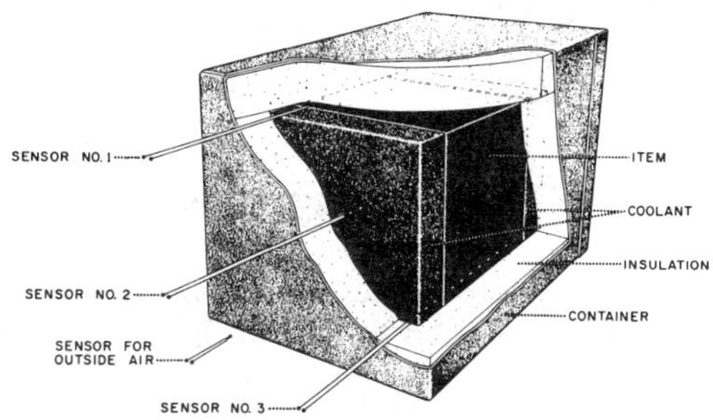

b—Containers Having a Usable Volume of More Than 1 ft^3 (0.03 m^3)

FIG. 1 Thermally Insulated Packages with Sensors Positioned

4.2 *exterior atmosphere*—the atmosphere in contact or near the exterior surface of a package.

4.3 *exterior package*—the outermost container of a package.

4.4 *interior atmosphere*—the atmosphere in contact or near the packaged item.

4.5 *interior package*—a package (often of corrugated fiberboard) located within another.

4.6 *thermal conductivity, homogeneous material*—the rate of heat flow, under steady conditions, through unit area, per unit temperature gradient in the direction perpendicular to the area.

5. Test Conditions and Apparatus

5.1 *Temperature of Exterior Atmosphere*—Draft-free environments large enough to accommodate the packages, but not necessarily the recorder, are required. Usual evaluation of the packaging materials will involve simple refrigerated or normal ambient exterior atmospheric conditions. These may be established as 4 ± 2°C (39 ± 4°F) and 23 ± 2°C (73 ± 4°F). Additionally, other temperature extremes expected during shipment might be used as an exterior atmosphere.

5.2 *Temperature of Interior Atmosphere of Package*—When no refrigerant is used, interior package temperature

[1] This test method is under the jurisdiction of ASTM Committee D-10 on Packaging and is the direct responsibility of Subcommittee D10.23 on Storage Test Methods.

Current edition approved July 27, 1984. Published September 1984. Originally published as D 3103 – 72. Last previous edition D 3103 – 72 (1977).

[2] Discontinued; see *1977 Annual Book of ASTM Standards*, Part 20.

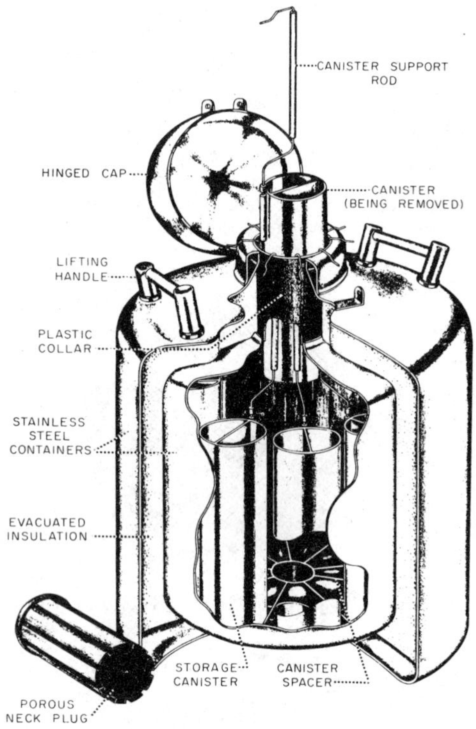

FIG. 2 Liquid Nitrogen Refrigerator

may serve as a measure of the thermal insulation quality of the package. If it is desired to control the interior temperature of the atmosphere, it should be done by one of the following:

5.2.1 *Refrigerant Gel or Water Ice Cans,* for ambient temperatures above 0°C (32°F),

5.2.2 *Dry Ice* (*Solid Carbon Dioxide* (CO_2)), kraft-paper wrapped, for temperatures to −73°C (−108.4°F), or

5.2.3 *Liquid Nitrogen* for temperatures to −195°C (−319°F).

5.2.3.1 **Caution**—Gaseous CO_2 and nitrogen are colorless, odorless, and noncombustible. In well-ventilated uses they present few problems, but evaporation or sublimation in airtight enclosures for prolonged periods (for example, 12 h) can produce sprung doors and asphyxiation of operating personnel. Usually these refrigerants can be used if provisions are made to evacuate the built-up gas periodically.

5.3 *Temperature Indicators:*

5.3.1 *Thermocouple-Potentiometer Recorders*—The recorder shall consist of a device capable of moving a chart at a constant speed past a pen or printer for the duration of the test. The result is a record produced either continuously or at short time intervals. The pen or printer shall be activated by a voltage from an insulated pair of copper-constantan or other suitable wires 30 gage B&S (0.255 mm) or less in diameter (not including the thickness of insulation). The recorder and chart shall have a capability that extends beyond the temperature values encountered in the test. Apparatus capable of producing from four to eight simultaneous records is preferable.

5.3.2 *Thermocouple-Potentiometer* (*Without Recorders*)—The thermocouple shall be copper-constantan or other suitable wires, 30 gage B&S or less in diameter. The potentiometer, scaled for temperature measurement, shall have a readable sensitivity of ±1°C.

5.3.3 *Thermistor-Recorder*—A thermistor sensor may be used, instead of a thermocouple, for sensing interior temperatures of the package. In general the recording equipment, as described in 5.3.1, shall suffice, but supplementary electrical circuitry will be required. The diameter of the extension wires from the package shall be the same as the thermocouple leads described in 5.3.1.

5.3.4 *Thermometer(s)*—Although less accurate, temperature can be recorded within the complete pack with one or more thermometers sealed in a hole(s) extending to the item. Between readings the thermometer(s) should be sealed in position with wax.

6. Sampling

6.1 Experimental package designs shall be made in accordance with the specifications and methods that will be used during actual production. When possible, the test packages shall be chosen by random sampling.

7. Test Specimens

7.1 A single test specimen shall consist of a package enclosing the actual item or a dummy load simulating the item. The package shall be closed, taped, or sealed in the same manner as will be used for actual shipment.

7.2 The mass, configuration, and location of refrigerant, if used, must be the same in each pack.

7.3 For exploratory evaluation of the overall insulation effectiveness of the container and insulation material, the interior cavity and package wall thickness shall be kept constant (for example, a 12 by 12 by 12-in. (305 by 305 by 305-mm)) interior cavity surrounded by a 1-in. (25-mm) thickness of insulation and a 14 by 14 by 14-in. (356 by 356 by 356-mm) container. A minimum of three such packages, shall be tested at each exposure listed in 5.1 to obtain an average result and range of performance.

7.4 When packages having previously established dimensions are evaluated, a minimum of five packages shall be exposed to the atmosphere given in 5.1 to obtain an average result and range of performance.

8. Conditioning

8.1 Materials shall be conditioned in accordance with Method D 641 or for 24 h at the conditions expected during actual production packing.

9. Procedures

9.1 *Packages Enclosing Solid Refrigerants:*

9.1.1 Calibrate the temperature-monitoring equipment.

9.1.2 Place the contents in the conditioned package.

9.1.3 Insert the sensors in the package so that they are in contact with the item(s) as follows:

9.1.3.1 For packages having an inside volume of 1 ft^3 (0.03 m^3) or less, place one sensor at or near the diagonally opposite corners of the item as shown in Fig. 1*a*. Additionally, the sensors and at least ½ in. (13 mm) of wire immediately behind them should be in contact with the item. Wrapping the wires around the item, if practical, is advantageous to minimize heat flow from the outside.

9.1.3.2 For packages having usable inside volume of more than 1 ft^3 (0.03 m^3), use the same sensor arrangement as described in 9.1.3.1 with an additional one, if practicable, at the geometrical center of the load.

9.1.4 Place a weighed and measured quantity of refrigerant in the package. When water ice or refrigerant gel are used, measure and record their temperature at the start of the test. Do this by securely taping a sensor to the outside of the refrigerant container or by inserting a sensor into the water prior to freezing.

NOTE—The quantity may be varied. However, when materials and designs of similar inside dimensions are being evaluated, the quantity by mass, configuration, and total surface area must be constant for the series for accurate evaluation of performance. When crushed or chopped dry ice is used, the fragments should be of the same general size for each package tested and should be free of dry ice dust.

9.1.5 Place the cover on the package with the sensor wires arranged so that the least possible heat transfer occurs through the joint.

9.1.6 Tape or otherwise secure the cover to the package in the normal manner.

9.1.7 Place the package in the test environment on a wooden platform or shelf in such a manner that no part of the package contacts the chamber walls or floor. If more than one package is tested at one time, they shall be separated from each other by at least 6 in. (152 mm). Mark the chart with date and time and start the recorder.

9.1.8 Allow the recorder to run until the indicated temperatures are above the maximum or below the minimum dictated by the product.

9.1.9 Check the calibration of the temperature-monitoring equipment.

9.1.10 Mark the date and hour on the chart.

9.2 *Containers with Liquid Nitrogen Refrigerant* (Fig. 1):

9.2.1 Detach the necessary sensor wires from the temperature recorder, calibrate, and place the sensing end(s) within the container adjacent to the contents.

9.2.2 Condition the container and contents by immersing them in liquid nitrogen for the length of time specified by the manufacturer, or by another recommended method.

9.2.2.1 **Caution**—The temperature of liquid nitrogen is about −195°C (−319°F). Oxygen will condense from air at normal barometric pressures at −183°C (−297°F). Therefore, when liquid nitrogen is used as the cooling agent, extreme care must be used during purging to maintain a nitrogen atmosphere adjacent to surfaces having a temperature near that of liquid nitrogen. This is necessary to prevent oxygen condensation and the related explosive fire danger.

9.2.3 With the contents in the conditioned container, add the liquid nitrogen as recommended by the container manufacturer.

9.2.4 Attach the cap or closure to the container with the sensor wire or wires extending between the closure and container in such a manner that the least possible nitrogen leakage occurs.

9.2.5 Attach the ends of the sensor wires to the recorder and proceed in accordance with 9.1.7.

9.2.6 Allow the recorder to run until the indicated temperatures are above the maximum established for the particular test.

9.2.7 Check the calibration of the temperature-monitoring equipment and mark the date and hour on the chart.

9.3 *Packages Without Refrigerants:*

9.3.1 When packages are tested without refrigerants, the thermal insulation quality is determined by observing the temperature change between the interior atmosphere or product and the controlled atmosphere outside the package.

9.3.2 Calibrate the temperature-monitoring equipment.

9.3.3 If the package (with or without interior packaging), is to be tested empty, place at least one sensor on the inside surface of the package, and at least one additional sensor near the center of the package.

9.3.4 If the package is to be tested with the product, embed at least two sensors in or in contact with the product, or as many sensors as are needed to test the entire interior.

9.3.5 Suspend one sensor in the controlled atmosphere outside and in close proximity of the package(s) to serve as a check against the usual recording controls of the climate chamber.

10. Report

10.1 The report shall include the following:

10.1.1 Dimensions of the package under test, its structural specifications, the materials of construction, the name of manufacturer, and the method of securing the lid,

10.1.2 Description of item(s) and secondary packaging materials,

10.1.3 Type, quantity, and location of refrigerant and devices within the package (in the case of water ice and refrigerant gel the initial ice and gel temperatures shall be recorded),

10.1.4 Type of temperature recording system used, (if a thermocouple-potentiometer system was used, indicate the location of thermocouple leads and junctions within the package),

10.1.5 Method of conditioning package and contents prior to test,

10.1.6 Temperature and relative humidity of the test environment,

10.1.7 Number of hours recorded on the chart by each sensor from the start of the test to the point where the temperature starts to exceed the maximum or minimum temperature established for the test.

10.1.8 Quantity, temperature, and state (if pertinent) of refrigerant remaining at the end of the test,

10.1.9 Description of the physical state of the contents at the end of the test,

10.1.10 Evaluation of the test with respect to the effective storage time in comparison to tests of other packaging materials or other designs, and

10.1.11 A statement that all tests were made in accordance with this method.

11. Precision and Bias

11.1 The precision and bias of this test method are limited by the precision and bias of the temperature-monitoring instruments (5.3).

11.1.1 *Precision* will be influenced by the construction of the container. The lower the insulating quality of the container, the more variation may be observed in the results for the type of containers described in 9.1.

11.1.2 *Bias*—In general, under properly controlled conditions, a variation of ±10 % among packages of similar construction tested in the same laboratory may be expected.

Standard Test Method for Water Vapor Transmission Through Screw-Cap Closure Liners[1]

This standard is issued under the fixed designation D 3199; the number immediately following the designation indicates the year of original adoption or, in the case of revision, the year of last revision. A number in parentheses indicates the year of last reapproval. A superscript epsilon (ϵ) indicates an editorial change since the last revision or reapproval.

1. Scope

1.1 This test method covers the measurement of the barrier efficiency of screw-cap liners against water vapor transmission either into or outward from a container capped under standard conditions.

1.2 *This standard does not purport to address the safety problems associated with its use. It is the responsibility of whoever uses this standard to consult and establish appropriate safety and health practices and determine the applicability of regulatory limitations prior to use.*

2. Referenced Document

2.1 *ASTM Standard:*

D 3198 Test Method for Application and Removal Torque of Threaded or Lug-Style Closures[2]

3. Summary of Test Method

3.1 This test method involves periodic weighing of containers sealed with screw caps containing the liners to be evaluated to determine moisture gain or loss under controlled conditions of temperature and humidity.

4. Significance and Use

4.1 This test method may be used to compare or screen new or existing closure liner materials by their performance in a standard test.

4.2 It also may be used to indicate whether the liner material will be satisfactory for use with moist products, dry products, or both types of products. When this method is used to assist in the choice of liner material for a specific use, the choice of containers and closures should simulate the conditions of use as closely as practicable.

4.3 This test method may be used to establish performance specifications.

5. Apparatus

5.1 *Bottles*, glass, ½ to 1 oz (approximately 15 to 30 cm^3) with a 28-mm Glass Packaging Institute (GPI) 400 or 405 finish. Select test bottles for uniformity of flatness of finish, width of lip sealing surface, and freedom from splits, chips, or cracks.

5.2 *Screw Caps*, 28 mm continuous thread (CT) of either metal or plastic.

5.3 *Torque Tester*[3], to determine application torque, accurate to within ±½ lbf·in. (0.06 N·m).

5.4 *Humidity Cabinet*, capable of control to ±1.8°F (±1°C) and ±2 % relative humidity at 100°F (38°C). Ranges required are 25 and 75 % relative humidity.

5.5 *Analytical Balance*, with an accuracy to at least 0.001 g.

5.6 *Pipet or Graduate*, capable of delivering 10 mL of water with an accuracy of ±0.1 mL.

6. Reagent

6.1 Activated silica gel (Grade H, Type 46, 16 mesh) anhydrous calcium chloride, or other suitable desiccant material.

7. Test Specimens

7.1 The closure-liner test specimen should be die-cut by either standard manufacturers' screw-cap lining equipment or by a single punch, using care that the liner is made uniform and without distortion of the liner facing.

7.2 No torn liners or wrinkled liners should be used. Prior to use, the liner backing combination must be laminated together and adhered to the cap, if plastic; or forced into the cap, if metal. At least ten test specimens and ten control specimens of each liner to be evaluated should be used for this test.

8. Procedure

8.1 *Water Vapor Transmission Outward from a Container:*

8.1.1 Tare weigh each ½-oz (28-mm) glass container together with its respective lined closure. Use five containers for this portion of the test.

8.1.2 Place 10.0 mL of distilled water in each tared glass container.

8.1.3 Apply the screw caps carrying the test liners with an application of 15 lbf·in. (approximately 1.7 N·m).

8.1.4 Assemble with the caps as in accordance with 8.1.1 and 8.1.3 and weigh five empty bottles for use as controls. Correct the test results by subtracting the average mass loss observed in the control containers from the average mass loss in the test containers.

[1] This test method is under the jurisdiction of ASTM Committee D-10 on Packaging and is the direct responsibility of Subcommittee D10.31 on Child Resistant Packaging.

Current edition approved Feb. 24, 1984. Published June 1984. Originally published as D 3199 – 73. Last previous edition D 3199 – 79.

[2] *Annual Book of ASTM Standards*, Vol 15.09.

[3] Owens-Illinois torque tester, available from Owens-Illinois, Inc., Closure Div., One Seagate, Toledo, OH 43666, or equivalent, has been found satisfactory for this method.

8.1.5 Place the containers upright in the humidity chamber at 100 ± 1.8°F (38 ± 1°C) and 25 ± 2 % relative humidity.

8.1.6 Seven days after starting the test, begin weighing each test container to the nearest 0.001 g at weekly intervals for 4 weeks. Allow the containers to come to room temperature before weighing.

8.1.7 Replace the test containers in the humidity cabinet immediately after each weighing.

8.2 *Water Vapor Transmission Into a Container:*

8.2.1 Tare weigh each ½-oz (28-mm) glass container together with its respective lined closure. Use five containers for this portion of the test.

8.2.2 Place 10.0 g of activated silica gel or other suitable desiccant in each container.

8.2.3 Apply the screw caps carrying the test liners with an application torque of 15 lbf·in. (approximately 1.7 N·m).

8.2.4 Assemble with the caps in accordance with 8.2.1 and 8.2.3 and weigh five empty bottles for use as controls. Correct the test results by subtracting the average mass gain observed in control containers from the average mass gain in the test containers.

8.2.5 Place the containers upright in the humidity chamber at 100 ± 1.8°F (38 ± 1°C) and 75 ± 2 % relative humidity.

8.2.6 Seven days after starting the test, begin weighing each test container at weekly intervals for 4 weeks. Allow the packages to come to room temperature before weighing.

8.2.7 Replace the test containers in the humidity cabinet immediately after each weighing.

9. Report

9.1 The report shall include the following:

9.1.1 Material of construction of the liner and whether adhered to or only forced into the cap,

9.1.2 Material of construction of the cap,

9.1.3 Name of the desiccant used,

9.1.4 Number of replicates and controls,

9.1.5 Time interval or intervals of the test,

9.1.6 Temperature and humidity of the test,

9.1.7 Mass losses and gains in tabular form, and their averages,

9.1.8 Average mass gain and loss, the range, and the standard deviation in milligrams corrected for the average mass gain and loss of the controls,

9.1.9 Calculation of daily gain or loss (optional),

9.1.10 Graph on arithmetical paper of mass gain or loss versus time (optional),

9.1.11 Evaluation of comparative results if more than one liner construction was tested,

9.1.12 Statement that this test was done in accordance with Test Method D 3199.

10. Precision and Bias

10.1 The precision of this method is to be determined. Test results that might allow statistical evaluation for this statement are herewith solicited.

10.2 The bias of this method includes quantitative estimates of the uncertainties of the dimensional measuring devices, the calibration of test equipment, and the skill of the operator. At this time, statements on bias should be limited to the developmental performance of particular laboratories.

Standard Test Methods for Peel Adhesion of Pressure-Sensitive Tape at 180° Angle[1]

This standard is issued under the fixed designation D 3330; the number immediately following the designation indicates the year of original adoption or, in the case of revision, the year of last revision. A number in parentheses indicates the year of last reapproval. A superscript epsilon (ε) indicates an editorial change since the last revision or reapproval.

1. Scope

1.1 These test methods cover the measurement of the peel adhesion of pressure-sensitive tapes.

1.1.1 Test Method A gives a measure of the adherence to a standard steel substrate or to other surface of interest for a single-coated tape (that is, adhesion to tape's own backing).

1.1.2 Test Method B gives a measure of the adherence to a standard steel substrate for double-coated tapes.

1.1.3 Test Method C gives a measure of the adherence of the release liner to the adhesive of either single-coated or double-coated tapes having a release liner.

1.2 These test methods provide a means of assessing the uniformity of the adhesion of a given type of pressure-sensitive tape. The assessment may be within a roll of tape, between rolls, or between production lots.

1.3 Variations in the tape's backing and adhesive affect the response. Therefore, these test methods cannot be used to pinpoint the specific cause(s) of nonuniformity.

1.4 These test methods may not be appropriate to test tapes having either relatively stiff backings, or stiff liners, or backings showing high stretch at low forces. These characteristics will result in a high variability of the test response which is not a true indication of the real nature of the adhesive bond.

NOTE 1—A complete metric companion to Test Methods D 3330 has been developed—D 3330M; therefore, no metric equivalents are presented in these test methods.

1.5 *This standard does not purport to address the safety problems associated with its use. It is the responsibility of the user of this standard to establish appropriate safety and health practices and determine the applicability of regulatory limitations prior to use.*

2. Referenced Documents

2.1 *ASTM Standards:*

A 666 Specification for Austenitic Stainless Steel, Sheet, Strip, Plate, and Flat Bar[2]

D 996 Terminology of Packaging and Distribution Environments[3]

D 2904 Practice for Interlaboratory Testing of a Textile Test Method that Produces Normally Distributed Data[4]

D 2906 Practice for Statements on Precision and Bias for Textiles[4]

D 3715 Practice for Quality Assurance of Pressure-Sensitive Tapes[3]

D 4332 Practice for Conditioning Containers, Packages or Packaging Components for Testing[3]

3. Terminology

3.1 *Definitions*—Terminology found in Terminology D 996 shall apply.

4. Summary of Test Methods

4.1 *Test Method A—Single-Coated Tapes*—A strip of tape is applied to a standard test panel (or other surface of interest) with controlled pressure. The tape is peeled from the panel at a specified rate, during which time the force required to effect peeling is measured.

4.2 *Test Method B—Double-Coated Tapes*—A strip of thin film is applied as a backing (substituted for the liner) and this composite is treated as a single-coated tape in accordance with Test Method A.

4.3 *Test Method C*—The tape is adhered to a rigid panel (with double-coated tape if single-coated) with the liner side up. The liner is peeled from the adhesive in the same manner as in peeling a single-coated tape from a standard panel in Test Method A.

5. Significance and Use

5.1 These test methods are tools for quality assurance use. Given a specific pressure-sensitive tape and a requirement in terms of the minimum or maximum peel value expected for this tape, the data from the test can be used in conjunction with acceptance criteria.

5.2 Test Methods A or B can show the relative bond strength of a given tape to one or more surfaces (material and texture) as compared to the standard stainless steel panel. Substitution of representative samples of the materials in question for the standard steel panel would suffice to do this.

5.3 Test Methods A or B cannot be used to compare two pressure-sensitive tapes of the same type but of different manufacture for their ability to adhere to a surface. This is because the measured peel force is not normalized for a fixed area of stress. The area under stress varies with backing stiffness and adhesive rheology (firmness). Two different tapes seldom agree in these properties.

5.4 Test Method C can show the amount of pull required to remove a liner that covers the adhesive side of a tape at a specified peel rate. The pull will be different at other peel rates.

5.5 These test methods may not provide design information as there is usually no direct relationship between peel adhesion and any functional requirement.

---

[1] These test methods are under the jurisdiction of ASTM Committee D-10 on Packaging and are the direct responsibility of Subcommittee D10.14 on Closure and Reinforcement.

Current edition approved June 29, 1990. Published August 1990. Originally published as D 3330 – 76. Last previous edition D 3330 – 87.

[2] *Annual Book of ASTM Standards*, Vols 01.03, 01.04 and 01.05.

[3] *Annual Book of ASTM Standards*, Vol 15.09.

[4] *Annual Book of ASTM Standards*, Vol 07.01.

6. Apparatus

6.1 *Specimen Cutter*[5]—The specimen cutter shall hold two single-edged razor blades in parallel planes, a precise distance apart, to form a cutter of exact specimen widths. Two cutters, 0.5 and 1-in. cutting width, shall be available.

NOTE 2—The 0.5-in. cutter shall consist of a 0.5 in. thick by 8-in. length of aluminum bar stock 0.5 in. wide. The edges for about 5 in. from one end shall be rounded slightly to form a handle. The width of the bar for 3 in. from the opposite end shall be narrowed to exactly 0.500 in. minus the thickness of a single razor blade (one or two used as cutting edges). The razor blades shall be held in position using side plates. The end of the cutter shall be cut away at a 45° angle to expose the cutting edge at one end of the blades. The edges shall be separated by 0.500 ± 0.005 in. A 1-in. cutter shall follow the same description, except that the barstock width shall be 1 in. and shall be narrowed to exactly 1.00 in. minus the thickness of a single razor blade.

6.2 *Dispensing System*, for solvents, such as a wash bottle.

6.3 *Panel*[5]—A 2 by 5 by no less than 0.043-in. panel of stainless steel 302 or 304 in accordance with Specification A 666 having a bright annealed finish. The surface roughness height shall be 2.0 ± 1.0 μin. arithmetical average deviation from the mean line. Panels showing stains, discoloration, or numerous scratches are not acceptable.

6.4 *Roller,*[5] mechanically operated:

6.4.1 A steel roller, 3.25 ± 0.1 in. in diameter and 1.75 ± 0.05 in. in width, covered with rubber approximately 0.25 in. in thickness, having a Shore scale A durometer hardness of 80 ± 5. The surface of the roller shall be a true cylinder void of any concave or convex deviations. The mass of the roller shall be 4.5 ± 0.1 lb.

6.4.2 No part of the apparatus shall increase the mass of the roller during use. The roller shall move mechanically at the rate of 12.0 ± 0.5 in./min.

NOTE 3—A simple check to determine if the rubber surface is cylindrical is to wrap the roller in a very thin paper (onionskin) and drag it across a flat glass plate on which is placed a carbon paper, face up. The carbon rubs off onto the thin paper wrapper to reveal high spots or hollows on the rubber surface.

6.5 *Adhesion Tester*—A constant-rate-of-extension (CRE) tension tested shall be used. The tester shall have two clamps with centers in the same plane, parallel with the direction of the motion of the stressing clamp, and so aligned that they will hold the specimen wholly in the same plane; a means of moving the stressing clamp at a uniform rate of 12.0 ± 0.5 in./min; and an autographic device for recording load. The instrument shall be calibrated to an accuracy of 0.5 % of full scale and the scale range used for any test shall be such that the mean test level falls within 20 to 50 % of full scale.

7. Reagents and Materials

7.1 *Purity of Reagents*—Reagent grade chemicals shall be used in all tests. Unless otherwise indicated, it is intended that all reagents shall conform to the specifications of the Committee on Analytical Reagents of the American Chemical Society, where such specifications are available.[6] Other grades may be used, provided it is first ascertained that the reagent is of sufficiently high purity to permit its use without lessening the accuracy of the determination.

7.2 *Solvents:*

7.2.1 *Diacetone Alcohol*, non-residual, technical grade or better.

7.2.2 One of the following:

7.2.2.1 *n-Heptane.*

7.2.2.2 *Methyl Alcohol (95 %).*

7.2.2.3 *Methyl Ethyl Ketone (MEK).*

7.2.2.4 *Methyl Isobutyl Ketone (MIBK).*

7.2.3 Where toxicity and flammability requirements are paramount, a mixture of *n*-heptane and a fluorinated hydrocarbon such as a refrigerant may meet the requirements.

7.3 *Cleaning Material*, absorbent, either surgical gauze or tissue. To be suitable, materials must be lint-free during use, absorbent, contain no additives that are soluble in the solvents listed in 7.2, and be made exclusively from virgin materials.

8. Sampling

8.1 Sampling shall be in accordance with Practice D 3715.

9. Test Specimens

9.1 The specimen shall be ½ in. to 1 in. wide. A tolerance of ±1/64 in. shall be allowed. The length shall be approximately 12 in.

9.2 Discard at least three but no more than six outer wraps of tape from the sample roll before taking the specimen for testing.

9.3 Remove one specimen per sample roll for each test to be performed. Remove the specimen from a freely rotating roll at the rate of 20 to 30 in./s. Where width or other factor, causing a high adherence to backing, makes it impossible to remove the specimen at the prescribed rate, remove it at a rate as close to 20 in./s as possible.

9.4 When the tape is wider than 1 in., specimens of the widest specified width are to be cut from the center of a strip removed from the roll in accordance with 9.3.

10. Conditioning

10.1 Condition the sample rolls of tape in the standard conditioning atmosphere as described in Practice D 4332 for a period of not less than 24 h. Test at these conditions.

11. Test Method A—Single-Coated Tapes

11.1 Dispense diacetone alcohol onto the panel. Scrub the panel with a clean piece of absorbent cleaning material. Dry the panel with fresh absorbent cleaning material. Dispense one of the solvents listed in 7.2.2 onto the panel, wiping it to dryness with fresh absorbent cleaning material. Repeat for a total of three washes with this solvent.

NOTE 4—Discard cleaned panels showing stains, discoloration, or numerous scratches. Avoid contacting panel surface with fingers. During storage, panels should be protected by covering with protective tape.

11.2 Touch one end of the specimen to an end of the test panel. Hold the other end of the specimen so that it does not make contact with the panel but is positioned loosely above it. Roll the tape mechanically once in each lengthwise direction, causing the roller to apply the tape to the panel. This prevents entrapment of air between the adhesive and

[5] Available from the Pressure Sensitive Tape Council (PSTC), 104 Wilmot Rd., Suite 201, Deerfield, IL 60015.

[6] "Reagent Chemicals, American Chemical Society Specifications," Am. Chemical Soc., Washington, DC. For suggestions on the testing of reagents not listed by the American Chemical Society, see "Reagent Chemicals and Standards," by Joseph Rosin, D. Van Nostrand Co., Inc., New York, NY, and the "United States Pharmacopeia."

the panel. Should this occur, discard the specimen.

NOTE 5—Where the width of the specimen is less than 1 in., prior to applying test specimen, apply a strip or strips of the tape, to give an equivalent width of 1 in. for rolling purposes.

11.3 Individually prepare each specimen and test immediately (within 1 min).

NOTE 6—Longer dwell time will give different results. Peel adhesion increases with dwell time at different rates for various tapes. A longer dwell time may be chosen purposely.

11.4 Double back the free end of the tape at an angle of 180° and peel 1 in. of the tape from the panel at the folded end. Clamp that end of the panel into the lower jaw of the adhesion testing machine and the free end of the tape into the upper jaw. Operate the lower jaw at 12 in./min.

11.5 After the lower jaw is started in motion, disregard the values obtained while the first inch of tape is mechanically peeled. Use the average pull value obtained during peeling of the next 2 in. as the adhesion value.

12. Test Method B—Double-Coated Tapes

12.1 *Face-Side*—Follow 10.1 to 10.3, then, remove the liner and superimpose on the test strip a strip of 0.001 in. thick polyester film, as wide as or slightly wider than the double-coated tape. Apply this film in the manner of applying the double-coated test strip to the panel so that the roller makes the actual application of the film to the double-coated tape.

NOTE 7—The two passes of the roller used in applying polyester film may be made using a hand roller of the same size. The rolling rate may be increased to 2 in./s.

12.1.1 Continue in accordance with 10.4 and 10.5.

12.2 *Liner Side*—Adhere the face side of the specimen to a strip of 1-mil polyester film in the manner described in 11.2 so that the roller makes the actual application of the tape to the film. Trim the film to be as wide as or slightly wider than the tape. Remove the liner.

12.2.1 Continue in accordance with 11.1 to 11.5.

13. Test Method C—Adhesion of Liner of Double-Coated and Single-Coated Tapes

13.1 *Double-Coated Tapes*—Follow 11.1. Apply 5 in. of one end of the specimen with the adhesive side (face side) down, to the panel. Make two passes with the mechanical roller, once in each direction. Separate the liner from the tape at the free end and cut away the free tape. Do not disturb the liner adhered to the tape on the panel.

13.2 *Single-Coated Tapes*—Follow 11.1. Apply a strip of double-coated tape, as wide as the specimen, the full length of the panel. Remove the liner from the double-coated tape. Superimpose 5 in. of one end of the specimen, backing side down, against the double-coated tape on the panel. Make two passes with the mechanical roller, once in each direction. Separate the liner from the tape at the free end and cut away the free tape. Do not disturb the liner adhered to the tape on the panel.

13.3 Double back the liner and proceed in accordance with 11.4 and 11.5.

14. Calculation

14.1 If observed pull value is not in ounces, convert to ounces. Calculate peel adhesion in ounces per inch by dividing the pull value by the width of tape expressed in inches.

15. Report

15.1 The report shall include the following:

15.1.1 Statement that this test method was used and indicating any deviations from the method as written,

15.1.2 Complete identification of each roll of tape tested, including tape, source, manufacturer's code number, and form,

15.1.3 Anomalous behavior during the test (that is, adhesive transfer or splitting),

15.1.4 Peel adhesion value in ounces per inch of width to the nearest 1 oz/in.,

15.1.5 Which test method was used, A, B or C, and if B, whether face side or liner side, and

15.1.6 Dwell time, if greater than the standard 1 min maximum.

16. Precision and Bias

16.1 *Summary:* The difference between two single observations should not exceed 18.8 % of the average of the two observations in 95 out of 100 cases when both observations are taken by the same well-trained operator using the same piece of test equipment and specimens randomly drawn from the same sample of material. Larger differences may occur under all other circumstances. The true value of peel adhesion at 180° angle can only be defined in terms of a specific test method. Within this limitation, Test Methods D 3330 has no known bias. Sections 16.2 through 16.5 explain the basis for this summary and for evaluations made under other conditions.

NOTE 8—Of the three methods in this test method only Test Method A was used in determining the precision. It is believed that the precision for Test Method B would be similar. It would probably not apply to Test Method C.

NOTE 9—The calculations for coefficient of variation and other statistics found in subsequent sections of this statement are described in Practice D 2906 and Appendix A3 of Practice D 2904.

16.2 *Interlaboratory Test Data*[7]—An interlaboratory study was made in 1980 in which randomly drawn samples of two materials were tested in each of six laboratories. Two operators in each laboratory each tested 3 specimens from each of 3 rolls of each material. The components of variance for Peel Adhesion at 180° results expressed as coefficients of variation (Note 9) and were calculated to be as follows.

	Specimens of the Same Material	Specimens of Different Material
Single-operator component	4.8 % of the average	9.7 % of the average
Within-laboratory component	2.1 % of the average	0 % of the average
Between-laboratory component	9.0 % of the average	3.7 % of the average
Replication component	4.9 % of the average	4.9 % of the average

16.3 *Critical Differences*—For the components of variance reported in 16.2, two averages of observed values should be considered significantly different at the 95 %

[7] Supporting data are available from ASTM Headquarters. Request RR:D-10-1002.

probability level if the difference equals or exceeds the critical differences shown in Table 1.

NOTE 10—The tabulated values of the critical differences and confidence limits should be considered to be a general statement particularly with respect to between-laboratory precision. Before a meaningful statement can be made about two specific laboratories, the amount of statistical bias between them, if any, must be established with each comparison being based on recent data obtained on specimens randomly drawn from one sample of the material to be evaluated.

16.4 *Confidence Limits*—For the components of variance reported in 16.2, single averages of observed values have the following 95 % confidence limits (See Note 10 and Table 2).

16.5 *Bias*—No justifiable statement can be made on the bias of Test Methods D 3330 for testing peel adhesion at 180° angle since the true value cannot be established by an accepted referee method.

17. Keywords

17.1 peel adhesion at 180° angle; pressure-sensitive tape

TABLE 1 Critical Difference, Percent of Grand Average for the Conditions Noted[A,B]

Number of Observations in Each Average	Single-Operator Precision	Within-Laboratory Precision	Between-Laboratory Precision
Specimens of the Same Material:			
1	18.8	19.7	42.5
5	14.5	15.6	29.4
10	13.9	15.0	29.1
Specimens of Different Material:			
1	30.1	30.1	31.8
5	28.0	28.0	29.4
10	27.3	27.3	29.1

[A] The critical differences were calculated using $t = 1.960$ which is based on infinite degrees of freedom.

[B] To convert the values of the critical differences to units of measure, multiply the average of the two specific sets of data being compared by the critical differences expressed as a decimal fraction.

TABLE 2 Width of 95 % Confidence Limits, Percent of the Grand Average for the Conditions Noted[A,B]

Number of Observations in Each Average	Single-Operator Precision	Within-Laboratory Precision	Between-Laboratory Precision
Specimens of the Same Material:			
1	±13.3	±13.9	±30.0
5	±10.3	±11.0	±20.8
10	±9.8	±10.6	±20.6
Specimens of Different Material:			
1	±21.3	±21.3	±22.5
5	±19.8	±19.8	±20.8
10	±19.3	±19.3	±20.6

[A] The confidence limits are calculated using $t = 1.960$ which is based on infinite degrees of freedom.

[B] To convert the values of the confidence limits to units of measure, multiply the average of the specific set of data which is of interest by the confidence limits expressed as a decimal fraction.

Standard Test Methods for Peel Adhesion of Pressure-Sensitive Tape at 180° Angle [Metric][1]

1. Scope

1.1 These test methods cover the measurement of the peel adhesion of pressure-sensitive adhesive tapes.

1.1.1 Test Method A gives a measure of the adherence to a standard steel substrate or to other surface of interest for a single-coated tape (that is, adhesion to tape's own backing).

1.1.2 Test Method B gives a measure of the adherence to a standard steel substrate for double-coated tapes.

1.1.3 Test Method C gives a measure of the adherence of the release liner to the adhesive of either single-coated or double-coated tapes having a release liner.

1.2 These methods provide a means of assessing the uniformity of the adhesion of a given type of pressure-sensitive adhesive tape. The assessment may be within a roll of tape, between rolls, or between production lots.

1.3 Variations in the tape's backing and adhesive affect the response. Therefore, these methods cannot be used to pinpoint the specific cause(s) of nonuniformity.

1.4 These test methods may not be appropriate to test tapes having either relatively stiff backings, or stiff liners, or backings showing high stretch at low forces. These characteristics will result in a high variability of the test response which is not a true indication of the real nature of the adhesive bond.

1.5 *This standard does not purport to address the safety problems associated with its use. It is the responsibility of the user of this standard to establish appropriate safety and health practices and determine the applicability of regulatory limitations prior to use.*

NOTE 1—These test methods are the metric counterpart of Test Methods D 3330.

2. Referenced Documents

2.1 *ASTM Standards:*

A 666 Specification for Austenitic Stainless Steel, Sheet, Strip, Plate, and Flat Bar[2]

D 996 Terminology of Packaging and Distribution Environments[3]

D 2904 Practice for Interlaboratory Testing of a Textile Test Method that Produces Normally Distributed Data[4]

D 2906 Practice for Statements on Precision and Bias for Textiles[4]

D 3715 Practice for Quality Assurance of Pressure-Sensitive Tapes[3]

D 4332 Practice for Conditioning Containers, Packages or Packaging Components for Testing[3]

3. Terminology

3.1 *Definitions*—Terminology found in Terminology D 996 shall apply.

4. Summary of Test Methods

4.1 *Test Method A—Single-Coated Tapes*—A strip of tape is applied to a standard test panel (or other surface of interest) with controlled pressure. The tape is peeled from the panel at a specified rate, during which time the force required to effect peeling is measured.

4.2 *Test Method B—Double-Coated Tapes*—A strip of thin film is applied as a backing (substituted for the liner) and this composite is treated as a single-coated tape in accordance with Test Method A.

4.3 *Test Method C*—The tape is adhered to a rigid panel (with double-coated tape if single coated) with the liner side up. The liner is peeled from the adhesive in the same manner as in peeling a single-coated tape from a standard panel in Test Method A.

5. Significance and Use

5.1 These test methods are tools for quality assurance use. Given a specific pressure-sensitive tape and a requirement in terms of the minimum or maximum peel value expected for this tape, the data from the test can be used in conjunction with acceptance criteria.

5.2 Test Methods A or B can show the relative bond strength of a given tape to one or more surfaces (material and texture) as compared to the standard stainless steel panel. Substitution of representative samples of the materials in question for the standard steel panel would suffice to do this.

5.3 Test Methods A or B cannot be used to compare two pressure-sensitive tapes of the same type but of different manufacture for their ability to adhere to a surface. This is because the measured peel force is not normalized for a fixed area of stress. The area under stress varies with backing stiffness and adhesive rheology (firmness). Two different tapes seldom agree in these properties.

5.4 Test Method C can show the amount of pull required to remove a liner that covers the adhesive side of a tape at a specified peel rate. The pull will be different at other peel rates.

5.5 These test methods may not provide design information as there is usually no direct relationship between peel adhesion and any functional requirement.

---

[1] These test methods are under the jurisdiction of ASTM Committee D-10 on Packaging and are the direct responsibility of Subcommittee D10.14 on Closure and Reinforcement.

Current edition approved June 29, 1990. Published August 1990. Originally published as D 3330M – 79. Last previous edition D 3330M – 87.

[2] *Annual Book of ASTM Standards*, Vols 01.03, 01.04 and 01.05.

[3] *Annual Book of ASTM Standards*, Vol 15.09.

[4] *Annual Book of ASTM Standards*, Vol 07.01.

6. Apparatus

6.1 *Specimen Cutter*[5]—The specimen cutter shall hold two single-edged razor blades in parallel planes, a precise distance apart, to form a cutter of exact specimen widths. Two cutters, 12 and 24-mm cutting width, shall be available.

NOTE 2—The 12-mm cutter shall consist of a 12 mm thick by 200 mm length of aluminum bar stock 12 mm wide. The edges for about 125 mm from one end shall be slightly rounded to form a handle. The width of the bar for 75 mm from the opposite end shall be narrowed to exactly 12 mm minus the thickness of a single razor blade (one of two used as cutting edges). The razor blades shall be held in position using side plates. The end of the cutter shall be cut away at 0.75 rad to expose the cutting edge at one end of the blades. The edges shall be separated by 12.0 ± 0.10 mm. The 24-mm cutter shall follow the same description, except that the barstock width shall be 24.0 mm and shall be narrowed exactly 24.0 mm minus the thickness of a single razor blade.

6.2 *Dispensing System*, for solvents, such as a wash bottle.

6.3 *Panel*[5]—A 50 by 125-mm panel no less than 1.1 mm thickness of stainless steel 302 or 304 in accordance with Specification A 666 having a bright annealed finish. The surface roughness height shall be 50 ± 5 nm arithmetical average deviation from the mean line. Panels showing stains, discoloration, or numerous scratches are not acceptable.

6.4 *Roller*,[5] mechanically operated:

6.4.1 A steel roller 85 ± 2.5 mm in diameter and 45 ± 1.5 mm in width, covered with rubber approximately 6 mm in thickness, having a Shore scale A durometer hardness of 80 ± 5. The surface of the roller shall be a true cylinder void of any concave or convex deviations. The mass of the roller shall be 2040 ± 45 g.

6.4.2 No part of the apparatus shall increase the mass of the roller during use. The roller shall move mechanically at the rate of 5.0 ± 0.2 mm/s.

NOTE 3—A simple check to determine if the rubber surface is cylindrical is to wrap the roller in a very thin paper (onionskin) and drag it across a flat glass plate on which is placed a carbon paper, face up. The carbon rubs off onto the thin paper wrapper to reveal high spots or hollows on the rubber surface.

6.5 *Adhesion Tester*—A constant-rate-of-extension (CRE) tension tester shall be used. The tester shall have two clamps with centers in the same plane, parallel with the direction of the motion of the stressing clamp, and so aligned that they will hold the specimen wholly in the same plane; a means of moving the stressing clamp at a uniform rate of 5.0 ± 0.2 mm/s; and an autographic device for recording load. The instrument shall be calibrated to an accuracy of 0.5 % of full scale and the scale range used for any test shall be such that the mean test level falls within 20 to 50 % of full scale.

7. Reagents and Materials

7.1 *Purity of Reagents*—Reagent grade chemicals shall be used in all tests. Unless otherwise indicated, it is intended that all reagents shall conform to the specifications of the Committee on Analytical Reagents of the American Chemical Society, where such specifications are available.[6] Other grades may be used, provided it is first ascertained that the reagent is of sufficiently high purity to permit its use without lessening the accuracy of the determination.

7.2 *Solvents:*

7.2.1 *Diacetone Alcohol*, non-residual, technical grade or better.

7.2.2 One of the following:

7.2.2.1 n-*Heptane.*

7.2.2.2 *Methyl Alcohol (95 %).*

7.2.2.3 *Methyl Ethyl Ketone (MEK).*

7.2.2.4 *Methyl Isobutyl Ketone (MIBK).*

7.2.3 Where toxicity and flammability requirements are paramount, a mixture of *n*-heptane and a fluorinated hydrocarbon such as refrigerant may meet the requirements.

7.3 *Cleaning Material*, absorbent, either surgical gauze or tissue. To be suitable, materials must be lint-free during use, absorbent, contain no additives that are soluble in the solvents listed in 7.2, and be made exclusively from virgin materials.

8. Sampling

8.1 Sampling shall be in accordance with Practice D 3715.

9. Test Specimens

9.1 The specimen shall be 12 to 24 mm wide. A tolerance of ±0.5 mm shall be allowed. The length shall be approximately 300 mm.

9.2 Discard at least three but no more than six outer wraps of tape from the sample roll before taking the specimen for testing.

9.3 Remove one specimen per sample roll for each test to be performed. Remove the specimen from a freely rotating roll at the rate of 500 to 750 mm/s. Where width or other factor, causing a high adherence to backing, makes it impossible to remove the specimen at the prescribed rate, remove it at a rate as close to 500 mm/s as possible.

9.4 When the tape is wider than 24 mm, specimens of the widest specified width are to be cut from the center of a strip removed from the roll as in 9.3.

10. Conditioning

10.1 Condition the sample rolls of tape in the standard conditioning atmosphere as described in Practice D 4332 for a period of not less than 24 h. Test at these conditions.

11. Test Method A—Single-Coated Tapes

11.1 Dispense diacetone alcohol onto the panel. Scrub the panel with a clean piece of absorbent cleaning material. Dry the panel with fresh absorbent cleaning material. Dispense one of the solvents listed in 7.2.2 onto the panel wiping it to dryness with fresh absorbent cleaning material. Repeat for a total of three washes with this solvent.

NOTE 4—Discard cleaned panels showing stains, discoloration, or numerous scratches. Avoid contacting panel surface with fingers. During storage, panels should be protected by covering with protective tape.

11.2 Touch one end of the specimen to an end of the test panel. Hold the other end of the specimen so that it does not make contact with the panel but is positioned loosely above it. Roll the tape mechanically once in each lengthwise direction, causing the roller to apply the tape to the panel. This prevents entrapment of air between the adhesive and

[5] Available from the Pressure-Sensitive Tape Council (PSTC), 104 Wilmot Rd., Suite 201, Deerfield, IL 60015.

[6] "Reagent Chemicals, American Chemical Society Specifications," Am. Chemical Soc., Washington, DC. For suggestions on the testing of reagents not listed by the American Chemical Society, see "Reagent Chemicals and Standards," by Joseph Rosin, D. Van Nostrand Co., Inc., New York, NY, and the "United States Pharmacopeia."

the panel. Should this occur, discard the specimen.

NOTE 5—Where the width of the specimen is less than 24 mm, prior to applying test specimen, apply a strip of strips of the tape, to give an equivalent width of 24 mm for rolling purposes.

11.3 Individually prepare each specimen and test immediately (within 1 min).

NOTE 6—Longer dwell time will give different results. Peel adhesion increases with dwell time at different rates for various tapes. A longer dwell time may be chosen purposely.

11.4 Double back the free end of the tape at an angle of 3.14 rad and peel 25 mm of the tape from the panel at the folded end. Clamp that end of the panel into the lower jaw of the adhesion testing machine and the free end of the tape into the upper jaw. Operate the lower jaw at 5 mm/s.

11.5 After the lower jaw is in motion, disregard the values obtained while the first 25 mm of tape is mechanically peeled. Use the average pull value obtained during peeling of the next 50 mm as the adhesion value.

12. Test Method B—Double-Coated Tapes

12.1 *Face Side*—Follow the procedure of 11.1 through 11.3, then remove the liner and superimpose on the test strip a strip of 0.025 mm thick polyester film, as wide as or slightly wider than the double-coated tape. Apply this film in the manner of applying the double-coated test strip to the panel so that the roller makes the actual application of the film to the double-coated tape.

NOTE 7—The two passes of the roller used in applying polyester film may be made using a hand roller of the same size. The roller rate may be increased to 50 mm/s.

12.1.1 Continue in accordance with 11.4 and 11.5.

12.2 *Liner Side*—Adhere the face side of the specimen to a strip of 0.025-mm thick polyester film in the manner described in 11.2 so that the roller makes the actual application of the tape to the film. Trim the film to be as wide as or slightly wider than the tape. Remove the liner.

12.2.1 Continue with the procedure of 11.1 through 11.5.

13. Test Method C—Adhesion of Liner of Double-Coated and Single-Coated Tapes

13.1 *Double-Coated Tapes*—Follow 11.1. Apply 125 mm of one end of the specimen with the adhesive side (face side) down, to the panel. Make two passes with the mechanical roller, once in each direction. Separate the liner from the tape at the free end and cut away the free tape. Do not disturb the liner adhered to the tape on the panel.

13.2 *Single-Coated Tapes*—Follow 11.1. Apply a strip of double-coated tape, as wide as the specimen, the full length of the panel. Remove the liner from the double-coated tape. Superimpose 125 mm of one end of the specimen, backing side down, against the double-coated tape on the panel. Make two passes with the mechanical roller, once in each direction. Separate the liner from the tape at the free end and cut away the free tape. Do not disturb the liner adhered to the tape on the panel.

13.3 Double back the liner and proceed in accordance with 11.4 and 11.5.

14. Calculation

14.1 If observed value is not in newtons, convert to newtons. Calculate peel adhesion in newtons per 100 mm of width by dividing the pull value by the width of the tape and multiplying by 100.

15. Report

15.1 The report shall include the following:

15.1.1 Statement that this test method was used and indicating any deviations from the method as written,

15.1.2 Complete identification of each roll of tape tested, including tape, source, manufacturer's code number, and form,

15.1.3 Anomalous behavior during the test (that is, adhesive transfer or splitting),

15.1.4 Peel adhesion value in N/100 mm of width to the nearest 1 N/100 mm,

15.1.5 Which test method was used, A, B or C, and if B, whether face side or liner side, and

15.1.6 Dwell time, if greater than the standard 1 min maximum.

16. Precision and Bias

16.1 *Summary:* The difference between two single observations should not exceed 18.8 % of the average of the two observations in 95 out of 100 cases when both observations are taken by the same well-trained operator using the same piece of test equipment and specimens randomly drawn from the same sample of material. Larger differences may occur under all other circumstances. The true value of peel adhesion at 180° angle can only be defined in terms of a specific test method. Within this limitation, Test Method D 3330 has no known bias. Sections 16.2 through 16.5 explain the basis for this summary and for evaluations made under other conditions.

NOTE 8—Of the three methods in this test method only Test Method A was used in determining the precision. It is believed that the precision for Test Method B would be similar. It would probably not apply to Test Method C.

NOTE 9—The calculations for coefficient of variation and other statistics found in subsequent sections of this statement are described in Practice D 2906 and Appendix A3 of Practice D 2904.

16.2 *Interlaboratory Test Data*[7]—An interlaboratory study was made in 1980 in which randomly drawn samples of two materials were tested in each of six laboratories. Two operators in each laboratory each tested 3 specimens from each of 3 rolls of each material. The components of variance for Peel Adhesion at 180° results expressed as coefficients of variations (Note 9) and were calculated to be as follows.

	Specimens of the Same Material	Specimens of Different Material
Single-operator component	4.8 % of the average	9.7 % of the average
Within-laboratory component	2.1 % of the average	0 % of the average
Between-laboratory component	9.0 % of the average	3.7 % of the average
Replication component	4.9 % of the average	4.9 % of the average

16.3 *Critical Differences*—For the components of variance reported in 16.2, two averages of observed values should be considered significantly different at the 95 %

[7] Supporting data are available from ASTM Headquarters. Request RR: D10-1002.

probability level if the difference equals or exceeds the critical differences shown in Table 1.

NOTE 10—The tabulated values of the critical differences and confidence limits should be considered to be a general statement particularly with respect to between-laboratory precision. Before a meaningful statement can be made about two specific laboratories, the amount of statistical bias between them, if any, must be established with each comparison being based on recent data obtained on specimens randomly drawn from one sample of the material to be evaluated.

16.4 *Confidence Limits*—For the components of variance reported in 15.2, single averages of observed values have the following 95 % confidence limits (See Note 10 and Table 2).

16.5 *Bias*—No justifiable statement can be made on the bias of Test Method D 3330 for testing peel adhesion at 180° angle since the true value cannot be established by an accepted referee method.

17. Keywords

17.1 peel adhesion at 180° angle; pressure-sensitive tape

TABLE 1 Critical Difference, Percent of Grand Average for the Conditions Noted[A,B]

Number of Observations in Each Average	Single-Operator Precision	Within-Laboratory Precision	Between-Laboratory Precision
Specimens of the Same Material:			
1	18.8	19.7	42.5
5	14.5	15.6	29.4
10	13.9	15.0	29.1
Specimens of Different Material:			
1	30.1	30.1	31.8
5	28.0	28.0	29.4
10	27.3	27.3	29.1

[A] The critical differences were calculated using t = 1.960 which is based on infinite degrees of freedom.

[B] To convert the values of the critical differences to units of measure, multiply the average of the two specific sets of data being compared by the critical differences expressed as a decimal fraction.

TABLE 2 Width of 95 % Confidence Limits, Percent of the Grand Average for the Conditions Noted[A,B]

Number of Observations in Each Average	Single-Operator Precision	Within-Laboratory Precision	Between-Laboratory Precision
Specimens of the Same Material:			
1	±13.3	±13.9	±30.0
5	±10.3	±11.0	±20.8
10	±9.8	±10.6	±20.6
Specimens of Different Material:			
1	±21.3	±21.3	±22.5
5	±19.8	±19.8	±20.8
10	±19.3	±19.3	±20.6

[A] The confidence limits are calculated using t = 1.960 which is based on infinite degrees of freedom.

[B] To convert the values of the confidence limits to units of measure, multiply the average of the specific set of data which is of interest by the confidence limits expressed as a decimal fraction.

Standard Test Methods for Mechanical-Shock Fragility of Products, Using Shock Machines[1]

This standard is issued under the fixed designation D 3332; the number immediately following the designation indicates the year of original adoption or, in the case of revision, the year of last revision. A number in parentheses indicates the year of last reapproval. A superscript epsilon (ε) indicates an editorial change since the last revision or reapproval.

1. Scope

1.1 These test methods cover the determination of the shock fragility of products. This fragility information may be used in designing shipping containers for transporting the products. It may also be used to improve product ruggedness. Unit or consumer packages, which are transported within an outer container, are considered to be the product for purposes of these test methods. Two methods are outlined as follows:

1.1.1 *Method A* is used first to determine the product's critical velocity.

1.1.2 *Method B* is used second to determine the product's critical acceleration.

1.2 *This standard does not purport to address the safety problems associated with its use. It is the responsibility of the user of this standard to establish appropriate safety and health practices and determine the applicability of regulatory limitations prior to use.*

2. Referenced Documents

2.1 *ASTM Standards:*

D 996 Terminology of Packaging and Distribution Environments[2]

D 2463 Test Method for Drop Impact Resistance of Blow-Molded Thermoplastic Containers[3]

D 4332 Practice for Conditioning Containers, Packages, or Packaging Components for Testing[2]

E 122 Practice for Choice of Sample Size to Estimate the Average Quality of a Lot or Process[4]

E 680 Test Method for Drop Weight Impact Sensitivity of Solid Phase Hazardous Materials[4]

3. Definitions

3.1 *acceleration of gravity* (g)—386.1 in./s^2 (9.806 m/s^2).

3.2 *critical acceleration* (A_c)—the maximum-faired acceleration level (see 8.3) above which product failure (or damage) occurs. A product usually has a different critical acceleration for each direction in which it is tested.

3.3 *critical velocity* (V_c)—the velocity change (see 8.2) below which product failure is unaffected by shock-pulse maximum-faired acceleration or waveform. A product usually has a different critical velocity for each direction in which it is tested.

3.4 *cushioning material*—a material used to isolate or reduce the effect of externally applied shock or vibration forces, or both.

3.5 *damage*—product failure that occurs during a shock test. Damage can render the product unacceptable because it becomes inoperable, fails to meet performance specifications, when its appearance is unacceptably altered, or some combination of these failure modes.

3.6 *damage boundary*—see Annex A3.

3.7 *shock pulse programmer*—a device used to control the parameters of the acceleration versus time shock pulse generated by a shock test machine.

3.8 *shock test machine drop height*—the distance through which the carriage of the shock test machine free falls before striking the shock pulse programmer.

3.9 Other definitions are given in Terminology D 996.

4. Significance and Use

4.1 These test methods are intended to provide the user with data on product shock fragility that can be used in choosing optimum-cushioning materials for shipping containers or for product redesign.

5. Apparatus

5.1 *Shock Test Machine:*

5.1.1 The machine shall consist of a flat horizontal test surface (carriage) of sufficient strength and rigidity to remain flat and horizontal under the stresses developed during the test. The test surface shall be guided to fall vertically without rotation or translation in other directions.

5.1.2 The machine shall incorporate sufficient carriage drop height to produce the shock pulses of 8.2 and 8.3. Drop height control shall be provided to permit reproducibility within ±0.25 in. (±6 mm).

5.1.3 The machine shall be equipped to produce shock pulses at the carriage as specified in 8.2 and 8.3.

5.1.4 Means shall be provided to arrest the motion of the carriage after impact to prevent secondary shock.

5.2 *Instrumentation:*

5.2.1 *Acceleration*—An accelerometer, a signal conditioner, and a data storage apparatus are required to record acceleration-time histories. The accelerometer shall be rigidly attached to the base structure of the product, or to the fixture, at or near a point where the fixture is fastened to the carriage. If the fixture is rigid enough so as not to distort the shock pulse imparted to the product, then the accelerometer may be mounted on the carriage. In some cases, when a

[1] These test methods are under the jurisdiction of ASTM Committee D-10 on Packaging and are the direct responsibility of Subcommittee D10.15 on Fragility Assessment.

Current edition approved March 25, 1988. Published May 1988.

[2] *Annual Book of ASTM Standards*, Vol 15.09.

[3] *Annual Book of ASTM Standards*, Vol 08.02.

[4] *Annual Book of ASTM Standards*, Vol 14.02.

product contains heavy resiliently supported masses which will severely distort the shock pulses, it may be necessary to precalibrate the shock machine. In this case the accelerometer is fastened to the carriage and a rigid mass weighing the same as the product is subjected to a series of shock pulses. The instrumentation system shall have sufficient response to permit measurements in the following ranges:

5.2.1.1 *Method A*—5 Hz or less to at least 1000 Hz.

5.2.1.2 *Method B*—3 Hz or less to at least 330 Hz.

5.2.1.3 *Accuracy*—Reading to be within ±5 % of the actual value.

5.2.1.4 *Cross-Axis Sensitivity*—Less than 5 % of the actual value.

5.2.2 *Velocity*—Instrumentation to measure the shock table's velocity change is required. This may be a device which electronically integrates the area under the shock pulse waveform. Alternatively, it can be measured by photodiode-type devices which measure shock table impact and rebound velocity. Calculation which assumes the shock pulse to be a perfect geometric figure usually is grossly inaccurate and should not be used.

6. Sampling

6.1 The sampling and the number of test specimens depend on the specific purposes and needs of the testing. Sample size determination based on Practice E 122 or other established statistical procedures is recommended.

7. Conditioning

7.1 If temperature and humidity conditioning is required for the product being tested, refer to Practice D 4332 for standard conditioning procedures. Conduct all tests with the same conditions prevailing, unless otherwise specified.

8. Procedure

8.1 Mount the product to be tested on the carriage of the shock test machine. The product should be supported by a fixture similar in shape and configuration to the cushion which will support the product in its shipping container. The fixture should be as rigid as possible so as not to distort the shock pulse imparted to the product. Securely fasten the fixture and product to the carriage so that it will not leave the surface of the carriage during the shock test.

NOTE 1—The points at which the fixture supports the product are very important because the dynamic response of the product is strongly influenced by the location of these support points.

NOTE 2—If the orientation of the product can change during handling impacts, then a test may be required for each of the directions in which the input shock can occur. Multidirectional tests are recommended since most products have different fragilities in different orientations.

8.2 *Method A—Critical Velocity Shock Test:*

8.2.1 *Scope*—This test method is used to determine the critical-velocity (V_c) portion of the damage boundary plot of a product.

8.2.1.1 A shock pulse having any waveform and having a duration (T_p) not longer than 3 ms may be used to perform this test. Shock pulse waveform is not limited since the critical velocity portion of the damage boundary is unaffected by shock pulse shape. Normally, since they are relatively easy to control, shock pulses having a half sine shock waveform are used. Occasionally, when testing small very rigid products, pulse durations shorter than 3 ms may be required.

8.2.2 *Procedure:*

8.2.2.1 Set the shock test machine so that the shock pulse produced has a velocity change below the anticipated critical velocity of the product.

8.2.2.2 Perform one shock test.

8.2.2.3 Examine or functionally test the product to determine if damage due to shock has occurred.

8.2.2.4 If no damage has occurred, set the shock test machine for a higher velocity change and repeat the shock test. Acceptable increment size is strongly influenced by the product being tested. For example, an increment of 5 in./s (0.13 m/s) may be appropriate for most products, but unacceptable for high value products.

8.2.2.5 Repeat 8.2.2.2 to 8.2.2.4 with incrementally increasing velocity change until product damage occurs. This point is shown as "Test No. 7" of Fig. A3.1.

8.2.2.6 Common practice is to define the critical velocity (V_c) as the midpoint between the last successful test and the test which produced failure. Depending on the purpose of the test, use of the last successful test point before failure may be considered as a more conservative estimate of (V_c).

8.3 *Method B—Critical Acceleration Shock Test:*

8.3.1 *Scope*—This test method is used to determine the critical acceleration (A_c) portion of the damage boundary plot of a product.

8.3.1.1 When a product's critical acceleration is known, package cushioning materials can be chosen to protect it.

8.3.1.2 If no cushioning materials are to be used in the package, it may be unnecessary to perform this test. In this case only the critical velocity test may suffice.

8.3.1.3 Trapezoidal shock pulses are normally used to perform this test. Although, in theory, a true square wave shock pulse is most desirable, it is not possible to obtain infinitely short rise and fall times. On the basis of much testing experience, it has been determined that rise and fall times (see Fig. A2.1) of 1.8 ms, or less, are required. Longer rise and fall times cause the critical acceleration line of the damage boundary curve to deviate from the horizontal, introducing errors in the test results. For the same reason waveforms having faired shapes which are not trapezoidal should not be used for this test. Their use would cause the critical acceleration line of the damage boundary curve to vary widely as a function of velocity change. As an example, if a half sine shock pulse waveform is used, a deeply scalloped critical acceleration line is produced and the test data becomes meaningless.

8.3.2 *Procedure:*

8.3.2.1 Set the shock test machine so that it will produce a trapezoidal shock pulse having a velocity change of at least 1.57 times as great as the critical velocity determined in Method A (8.2). For an added safety margin, a factor of 2 or more is normally used. This is required to avoid the rounded intersection of the critical velocity and critical acceleration lines. Maximum-faired acceleration level of the first shock pulse should be below the anticipated failure level of the product.

8.3.2.2 Perform one shock test.

8.3.2.3 Examine the recorded shock pulse to be certain the desired maximum-faired acceleration and velocity change were obtained.

8.3.2.4 Examine or functionally test the product to determine if damage due to shock has occurred.

8.3.2.5 If no damage has occurred, set the shock test machine for a higher maximum-faired acceleration level. Be certain the velocity change of subsequent shock pulses is maintained at or above the level determined in 8.3.2.1. Acceptable increment size is strongly influenced by the product being tested. For example, an increment of 5 *g*'s may be appropriate for most products, but unacceptable for high-value products.

8.3.2.6 Repeat 8.3.2.2 to 8.3.2.5 with incrementally increasing maximum-faired acceleration until product damage occurs. This point is known as "Test No. 14" of Fig. A3.1. Common practice is to define the critical acceleration (A_c) as the midpoint between the last successful test and the test which produced failure. Depending on the purpose of the test, use of the last successful test point before failure may be considered as a more conservative estimate of (A_c).

9. Report

9.1 *The report shall include the following:*

9.1.1 Complete identification of the product being tested including type, manufacturer's code numbers, general description of configuration, and its pretest condition.

9.1.2 Method of mounting the product on the carriage of the shock test machine.

9.1.3 Type of instrumentation used and critical settings thereof.

9.1.4 Recordings of the shock pulses which caused product damage.

9.1.5 Record of shock test machine drop height for each shock pulse that caused product damage.

9.1.6 Record of damage including photograph of product damage, if visible.

9.1.7 Record waveform, maximum-faired acceleration, pulse duration, and velocity change of the shock pulses.

9.1.8 Record of conditioning used.

9.1.9 Plots of the product's damage boundaries.

9.1.10 If multiple products are used, record sampling methods, average or median test levels, and standard deviations.

10. Precision and Bias

10.1 *Precision*—The within-laboratory or repeatability standard deviation is largely dependent on the particular item being tested. A research report[5] describes an interlaboratory test program of three types of items (in packages) for a critical velocity shock test. The repeatability standard deviations were 6.7, 14.7, and 21.5 in./s (0.17, 0.37, and 0.55 m/s). Other items may have more or less variability. The between laboratory or reproducibility standard deviation was 5.7 in./s (0.15 m/s).

10.2 *Bias*—No justifiable statement can be made on the bias of these test methods since a true value cannot be established by an accepted referee method.

[5] Available from ASTM Headquarters. Request RR: D10-1004.

ANNEXES

(Mandatory Information)

A1. HALF-SINE SHOCK PULSE VELOCITY CHANGE, USING INTEGRATING INSTRUMENTATION

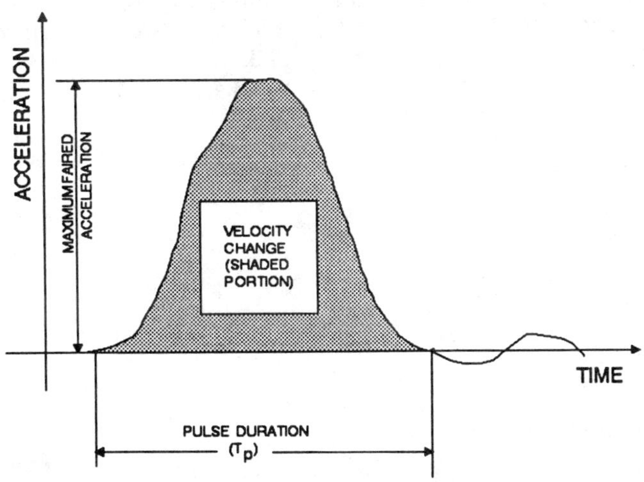

FIG. A1.1 Half-Sine Shock Pulse Diagram

A1.1 *Integrating Instrumentation*—Integrate the area under the curve from the point where the acceleration level first leaves the zero axis in a positive direction to the point where the acceleration next returns to zero (see Fig. A1.1).

A2. TRAPEZOIDAL SHOCK PULSE VELOCITY CHANGE USING INTEGRATING INSTRUMENTATION

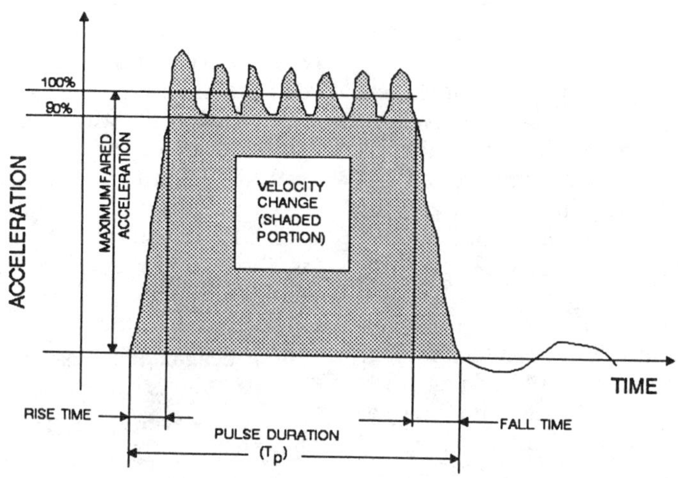

FIG. A2.1 Trapezoidal Shock Pulse Diagram

A2.1 *Integrating Instrumentation*—Integrate the area under the curve from the point where the acceleration level first leaves the zero axis in a positive direction to the point where the level next returns to zero (see Fig. A2.1).

A3. DAMAGE BOUNDARY

A3.1 Sensitivity to shock of a product is dependent on three parameters of the shock pulse: shock pulse shape, shock-pulse velocity change, and shock-pulse maximum-faired acceleration. For a given product, the interrelation of these three parameters is shown by damage boundary as plotted in Fig. A3.1.

A3.2 For shock pulses having peak acceleration and velocity-change values falling in the shaded area, product damage will occur. Shock pulses having values outside the shaded area will not damage the product. For most products the damage boundary will be different for each direction in which the shock occurs.

A3.3 The example plotted in Fig. A3.1 is based on tests conducted in accordance with Methods A and B. A sample of the product was subjected to half-sine shock pulses in accordance with Method A.

A3.3.1 Tests numbered 1 to 7 with both drop height and acceleration increasing successively, were performed. Failure occurred in the seventh test establishing the vertical critical-velocity line midway between the sixth and seventh test levels (see 8.2.2.6).

A3.3.2 Then another sample or a repaired sample of the product was subjected to trapezoidal shock pulses in accordance with Method B (8.3). Each trapezoidal shock pulse had a velocity change of more than two times the critical velocity (V_c) determined previously. Each trapezoidal shock pulse has a failed acceleration level incrementally higher than the previous shock pulse. Failure occurred in the fourteenth test, establishing the horizontal critical acceleration line (A_c) midway between the thirteenth and fourteenth test levels (see (8.3.2.6).

A3.4 When the damage boundary has plotted three things

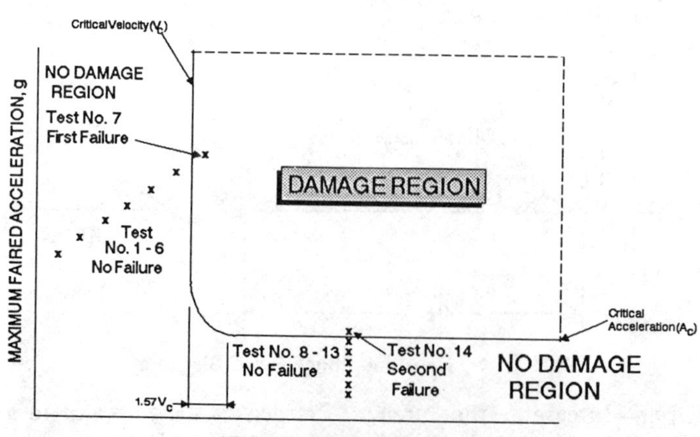

FIG. A3.1 Damage Boundary

can be learned from it:

A3.4.1 If the velocity change that the product will undergo in shipment is below the critical velocity (V_c), no cushioning is required.

A3.4.2 If the critical velocity (V_c) is below the velocity change which the product will be subjected to during unpackaged product handling, then the product should be modified to increase its critical velocity. Examples of unpackaged product handling are movement of the finished product on a production line, before packaging and customer handling and installation upon receipt. In these cases, the test will have shown that the unmodified product is too fragile to be handled in its normal production or in-use environment.

A3.4.3 If the velocity change that the product will undergo in shipment is above the critical velocity (V_c), package cushioning should be designed so that it transmits no more than the critical acceleration (A_c).

A3.4.4 The actual shape of the pulse transmitted to the product by the cushion is usually not known. The pulse shape depends on the dynamic force-versus-deflection characteristics of the cushion and will vary for different cushion materials, cushion deflections, etc. The damage boundary of a trapezoidal-shock pulse envelopes damage boundaries produced by other waveforms. For this reason shocks transmitted by some cushion materials will be less severe than those produced by the trapezoidal-shock pulse test. None will be more severe than those produced by the trapezoidal shock pulse. Therefore, the test in accordance with Method B (8.3) introduces a safety factor.

A3.4.5 As shown in Fig. A3.1, the corner where the critical velocity and critical acceleration lines intersect is rounded. To avoid inconclusive test results, the critical acceleration test is conducted at velocity changes at least two times the product's critical velocity. In this way the rounded region of the damage boundary is avoided.

A4. EFFECT OF MULTIPLE SHOCKS

A4.1 Methods A and B require that the product being tested be subjected to a series of shocks of incrementally increasing severity. Most products are not affected by this multiplicity of tests. However, some products will fail prematurely due to cumulative effects. When a second sample of such a product is subjected to a single shock pulse at the same level which caused the first sample to fail, it will not fail. Only when it is subjected to even higher level shocks will it fail. For a product of this type, it is important to determine the probable number of shocks which it will be subjected to in shipment. If significantly fewer shocks than those used in the test are anticipated, then the test data will have to be corrected. Usually multiple samples of such a product are tested.

A4.2 If only a few samples of the product are available, a simplified calculation technique may be used to determine the effect of multiple shocks. After the tests of the first sample, successive samples are tested at shock levels beginning near the failure level of the first sample. Three to five new or repaired test items are often used for each test orientation and for each part of the damage boundary (V_c and A_c). The failure level is then defined as the average (arithmetic mean) of the midpoints between the last tests and the test which produced failure (excluding the first sample, which failed prematurely due to cumulative effects). This procedure is less accurate than the procedure described in A4.3.

A4.3 A test procedure known as the "up-and-down" or "staircase" method is well suited for use in product fragility testing. Several specimens are tested sequentially with the test specimen being discarded or repaired after each individual shock test. The first specimen is tested at the estimated failure point. If it fails at that shock level, the next specimen is tested at a level which is a fixed increment lower. If it passes, the specimen is tested at a shock level which is incrementally higher. The shock input for each test is thus determined by the previous test result.

A4.3.1 At the completion of a fixed number of tests, often ten or more, an average or median value and the standard deviation are calculated. This procedure is repeated for each orientation and each part of the damage boundary (V_c and A_c) which is of interest. When possible analyze the data for normality (reasonable conformance with Gaussian probability distribution).

A4.3.2 Several texts[6,7,8] on experimental statistics, listed below, describe this procedure and computations in detail. In addition, Test Methods D 2463 and E 680 also describe this procedure.

A4.4 The effect of multiple shocks should be considered even if only a single sample of the product is available for testing. If the product is complex, usually some sub-element of the product will fail first. Frequently, even though the product may be a prototype, additional sub-elements are available to replace the one which was damaged. In this case the procedure of A4.2 may be used.

A4.4.1 If all parts of the product are one-of-a-kind and no more are available, then a correction factor allowing for the effects of multiple tests may have to be used. Such a factor will vary widely for different types of products. As more product samples become available, test results should be refined using the procedures of either A4.2 or A4.3.

[6] Dixon, W.J., and Massey, F.J., *Introduction to Statistical Analysis*, McGraw Hill, 1969, p. 377–393.

[7] Lipson, C., and Sheth, N.J., *Statistical Design and Analysis of Engineering Experiments*, McGraw Hill, 1973, pp. 270–274.

[8] Natrella, M.G., *Experimental Statistics*, NBS Handbook 91, U.S. Government Printing Office, 1963, p. 10–1, 10–22, and 10–23.

Standard Test Method of Vibration (Vertical Sinusoidal Motion) Test of Products[1]

This standard is issued under the fixed designation D 3580; the number immediately following the designation indicates the year of original adoption or, in the case of revision, the year of last revision. A number in parentheses indicates the year of last reapproval. A superscript epsilon (ε) indicates an editorial change since the last revision or reapproval.

1. Scope

1.1 This test method covers the determination of resonances of unpackaged products and components of unpackaged products by means of vertical linear sinusoidal motion at the surface on which the product is mounted for test.

1.2 This information may be used for the design of a container or interior package that will minimize transportation vibration inputs at these critical frequencies, when these products resonances are within the expected transportation environment frequency range. Since vibration damage is most likely to occur at product resonant frequencies, these resonances may be thought of as potential product fragility points.

1.3 Information obtained from this test method can be used for product modification. This may become necessary if a product's response would require design of an impractical or excessively costly shipping container.

1.4 This test method does not simulate the vibration effects that the product will encounter in its operational or in-use environment. Other, more suitable test procedures should be used for this purpose. See Section 2.

1.5 Test levels given in this test method represent the correlation of the best information currently available from research investigation and from experience in the use of this method. If more applicable or accurate data are or become available, they should be substituted.

1.6 *This standard does not purport to address all of the safety problems associated with its use. It is the responsibility of the user of this standard to establish appropriate safety and health practices and determine the applicability of regulatory limitations prior to use.* See Section 6 for specific precautionary statements.

2. Referenced Documents

2.1 *ASTM Standards:*

D 996 Terminology of Packaging and Distribution Environments[2]

D 4169 Practice for Performance Testing of Shipping Containers and Systems[2]

D 4332 Practice for Conditioning Containers, Packages, or Packaging Components for Testing[2]

E 122 Practice for Choice of Sample Size to Estimate the Average Quality of a Lot or Process[3]

2.1 *Military Standard:*

MIL-STD 810 D, Method 514 Vibration[4]

2.2 *International Electrotechnical Commission Recommendation Publication:*

68-2-6, Part 2, Test F Vibration, Basic Environmental Testing Procedures for Electronic Components and Electrical Equipment[5]

3. Terminology

3.1 *Definitions*—For definitions of terms used in this standard, see Terminology D 996.

3.2 *Descriptions of Terms Specific to This Standard:*

3.2.1 *amplification*—the ratio of measured acceleration level of the component to the measured input acceleration to the product.

3.2.2 *amplitude*—the maximum value of a sinusoidal quantity (zero to peak).

3.2.3 *octave*—the interval of two frequencies having a basic frequency ratio of 2 (1 octave/min = 0.301 decade/min).

3.2.4 *resonance*—for a system undergoing forced vibration, the frequency at which any change of the exciting frequency causes a decrease in the response of the system.

4. Significance and Use

4.1 Products are exposed to complex dynamic stresses in the transportation environment. The determination of the resonant frequencies of the product may aid the packaging designer in determining the proper packaging system to provide adequate protection for the product, as well as providing an understanding of the complex interactions between the components of the product as they relate to expected transportation vibration inputs.

5. Apparatus

5.1 *Vibration Test Machine*—The machine shall consist of a flat horizontal test surface of sufficient strength and rigidity such that the applied vibrations are essentially uniform over the entire test surface. The test surface shall be driven to move only vertically in sinusoidal motion (a rotary motion is not acceptable). The frequency and amplitude of the motion shall be variable, under control, to cover the range specified in 10.3.

[1] This test method is under the jurisdiction of ASTM Committee D-10 on Packaging and is the direct responsibility of Subcommittee D10.15 on Fragility Assessment.

Current edition approved June 29, 1990. Published August 1990. Originally published as D 3580 – 77T. Last previous edition D 3580 – 89.

[2] *Annual Book of ASTM Standards*, Vol 15.09.

[3] *Annual Book of ASTM Standards*, Vol 14.02.

[4] Available from Naval Publications and Forms Center, 5801 Tabor Ave., Philadelphia, PA 19120.

[5] Available from American National Standards Institute, Inc., 1430 Broadway, New York, NY 10018.

5.2 *Specimen-Mounting Devices*—Devices of sufficient strength and rigidity are required to attach the product securely to the test surface. The mounting devices shall not have resonances in the test frequency range. They shall support the product in a manner similar to the way in which it will be supported in its shipping container. Relative motion between the test surface and product shall not be permitted.

5.3 *Instrumentation:*

5.3.1 Sensors, signal conditioners, filters, and a data acquisition apparatus are required to monitor or record, or both, the accelerations and frequencies at the test surface of the apparatus and at points of interest in the product. The instrumentation system shall have a response accurate to within ±5 % over the test range.

5.3.2 For visual examination of the specimen under test, a stroboscope or video system may be beneficial.

6. Safety Precautions

6.1 **Precaution:** This test method may produce severe mechanical response in the product being tested. Therefore, the means used to fasten the product to the test surface must be of sufficient strength to keep it adequately secured. Operating personnel must remain alert to potential hazards and take necessary precautions for their safety. Stop the test immediately if a dangerous condition should develop.

7. Sampling

7.1 Test specimens and number of samples shall be chosen to permit an adequate determination of representative performance. Whenever sufficient products are available, five or more replicate samples should be tested to improve the statistical reliability of the data obtained (see Practice E 122).

8. Test Specimens

8.1 The product as intended for packaging shall constitute the test specimen. Sensor(s) may be applied as appropriate to measure data points of interest with the minimum possible alteration of the test specimen. Parts and surfaces of the specimen may be marked for identification and reference. When necessary to observe the interior components of the product during tests, holes may be cut in non-critical areas or noncritical panels may be removed.

9. Conditioning

9.1 Condition test specimens prior to test and maintain in accordance with any requirements. In the absence of other requirements, conditioning in accordance with Practice D 4332 is recommended (standard conditioning atmosphere of 23 ± 2°C (73.4 ± 3.6°F) and 50 ± 2 % relative humidity).

10. Procedure

10.1 Perform the tests in the conditioned environment or immediately upon removal from that environment.

10.2 Attach the test specimen to the test surface, near the center of the apparatus in a manner that will prevent the specimen from leaving or moving across the test surface during vibration. Caution is necessary to avoid excessive pressure or mounting methods that could influence the characteristics of the product.

10.3 Test intensities shall be sufficient to vibrate the product at acceleration and frequency levels that determine if product resonances exist in the expected transportation environment. Experience has shown that individual transportation environments may contain frequencies ranging from 0.2 Hz to above 100 Hz. Acceleration levels sufficient to excite resonance normally range from 0.1 to 0.5 g.

10.3.1 The following test procedure may be used. Sweep the frequency range from 3 to 100 Hz and return using automatic or manual sweep, while maintaining a nearly constant acceleration level.

10.3.2 Select an acceleration level between 0.1 and 0.5 g (zero to peak). Starting at 3 Hz, vary the frequency of vibration at a continuous logarithmic rate of .5 to 1 octave/min to 100 Hz and back to 3 Hz. Record any resonant responses of the product, repeat the cycle if necessary.

NOTE 1—For some specific product/environmental combinations higher frequencies may be required to produce product resonances. For an example see MIL-STD 810D.

10.4 Monitor the acceleration and frequency data sensed on test surface to ensure that the desired test conditions are produced. Mount the accelerometer, either top or bottom, as close to the center of the test surface as possible.

10.5 Monitor the test specimen and its components for any resonant vibrations. Use a stroboscope, sensors, visual, auditory or other means as applicable to determine these resonances. If necessary, the frequency sweep may be interrupted for short time periods in order to properly identify a resonating component.

10.6 Record the frequencies of any resonances and identify the product components that are resonating.

10.7 Test the product in each of the potential shipping orientations of concern.

10.8 *(Optional)* Dwell at each resonant frequency found in 10.3.1, if it is determined to be within the expected transportation environment, dwell time, acceleration level, and damage criteria to be specified by the user. Adjust the frequency of the vibration as necessary to maintain resonance.

NOTE 2—If no dwell time is specified, a dwell time of 15 min is recommended. Practice D 4169 may be referred to for test durations.

11. Report

11.1 The report shall include the following:

11.1.1 Description of the test product in sufficient detail for proper identification.

11.1.2 Identification of the purpose of the test.

11.1.3 Descriptions of the test sequence, the input acceleration level, frequency range swept, and sweep rate.

11.1.4 Descriptions of any deviations from the specified test method.

11.1.5 A statement of the number of test replications, if any.

11.1.6 Identification of apparatus and instrumentation used, including dates of last instrument calibrations, manufacturer's names and model numbers. Details of any modifications thereto, if known, shall be included.

11.1.7 Method of conditioning.

11.1.8 Results of any prior tests.

11.1.9 Components that displayed resonant vibration and

their corresponding frequencies. Any other significant data including measurements and observations shall be included.

11.1.10 If applicable, identification of components that could be redesigned to eliminate excessive resonant vibration or to change the resonant frequency.

11.1.11 Resonant amplifications (the ratio of measured acceleration level of component over the measured input acceleration to the product), when measured, shall be reported.

11.1.12 If applicable, a description of shipping container characteristics desirable for vibration protection of the product.

12. Precision and Bias

12.1 No statement is made about either the precision or bias of this test method since the results merely indicate the presence of responses from vertical linear sinusoidal vibration input.

Standard Test Method for Holding Power of Pressure-Sensitive Tapes[1]

This standard is issued under the fixed designation D 3654; the number immediately following the designation indicates the year of original adoption or, in the case of revision, the year of last revision. A number in parentheses indicates the year of last reapproval. A superscript epsilon (ε) indicates an editorial change since the last revision or reapproval.

[ε1] NOTE—The address in Footnote 5 was changed editorially in May 1990.

1. Scope

1.1 This test method provides procedures to determine the ability of pressure-sensitive tapes to remain adhered under constant load applied parallel to the surfaces of the tape. They may also be used to measure the displacement or slippage of such tapes under a standard load in a given or specified time period.

1.1.1 Procedure A gives a measure of holding power to a standard fiberboard surface.

1.1.2 Procedure B gives a measure of holding power of filament reinforced tapes to a standard fiberboard surface.

1.1.3 Procedure C gives a measure of holding power to a standard steel surface at an elevated temperature.

NOTE 1—A complete metric companion to Test Method D 3654 has been developed—D 3654M; therefore, no metric equivalents are presented in this method. Test Method D 3654M uses test areas of 12 by 12 mm and 12 by 75 mm, rather than 0.5 by 0.5 in. and 0.5 by 3 in. used here. This difference in test parameters will generally produce lower times to failure for Test Method D 3654M than for Test Method D 3654 but not necessarily by a constant factor.

1.2 *This standard does not purport to address the safety problems associated with its use. It is the responsibility of the user of this standard to establish appropriate safety and health practices and determine the applicability of regulatory limitations prior to use.*

2. Referenced Documents

2.1 *ASTM Standards:*

A 666 Specification for Austenitic Stainless Steel, Sheet, Strip, Plate, and Flat Bar[2]

D 3715 Practice for Quality Assurance of Pressure-Sensitive Tapes[3]

E 122 Practice for Choice of Sample Size to Estimate the Average Quality of a Lot or Process[4]

3. Summary of Test Method

3.1 A standard-size tape specimen is applied to a specified test surface with controlled pressure. The tape is subjected to a shear force by use of a specified mass acting parallel to the surfaces of the specimen.

3.1.1 *Procedure A*—A standard area of tape, generally a tape used for box closure applications, is applied to a vertical panel that is covered with a standard fiberboard surface. A standard mass is applied and allowed to act until failure.

3.1.2 *Procedure B*—A standard area of filament reinforced tape is applied to a horizontal panel covered with a standard fiberboard surface. A standard mass is applied to the end of the specimen after the specimen is formed around a 90° angle to the vertical plane, and allowed to act for a specific time.

3.1.3 *Procedure C*—A standard area is applied to a standard steel vertical surface. A standard mass is applied. The test is conducted at an elevated temperature.

4. Significance and Use

4.1 Procedure A may be used to determine the holding power of tapes generally used to close fiberboard boxes in packaging applications.

4.2 Procedure B may be used to determine the ability of a filament reinforced tape to hold to fiberboard when placed under a constant stress.

4.3 Procedure C may be used to compare the holding power of tapes when applied to a standard steel surface with the duration of the test reduced by conducting the test at an elevated temperature.

5. Apparatus

5.1 *Specimen Cutter*[5]—The specimen cutter shall hold two single-edged razor blades in parallel planes precisely 0.5 in. apart.

NOTE 2—The 0.5-in. cutter shall consist of a 0.5 in. thick by 8 in. length of aluminum bar stock 0.5 in. wide. The edges for about 5 in. for one end shall be rounded slightly to form a handle. The width of the bar for 3 in. from the opposite end shall be narrowed to exactly 0.500 in. minus the thickness of a single razor blade (one of two used as cutting edges). The razor blades shall be held in position using side plates. The end of the cutter shall be cut away at a 45° angle to expose the cutting edge at one end of the blades. The edges shall be separated by 0.500 ± 0.005 in.

5.2 *Solvents and Related Materials:*

5.2.1 *Diacetone Alcohol*, nonresidual, technical grade or better.

5.2.2 One of the following reagent grade materials:

5.2.2.1 *n*-heptane,

5.2.2.2 methyl alcohol,

[1] This test method is under the jurisdiction of ASTM Committee D-10 on Packaging and is the direct responsibility of Subcommittee D10.14 on Closure and Reinforcement of Packages.

Current edition approved Feb. 26, 1988. Published May 1988. Originally published as D 3654 – 78. Last previous edition D 3654 – 87.

[2] *Annual Book of ASTM Standards*, Vol 01.03.

[3] *Annual Book of ASTM Standards*, Vol 15.09.

[4] *Annual Book of ASTM Standards*, Vol 14.02.

[5] Available from the Pressure-Sensitive Tape Council (PSTC), 104 Wilmot Rd., Suite 201, Deerfield, IL 60015.

5.2.2.3 methyl ethyl ketone (MEK), or
5.2.2.4 methyl isobutyl ketone (MIBK).

NOTE 3—Where toxicity and flammability requirements are primary concern, a mixture of *n*-heptane and a fluorinated hydrocarbon may be used.

5.2.3 *Dispensing System for Solvents*, such as a wash bottle.

5.2.4 *Absorbent Cleaning Material*, such as surgical gauze or tissue. To be suitable, cleaning material must be free of lint during use, absorbent, and contain no additives that are soluble in the solvents listed in 5.2.2, and made exclusively of virgin materials.

5.3 *Panels*[5]*:*

5.3.1 *Procedures A and C*—A 2 by 5 by no less than 0.043-in. thick panel of stainless steel Type 302 and 304 in accordance with Specification A 666 and having a bright annealed finish. The surface roughness height shall be 2.0 ± 1.0 µin. arithmetical average deviation from the mean line. One or both of the panel ends shall be ground to form a 90° angle with the panel surface. Panels showing stains, discolorations, or numerous scratches are not acceptable.

5.3.2 *Procedure B*—A 2 by 5 by no less than 0.043 in. thick steel panel. A ½-in. length of one end of the panel shall be bent through an arc of 120° away from the test surface. The radius of curvature of the finished surface at the bend shall be 1/16 to ⅛ in.

5.4 *Roller,*[5] *Hand-Operated or Mechanically Operated:*

5.4.1 A steel roller, 3.25 ± 0.1 in. in diameter and 1.75 ± 0.05 in. in width, covered with rubber approximately 0.25 in. in thickness, having a Shore scale A durometer hardness of 80 ± 5. The surface of the roller shall be a true cylinder, void of concave or convex deviations. The mass of the roller shall be 4.5 ± 0.1 lb. If mechanically operated, rate shall be 12.0 ± 0.5 in./min.

5.4.2 The roller construction shall not allow the mass of the handle to increase the force of the roller during use.

5.5 *Test Stands and Ancillary Apparatus:*

5.5.1 *Procedures A and C*—A test stand, that will hold the test panel, with tape applied, at an angle of 0 to 2° with the vertical, so that when the mass is acting on the test specimen, no peel forces will be exerted on the tape.

5.5.2 *Procedure B*—A test stand that will support the test panel in a horizontal plane, approximately 12 in. above the work surface.

5.5.3 *A clamp*, that will allow the attachment of the mass to the specimen, distributing the load equally across the tape specimen width.

5.6 *Test Masses:*

5.6.1 *Procedures A and C*—The test mass shall be 1000 ± 5 g, or other mass as specified. The mass of the clamp described in 5.5.3 shall be included as part of the total mass.

5.6.2 *Procedure B*—The test mass shall be 10 ± 0.5 lb, or other mass, as specified. The mass of the clamp described in 5.5.2 shall be included as part of the total mass.

5.6.3 *A timing system*, to measure interval between the application of the load to the specimen and its separation from the panel.

6. Sampling

6.1 *Acceptance Sampling*—Sampling shall be in accordance with Practice D 3715. For Procedures A and C, five replicate specimens shall be averaged as required by Section 10 for each test result. No single specimen value shall be considered as representative of the roll under test.

6.2 *Sampling for Other Purposes*—The sampling and the number of test specimens depends on the purpose of the testing. Practice E 122 is recommended. It is common to test at least five specimens of a particular tape. Test specimens should be taken from several rolls of a tape and, whenever possible, among several production runs of a tape. Strong conclusions about a specific property of a tape cannot be based on test results of a single unit (roll) of product.

7. Conditioning

7.1 Condition the sample rolls of tape at 73.4 ± 3.6°F and 50 ± 5 % relative humidity for at least 24 h prior to testing. Place the rolls of tape and the test surface so that the conditioning atmosphere has free access to all normally exposed surfaces.

7.2 Unless otherwise specified, conduct all tests in the standard conditions.

8. Test Specimens

8.1 *Removal From Roll:*

8.1.1 Unwind and discard at least three but no more than six outer wraps of tape from the sample roll before taking a specimen for testing.

8.1.2 Remove one specimen (or more when specified), from random area(s) within the sample roll. Remove each specimen from a freely rotating roll at rate of 20 to 30 in./s. Where width or other factors cause high adherence to tape backing and makes removal at prescribed rates impossible, remove at a rate as close to 20 in./s as possible.

8.2 When the sample roll width is wider than the required specimen width, test specimens shall be cut from the center of a strip removed from the roll as described in 8.1.1 and 8.1.2 using the appropriate specimen cutter.

8.3 *Specimen Size:*

8.3.1 *Procedures A and C*—The specimen shall be 0.5 ± 0.016 in. wide. The length shall be approximately 6 in.

8.3.2 *Procedure B*—The specimen shall be 0.5 ± 0.016 in. wide and approximately 12 in. long.

9. Procedure

9.1 For Procedures A and B, apply by means of a double-coated pressure-sensitive tape a 1 by 4-in. piece of Standard Reference Material 1810A[6] with the 1-in. dimension centered at one end of the test panel. See 5.3.1 and 5.3.2.

NOTE 4—Take care that the 1810 material is applied with the proper side up and with the grain of the paper perpendicular to the 4-in. dimension of the paper. The 4-in. dimension should be parallel to the long edge of the test panel.

9.2 *Procedure A:*

9.2.1 Center the tape specimen on the 1-in. dimension of the test paper, see 9.1, at one end of the test panel and apply without added pressure to cover an area exactly 0.5 by 0.5 in. with the tape. The adhesive surface of the free end of the

[6] Standard Reference Material 1810A is available from the Office of Standard Reference Materials, National Institute of Standards and Technology, Washington, DC 20234.

specimen should be masked by applying a strip of paper approximately 5 in. by the width of the specimen.

9.2.2 To prevent cutting the specimen by the end of the panel, place another panel of the same or slightly lesser thickness under the free masked end of the specimen, and in contact with the end of the test panel prior to roll down. Roll down the applied test area twice, once in each lengthwise direction, at a velocity of 12 ± 0.5 in./min.

9.2.3 Place the clamp on the masked free end of the specimen, ensuring that the clamp extends completely across the width of the specimen and is aligned to uniformly distribute the load.

9.2.4 Place the test assembly in the test stand so that the free end of the test specimen is vertical, ensuring that no peel forces act on the specimen.

9.2.5 Apply the 1000-g mass to the clamp gently so as not to cause any shear impact force on the tape specimen. Record the time elapsed in which the tape specimen has completely separated from the test panel.

9.3 *Procedure B:*

9.3.1 Apply one end of the specimen, about 4 in. in length, adhesive side down, to the longitudinal surface of the test panel, see 9.1. The tape must be at a true right angle to the bent edge of the panel. Allow the remaining 8 in. to extend over and beyond the bent edge of the panel.

9.3.2 Using a square, cut across and through the width of the tape specimen 3.0 in. back from the front of the bend in the horizontal plane of the test panel surface.

9.3.3 Pass the rubber roller four times over the surface of the test specimen, twice in each lengthwise direction, at a velocity of 12 ± 0.5 in./min.

9.3.4 Place the clamp on the free end of the specimen, ensuring that the clamp extends completely across the width, and is aligned to distribute the load uniformly.

9.3.5 Place the test assembly in the test stand so that the free end of the test specimen is vertical. Apply the specified 10-lb. mass to the clamp gently so as not to cause any impact force on the specimen.

9.3.6 At the end of 48 h under load, examine the specimen for evidence of slippage. Measure any slippage that has occurred to the nearest 1⁄32 in.

9.4 *Procedure C:*

9.4.1 Dispense diacetone alcohol onto the panel. **Caution:** All operations with solvents should be conducted in a well-ventilated hood. Scrub the panel with a clean piece of absorbent cleaning material. Dry the panel with fresh absorbent cleaning material. Dispense one of the solvents listed in 5.2.2 onto the panel, wiping it to dryness with fresh absorbent cleaning material. Repeat for a total of three washes with this solvent. Do not touch the cleaned panel surface with fingers.

9.4.2 Apply test specimen to test surface as described in 9.2.1, except test panel shall be steel as described in 5.3.1. Prepare test assembly as described in 9.2.2, 9.2.3, and 9.2.4.

9.4.3 Place the test strand in an oven maintained at 120 ± 2°F.

9.4.4 Allow to condition at 120°F for 10 min, then apply the 1000-g mass to the clamp gently so as not to cause any shear impact force on the tape specimen. Record the time elapsed in which the tape specimen has completely separated from the test panel.

10. Calculation

10.1 *Acceptance Sampling:*

10.1.1 *Procedures A and C*—To determine the test result for each roll of tape, convert each of the five specimen test results (times to failure) to its common or natural logarithm. Obtain the arithmetic mean of all logarithms and then convert back to time by obtaining the appropriate antilogarithm. This gives the test result for the roll of tape under consideration in the sampling plan.

10.1.2 *Procedure B*—It is common to interpret the results of Procedure B as passing or failing a pre-set acceptance criterion.

10.2 *Other Purposes*—Convert the test result (time to failure) from each specimen to its common or natural logarithm. Conduct an analysis of the logarithms of the data to determine the desired descriptive or comparative statistics. The resulting means and confidence intervals may be converted back to the original time units by obtaining the appropriate antilogarithm.

11. Report

11.1 The report shall include the following:

11.1.1 Method designation.

11.1.2 Manufacturer's name and tape designation.

11.1.3 Procedure used.

11.1.4 *For Procedure A or C*, amount of time for the tape to separate completely from the panel.

11.1.5 *For Procedure B*, distance slipped to nearest 1⁄32 in. in 48 h.

11.1.6 Any observations of tape behavior as called for in the material specification.

11.1.7 Individual roll results shall be reported unless otherwise specified.

11.1.8 When using sampling as described in 6.2, report average calculated in 10 and the number of observations.

12. Precision and Bias

12.1 *Precision:*

12.1.1 *Procedure A*—Repeatability (within laboratory variability) information is available from one laboratory; several types of tape were evaluated using the time to failure in minutes converted to natural logarithms for analysis. Examples of averages and within roll standard deviations (including test error) are:

Tape	X	Converted X, min	Within Roll σ
A	8.74	6248	0.32
B	8.27	3905	0.23
C	2.43	11.4	0.27
D	6.50	665	0.45
E	5.98	395	0.82

The remaining components of variance are expected to be similar to those reported for Procedure C. The between roll standard deviation and the residual (including within roll) standard deviation will vary, depending on the tape type and the manufacturer.

12.1.2 *Procedure B*—No statement is made about the precision of Procedure B since the result merely states whether there is conformance to the specified criteria for success.

12.1.3 *Procedure C*—An interlaboratory evaluation of two types of pressure sensitive tapes by three laboratories has been conducted for Procedure C: the results are available.[7] It is based on an evaluation using the lognormal distribution; the times to failure in minutes were converted to their natural (base e) logarithms for the analysis. The tapes tested in this study had a grand mean of 1.2 which converts to 3.3 min. The components of variance for the following factors were estimated: between laboratory, between operators in a laboratory, between rolls of a tape, and residual (including test error and within-roll variation).

A summary of the pooled standard deviation is:

Between laboratories	0.46
Between testers	0.247
Between rolls	0.185
Residual	0.411

These components may be combined in several ways to obtain the desired repeatability (within laboratory) and reproducibility (between laboratory) estimates of precision. One company reports that, with care, the between laboratory standard deviation can be reduced to about 0.2. The between roll standard deviation and the residual (including within roll) standard deviation may be higher or lower than reported here, depending on the tape type and the manufacturer.

12.2 *Bias*—No measure of bias is possible with this test method because an accepted reference or referee value is not available.

[7] Supporting data are available from ASTM Headquarters. Request RR: D-10-1002, Report 2.

Standard Test Method for
Holding Power of Pressure-Sensitive Tapes [Metric][1]

This standard is issued under the fixed designation D 3654M; the number immediately following the designation indicates the year of original adoption or, in the case of revision, the year of last revision. A number in parentheses indicates the year of last reapproval. A superscript epsilon (ε) indicates an editorial change since the last revision or reapproval.

[ε1] NOTE—The address in Footnote 5 was changed editorially in May 1990.

1. Scope

1.1 This test method provides procedures to determine the ability of pressure-sensitive tapes to remain adhered under constant load applied parallel to the surfaces of the tape. They may also be used to measure the displacement or slippage of such tapes under a standard load in a given or specified time period.

1.1.1 Procedure A gives a measure of holding power to a standard fiberboard surface.

1.1.2 Procedure B gives a measure of holding power of filament reinforced tapes to a standard fiberboard surface.

1.1.3 Procedure C gives a measure of holding power to a standard steel surface at an elevated temperature.

NOTE 1—This test method is a metric companion to Test Method D 3654 and uses test areas of 12 by 12 mm and 12 by 75 mm, rather than 0.5 by 0.5 in. and 0.5 by 3 in. used in Test Method D 3654. This difference in test parameters will generally produce lower times to failure for Test Method D 3654M than for Test Method D 3654 but not necessarily by a constant factor.

1.2 *This standard does not purport to address all of the safety problems associated with its use. It is the responsibility of the user of this standard to establish appropriate safety and health practices and determine the applicability of regulatory limitations prior to use.* For a specific cautionary statement, see Note 5.

2. Referenced Documents

2.1 *ASTM Standards:*

A 666 Specification for Austenitic Stainless Steel, Sheet, Strip, Plate, and Flat Bar[2]

D 3715 Practice for Quality Assurance of Pressure-Sensitive Tapes[3]

E 122 Practice for Choice of Sample Size to Estimate the Average Quality of a Lot or Process[4]

3. Summary of Test Method

3.1 A standard-size tape specimen is applied to a specified test surface with controlled pressure. The tape is subjected to a shear force by use of a specified mass acting parallel to the surfaces of the specimen.

3.1.1 *Procedure A*—A standard area of tape, generally a tape used for box closure applications, is applied to a vertical panel that is covered with a standard fiberboard surface. A standard mass is applied and allowed to act until failure.

3.1.2 *Procedure B*—A standard area of filament reinforced tape is applied to a horizontal panel covered with a standard fiberboard surface. A standard mass is applied to the end of the specimen after the specimen is formed around a 90° angle to the vertical plane, and allowed to act for a specified time.

3.1.3 *Procedure C*—A standard area is applied to a standard steel vertical surface. A standard mass is applied. The test is conducted at an elevated temperature.

4. Significance and Use

4.1 Procedure A may be used to determine the holding power of tapes generally used to close fiberboard boxes in packaging applications.

4.2 Procedure B may be used to determine the ability of a filament reinforced tape to hold to fiberboard when placed under a constant stress.

4.3 Procedure C may be used to compare the holding power of tape when applied to a standard steel surface with the duration of the test reduced by conducting the test at an elevated temperature.

5. Apparatus

5.1 *Specimen Cutter*[5]—The specimen cutter shall hold two single-edged razor blades in parallel planes precisely 12 mm apart.

NOTE 2—The 12-mm cutter shall consist of a 12 mm thick by 200 mm length of aluminum bar stock 12 mm wide. The edges for about 125 mm from one end shall be rounded slightly to form a handle. The width of the bar for 75 mm from the opposite end shall be narrowed to exactly 12 mm minus the thickness of a single razor blade (one of two used as cutting edges). The razor blades shall be held in position using side plates. The end of the cutter shall be cut away at a 45° angle to expose the cutting edge at one end of the blades. The edges shall be separated by 12 ± 0.1 mm.

5.2 *Solvents and Related Materials:*

5.2.1 *Diacetone Alcohol*, nonresidual, technical grade or better.

5.2.2 One of the following reagent grade materials:

5.2.2.1 *n*-heptane,

5.2.2.2 methyl alcohol,

[1] This test method is under the jurisdiction of ASTM Committee D-10 on Packaging and is the direct responsibility of Subcommittee D 10.14 on Closure and Reinforcement of Packages.

Current edition approved Feb. 26, 1988. Published May 1988. Originally published as D 3654M – 82. Last previous edition D 3654M – 87.

[2] *Annual Book of ASTM Standards*, Vol 01.03.

[3] *Annual Book of ASTM Standards*, Vol 15.09.

[4] *Annual Book of ASTM Standards*, Vol 14.02.

[5] Available from the Pressure Sensitive Tape Council (PSTC), 104 Wilmot Rd., Suite 201, Deerfield, IL 60015.

5.2.2.3 methyl ethyl ketone (MEK), or
5.2.2.4 methyl isobutyl ketone (MIBK).

NOTE 3—Where toxicity and flammability requirements are primary concern, a mixture of *n*-heptane and a fluorinated hydrocarbon may be used.

5.2.3 *Dispensing System for Solvents*, such as a wash bottle.

5.2.4 *Absorbent Cleaning Material*, such as surgical gauze or tissue. To be suitable, cleaning material must be free of lint during use, absorbent, and contain no additives that are soluble in the solvents listed in 5.2.2, and made exclusively of virgin materials.

5.3 *Panels*[5]*:*

5.3.1 *Procedures A and C*—A 50 by 125 by no less than 1.1 mm thick panel of stainless steel Type 302 or 304 in accordance with Specification A 666 and having a bright annealed finish. The surface roughness height shall be 50 nm ± 5 nm arithmetical average deviation from the mean line. One or both of the panel ends shall be ground to form a 90° angle with the panel surface. Panels showing stains, discolorations, or numerous scratches are not acceptable.

5.3.2 *Procedure B*—A 50 by 125 by no less than 1.1 mm thick steel panel. A 12-mm length of one end of the panel shall be bent through an arc of 120° away from the test surface. The radius of curvature of the finished surface at the bend shall be 1.5 to 3 mm.

5.4 *Roller,*[5] *Hand-Operated or Mechanically Operated:*

5.4.1 A steel roller, 82.5 ± 2.5 mm in diameter and 45 ± 1.5 mm in width, covered with rubber approximately 6 mm in thickness, having a Shore scale A durometer hardness of 80 ± 5. The surface of the roller shall be a true cylinder, void of concave or convex deviations. The mass of the roller shall be 2040 ± 45 g. If mechanically operated, rate shall be 5 ± 0.2 mm/s.

5.4.2 The roller construction shall not allow the mass of the handle to increase the force of the roller during use.

5.5 *Test Stands and Ancillary Apparatus:*

5.5.1 *Procedures A and C*—A test stand, that will hold the test panel, with tape applied, at an angle of 0 to 2° with the vertical, so that when the mass is acting on the test specimen, no peel forces will be exerted on the tape.

5.5.2 *Procedure B*—A test stand that will support the test panel in a horizontal plane, approximately 300 mm above the work surface.

5.5.3 *A clamp*, that will allow the attachment of the mass to the specimen, distributing the load equally across the tape specimen width.

5.6 *Test Masses:*

5.6.1 *Procedures A and C*—The test mass shall be 1000 ± 5 g, or other mass as specified. The mass of the clamp described in 5.5.3 shall be included as part of the total mass.

5.6.2 *Procedure B*—The test mass shall be 4.5 ± 0.2 kg, or other mass, as specified. The mass of the clamp described in 5.5.3 shall be included as part of the total mass.

5.6.3 *A timing system*, to measure interval between the application of the load to the specimen and its separation from the panel.

6. Sampling

6.1 *Acceptance Sampling*—Sampling shall be in accordance with Practice D 3715. For Procedures A and C, five replicate specimens shall be averaged as required by Section 10 for each test result. No single specimen value shall be considered as representative of the roll under test.

6.2 *Sampling for Other Purposes*—The sampling and the number of test specimens depends on the purpose of the testing. Practice E 122 is recommended. It is common to test at least five specimens of a particular tape. Test specimens should be taken from several rolls of a tape and, whenever possible, among several production runs of a tape. Strong conclusions about a specific property of a tape cannot be based on test results of a single unit (roll) of product.

7. Conditioning

7.1 Condition the sample rolls of tape at 23 ± 2°C and 50 ± 5 % relative humidity for at least 24 h prior to testing. Place the rolls of tape and the test surface so that the conditioning atmosphere has free access to all normally exposed surfaces.

7.2 Unless otherwise specified, conduct all tests in the standard conditions.

8. Test Specimens

8.1 *Removal From Roll:*

8.1.1 Unwind and discard at least three but no more than six outer wraps of tape from the sample roll before taking a specimen for testing.

8.1.2 Remove one specimen (or more when specified), from random area(s) within the sample roll. Remove each specimen from a freely rotating roll at rate of 500 to 750 mm/s. Where width or other factors cause high adherence to tape backing and makes removal at prescribed rates impossible, remove at a rate as close to 500 mm/s as possible.

8.2 When the sample roll width is wider than the required specimen width, test specimens shall be cut from the center of a strip removed from the roll as described in 8.1.1 and 8.1.2 using the appropriate specimen cutter.

8.3 *Specimen Size:*

8.3.1 *Procedures A and C*—The specimen shall be 12 ± 0.5 mm wide. The length shall be approximately 150 mm.

8.3.2 *Procedure B*—The specimen shall be 12 ± 0.5 mm wide and approximately 300 mm long.

9. Procedure

9.1 For Procedures A and B, apply by means of a double-coated pressure-sensitive tape a 24 by 100-mm piece of Standard Reference Material 1810A[6] with the 24-mm dimension centered at one end of the test panel. See 5.3.1 and 5.3.2.

NOTE 4—Take care that the 1810 material is applied with the proper side up and with the grain of the paper perpendicular to the 100-mm dimension of the paper. The 100-mm dimension should be parallel to the long edge of the test panel.

9.2 *Procedure A:*

9.2.1 Center the tape specimen on the 24-mm dimension of the test paper, see 9.1, at one end of the test panel and apply without added pressure to cover the area exactly 12 by

[6] Standard Reference Material 1810A is available from the Office of Standard Reference Materials, National Institute of Standards and Technology, Washington, DC 20234.

12 mm with the tape. The adhesive surface of the free end of the specimen should be masked by applying a strip of paper approximately 125 mm by the width of the specimen.

9.2.2 To prevent cutting the specimen by the end of the panel, place another panel of the same or slightly lesser thickness under the free masked end of the specimen, and in contact with the end of the test panel prior to roll down. Roll down the applied test area twice, once in each lengthwise direction, at a velocity of 5 ± 0.2 mm/s.

9.2.3 Place the clamp on the masked free end of the specimen, ensuring that clamp extends completely across the width of the specimen and is aligned to uniformly distribute the load.

9.2.4 Place the test assembly in the test strand so that the free end of the test specimen is vertical, ensuring that no peel forces act on the specimen.

9.2.5 Apply the 1000-g mass to the clamp gently so as not to cause any shear impact force on the tape specimen. Record the time elapsed in which the tape specimen has completely separated from the test panel.

9.3 *Procedure B:*

9.3.1 Apply one end of the specimen, about 100 mm in length, adhesive side down, to the longitudinal surface of the test panel, see 9.1. The tape must be at a true right angle to the bent edge of the panel. Allow the remaining 200 mm to extend over and beyond the bent edge of the panel.

9.3.2 Using a square, cut across and through the width of the tape specimen 75 mm back from the front of the bend in the horizontal plane of the test panel surface.

9.3.3 Pass the rubber roller four times over the surface of the test specimen, twice in each lengthwise direction, at a velocity of 5 ± 0.2 mm/s.

9.3.4 Place the clamp on the free end of the specimen, ensuring that the clamp extends completely across the width, and is aligned to distribute the load uniformly.

9.3.5 Place the test assembly in the test stand so that the free end of the test specimen is vertical. Apply the specified 4.5-kg mass to the clamp gently so as not to cause any impact force on the specimen.

9.3.6 At the end of 48 h under load, examine the specimen for evidence of slippage. Measure any slippage that has occurred to the nearest 1 mm.

9.4 *Procedure C:*

9.4.1 Dispense diacetone alcohol onto the panel.

NOTE 5—**Caution**: All operations with solvents should be conducted in a well-ventilated hood. Scrub the panel with a clean piece of absorbent cleaning material. Dry the panel with fresh absorbent cleaning material. Dispense one of the solvents listed in 5.2.2 onto the panel, wiping it to dryness with fresh absorbent cleaning material. Repeat for a total of three washes with this solvent. Do not touch the cleaned panel surface with fingers.

9.4.2 Apply test specimen to test surface as described in 9.2.1, except test panel shall be steel as described in 5.3.1. Prepare test assembly as described in 9.2.2, 9.2.3, and 9.2.4.

9.4.3 Place the test stand in an oven maintained at 49 ± 1°C.

9.4.4 Allow to condition at 49°C for 10 min, then apply the 1000-g mass to the clamp gently so as not to cause any shear impact force on the tape specimen. Record the time elapsed in which the tape specimen has completely separated from the test panel.

10. Calculation

10.1 *Acceptance Sampling:*

10.1.1 *Procedures A and C*—To determine the test result for each roll of tape, convert each of the five specimen test results (times to failure) to its common or natural logarithm. Obtain the arithmetic mean of all logarithms and then convert back to time by obtaining the appropriate antilogarithm. This gives the test result for the roll of tape under consideration in the sampling plan.

10.1.2 *Procedure B*—It is common to interpret the results of Procedure B as passing or failing a pre-set acceptance criterion.

10.2 *Other Purposes*—Convert the test result (time to failure) from each specimen to its common or natural logarithm. Conduct an analysis of the logarithms of the data to determine the desired descriptive or comparative statistics. The resulting means and confidence intervals may be converted back to the original time units by obtaining the appropriate antilogarithm.

11. Report

11.1 The report shall include the following:

11.1.1 Method designation.

11.1.2 Manufacturer's name and tape designation.

11.1.3 Procedure used.

11.1.4 *For Procedure A or C*, amount of time for the tape to separate completely from the panel.

11.1.5 *For Procedure B*, distance slipped to nearest 1 mm in 48 h.

11.1.6 Any observations of tape behavior as called for in the material specification.

11.1.7 Individual roll results shall be reported unless otherwise specified.

11.1.8 When using sampling as described in 6.2, report average calculated in 10 and the number of observations.

12. Precision and Bias

12.1 *Bias*—No measure of bias is possible with D 3654M because an accepted reference or referee value is not available.

12.2 *Precision:*

12.2.1 *Procedure A*—Repeatability (within laboratory variability) information is available from one laboratory; several types of tape were evaluated using the time to failure in minutes converted to natural logarithms for analysis. Examples of averages and within roll standard deviations (including test error) are:

Tape	X	Converted X, min	Within Roll σ
A	8.74	6248	0.32
B	8.27	3905	0.23
C	2.43	11.4	0.27
D	6.50	665	0.45
E	5.98	395	0.82

The remaining components of variance are expected to be similar to those reported for Procedure C. The between roll standard deviation and the residual (including within roll) standard deviation will vary, depending on the tape type and the manufacturer.

12.2.2 *Procedure B*—No statement is made about the precision of Procedure B since the result merely states whether there is conformance to the specified criteria for success.

12.2.3 *Procedure C*—An interlaboratory evaluation of two types of pressure sensitive tapes by three laboratories has been conducted for Procedure C: the results are available.[7] It is based on an evaluation using the lognormal distribution; the times to failure in minutes were converted to their natural (base e) logarithms for the analysis. The tapes tested in this study had a grand mean of 1.2 which converts to 3.3 min. The components of variance for the following factors were estimated: between laboratory, between operators in a laboratory, between rolls of a tape, and residual (including test error and within roll variation).

A summary of the pooled standard deviations is:

Between laboratories	0.46
Between testers	0.247
Between rolls	0.185
Residual	0.411

These components may be combined in several ways to obtain the desired repeatability (within laboratory) and reproducibility (between laboratory) estimates of precision. One company reports that, with care, the between laboratory standard deviation can be reduced to about 0.2. The between roll standard deviation and the residual (including within roll) standard deviation may be higher or lower than reported here, depending on the tape type and the manufacturer.

[7] Supporting data are available from ASTM Headquarters. Request RR: D-10-1002, Report 2.

Designation: D 3662 – 88

Standard Test Method for Bursting Strength of Pressure-Sensitive Tapes[1]

This standard is issued under the fixed designation D 3662; the number immediately following the designation indicates the year of original adoption or, in the case of revision, the year of last revision. A number in parentheses indicates the year of last reapproval. A superscript epsilon (ε) indicates an editorial change since the last revision or reapproval.

This standard has been approved for use by agencies of the Department of Defense. Consult the DoD Index of Specifications and Standards for the specific year of issue which has been adopted by the Department of Defense.

1. Scope

1.1 This test method covers the measurement of the bursting strength of pressure-sensitive tape used in the closure of fiberboard containers.

1.2 *This standard does not purport to address the safety problems associated with its use. It is the responsibility of the user of this standard to establish appropriate safety and health practices and determine the applicability of regulatory limitations prior to use.*

NOTE 1—A complete metric companion to Test Method D 3662 has been developed—D 3662M; therefore, no metric equivalents are presented in this test method.

2. Referenced Documents

2.1 *ASTM Standards:*

D 996 Terminology of Packaging and Distribution Environments[2]

D 2529 Test Method for Bursting Strength of Paperboard and Linerboard[3]

D 2904 Practice for Interlaboratory Testing of a Textile Test Method that Produces Normally Distributed Data[4]

D 2906 Practice for Statements on Precision and Bias for Textiles[4]

D 3715 Practice for Quality Assurance of Pressure-Sensitive Tapes[2]

E 122 Practice for Choice of Sample Size to Estimate the Average Quality of a Lot or Process[5]

3. Terminology

3.1 Terminology found in Terminology D 996 shall apply.

4. Significance and Use

4.1 The bursting strength of pressure-sensitive tapes is an indication of the character of the backing of the product. It gives a measure of the relative resistance to rupture of the tape applied to seams and panels when under stress from inside and outside the container.

5. Apparatus

5.1 *Bursting Tester,* consisting of the following:

5.1.1 Means for clamping the test specimen between two annular, plane surfaces having fine concentric tool marks to minimize slippage. The upper clamping platen shall have a diameter of 3.75 in., a thickness of 0.375 in., and a circular opening 1.240 ± 0.001 in. in diameter. The lower edge of the opening (side in contact with the tape) shall have a 0.0025-in. radius. The lower clamping surface (diaphragm plate) shall have a thickness of 0.219 ± 0.003 in. with an opening 1.240 ± 0.001 in. in diameter. The upper edge of the opening (in contact with the tape) shall have a 0.016-in. radius, and the lower edge of the opening (in contact with the rubber diaphragm) a radius of 0.125 in. to prevent cutting the rubber when pressure is applied. The upper clamping ring shall be connected to the clamping mechanism through a swivel joint to facilitate an even clamping pressure. The openings in the two clamping plates shall be concentric to within 0.005 in. and their clamping faces shall be flat and parallel (see Annex A1.1 of Test Method D 2529).

5.1.2 A molded (disk-shaped) diaphragm requiring a pressure of not less than 23 psi nor more than 30 psi to distend it to a height of 0.375 in. above the diaphragm plate (see Annex A1.2 of Test Method D 2529).

5.1.3 Means of forcing liquid into the pressure chamber below the diaphragm at a steady rate of 170 ± 15 mL/min (see Annex A1.3 of Test Method D 2529). This pressure shall be generated by a motor-driven piston forcing a liquid (glycerin) into the pressure chamber of the apparatus.

5.1.4 A pressure gage of the maximum reading or the lazy-hand type. The scale should have a radius of 1.875 in. with graduations extending over a minimum arc of 270° indicating bursting pressure in pounds-force per square inch, with an accuracy of ½ % of full scale, and have sufficient capacity so that all readings can be maintained in the middle half of the scale. In its operating position, have the gage inclined between horizontal and not over 30 % from the horizontal. When more than one gage is mounted on a single apparatus, only the gage on which the measurement is being made shall be open to the hydraulic system so as not to reduce the rate of distention of the specimen (Note 2). Calibrate the gages frequently enough to ensure their accuracy (once a week of in constant use, or before using if they have been idle for a month or more). Calibrate the gages while inclined at the same angle as used during a test by means of a dead-weight gage tester. During calibration, apply the pressure so that the rate of travel of the needle is similar to that when actually testing the tape. If a gage is accidently subjected to a pressure beyond its capacity, recalibrate it

[1] This test method is under the jurisdiction of ASTM Committee D-10 on Packaging and is the direct responsibility of Subcommittee D10.14 on Closure and Reinforcement of Packages.

Current edition approved Aug. 26, 1988. Published October 1988. Originally published as D 3662 – 78. Last previous edition D 3662 – 83.

[2] *Annual Book of ASTM Standards,* Vol 15.09.

[3] Discontinued; see *1985 Annual Book of ASTM Standards,* Vol 15.09.

[4] *Annual Book of ASTM Standards,* Vol 07.01.

[5] *Annual Book of ASTM Standards,* Vol 14.02.

before it is used again (see Annex A1.4 of Test Method D 2529).[6]

NOTE 2—Calibrate the apparatus in accordance with Annex A1 of Test Method D 2529.

6. Sampling

6.1 *Acceptance Sampling*—Sampling shall be in accordance with Practice D 3715.

6.2 *Sampling for Other Purposes*—The sampling and the number of test specimens depends on the purpose of the testing. Practice E 122 is recommended. It is common to test at least five specimens of a particular tape. Test specimens should be taken from several rolls of tape and whenever possible, among several production runs of a tape. Strong conclusions about a specific property of a tape cannot be based on test results of a single unit (roll) of a product.

7. Test Specimens

7.1 The specimen shall be a minimum of 1½ in. wide. The length shall be approximately 6 in. It shall be free of wrinkles and creases.

7.2 Unwind and discard at least three, but no more than six, outer wraps of tape from the sample roll before taking specimens for testing.

7.3 Unless otherwise specified, test a minimum of three specimens per sample roll. Remove the specimen from a freely rotating roll at the rate of 20 to 30 in./s. Where width or other factors causing a high adherence to backing make it impossible to remove the specimen at the prescribed rate, remove it at a rate as close to 20 in./s as possible.

8. Conditioning

8.1 Condition the sample rolls in an atmosphere uniformly maintained at 73.4 ± 3.6°F and 50 ± 5 % relative humidity for a period not less than 24 h. Testing shall also be done at these conditions.

9. Procedure

9.1 Center a specimen on the diaphragm plate with the adhesive side down. Sweep a finger around the plate to ensure the tape is adhered smoothly and without wrinkles or folds.

NOTE 3—This step can be facilitated by either fully raising the upper clamping platen or removing it temporarily.

9.2 Lower the upper clamping platen onto the tape backing. If the tester has a handscrew-type clamp, use care so as not to apply excessive pressure and cause tape backing breakage and subsequent pull out of the specimen from the clamp.

9.3 Apply the bursting pressure by driving the piston forward until the diaphragm ruptures the specimen; then return piston to its original position to prevent rupture of the diaphragm. Record the maximum pressure registered by the gage and gently return the indicator of the gage to zero.

9.4 Remove the specimen remains from the diaphragm; this adhesive can be removed with a series of pats and swipes with fresh tape. After ensuring that the diaphragm is free of adhesive residue, proceed with the testing of the remaining specimens.

10. Calculation

10.1 Determine an average value (arithmetic mean) of bursting pressure for each sample roll tested. This is the sum of individual values for each roll divided by the number of values.

10.2 Where applicable, determine a lot or batch average. This is the sum of average for each sample roll (see 10.1) divided by the number of sample rolls.

11. Report

11.1 Report the bursting strength in pounds-force per square inch to three significant figures.

11.2 The report shall include the following:

11.2.1 Manufacturer's name and designation for the tape,

11.2.2 Sample roll average(s) (see 10.1), and

11.2.3 Lot or batch average where appropriate (see 10.2).

12. Precision and Bias

12.1 *Summary*—The difference between two single observations should not exceed 14.8 % of the average of the two observations in 95 out of 100 cases when both observations are taken by the same well-trained operator using the same sample of material. Larger differences may occur under all other circumstances. The true value of bursting strength can only be defined in terms of a specific test method. Within this limitation, Test Method D 3662 has no known bias. Sections 12.2 through 12.5 explain the basis for this summary and for evaluations made under other conditions.

12.2 *Interlaboratory Test Data*[7]—An interlaboratory study was made in 1980 in which randomly drawn samples of two materials were tested in each of four laboratories. Two operators in each laboratory each tested three specimens from each of three rolls of each material. The components of variance for bursting strength results expressed as coefficients of variation were calculated to be as follows:

	Specimens of the Same Material	Specimens of Different Material
Single-operator component	3.1 % of the average	7.3 % of the average
Within-laboratory component	2.7 % of the average	0 % of the average
Between-laboratory component	8.3 % of the average	5.7 % of the average
Replication component	4.3 % of the average	4.3 % of the average

NOTE 4—The calculations for coefficient of variation and other statistics found in subsequent sections of this statement are described in Practice D 2906 and Annex A3 of Practice D 2904.

12.3 *Critical Differences*—For the components of variance reported in 12.2, two averages observed values should be considered significantly different at the 95 % probability level if the difference equals or exceeds the critical differences shown in Table 1.

12.4 *Confidence Limits*—For the components of variance reported in 12.2, single averages of observed values have the following 95 % confidence limits shown in Table 2.

[6] A test instrument that meets these requirements is the Model "A" Mullen tester manufactured by B. F. Perkins Co., P. O. Box 366-TR, Chicopee, MA 01021.

[7] Supporting data are available from ASTM Headquarters. Request RR:D-10-1002.

NOTE 5—The tabulated values of the critical differences and confidence limits should be considered to be a general statement particularly with respect to between-laboratory precision. Before a meaningful statement can be made about two specific laboratories, the amount of statistical bias between them, if any, must be established with each comparison being based on recent data obtained on specimens randomly drawn from one sample of the material to be evaluated.

12.5 *Bias*—No justifiable statement can be made on the bias of Test Method D 3662 for testing bursting strength since the true value cannot be established by an accepted referee method.

TABLE 1 Critical Difference, Percent of Grand Average for the Conditions Noted[A,B]

Number of Observations in Each Average	Single-Operator Precision	Within-Laboratory Precision	Between-Laboratory Precision
Specimens from the Same Material:			
1	14.8	16.6	28.3
5	10.1	12.6	26.2
10	9.4	12.0	25.9
Specimens from Different Material:			
1	23.6	23.6	28.3
5	20.9	20.9	26.2
10	20.6	20.6	25.9

[A] The critical differences were calculated using $t = 1.960$ which is based on infinite degrees of freedom.

[B] To convert the values of the critical differences to units of measure, multiply the average of the two sets of data being compared by the critical differences expressed as a decimal fraction.

TABLE 2 Width of 95 % Confidence Limits, Percent of the Grand Average for the Conditions Noted[A,B]

Number of Observations in Each Average	Single-Operator Precision	Within-Laboratory Precision	Between-Laboratory Precision
Specimens from the Same Material:			
1	±10.4	±11.7	±20.0
5	±7.2	±8.9	±18.5
10	±6.6	±8.5	±18.3
Specimens from Different Material:			
1	±16.7	±16.7	±20.0
5	±14.8	±14.8	±18.5
10	±14.6	±14.6	±18.3

[A] The confidence limits are calculated using $t = 1.960$ which is based on infinite degrees of freedom.

[B] To convert the values of confidence limits to units of measure, multiply the average of the specific set of data which is of interest by the confidence limits expressed as a decimal fraction.

Standard Test Method for Bursting Strength of Pressure-Sensitive Tapes [Metric][1]

This standard is issued under the fixed designation D 3662M; the number immediately following the designation indicates the year of original adoption or, in the case of revision, the year of last revision. A number in parentheses indicates the year of last reapproval. A superscript epsilon (ϵ) indicates an editorial change since the last revision or reapproval.

1. Scope

1.1 This test method covers the measurement of the bursting strength of pressure-sensitive tape used in the closure of fiberboard containers.

1.2 *This standard does not purport to address the safety problems associated with its use. It is the responsibility of the user of this standard to establish appropriate safety and health practices and determine the applicability of regulatory limitations prior to use.*

NOTE 1—This test method is the metric companion to Test Method D 3662.

2. Referenced Documents

2.1 *ASTM Standards:*

D 996 Terminology of Packaging and Distribution Environments[2]

D 2529 Test Method for Bursting Strength of Paperboard and Linerboard[3]

D 2904 Practice for Interlaboratory Testing of a Textile Test Method that Produces Normally Distributed Data[4]

D 2906 Practice for Statements on Precision and Bias for Textiles[4]

D 3715 Practice for Quality Assurance of Pressure-Sensitive Tapes[2]

E 122 Practice for Choice of Sample Size to Estimate the Average Quality of a Lot or Process[5]

3. Terminology

3.1 Terminology found in Terminology D 996 shall apply.

4. Significance and Use

4.1 The bursting strength of pressure-sensitive tapes is an indication of the character of the backing of the product. It gives a measure of the relative resistance to rupture of the tape applied to seams and panels when under stress from inside and outside the container.

5. Apparatus

5.1 *Bursting Tester,* consisting of the following:

5.1.1 Means for clamping the test specimen between two annular, plane surfaces having fine concentric tool marks to minimize slippage. The upper clamping platen shall have a diameter of 95 mm, a thickness of 9.5 mm, and a circular opening 31.50 ± 0.025 mm in diameter. The lower edge of the opening (side in contact with the tape) shall have a 0.064-mm radius. The lower clamping surface (diaphragm plate) shall have a thickness of 5.56 ± 0.08 mm with an opening 31.50 ± 0.02 mm in diameter. The upper edge of the opening (in contact with the tape) shall have a 0.4-mm radius, and the lower edge of the opening (in contact with the rubber diaphragm) a radius of 3.2 mm to prevent cutting the rubber when pressure is applied. The upper clamping ring shall be connected to the clamping mechanism through a swivel joint to facilitate an even clamping pressure. The openings in the two clamping plates shall be concentric to within 0.13 mm and their clamping faces shall be flat and parallel (see Appendix A1.1 of Test Method D 2529).

5.1.2 A molded (disk-shaped) diaphragm requiring a pressure of not less than 160 kPa nor more than 210 kPa to distend it to a height of 9.5 mm above the diaphragm plate (see Appendix A1.2 of Test Method D 2529).

5.1.3 Means of forcing liquid into the pressure chamber below the diaphragm at a steady rate of 170 ± 15 mL/min (see Appendix A1.3 of Test Method D 2529). This pressure shall be generated by a motor-driven piston forcing a liquid (glycerin) into the pressure chamber of the apparatus.

5.1.4 A pressure gage of the maximum reading or the lazy-hand type. The scale should have a radius of 48 mm with graduations extending over a minimum arc of 270° indicating bursting pressure in kilopascals with an accuracy of ½ % of full scale, and have sufficient capacity so that all readings can be maintained in the middle half of the scale. In its operating position, have the gage inclined between horizontal and not over 30° from the horizontal. When more than one gage is mounted on a single apparatus, only the gage on which the measurement is being made shall be open to the hydraulic system so as not to reduce the rate of distention of the specimen (Note 2). Calibrate the gages frequently enough to ensure their accuracy (once a week if in constant use, or before using if they have been idle for a month or more). Calibrate the gages while inclined at the same angle as used during a test by means of a dead-weight gage tester. During calibration, apply the pressure so that the rate of travel of the needle is similar to that when actually testing the tape. If a gage is accidently subjected to a pressure

---

[1] This test method is under the jurisdiction of ASTM Committee D-10 on Packaging and is the direct responsibility of Subcommitte D10.14 on Closure and Reinforcement of Packages.

Current edition approved Aug. 26, 1988. Published October 1988. Originally published as D 3662M – 83. Last previous edition D 3662M – 83.

[2] *Annual Book of ASTM Standards,* Vol 15.09.

[3] Discontinued; see *1985 Annual Book of ASTM Standards,* Vol 15.09.

[4] *Annual Book of ASTM Standards,* Vol 07.01.

[5] *Annual Book of ASTM Standards,* Vol 14.02.

beyond its capacity, recalibrate it before it is used again (see Appendix A1.4 of Test Method D 2529).[6]

NOTE 2—Calibrate the apparatus in accordance with Appendix A1 of Test Method D 2529.

6. Sampling

6.1 *Acceptance Sampling*—Sampling shall be in accordance with Practice D 3715.

6.2 *Sampling for Other Purposes*—The sampling and the number of test specimens depends on the purpose of the testing. Practice E 122 is recommended. It is common to test at least five specimens of a particular tape. Test specimens should be taken from several rolls of a tape and, whenever possible, among several production runs of a tape. Strong conclusions about a specific property of a tape cannot be based on test results of a single unit (roll) of a product.

7. Test Specimens

7.1 The specimen shall be a minimum of 48 mm wide. The length shall be approximately 150 mm. It shall be free of wrinkles and creases.

7.2 Unwind and discard at least three, but no more than six, outer wraps of tape from the sample roll before taking specimens for testing.

7.3 Unless otherwise specified, test a minimum of three specimens per sample roll. Remove the specimen from a freely rotating roll at the rate of approximately 500 to 750 mm/s. Where width or other factors causing a high adherence to backing make it impossible to remove the specimen at the prescribed rate, remove it at a rate as close to 500 mm/s as possible.

8. Conditioning

8.1 Condition the sample rolls in an atmosphere uniformly maintained at 23 ± 2°C and 50 ± 5 % relative humidity for a period not less than 24 h. Testing shall also be done at these conditions.

9. Procedure

9.1 Center a specimen on the diaphragm plate with the adhesive side down. Sweep a finger around the plate to ensure the tape is adhered smoothly and without wrinkles or folds.

NOTE 3—This step can be facilitated by either fully raising the upper clamping platen or removing it temporarily.

9.2 Lower the upper clamping platen onto the tape backing. If the tester has a handscrew-type clamp, use care so as not to apply excessive pressure and cause tape backing breakage and subsequent pull out of the specimen from the clamp.

9.3 Apply the bursting pressure by driving the piston forward until the diaphragm ruptures the specimen; then return piston to its original position to prevent rupture of the diaphragm. Record the maximum pressure registered by the gage and gently return the indicator of the gage to zero.

9.4 Remove the specimen remains from the diaphragm; this adhesive can be removed with a series of pats and swipes with fresh tape. After ensuring that the diaphragm is free of adhesive residue, proceed with the testing of the remaining specimens.

10. Calculation

10.1 Determine an average value (arithmetic mean) of bursting pressure for each sample roll tested. This is the sum of individual values for each roll divided by the number of values.

10.2 Where applicable, determine a lot or batch average. This is the sum of average for each sample roll (see 10.1) divided by the number of sample rolls.

11. Report

11.1 Report the bursting strength in kilopascals to three significant figures.

11.2 The report shall include the following:

11.2.1 Manufacturer's name and designation for the tape,

11.2.2 Sample roll average(s) (see 10.1), and

11.2.3 Lot or batch average where appropriate (see 10.2).

12. Precision and Bias

12.1 *Summary*—The difference between two single observations should not exceed 14.8 % of the average of the two observations in 95 out of 100 cases when both observations are taken by the same well-trained operator using the same sample of material. Larger differences are likely to occur under all other circumstances. The true value of bursting strength can only be defined in terms of a specific test method. Within this limitation, Test Method D 3662M has no known bias. Sections 12.2 through 12.5 explain the basis for this summary and for evaluations made under other conditions.

12.2 *Interlaboratory Test Data*[7]—An interlaboratory study was made in 1980 in which randomly drawn samples of two materials were tested in each of four laboratories. Two operators in each laboratory each tested three specimens from each of three rolls of each material. The components of variance for bursting strength results expressed as coefficients of variation were calculated to be as follows:

	Specimens of the Same Material	Specimens of Different Material
Single-operator component	3.1 % of the average	7.3 % of the average
Within-laboratory component	2.7 % of the average	0 % of the average
Between-laboratory component	8.3 % of the average	5.7 % of the average
Replication component	4.3 % of the average	4.3 % of the average

NOTE 4—The calculations for coefficient of variation and other statistics found in subsequent sections of this statement are described in Practice D 2906 and Appendix A3 of Practice D 2904.

12.3 *Critical Differences*—For the components of variance reported in 12.2, two averages observed values should be considered significantly different at the 95 % probability level if the difference equals or exceeds the critical difference shown in Table 1.

[6] A test instrument that meets these requirements is the Model "A" Mullen Tester manufactured by B. F. Perkins Co., P.O. Box 366-TR, Chicopee, MA 01021.

[7] Supporting data are available from ASTM Headquarters. Request RR: D-10-1002.

TABLE 1 Critical Difference, Percent of Grand Average for the Conditions Noted[A, B]

Number of Observations in Each Average	Single-Operator Precision	Within-Laboratory Precision	Between-Laboratory Precision
Specimens from the Same Material:			
1	14.8	16.6	28.3
5	10.1	12.6	26.2
10	9.4	12.0	25.8
Specimens from Different Material:			
1	23.6	23.6	28.3
5	20.9	20.9	26.2
10	20.6	20.6	25.9

[A] The critical differences were calculated using $t = 1.960$ which is based on infinite degrees of freedom.

[B] To convert the values of the critical differences to units of measure, multiply the average of the two specific sets of data being compared by the critical differences expressed as a decimal fraction.

TABLE 2 Width of 95 % Confidence Limits, Percent of the Grand Average for the Conditions Noted[A, B]

Number of Observations in Each Average	Single-Operator Precision	Within-Laboratory Precision	Between-Laboratory Precision
Specimens from the Same Material:			
1	±10.4	±11.7	±20.0
5	±7.2	±8.9	±18.5
10	±6.6	±8.5	±18.3
Specimens from Different Material:			
1	±16.7	±16.7	±20.0
5	±14.8	±14.8	±18.5
10	±14.6	±14.6	±18.3

[A] The confidence limits are calculated using $t = 1.960$ which is based on infinite degrees of freedom.

[B] To convert the values of the confidence limits to units of measure, multiply the average of the specific set of data which is of interest by the confidence limits expressed as a decimal fraction.

12.4 *Confidence Limits*—For the components of variance reported in 12.2, single averages of observed values have the following 95 % confidence limits shown in Table 2.

NOTE 5—The tabulated values of the critical differences and confidence limits should be considered to be a general statement particularly with respect to between-laboratory precision. Before a meaningful statement can be made about two specific laboratories, the amount of statistical bias between them, if any, must be established with each comparison being based on recent data obtained on specimens randomly drawn from one sample of the material to be evaluated.

12.5 *Bias*—No justifiable statement can be made on the bias of Test Method D 3662M for testing bursting strength since the true value cannot be established by an accepted referee method.

Standard Test Method for
Tensile Strength and Elongation of Pressure-Sensitive Tapes[1]

This standard is issued under the fixed designation D 3759; the number immediately following the designation indicates the year of original adoption or, in the case of revision, the year of last revision. A number in parentheses indicates the year of last reapproval. A superscript epsilon (ε) indicates an editorial change since the last revision or reapproval.

ε1 NOTE—The address in Footnote 5 was changed editorially in May 1990.

1. Scope

1.1 This test method covers the measurement of tensile strength (breaking strength) and stretch properties (elongation and "F" value) for pressure-sensitive tapes. It includes procedures for machine direction and cross-direction tests and tests for tapes with low and high stretch and reinforced backings. It also includes a procedure for obtaining stretch force ("F" value) in conjunction with a determination of tensile strength. These procedures employ a constant-rate-of-extension (CRE)-type testing machine. They apply to the principle of stretching the specimen at a fixed strain rate of one inch per inch of specimen length per minute with the exception of the procedure for reinforced tapes.

1.2 *This standard does not purport to address the safety problems associated with its use. It is the responsibility of the user of this standard to establish appropriate safety and health practices and determine the applicability of regulatory limitations prior to use.*

NOTE 1—A complete metric companion to Test Method D 3759 has been developed—D 3759M; therefore, no metric equivalents are presented in this test method. D 3759M uses specimen widths of 12 mm and 24 mm rather than ½ in. and 1 in. used in Test Method D 3759. This difference in test parameters will generally produce lower values for Test Method D 3759M.

2. Referenced Documents

2.1 *ASTM Standards:*

D 996 Terminology of Packaging and Distribution Environments[2]

D 2904 Practice for Interlaboratory Testing of a Textile Test Method that Produces Normally Distributed Data[3]

D 2906 Practice for Statements on Precision and Bias for Textiles[3]

D 3715 Practice for Quality Assurance of Pressure-Sensitive Tapes[2]

E 122 Practice for Choice of Sample Size to Estimate Quality of a Lot or Process[4]

3. Terminology

3.1 Terminology found in Terminology D 996 shall apply.

4. Significance and Use

4.1 This test method provides information that can be used in material specifications for product design and quality assurance applications. It can be used in comparing different products.

4.2 The use of this test method must be related to the purpose for which the test is performed. One purpose is for determining the relative strength of the tape in the size in which it is purchased or used. Another purpose is to identify or characterize a particular backing material.

4.2.1 When relative strength is of interest, the test should be performed on the tape as-received, that is, without cutting the material to a specimen width less than the as-received width.

4.2.1.1 Usually tapes wider than 2 in. are not tested due principally to the limitations of equipment. Tapes as narrow as approximately 0.125 in. can be tested.

4.2.1.2 Comparison of materials by different test methods should be avoided because the test parameters of specimen dimensions and crosshead velocity determine the outcome. Changes in the parameter levels will produce different results for the same material.

NOTE 2—It is usual to find the tensile strength increasing significantly with increasing crosshead velocity and therefore, strain rate.

4.2.2 When identity or material characterization is of interest, the test should be performed on a specimen cut from within the sample material boundaries using a sharp razor cutter, such as that defined in Section 5.

NOTE 3—Some of the traditional tools for specimen preparation must be avoided when the backing is comprised of thin plastic sheeting. These include chopping dies and sample cutters operating on a shearing principle. The reason for this restraint is that edges sufficiently ragged and damaged resulting from chopping or shearing cause tearing to occur before the true tensile strength level is reached. Tapes with fibrous backings may be cut to satisfactory specimens with these tools.

4.2.3 Stretch characteristics can be related to the tape's intended use or for identifying or characterizing a material.

NOTE 4—Elongation measurements become difficult to perform on stretchy materials (greater than 25% ultimate elongation) when the ratio of specimen length to width is small (approaching 2). The results show high variability and do not allow for practical use of this information except when one wishes to demonstrate large differences between material.

---

[1] This test method is under the jurisdiction of ASTM Committee D-10 on Packaging and is the direct responsibility of Subcommittee D10.14 on Closure and Reinforcement of Packages.

Current edition approved Aug. 26, 1988. Published October 1988. Originally published as D 3759 – 83. Last previous edition D 3759 – 83.

[2] *Annual Book of ASTM Standards*, Vol 15.09.

[3] *Annual Book of ASTM Standards*, Vol 07.01.

[4] *Annual Book of ASTM Standards*, Vol 14.02.

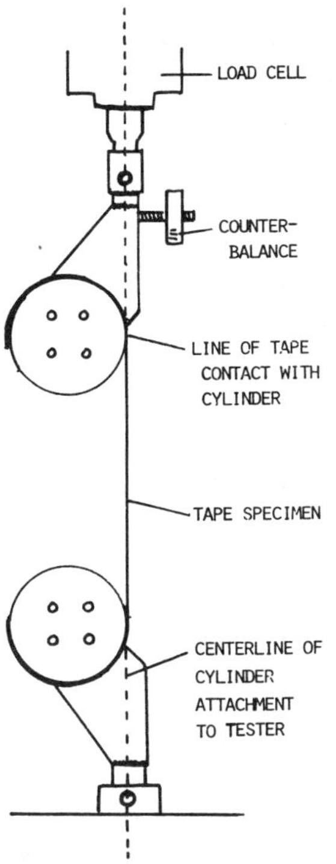

FIG. 1 Test Configuration for Reinforced Tapes

TABLE 1 Tester Preparation and Specimen Dimensions

	Gage Length, in.	Cross-head Veloc-ity, in./min	Chart Veloc-ity,[A] in./min	Speci-men Width,[B] in.	Length, in.
Tapes with ultimate elongation of:					
Up to 200 %:					
Machine direction	5	5	5	½–1	9
Cross direction[C,D,E]	1	1	5	¼–½	5
200 % and up:					
Machine direction	2	2	5	½	6
Reinforced tapes	10	5	5	½–1	28

[A] The chart velocity may be set at other velocities. It should not be slower than the crosshead velocity.

[B] The specimen widths shown are for tests in which the specimen is cut from within the sample dimension. See 4.2.1.1.

[C] Cross-direction (C.D.) tests are limited to sample rolls of tape at least 2 in. in width.

[D] It is unusual to test C.D. tensile strength of tapes having ultimate elongations greater than 150 %. Therefore no reference to this is made in Table 1. However C.D. tests could be made under that category on the high stretch materials.

[E] If the sample provides ample material, C.D. tests should preferably be made in the same way machine direction (M.D.) tests are. This would occur with web material or sufficiently wide rolls.

5. Apparatus

5.1 *Tension Tester*—A constant-rate-of-extension (CRE) type with load cell capacity such that the maximum expected specimen strength does not exceed 90 % of its normal limit.

5.1.1 *Test Information* should be displayed in at least an alphanumeric digital display or a load-elongation curve plotted by a pen or stylus responding to load and with a chart driven synchronously with the crosshead.

5.1.2 *Load-Elongation Curve,* plotted by a pen or stylus responding to load and with a servo-chart drive or *x-y* recorder driven by an extensometer.

5.1.3 *Clamps,* preferably the pneumatic action type.

5.1.3.1 Clamp faces at least 2 in. wide by 1½ in. deep. Faces shall have a light cross-hatch serration.

NOTE 5—Plastic materials are reduced in width and thickness while being stretched. This causes them to be drawn out of the clamps. Pneumatic clamps minimize this effect. It can be further reduced by the appropriate choice of surface of the clamps. The greatest improvement, both with respect to the above mentioned shrinkage problem and simple slippage, may be found from the use of urethane film which can be obtained as a pressure-sensitive tape approximately 20 mils thick. This material has a very high coefficient of friction, is somewhat malleable, and is easily replaced. Alternative materials are coated abrasive, rubber (neoprene or other synthetic type), or other tape.

5.1.4 *Cylinders,* in place of clamps for testing reinforced tapes. Each of two cylinders shall be 4-in. diameter by 1.5-in. thick held in the position ordinarily occupied by the clamps so that the tape, when applied to the cylinders and extending between them, falls in the line of stress otherwise occupied by the specimens when clamps are used. See Fig. 1.

5.1.5 *Scale,* approximately 1 in. in length divided into 0.1-in. increments attached to each cylinder. The zero point (origin) shall be at the point of tangency of the tape with the cylinder during the test and the scale shall increase upward on the lower cylinder and downward on the upper cylinder.

NOTE 6—These scales will be used to observe and measure the tape slippage during the tension test for reinforced tapes.

5.2 *Cutter,*[5] holding two single-edged razor blades in parallel planes, a precise distance apart, to form a cutter of exact specimen width. Appropriate widths shall be available (refer to specimen width in Table 1) patterned after the ½-in. cutter in 5.2.1. The differences between cutters of various widths is in the final width of the bar after removing the thickness of one razor blade.

5.2.1 The cutter shall consist of a ½-in. thick by approximately 8-in. length of aluminum bar stock ½-in. wide. The edges, for about 5 in. from one end shall be slightly rounded to form a handle. The width of the bar, for approximately 3 in. from the opposite end, shall be narrowed to exactly 0.500 in. minus the thickness of a single razor blade (one of two used as cutting edges). The razor blades shall be held in position using side plates. The end of the cutter shall be cut away at a 45° angle to expose the cutting edges at one end of the blades.

6. Sampling

6.1 *Acceptance Sampling*—Acceptance sampling shall be in accordance with Practice D 3715.

6.2 *Sampling for Other Purposes*—The sampling and the number of test specimens depends on the purpose of the testing. Practice E 122 is recommended. It is common to test at least five specimens of a particular tape. Test specimens should be taken from several rolls of tape, and whenever possible, among several production runs of tape. Strong conclusions about a specific property of a tape cannot be based on tests of a single unit (roll) of a product.

[5] Available from The Pressure-Sensitive Tape Council (PSTC), 104 Wilmot Rd., Suite 201, Deerfield, IL 60015.

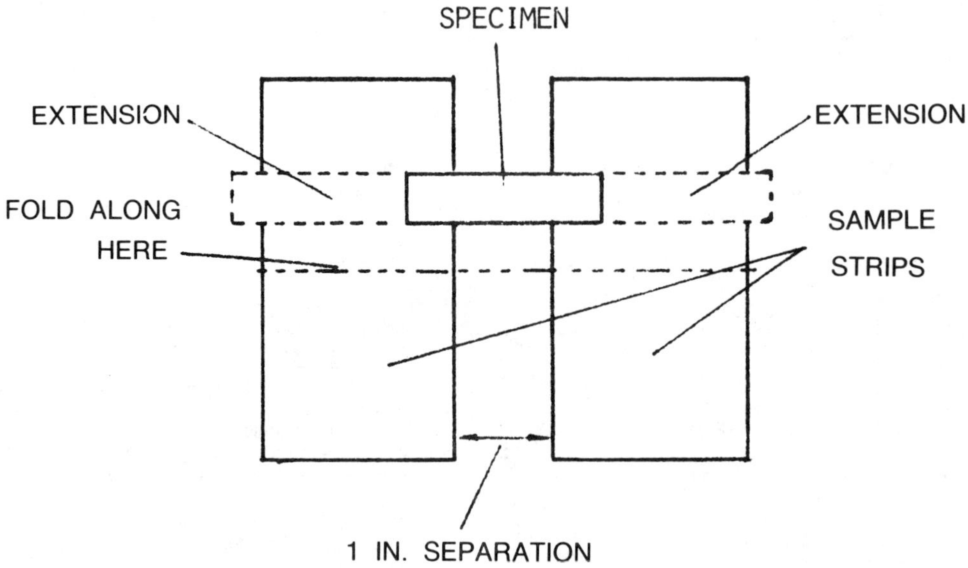

FIG. 2 Cross Direction (C.D.) Specimen Preparation

7. Test Specimens

7.1 Specimens shall have the dimensions shown in Table 1.

7.2 Unwind and discard at least three, but no more than six, outer wraps of tape from the sample roll before taking specimens for testing.

7.3 Test one specimen per sample roll, unless otherwise specified.

7.4 The following applies to nonreinforced tapes:

7.4.1 Specimen ends, that are clamped shall be prepared by covering the adhesive with paper, some other tape, or an extension of the specimen. In the latter case the specimen must be cut at least 4 in. longer than defined in Table 1.

7.4.2 The covering shall be free of wrinkles, leaving the gage-length area uncovered and completely cover the rest of the specimen so that the clamps will apply uniform pressure against the specimen.

7.4.3 A special specimen preparation is required for cross direction (C.D.) specimens from rolls less than 4 in. in width. Lay two rectangular sample strips on a flat surface with the adhesive side facing up. See Fig. 2. Each strip shall be as wide as the sample roll and approximately 5 in. in length. Position these strips side by side with one long edge of one strip parallel to and 1.0 in. separated from one long edge of the second strip.

7.4.3.1 Cut a specimen from the sample roll to have the width specified in Table 1 and length equal to the width of the roll.

7.4.3.2 Lay this specimen adhesive side up across the 1.0-in. separation of the strips. Position it toward one end of the sample strips so that it rests equally on both strips and at a right angle to their parallel edges.

7.4.3.3 Cut two additional strips from the sample roll having the same width as the specimen. *Butt* the end of one of these at one end to form a continuation of the specimen across the remainder of the sample strip. Use the second strip to butt against the other end of the specimen in like manner.

7.4.3.4 Fold each of the original sample strips over onto itself to form a three-ply tab that will be gripped by the clamps during the test.

7.4.3.5 Trim off any excess (single ply of tape) of either the sample strips or the extension strips extending beyond the two- or three-ply parts of the assembly.

NOTE 7—The extension serves to keep the clamping pressure uniform over the whole area of the specimen. This is an imperative factor to a successful test.

7.5 For reinforced tapes the specimen requires no further preparation than to have the appropriate dimensions (Table 1) and ensure that the adhesive is not contaminated so it will adhere well to the cylinders.

8. Preparation of Apparatus

8.1 Table 1 shows the tension tester settings for use with the specified test categories.

8.2 For testing reinforced tapes, set the cylinders 6 in. apart so that at the start of a test 10 in. of tape will extend between and without contact with the cylinders.

NOTE 8—The upper cylinder should be counterbalanced in order that the line of tape contact on the cylinders intersects an imaginary line running between the points of cylinder attachment to the tester and no side forces are exerted during the test. See Fig. 1.

9. Conditioning

9.1 Condition the sample to equilibrium in an atmosphere uniformly maintained at 73.4 ± 3.6°F and 50 ± 5 % relative humidity for 24 h. Testing shall be performed at the same conditions.

10. Procedure

10.1 *Nonreinforced Tapes:*

10.1.1 Clamp the specimen in the grips of the testing machine. Take care to align the long axis of the specimen with an imaginary line running between the points of attachment of the grips and including the center of the grips. Apply no more tension to the specimen during clamping than is necessary to remove slack.

10.1.2 Start the crosshead in motion at the specified velocity (Table 1) and ensure that the mechanism that displays the response (that is charts, plotter, and digital display) is operating. Continue until the specimen ruptures.

10.1.3 Record a numerical display of the results.

10.2 *Reinforced Tapes:*

10.2.1 Adhere approximately 9 in. of the specimen on the upper cylinder beginning at the line of tape contact (see 8.2), and wrap the specimen around the top surface of the cylinder. Repeat this with the free end of the specimen on the lower cylinder, except wrap the specimen around the bottom surface of the cylinder. The applied specimen must be centered on the center line around the cylinder surface. This elimination of skewness prevents nonuniform stress loading across the width of the specimen. The specimen shall also be sufficiently taut to remove slack.

10.2.2 Mark the specimen (and cylinder if not already done), with a marking pen making a line approximately 1⁄32 in. wide at the line where the tape contacts each cylinder. These bench marks will be 10 in. apart and shall be checked to ensure this.

10.2.3 Start the crosshead in motion at the specified velocity and ensure that the response-indicator mechanism is operating to indicate both load and elongation, if the latter is required.

10.2.4 Observe the bench marks on the specimen to determine their change in position relative to the marks on the cylinders. Use the scales appended to the cylinders.

10.2.5 When the specimen breaks, record the sum of the upper and lower bench mark changes to the nearest 0.1 in. This will be the correction for the elongation.

10.2.6 Also record the indicated responses for tensile strength and elongation when the tester provides a numerical display of this information.

11. Calculation

11.1 *Tensile Strength:*

11.1.1 When the recorded load-elongation curve results from the testing, calculate tensile strength as follows:

11.1.1.1 Find the farthest advance of the plotting pen or stylus from the origin in the direction representing increasing force. Record this advance as a percent of the chart scale.

11.1.1.2 Multiply the full-scale load range used during the test by the percent found in 11.1.1.1.

NOTE 9—For this calculation express the percent as a decimal fraction, that is, use 0.87 for 87 %.

11.1.1.3 Convert the value found in 11.1.1.2 to pounds-force per the desired basis of dimension for the final step. For lbf/in. of width, divide by the specimen width in inches.

11.2 *Ultimate Elongation:*

11.2.1 When the recorded load-elongation curve results from the use of a synchronous chart drive, calculate elongation as follows:

11.2.1.1 Measure the distance from the start of the plot to the indicated breaking point along the motion (time) axis of the chart.

11.2.1.2 When the chart velocity equals the crosshead velocity, the distance from 11.2.1.1 divided by the gage length and multiplied by 100 equals the ultimate elongation.

11.2.1.3 When the chart and crosshead velocities are different, multiply the calculation completed in 11.2.1.2 by the ratio of the crosshead velocity to the chart velocity to obtain the correct ultimate elongation.

11.2.2 When the recorded load-elongation curve results from use of a servo-chart drive or *x-y* recorder driven by an extensometer, calculate elongation as follows:

11.2.2.1 Measure the distance from the start of the plot to the indicated breaking point. Divide this by the magnification ratio (to get the true dimension of elongation). Divide this by the gage length. Next, multiply by 100 to convert the value to percent ultimate elongation.

NOTE 10—The magnification ratio is the chart motion over the actual extensometer motion.

11.2.3 When testing reinforced tapes, the correction found in 10.2.5 shall first be subtracted from the amount of extension indicated or measured along the time axis of the recording chart and the difference divided by the original gage length. Multiply this quotient by 100 to convert it into percent ultimate elongation.

11.3 "*F*" Value:

11.3.1 When the recorded load-elongation curve results from the use of a synchronous chart drive, calculate the "F" value as follows:

11.3.1.1 Find the chart scale equivalent to the desired elongation point. (If one were determining the F-3 value, that is, the force required to stretch the specimen 3 %, one would be finding the chart scale equivalent to 3 % elongation). This equivalent is the distance measured from the start of the plot to the prescribed elongation point on the chart. To calculate the distance in terms of the chart scale, use the equation:

$$d = \frac{E \times G \times C}{100\,(H)}$$

where:
d = chart scale equivalent distance,
E = percent elongation where the "F" value is to be determined,
G = gage length,
C = chart velocity, and
H = crosshead velocity.

The units of these measurements shall be inches.

11.3.1.2 Using the calculated *d*, measure along the chart time axis (same as stretch, extension, or elongation axis) from the origin of the recorded plot to a point distance, *d*. This point shall be at zero force as was the plot origin. Draw a line from this point parallel to the force axis to intersect with the recorded plot. Measure the scale length of this line from zero force and convert this length into percent of chart scale and then into force. See 11.1.1.1, 11.1.1.2, and Note 9.

11.3.1.3 Multiply the full-scale load range used during the test by the percent found in 11.3.1.2. Express the percent as a decimal fraction. The product is the "F" value in pounds-force or newtons for the specimen width.

11.3.1.4 The final step is the same as 11.1.1.3, except substitute 11.3.1.3 for 11.1.1.2.

11.3.2 When the recorded load-elongation curve results from the use of a servo-chart drive or *x-y* recorder driven by an extensometer, calculate the "F" value as follows:

11.3.2.1 Find the scale equivalent to the desired elongation point using this equation:

$$d = \frac{E \times G \times M}{100}$$

TABLE 2 Components of Variation as Coefficient of Variation Percentage Point

Names of Properties	Single-Operator Component	Within-Laboratory Component	Between-Laboratory Component	Replication Component
Specimens of the Same Material:				
M.D. Tensile (nonreinforced)	4.2	2.9	12.4	5.7
C.D. Tensile (nonreinforced)	2.6	15.5	39.1	7.0
Elongation (nonreinforced)	6.6	4.4	28.3	12.6
Tensile (reinforced)	7.1	2.4	4.2	3.3
Elongation (reinforced)	3.3	4.9	27.1	8.3
Specimens of Different Material:				
M.D. Tensile (nonreinforced)	10.0	2.6	7.3	5.7
C.D. Tensile (nonreinforced)	31.8	11.4	25.1	7.0
Elongation (nonreinforced)	27.3	6.4	10.9	12.6
Tensile (reinforced)	7.5	0.0	4.2	3.3
Elongation (reinforced)	5.3	4.8	26.7	8.3

where:

d = scale equivalent,
E = percent elongation where the "F" value is to be determined,
G = gage length, and
M = magnification ratio (see Note 10).

11.3.2.2 Using the calculated value of d from the 11.3.2.1, perform the steps in 11.3.1.2 and 11.3.1.3.

12. Report

12.1 Report the tensile strength in pounds-force per inch of width to three significant places.

12.2 Report the ultimate elongation in percent to two significant places.

12.3 Report the "F" value in the same manner as the tensile strength (12.1).

12.4 Include the manufacturer's name and designation for the tape.

12.5 The following test parameters are to be reported:

12.5.1 crosshead velocity,

12.5.2 gage length, and

12.5.3 specimen width.

12.6 When the desired test response includes ultimate elongation, indicate whether slippage of material from within the clamps occurred and estimate the amount.

12.7 Any anomalous behavior noticed besides slippage.

13. Precision and Bias

13.1 *Summary*—The difference between two single observations should not exceed the following critical differences in 95 out of 100 when all of the observations are taken by the same well-trained operator using the same piece of test equipment and specimens randomly drawn from the same sample of material.

M.D. Tensile (nonreinforced)	19.5 % of the average
C.D. Tensile (nonreinforced)	20.8 % of the average
Elongation (nonreinforced)	39.5 % of the average
Tensile (reinforced)	21.6 % of the average
Elongation (reinforced)	24.7 % of the average

The size of the differences is likely to be affected adversely by different circumstances. The true values of M.D. tensile (nonreinforced), C.D. tensile (nonreinforced), elongation (nonreinforced), tensile (reinforced) and elongation (reinforced) can be defined only in terms of specific test methods. Within this limitation, the procedures in Test Method D 3759 for determining these properties have no known bias. Sections 13.2 through 13.5 explain the basis for this summary and for evaluations made under other conditions.

13.2 *Interlaboratory Test Data*[6]—An interlaboratory study was made in 1980 in which randomly drawn samples of two materials were tested in six laboratories. Two operators in each laboratory each tested three specimens from each of three rolls of each material. The components of variance expressed as coefficient of variation were calculated to be the values listed in Table 2.

NOTE 11—The calculations for coefficient of variation and other statistics found in subsequent sections of this statement are described in Practice D 2906 and Annex A3 of Practice D 2904.

13.3 *Critical Differences*—For the components of variance listed in Table 2, two averages of observed values should be considered significantly different at the 95 % probability level if the difference equals or exceeds the critical differences listed in Table 3.

13.4 *Confidence Limits*—For the components of variance listed in Table 2, single averages of observed values have the 95 % confidence limits listed in Table 4.

NOTE 12—The tabulated values of the critical differences and confidence limits should be considered to be a general statement particularly with respect to between-laboratory precision. Before a meaningful statement can be made about two specific laboratories, the amount of statistical bias between them, if any, must be established with each comparison being based on recent data obtained on specimens randomly drawn from one sample of the material to be evaluated.

13.5 *Bias*—No justifiable statement can be made on the bias of Test Method D 3759 for testing tensile strength and elongation since the true value cannot be established by an accepted referee method.

[6] Supporting data are available from ASTM Headquarters. Request RR:D-10-1002.

TABLE 3 Critical Differences for the Properties Noted, Percentage Points[A]

Names of Properties	Number of Observation in Each Average	Single-Operator Precision Specimens		Within-Laboratory Precision Specimens		Between-Laboratories Precision Specimens	
		Same Material	Different Material	Same Material	Different Material	Same Material	Different Material
M.D. Tensile (nonreinforced)	1	19.5	32.1	34.8	32.9	40.3	38.6
	5	13.6	28.8	15.8	29.7	37.8	36.0
	10	12.6	28.4	15.0	29.3	37.5	35.6
C.D. Tensile (nonreinforced)	1	20.8	90.3	47.9	95.7	118.0	118.3
	5	11.3	88.6	44.3	94.1	117.0	117.0
	10	9.4	88.4	44.1	93.8	117.0	116.9
Elongation (nonreinforced)	1	39.5	83.3	41.3	85.2	88.7	90.4
	5	24.0	77.2	26.9	79.2	83.0	84.8
	10	21.3	76.4	24.5	78.4	82.3	84.1
Tensile (reinforced)	1	21.6	22.6	22.5	22.6	25.4	25.4
	5	20.0	21.1	21.0	21.1	24.1	24.1
	10	19.8	20.9	20.9	20.9	23.9	23.9
Elongation (reinforced)	1	24.7	27.2	28.1	30.4	80.7	80.1
	5	13.7	17.9	19.2	22.4	77.4	77.5
	10	11.6	16.4	17.8	21.2	77.1	77.1

[A] Critical differences were calculated using $t = 1.96$ which is based on infinite degrees of freedom.

TABLE 4 Width of 95 % Confidence Limits for the Properties Noted, Percentage Points[A]

Names of Properties	Number of Observation in Each Average	Single-Operator Precision Specimens		Within-Laboratory Precision Specimens		Between-Laboratories Precision Specimens	
		Same Material	Different Material	Same Material	Different Material	Same Material	Different Material
M.D. Tensile (nonreinforced)	1	±13.8	±22.7	±24.6	±23.2	±28.5	±27.3
	5	±9.6	±20.4	±11.1	±21.0	±26.8	±25.4
	10	±8.9	±20.1	±10.6	±20.7	±26.5	±25.2
C.D. Tensile (nonreinforced)	1	±14.7	±63.9	±33.9	±67.7	±83.7	±83.7
	5	±8.0	±62.7	±31.4	±66.5	±82.7	±82.8
	10	±6.7	±62.5	±31.2	±66.4	±82.7	±82.7
Elongation (nonreinforced)	1	±27.9	±58.9	±29.2	±60.2	±62.7	±63.9
	5	±17.0	±54.6	±19.0	±56.0	±58.7	±60.0
	10	±15.1	±54.1	±17.3	±54.5	±58.2	±59.9
Tensile (reinforced)	1	±15.3	±16.0	±16.0	±16.0	±18.0	±18.0
	5	±14.2	±14.9	±14.9	±14.9	±17.0	±17.0
	10	±14.0	±14.8	±14.8	±14.8	±16.9	±16.9
Elongation (reinforced)	1	±17.5	±19.3	±19.9	±21.5	±56.7	±56.7
	5	±9.7	±12.7	±13.6	±15.9	±54.8	±54.8
	10	±8.2	±11.6	±12.6	±15.0	±54.5	±54.5

[A] Critical differences were calculated using $t = 1.96$ which is based on infinite degrees of freedom.

Standard Test Method for Tensile Strength and Elongation of Pressure-Sensitive Tapes [Metric][1]

This standard is issued under the fixed designation D 3759M; the number immediately following the designation indicates the year of original adoption or, in the case of revision, the year of last revision. A number in parentheses indicates the year of last reapproval. A superscript epsilon (ε) indicates an editorial change since the last revision or reapproval.

ε1 NOTE—The address in Footnote 5 was editorially changed in May 1990.

1. Scope

1.1 This test method covers the measurement of tensile strength (breaking strength) and stretch properties (elongation and "F" value) for pressure-sensitive tapes. It includes procedures for machine direction and cross-direction tests and tests for tapes with low and high stretch and reinforced backings. It also includes a procedure for obtaining stretch force ("F" value) in conjunction with a determination of tensile strength. These procedures employ a constant-rate-of-extension (CRE)-type testing machine. They apply to the principle of stretching the specimen at a fixed strain rate of 25 mm per 25 mm of specimen length per minute with the exception of the procedure for reinforced tapes.

1.2 *This standard does not purport to address the safety problems associated with its use. It is the responsibility of the user of this standard to establish appropriate safety and health practices and determine the applicability of regulatory limitations prior to use.*

NOTE 1—This test method is the metric companion to Test Method D 3759 and uses specimen widths of 12 mm and 24 mm rather than ½ in. and 1 in. used in Test Method D 3759. This difference in test parameters will generally produce lower values for Test Method D 3759M.

2. Referenced Documents

2.1 *ASTM Standards:*

D 996 Terminology of Packaging and Distribution Environments[2]

D 2904 Practice for Interlaboratory Testing of a Textile Test Method that Produces Normally Distributed Data[3]

D 2906 Practice for Statements on Precision and Bias for Textiles[3]

D 3715 Practice for Quality Assurance of Pressure-Sensitive Tapes[2]

E 122 Practice for Choice of Sample Size to Estimate the Average Quality of a Lot or Process[4]

3. Terminology

3.1 Terminology found in D 996 shall apply.

4. Significance and Use

4.1 This test method provides information that can be used in material specifications for product design and quality assurance applications. It can be used in comparing different products.

4.2 The use of this test method must be related to the purpose for which the test is performed. One purpose is for determining the relative strength of the tape in the size in which it is purchased or used. Another purpose is to identify or characterize a particular backing material.

4.2.1 When relative strength is of interest, the test should be performed on the tape as-received, that is, without cutting the material to a specimen width less than the as-received width.

4.2.1.1 Usually tapes wider than 48 mm are not tested due principally to the limitations of equipment. Tapes as narrow as approximately 3 mm can be tested.

4.2.1.2 Comparison of materials by different methods should be avoided because the test parameters of specimen dimension and crosshead velocity determine the outcome. Changes in the parameter levels will produce different results for the same material.

NOTE 2—It is usual to find the tensile strength increasing significantly with increasing crosshead velocity and therefore, strain rate.

4.2.2 When identity of material characterization is of interest, the test should be performed on a specimen cut from within the sample material boundaries using a sharp razor cutter, such as that defined in Section 5.

NOTE 3—Some of the traditional tools for specimen preparation must be avoided when the backing is comprised of thin plastic sheeting. These include chopping dies and sample cutters operating on a shearing principle. The reason for this restraint is that edges sufficiently ragged and damaged resulting from chopping or shearing cause tearing to occur before the true tensile strength level is reached. Tapes with fibrous backings may be cut to satisfactory specimens with these tools.

4.2.3 Stretch characteristics can be related to the tape's intended use or for identifying or characterizing a material.

NOTE 4—Elongation measurements become difficult to perform on stretchy materials (greater than 25 % ultimate elongation) when the ratio of specimen length to width is small (approaching 2). The results show high variability and do not allow for practical use of this information except when one wishes to demonstrate large differences between material.

[1] This test method is under the jurisdiction of ASTM Committee D-10 on Packaging and is the direct responsibility of Subcommittee D10.14 on Closure and Reinforcement of packages.

Current edition approved Sept. 30, 1988. Published November 1988. Originally published as D 3759M – 83. Last previous editioin D 3759M – 83.

[2] *Annual Book of ASTM Standards,* Vol 15.09.

[3] *Annual Book of ASTM Standards,* Vol 07.01.

[4] *Annual Book of ASTM Standards,* Vol 14.02.

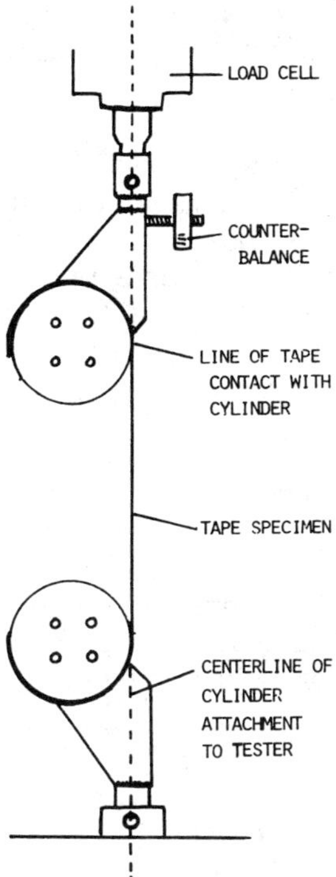

FIG. 1 Test Configuration for Reinforced Tapes

5. Apparatus

5.1 *Tension Tester*—A constant-rate-of-extension (CRE) type with load cell capacity such that the maximum expected specimen strength does not exceed 90 % of its normal limit.

5.1.1 *Test Information* should be displayed in at least an alphanumeric digital display or a load-elongation curve plotted by a pen or stylus responding to load and with a chart driven synchronously with the crosshead.

5.1.2 *Load-Elongation Curve,* plotted by a pen or stylus reponding to load and with a servo-chart drive or *x-y* recorder driven by an extensometer.

5.1.3 *Clamps,* preferably the preumatic action type.

5.1.3.1 Clamp faces at least 50 mm wide by 38 mm deep. Faces shall have a light cross-hatch serration.

NOTE 5—Plastic materials are reduced in width and thickness while being stretched. This causes them to be drawn out of the clamps. Pneumatic clamps minimize this effect. It can be further reduced by the appropriate choice of surface of the clamps. The greatest improvement, both with respect to the above mentioned shrinkage problem and simple slippage, may be found from the use of urethane film which can be obtained as a pressure-sensitive tape approximately 0.5 mm thick. This material has a very high coefficient of friction, is somewhat malleable, and is easily replaced. Alternative materials are coated abrasive, rubber (neoprene or other synthetic type), or other tape.

5.1.4 *Cylinders,* in place of clamps for testing reinforced tapes. Each of two cylinders shall be approximately 100 mm diameter by approximately 38 mm thick held in the position ordinarily occupied by the clamps so that the tape, when applied to the cylinders and extending between them, falls in the line of stress otherwise occupied by the specimens when clamps are used. See Fig. 1.

5.1.5 *Scale,* approximately 22 mm in length divided into 2-mm increments attached to each cylinder. The zero point (origin) shall be at the point of tangency of the tape with the cylinder during the test and the scale shall increase upward on the lower cylinder and downward on the upper cylinder.

NOTE 6—These scales will be used to observe and measure the tape slippage during the tension test for reinforced tapes.

TABLE 1 Tester Preparation and Specimen Dimensions

	Gage Length, mm	Cross-head Veloc-ity, min, mm/min	Chart Veloc-ity,[A] min, mm/min	Spe-cimen Width,[B] mm	Length, mm
Tapes with ultimate elongation of:					
Up to 200 %:					
Machine direction	125	125	125	12–24	230
Cross direction[C,D,E]	25	25	125	12	125
200 % and up:					
Machine direction	50	50	125	12	150
Reinforced tapes	250	125	125	12–24	710

[A] The chart velocity may be set at other velocities. It should be slower than the crosshead velocity.

[B] The specimen widths shown are for tests in which the specimen is cut from within the sample dimension. See 4.2.1.1.

[C] Cross-direction (C.D.) tests are limited to sample rolls of tape at least 48 mm in width.

[D] It is unusual to test C.D. tensile strength of tapes having ultimate elongations greater than 200 %. Therefore no reference to this is made in Table 1. However C.D. tests could be under that category on the high stretch materials.

[E] If the sample provides ample material, C.D. tests should preferably be made in the same way machine direction (M.D.) tests are. This would occur with web material or sufficiently wide rolls.

5.2 *Cutter,*[5] holding two single-edged razor blades in parallel planes, a precise distance apart, to form a cutter of exact specimen width. Appropriate widths shall be available (refer to specimen width in Table 1) patterned after the 12-mm cutter in 5.2.1. The differences between cutters of various widths is in the final width of the bar after removing the thickness of one razor blade.

5.2.1 The cutter shall consist of 12 mm thick by approximately 200 mm length of aluminum bar stock 125 mm from one end shall be slightly rounded to form a handle. The width of the bar, for approximately 75 mm from the opposite end, shall be narrowed to exactly 12 mm minus the thickness of a single razor blade (one of two used as cutting edges). The razor blades shall be held in position using side plates. The end of the cutter shall be cut away at a 45° angle to expose the cutting edges at one end of the blades.

6. Sampling

6.1 *Acceptance Sampling*—Sampling shall be in accordance with Practice D 3715.

6.2 *Sampling for Other Purposes*—The sampling and the number of test specimens depends on the purpose of the testing. Practice E 122, is recommended. It is common to test at least five specimens of a particular tape. Test specimens should be taken from several rolls of a tape and, whenever possible, among several production runs of a tape. Strong conclusions about a specific property of a tape cannot be based on test results of a single unit (roll) of product.

[5] Available from the Pressure-Sensitive Tape Council (PSTC), 104 Wilmot Rd., Suite 201, Deerfield, IL 60015.

7. Test Specimens

7.1 Specimens shall have the dimensions shown in Table 1.

7.2 Unwind and discard at least three, but no more than six, outer wraps of tape from the sample roll before taking specimens for testing.

7.3 Test one specimen per sample roll, unless otherwise specified.

7.4 The following applies to nonreinforced tapes:

7.4.1 Specimen ends that are clamped shall be prepared by covering the adhesive with paper, some other tape, or an extension of the specimen. In the latter case the specimen must be cut at least 100 mm longer than defined in Table 1.

7.4.2 The covering shall be free of wrinkles, leaving the gage-length area uncovered and completely cover the rest of the specimen so that the clamps will apply uniform pressure against the specimen.

7.4.3 A special specimen preparation is required for cross direction (C.D.) specimens from rolls less than 96 mm in width. Lay two rectangular sample strips on a flat surface with the adhesive side facing up. See Fig. 2. Each strip shall be as wide as the sample roll and approximately 125 mm in length. Position these strips side by side with one long edge of one strip parallel to and 25 mm separated from one long edge of the second strip.

7.4.3.1 Cut a specimen from the sample roll to have the width specified in Table 1 and length equal to the width of the roll.

7.4.3.2 Lay this specimen adhesive side up across the 25 mm separation of the strips. Position it toward one end of the sample strips so that it rests equally on both strips and at a right angle to their parallel edges.

7.4.3.3 Cut two additional strips from the sample roll having the same width as the specimen. Butt the end of one of these at one end to form a continuation of the specimen across the remainder of the sample strip. Use the second strip to butt against the other end of the specimen in like manner.

7.4.3.4 Fold each of the original sample strips over onto itself to form a three-ply tab that will be gripped by the clamps during the test.

7.4.3.5 Trim off any excess (single ply of tape) of either the sample strips or the extension strips extending beyond the two- or three-ply parts of the assembly.

NOTE 7—The extension serves to keep the clamping pressure uniform over the whole area of the specimen. This is an imperative factor to a successful test.

7.5 For reinforced tapes the specimen requires no further preparation than to have the appropriate dimensions (Table 1) and ensure that the adhesive is not contaminated so it will adhere well to the cylinders.

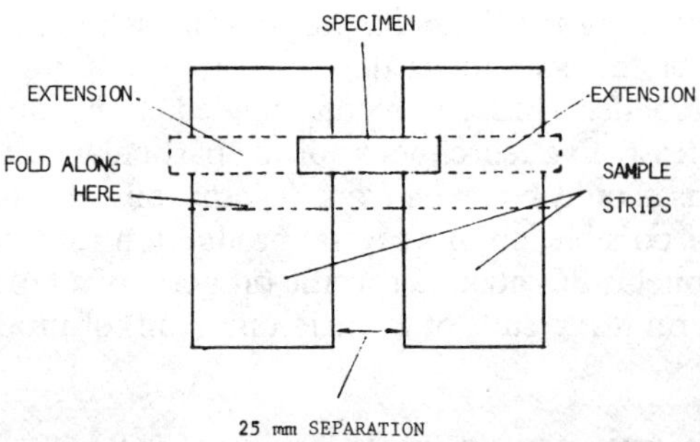

FIG. 2 Cross Direction (C.D.) Specimen Preparation

8. Preparation of Apparatus

8.1 Table 1 shows the tension tester settings for use with the specified test categories.

8.2 For testing reinforced tapes, set the cylinders 150 mm apart so that at the start of a test 25 mm of tape will extend between and without contact with the cylinders.

NOTE 8—The upper cylinder should be counter-balanced in order that the line of tape contact on the cylinders intersects an imaginary line running between the points of cylinder attachment to the tester and no side forces are exerted during the test. See Fig. 1.

9. Conditioning

9.1 Condition the sample to equilibrium in an atmosphere uniformly maintained at 23 ± 2°C and 50 ± 5 % relative humidity for 24 h. Testing shall be performed at the same conditions.

10. Procedure

10.1 *Nonreinforced Tapes:*

10.1.1 Clamp the specimen in the grips of the testing machine. Take care to align the long axis of the specimen with an imaginary line running between the points of attachment of the grips and including the center of the grips. Apply no more tension to the specimen during clamping than is necessary to remove slack.

10.1.2 Start the crosshead in motion at the specified velocity (Table 1) and ensure that the mechanism that displays the response (that is charts, plotter, and digital display) is operating. Continue until the specimen ruptures.

10.1.3 Record a numerical display of the results.

10.2 *Reinforced Tapes:*

10.2.1 Adhere approximately 230 mm of the specimen on the upper cylinder beginning at the line of tape contact (see 8.2), and wrap the specimen around the top surface of the cylinder. Repeat this with the free end of the specimen on the lower cylinder, except wrap the specimen around the bottom surface of the cylinder. The applied specimen must be centered on the center line around the cylinder surface. This elimination of skewness prevents nonuniform stress loading across the width of the specimen. The specimen shall also be sufficiently taut to remove slack.

10.2.2 Mark the specimen (and cylinder if not already done), with a marking pen making a line approximately 1 mm wide at the line where the tape contacts each cylinder. These bench marks will be 25 mm apart and shall be checked to ensure this.

10.2.3 Start the crosshead in motion at the specified velocity and ensure that the response-indicator mechanism is operating to indicate both load and elongation, if the latter is required.

10.2.4 Observe the bench marks on the specimen to determine their change in position relative to the marks on the cylinders. Use the scales appended to the cylinders.

10.2.5 When the specimen breaks, record the sum of the upper and lower bench mark changes to the nearest 2 mm. This will be the correction for the elongation.

10.2.6 Also record the indicated responses for tensile

strength and elongation when the tester provides a numerical display of this information.

11. Calculation

11.1 *Tensile Strength:*

11.1.1 When the recorded load-elongation curve results from the testing, calculate tensile strength as follows:

11.1.1.1 Find the farthest advance of the plotting pen or stylus from the origin in the direction representing increasing force. Record this advance as a percent of the chart scale.

11.1.1.2 Multiply the full-scale load range used during the test by the percent found in 11.1.1.1.

NOTE 9—For this calculation express the percent as a decimal fraction, that is, use 0.87 for 87 %.

11.1.1.3 Convert the value found in 11.1.1.2 to newtons per the desired basis of dimension for the final step. For N/100 mm of width, divide by the specimen width in millimetres and multiply by 100.

11.2 *Ultimate Elongation:*

11.2.1 When the recorded load-elongation curve results from the use of a synchronous chart drive, calculate elongation as follows:

11.2.1.1 Measure the distance from the start of the plot to the indicated breaking point along the motion (time) axis of the chart.

11.2.1.2 When the chart velocity equals the crosshead velocity, the distance from 11.2.1.1 divided by the gage length and multiplied by 100 equals the ultimate elongation.

11.2.1.3 When the chart and crosshead velocities are different, multiply the calculation completed in 11.2.1.2 by the ratio of the crosshead velocity to the chart velocity to obtain the correct ultimate elongation.

11.2.2 When the recorded load-elongation curve results from use of a servo-chart drive or *x-y* recorder driven by an extensometer, calculate elongation as follows:

11.2.2.1 Measure the distance from the start of the plot to the indicated breaking point. Divide this by the magnification ratio (to get the true dimension of elongation). Divide this by the gage length. Next, multiply by 100 to convert the value to percent ultimate elongation.

NOTE 10—The magnification ratio is the chart motion over the actual extensometer motion.

11.2.3 When testing reinforced tapes, the correction found in 10.2.5 shall first be subtracted from the amount of extension indicated or measured along the time axis of the recording chart and the difference divided by the original gage length. Multiply this quotient by 100 to convert it into percent ultimate elongation.

11.3 *"F" Value:*

11.3.1 When the recorded load-elongation curve results from the use of a synchronous chart drive, calculate the "F" value as follows:

11.3.1.1 Find the chart scale equivalent to the desired elongation point. (If one were determining the F-3 value, that is, the force required to stretch the specimen 3 %, one would be finding the chart scale equivalent to 3 % elongation). This equivalent is the distance measured from the start of the plot to the prescribed elongation point on the chart. To calculate the distance in terms of the chart scale, use this equation:

$$d = \frac{E \times G \times C}{100\,(H)}$$

where:
d = chart scale equivalent distance,
E = percent elongation where the "F" value is to be determined,
G = gage length,
C = chart velocity, and
H = crosshead velocity.
The units of these measurements shall be millimetres.

11.3.1.2 Using the calculated d, measure along the chart time axis (same as stretch, extension, or elongation axis) from the origin of the recorded plot to a point distance, d. This point shall be at zero force as was the plot origin. Draw a line from this point parallel to the force axis to intersect with the recorded plot. Measure the scale length of this line from zero force and convert this length into percent of chart scale and then into force. See 11.1.1.1, 11.1.1.2, and Note 9.

11.3.1.3 Multiply the full-scale load range used during the test by the percent found in 10.3.1.2. Express the percent as a decimal fraction. The product is the "F" value in newtons for the specimen width.

11.3.1.4 The final step is the same as 11.1.1.3, except substitute 11.3.1.3 for 11.1.1.2.

11.3.2 When the recorded load-elongation curve results from the use of a servo-chart drive or *x-y* recorder driven by an extensometer, calculate the "F" value as follows:

11.3.2.1 Find the scale equivalent to the desired elongation point using this equation:

$$d = \frac{E \times G \times M}{100}$$

where:
d = scale equivalent,
E = percent elongation where the "F" value is to be determined,
G = gage length, and
M = magnification ratio (see Note 10).

11.3.2.2 Using the calculated value of d from the 11.3.2.1, perform the steps in 11.3.1.2 and 11.3.1.3.

12. Report

12.1 Report the tensile strength in newtons per 100 mm of width to three significant places.

12.2 Report the ultimate elongation in percent to two significant places.

12.3 Report the "F" value in the same manner as the tensile strength (11.1).

12.4 Include the manufacturer's name and designation for the tape.

12.5 The following test parameters are to be reported:

12.5.1 crosshead velocity,

12.5.2 gage length, and

12.5.3 specimen width.

12.6 When the desired test response includes ultimate elongation, indicate whether slippage of material from within the clamps occurred and estimate the amount.

12.7 Any anomalous behavior noticed besides slippage.

TABLE 2 Components of Variation as Coefficient of Variation Percentage Point

Names of Properties	Single-Operator Component	Within-Laboratory Component	Between-Laboratory Component	Replication Component
Specimens of the Same Material:				
M.D. Tensile (nonreinforced)	4.2	2.9	12.4	5.7
C.D. Tensile (nonreinforced)	2.6	15.5	39.1	7.0
Elongation (nonreinforced)	6.6	4.4	28.3	12.6
Tensile (reinforced)	7.1	2.4	4.2	3.3
Elongation (reinforced)	3.3	4.9	27.1	8.3
Specimens of Different Material:				
M.D. Tensile (nonreinforced)	10.0	2.6	7.3	5.7
C.D. Tensile (nonreinforced)	31.8	11.4	25.1	7.0
Elongation (nonreinforced)	27.3	6.4	10.9	12.6
Tensile (reinforced)	7.5	0.0	4.2	3.3
Elongation (reinforced)	5.3	4.8	26.7	8.3

TABLE 3 Critical Differences for the Properties Noted, Percentage Points[A]

Names of Properties	Number of Observation in Each Average	Single-Operator Precision		Within-Laboratory Precision		Between-Laboratory Precision	
		Specimens		Specimens		Specimens	
		Same Material	Different Material	Same Material	Different Material	Same Material	Different Material
M.D. Tensile (nonreinforced)	1	19.5	32.1	34.8	32.9	40.3	38.6
	5	13.6	28.8	15.8	29.7	37.8	36.0
	10	12.6	28.4	15.0	29.3	37.5	35.6
C.D. Tensile (nonreinforced)	1	20.8	90.3	47.9	95.7	118.0	118.3
	5	11.3	88.6	44.3	94.1	117.0	117.0
	10	9.4	88.4	44.1	93.8	117.0	116.9
Elongation (nonreinforced)	1	39.5	83.3	41.3	85.2	88.7	90.4
	5	24.0	77.2	26.9	79.2	83.0	84.8
	10	21.3	76.4	24.5	78.4	82.3	84.1
Tensile (reinforced)	1	21.6	22.6	22.5	22.6	25.4	25.4
	5	20.0	21.1	21.0	21.1	24.1	24.1
	10	19.8	20.9	20.9	20.9	23.9	23.9
Elongation (reinforced)	1	24.7	27.2	28.1	30.4	80.7	80.1
	5	13.7	17.9	19.2	22.4	77.4	77.5
	10	11.6	16.4	17.8	21.2	77.1	77.1

[A] Critical differences were calculated using $t = 1.96$ which is based on infinite degrees of freedom.

13. Precision and Bias

13.1 *Summary*—The difference between two single observations should not exceed the following critical differences in 95 out of 100 cases when all of the observations are taken by the same well-trained operator using the same piece of test equipment and specimens randomly drawn from the same sample of material.

M.D. Tensile (nonreinforced)	19.5 % of the average
C.D. Tensile (nonreinforced)	20.8 % of the average
Elongation (nonreinforced)	39.5 % of the average
Tensile (reinforced)	21.6 % of the average
Elongation (reinforced)	24.7 % of the average

The size of the differences is likely to be affected adversely by different circumstances. The true values of M.D. tensile (nonreinforced), C.D. tensile (nonreinforced), elongation (nonreinforced), tensile (reinforced) and elongation (reinforced) can be defined only in terms of specific test methods. Within this limitation, the procedures in Test Method D 3759M for determining these properties have no known bias. Sections 13.2 through 13.5 explain the basis for this summary and for evaluations made under other conditions.

13.2 *Interlaboratory Test Data*[6]—An interlaboratory study was made in 1980 in which randomly drawn samples of two materials were tested in six laboratories. Two operators in each laboratory each tested three specimens from each of three rolls of each material. The components of variance expressed as coefficient of variation were calculated to be the values listed in Table 2.

NOTE 11—The calculations for coefficient of variation and other statistics found in subsequent sections of this statement are described in Practice D 2906 and Appendix A3 of Practice D 2904.

13.3 *Critical Differences*—For the components of variance listed in Table 2, two averages of observed values should be considered significantly different at the 95 % probability level if the difference equals or exceeds the critical differences listed in Table 3.

13.4 *Confidence Limits*—For the components of variance listed in Table 2, single averages of observed values have the 95 % confidence limits listed in Table 4.

NOTE 12—The tabulated values of the critical differences and confidence limits should be considered to be a general statement particularly with respect to between-laboratory precision. Before a meaningful statement can be made about two specific laboratories, the amount of statistical bias between them, if any, must be established with each comparison being based on recent data obtained on specimens randomly drawn from one sample of the material to be evaluated.

13.5 *Bias*—No justifiable statement can be made on the bias of Test Method D 3759M for testing tensile strength and elongation since the true value cannot be established by an accepted referee method.

[6] Supporting data are available from ASTM Headquarters. Request RR: D-10-1002.

TABLE 4 Width of 95 % Confidence Limits for the Properties Noted, Percentage Points [A]

Names of Properties	Number of Observations in Each Average	Single-Operator Precision		Within-Laboratory Precision		Between-Laboratory Precision	
		Same Material	Different Material	Same Material	Different Material	Same Material	Different Material
M.D. Tensile (nonreinforced)	1	±13.8	±22.7	±24.6	±23.2	±28.5	±27.3
	5	±9.6	±20.4	±11.1	±21.0	±26.8	±25.4
	10	±8.9	±20.1	±10.6	±20.7	±26.5	±25.2
C.D. Tensile (nonreinforced)	1	±14.7	±63.9	±33.9	±67.7	±83.7	±83.7
	5	±8.0	±62.7	±31.4	±66.5	±82.7	±82.8
	10	±6.7	±62.5	±31.2	±66.4	±82.7	±82.7
Elongation (nonreinforced)	1	±27.9	±58.9	±29.2	±60.2	±62.7	±63.9
	5	±17.0	±54.6	±19.0	±56.0	±58.7	±60.0
	10	±15.1	±54.1	±17.3	±54.5	±58.2	±59.9
Tensile (reinforced)	1	±15.3	±16.0	±16.0	±16.0	±18.0	±18.0
	5	±14.2	±14.9	±14.9	±14.9	±17.0	±17.0
	10	±14.0	±14.8	±14.8	±14.8	±16.9	±16.9
Elongation (reinforced)	1	±17.5	±19.3	±19.9	±21.5	±56.7	±56.7
	5	±9.7	±12.7	±13.6	±15.9	±54.8	±54.8
	10	±8.2	±11.6	±12.6	±15.0	±54.5	±54.5

[A] Critical differences were calculated using $t = 1.96$ which is based on infinite degrees of freedom.

Standard Test Method for Adherence to Linerboard of Pressure-Sensitive Tape at Low Temperature[1]

This standard is issued under the fixed designation D 3889; the number immediately following the designation indicates the year of original adoption or, in the case of revision, the year of last revision. A number in parentheses indicates the year of last reapproval. A superscript epsilon (ε) indicates an editorial change since the last revision or reapproval.

This standard has been approved for use by agencies of the Department of Defense. Consult the DoD Index of Specifications and Standards for the specific year of issue which has been adopted by the Department of Defense.

[ε1] NOTE—Footnotes 4 and 5 were changed editorially in May 1990.

1. Scope

1.1 This test method provides one procedure for determining the adherence of pressure-sensitive tapes to a standard linerboard or other similar surface.

1.2 This test method provides for an attribute response (pass or fail) and does not yield numerical data on a variable or continuous scale.

1.3 *This standard does not purport to address the safety problems associated with its use. It is the responsibility of the user of this standard to establish appropriate safety and health practices and determine the applicability of regulatory limitations prior to use.*

2. Referenced Documents

2.1 *ASTM Standards:*

D 996 Terminology of Packaging and Distribution Environments[2]

D 3715 Practice for Quality Assurance of Pressure-Sensitive Tapes[2]

D 4332 Practice for Conditioning Containers, Packages, or Packaging Components for Testing[2]

E 122 Practice for Choice of Sample Size to Estimate a Measure of Quality of a Lot or Process[3]

3. Terminology

3.1 *Definitions*—General terms in this test method are defined in Terminology D 996.

4. Summary of Test Method

4.1 The pressure-sensitive tape to be tested is adhered to the standard linerboard or other material of interest. Following a dwell period at an elevated temperature, the specimen assembly is cooled to a specific temperature. After a specified time at the low temperature the specimen assembly is flexed around a small mandrel and observed for indications of bond failure.

5. Significance and Use

5.1 Satisfactory tape performance at the test temperature implies satisfactory cold temperature performance at or above the test temperature.

5.2 This procedure would be applicable for questions of adherence to 100 % kraft fiberboards similar to the standard linerboard surface without substitution but for direct information on any board or surface the actual surface in question should be substituted.

6. Apparatus

6.1 *Standard Reference Material 1810A,*[4] length approximately 11 in., width approximately ½ in. wider than the specimen width. See Section 8.

NOTE 1—When agreed upon by the parties involved or when called for by the material specification, other flexible material may be substituted for the 1810 linerboard.

6.2 *Rubber-Covered Steel Roller,*[5] 3.25 ± 0.1 in. in diameter and 1.75 ± 0.05 in. wide, covered with rubber approximately 0.25 in. thick having a Shore A durometer hardness of 80 ± 5. The surface of the rubber shall form a true cylinder devoid of any convex or concave deviations. The mass of the roller shall be 4.5 ± 0.1 lb. The mass of the handle shall not add to the mass of the roller during use.

6.3 *Oven,* air circulating, maintained at 150 ± 2°F.

6.4 *Cold Chamber,* maintained at a specified temperature −65 ± 2°F if no other temperature specified.

6.5 *Nonrotating, Smooth Surface, Metal Cylinder,* having a diameter of 1.5 ± 0.25 in. and minimum length of 2 in. fixed to a base or support for use as a mandrel around which to flex the tape specimen assembly.

7. Sampling

7.1 *Acceptance Sampling*—Sampling shall be in accordance with Practice D 3715.

7.2 *Sampling for Other Purposes*—The sampling and the number of specimens depends on the purpose of the testing. Practice E 122 is recommended. It is common to test at least

[1] This test method is under the jurisdiction of ASTM Committee D-10 on Packaging and is the direct responsibility of Subcommittee D10.14 on Closure and Reinforcement.

Current edition approved Dec. 30, 1988. Published February 1989. Originally published as D 3889 – 80. Last previous edition D 3889 – 85.

[2] *Annual Book of ASTM Standards,* Vol 15.09.

[3] *Annual Book of ASTM Standards,* Vol 14.02.

[4] Standard Reference Material 1810A is available from the Office of Standard Reference Materials, National Institute of Standards and Technology, Washington, DC 20234.

[5] Available from the Pressure-Sensitive Tape Council (PSTC), 104 Wilmot Rd., Suite 201, Deerfield, IL 60015.

five specimens of a particular tape. Test specimens should be taken from several rolls of tape and wherever possible, among several production runs of tape. Strong conclusions about a specific property of a tape cannot be based on test results of a single unit (roll) of a product.

8. Test Specimens

8.1 The specimens shall be 0.5 in. to 1.0 in. wide. The length shall be a maximum of 6.0 in.

8.2 Unwind and discard at least three, but no more than six, outer wraps of tape from the sample roll before taking specimens for testing.

8.3 Remove specimens from a freely rotating roll at the rate of 20 to 30 in./s. Wherever width or other factors causing high adherence to backing make it impossible to remove the specimen at the prescribed rate, remove it at a rate as close to 20 in./s as possible.

9. Conditioning

9.1 Except when stated elsewhere in the method, prepare and condition the sample materials in the Standard Conditioning Atmosphere as described in Practice D 4332 for a period not less than 24 h.

10. Procedure

10.1 Apply the specimen centrally to the linerboard strip. Roll over the tape against a flat smooth surface using the rubber-covered roller. Make two passes, one in each lengthwise direction at the rate of 12 in./min.

10.2 Expose the specimen assembly to a temperature of 150 ± 2°F for 24 h. Within the next 24 h expose the specimen assembly together with the 1.5-in. mandrel, to the specified temperature for 2 h. If no temperature is specified, specimen should be conditioned at −65°F in the cold chamber for 2 h.

10.3 Without removing the specimen assembly from the cold environment nor effecting any change in that temperature, hold an extreme end (linerboard only) in each hand. With the linerboard side of the assembly against the mandrel draw first one end and then the other forward so that both ends of tape on the linerboard will have been formed into the curve of the mandrel. Draw the ends forward in turn at the rate of 24 in./s. Keep the ends of the assembly taut and parallel forming a 180° bend. End the drawing with the assembly in the same position as when starting having completed one complete cycle.

10.4 Observe the tape and linerboard during and following the drawing for any evidence of release of tape from the linerboard.

11. Report

11.1 The report shall include the following:

11.1.1 Statement that this test method was used. Indicate any deviations from the test method as written,

11.1.2 Temperature at which the test was conducted.

11.1.3 Manufacturer's name and designation for the tape.

11.1.4 The observation made in 10.4, that is, whether the tape did or did not release from the linerboard.

12. Precision and Bias

12.1 No statement is made about either the precision or bias of this test method since the results merely state whether there is conformance to the criteria for success specified in the procedure.

Standard Test Method for Adherence to Linerboard of Pressure-Sensitive Tape at Low Temperature [Metric][1]

This standard is issued under the fixed designation D 3889M; the number immediately following the designation indicates the year of original adoption or, in the case of revision, the year of last revision. A number in parentheses indicates the year of last reapproval. A superscript epsilon (ε) indicates an editorial change since the last revision or reapproval.

ε1 NOTE—Footnote 4 was added and Footnote 5 was changed editorially in May 1990.

1. Scope

1.1 This test method provides one procedure for determining the adherence of pressure-sensitive tapes to a standard linerboard or other similar surface.

1.2 This test method provides for an attribute response (pass or fail) and does not yield numerical data on a variable or continuous scale.

1.3 This test method is the metric companion to Test Method D 3889.

1.4 *This standard does not purport to address the safety problems associated with its use. It is the responsibility of the user of this standard to establish appropriate safety and health practices and determine the applicability of regulatory limitations prior to use.*

2. Referenced Documents

2.1 *ASTM Standards:*

D 996 Terminology of Packaging and Distribution Environments[2]

D 3715 Practice for Quality Assurance of Pressure-Sensitive Tapes[2]

D 4332 Practice for Conditioning Containers, Packages, and Packaging Components for Testing[2]

E 122 Practice for Choice of Sample Size to Estimate a Measure of Quality of a Lot or Process[3]

3. Terminology

3.1 *Definitions*—General terms in this test method are defined in Terminology D 996.

4. Summary of Test Method

4.1 The pressure-sensitive tape to be tested is adhered to the standard linerboard or other material of interest. Following a dwell period at an elevated temperature, the specimen assembly is cooled to a specific temperature. After a specified time at the low temperature, the specimen assembly is flexed around a small mandrel and observed for indications of bond failure.

5. Significance and Use

5.1 Satisfactory tape performance at the test temperature implies satisfactory cold temperature performance at or above the test temperature.

5.2 This procedure would be applicable for questions of adherence to 100 % kraft fiberboards similar to the standard linerboard surface without substitution. However, for direct information on any board or surface, the actual surface in question should be substituted.

6. Apparatus

6.1 *Standard Reference Material 1810A,*[4] length approximately 280 mm, width approximately 13 mm wider than the specimen width. See Section 8.

NOTE—When agreed upon by the parties involved or when called for by the material specification, other flexible material may be substituted for the 1810 linerboard.

6.2 *Rubber-Covered Steel Roller,*[5] 82.5 ± 1.3 mm in diameter and 44.5 ± 1.3 mm wide, covered with rubber approximately 6 mm thick having a Shore A durometer hardness of 80 ± 5. The surface of the rubber shall form a true cylinder devoid of any convex or concave deviations. The mass of the roller shall be 2040 ± 45 g. The mass of the handle shall not add to the mass of the roller during use.

6.3 *Oven,* air circulating, maintained at 66 ± 1°C.

6.4 *Cold Chamber,* maintained at a specified temperature of −54 ± 1°C if no other temperature specified.

6.5 *Nonrotating, Smooth, Surface, Metal Cylinder,* having a diameter of 40 ± 6 mm and minimum length of 50 mm fixed to a base or support for use as a mandrel around which to flex the tape specimen assembly.

7. Sampling

7.1 *Acceptance Sampling*—Sampling shall be in accordance with Practice D 3715.

7.2 *Sampling for Other Purposes*—The sampling and the number of specimens depends on the purpose of the testing. Practice E 122 is recommended. It is common to test at least five specimens of the particular tape. Test specimens should be taken from several rolls of tape and wherever possible,

---

[1] This test method is under the jurisdiction of ASTM Committee D-10 on Packaging and is the direct responsibility of Subcommittee D10.14 on Closure and Reinforcement.

Current edition approved Dec. 30, 1988. Published February 1989. Originally published as D 3889M – 85. Last previous edition D 3889M – 85.

[2] *Annual Book of ASTM Standards,* Vol 15.09.

[3] *Annual Book of ASTM Standards,* Vol 14.02.

[4] Standard Reference Material 1810A is available from the Office of Standard Reference Materials, National Institute of Standards and Technology, Washington, DC 20234.

[5] Available from the Pressure-Sensitive Tape Council (PSTC), 104 Wilmot Rd., Suite 201, Deerfield, IL 60015.

among several productions runs of a tape. Strong conclusions about a specific property of a tape cannot be based on the test results of a single unit (roll) of a product.

8. Test Specimen

8.1 The specimen shall be 12 to 24 mm wide. The length shall be a maximum of 150 mm.

8.2 Unwind and discard at least three, but no more than six, outer wraps of tape from the sample roll before taking specimens for testing.

8.3 Remove specimens from a freely rotating roll at the rate of 500 to 750 mm/s. Where width or other factors causing high adherence to backing make it impossible to remove the specimen at the prescribed rate, remove it at a rate as close to 500 mm/s as possible.

9. Conditioning

9.1 Except when stated elsewhere in this test method, prepare and condition the sample materials in the Standard Conditioning Atmosphere as described in Practice D 4332 for a period of not less than 24 h.

10. Procedure

10.1 Apply the specimen centrally to the linerboard strip. Roll over the tape against a flat smooth surface using the rubber-covered roller. Make two passes, one in each lengthwise direction at the rate of 5 mm/s.

10.2 Expose the specimen assembly to a temperature of 66 ± 1°C for 24 h. Within the next 24 h, expose the specimen assembly together with the 40-mm mandrel to the specified temperature for 2 h. If no temperature is specified, specimen should be conditioned at −54°C in the cold chamber for 2 h.

10.3 Without removing the specimen assembly from the cold environment, nor effecting any change in the temperature, hold an extreme end (linerboard only) in each hand. With the linerboard side of the assembly against the mandrel, draw first one end and then the other forward so that both ends of the tape on the linerboard will have been formed into the curve of the mandrel. Draw the ends forward in turn at the rate of 0.8 m/s. Keep the ends of the assembly taut and parallel forming a 180° bend. End the drawing with the assembly in the same position as when starting, having completed one complete cycle.

10.4 Observe the tape and linerboard during and following the drawing for any evidence of release of tape from the linerboard.

11. Report

11.1 The report shall include the following information:

11.1.1 Statement that this test method was used. Indicate any deviations from the test method as written.

11.1.2 Temperature at which the test was conducted.

11.1.3 Manufacturer's name and designation for the tape.

11.1.4 The observation made in 10.4, that is, whether the tape did or did not release from the linerboard.

12. Precision and Bias

12.1 No statement is made about either the precision or bias of Test Method D 3889M since the results merely state whether there is conformance to the criteria for success specified in the procedure.

Designation: D 3950 – 90

Standard Specification for Strapping, Nonmetallic (and Joining Methods)[1]

This standard is issued under the fixed designation D 3950; the number immediately following the designation indicates the year of original adoption or, in the case of revision, the year of last revision. A number in parentheses indicates the year of last reapproval. A superscript epsilon (ε) indicates an editorial change since the last revision or reapproval.

1. Scope

1.1 This specification covers nonmetallic strapping and joining methods intended for use in closing, reinforcing, and bundling articles for shipment, unitizing, palletizing, and bracing for car loading and truck loading.

1.2 The values stated in inch-pound units are to be regarded as standard.

1.3 The following safety hazards caveat pertains only to the test method portion, Section 12, of this specification: *This standard does not purport to address the safety problems associated with its use. It is the responsibility of the user of this standard to establish appropriate safety and health practices and determine the applicability of regulatory limitations prior to use.*

2. Referenced Documents

2.1 *ASTM Standards:*

D 996 Terminology of Packaging and Distribution Environments[2]

D 3951 Practice for Commercial Packaging[2]

D 4332 Practice for Conditioning Containers, Packages, or Packaging Components for Testing[2]

D 4675 Guide for Selection and Use of Flat Strapping Materials[2]

2.2 *Military Standard:*

MIL-STD-105 Sampling Procedures and Tables for Inspection by Attributes[3]

3. Terminology

3.1 For general definitions of packaging and distribution environments, see Definitions D 996.

4. Classification

4.1 *Type and Grades:*

Type — Strapping, bonded rayon cord.
- *Grade 1*—Light duty.
- *Grade 2*—Regular duty.
- *Grade 3*—Heavy duty.

Type IA— Strapping, bonded, or woven polyester cord.
- *Grade 1*—Light duty.
- *Grade 2*—Regular duty.
- *Grade 3*—Heavy duty.
- *Grade 4*—Extra heavy duty.
- *Grade 5*—Special duty.

Type II — Strapping, polyolefin plastic.

Type III— Strapping, nylon plastic.

Type IV— Strapping, polyester plastic.

5. Ordering Information

5.1 The inquiry and order shall indicate the following:

5.1.1 Type, grade, and dimensions required (see 4.1 and 7.1),

5.1.2 Length per coil (see 8.1),

5.1.3 Joining method (see 6.2), type and size required (if needed),

5.1.4 If an embossed finish on strapping is desired or allowed (see Footnote in Table 3),

5.1.5 Make and model of strapping equipment that the strapping and joining method must work in, if applicable,

5.1.6 Coil dimensions (see 8.1),

5.1.7 Level of packaging and packing if other than commercial (see Section 15), and

5.1.8 ASTM designation and date of issue.

6. Materials and Manufacture

6.1 Materials shall be of the quality necessary to meet the physical requirements within the allowable dimensions.

6.1.1 *Type I*—Strapping shall consist of longitudinal rayon cords bonded with a plastic binder so that a nonwoven material is formed.

6.1.1.1 *Type IA*—Strapping shall consist of longitudinal polyester cords bonded with a plastic binder to form a nonwoven material or longitudinal polyester cords woven with a weft thread and treated with a plastic binder to form a woven material.

6.1.2 *Type II*—Strapping shall be an extruded, oriented polyolefin.

6.1.3 *Type III*—Strapping shall be an extruded, oriented nylon.

6.1.4 *Type IV*—Strapping shall be an extruded, oriented polyester.

6.2 *Joining Methods*—If seals or buckles are to be used, they shall be steel and have a coating of zinc, black iron oxide, or equivalent protection from corrosion, or buckles may be made of plastic.

7. Mechanical Properties

7.1 *Breaking Strength and Elongation* (see 12.2):

7.1.1 Type I and Type IA strapping shall conform to the breaking strengths and elongations prescribed in Tables 1 and 2.

7.1.2 Type II strapping shall conform to the breaking strengths and elongations prescribed in Table 3.

---

[1] This specification is under the jurisdiction of ASTM Committee D-10 on Packaging and is the direct responsibility of Subcommittee D10.14 on Closure and Reinforcement.

Current edition approved July 27, 1990. Published September 1990. Originally published as D 3950 – 80. Last previous edition D 3950 – 87.

[2] *Annual Book of ASTM Standards*, Vol 15.09.

[3] Available from Naval Publications and Forms Center, 5801 Tabor Ave., Philadelphia, PA 19120.

TABLE 1 Breaking Strengths and Elongation[A] of Type I Bonded Rayon Cord Strapping

Nominal Width of Strapping, in. (mm)	Grade	Minimum Breaking Strength, lbf (N)
¼ (6.4)	2	235 (1 045)
⅜ (9.5)	1	290 (1 290)
	2	350 (1 555)
½ (12.7)	1	410 (1 820)
	2	465 (2 070)
⅝ (15.9)	1	525 (2 335)
	2	585 (2 600)
	3	765 (3 400)
¾ (19.1)	1	640 (2 845)
	2	700 (3 110)
	3	900 (4 000)
1¼ (31.8)	3	1 575 (7 005)

[A] Average range of elongation at break is 10 to 15 %. Minimum transverse strength is 2.75 lbf (12.2 N).

7.1.3 Type III strapping shall conform to the breaking strengths and elongations prescribed in Table 4.

7.1.4 Type IV strapping shall conform to the breaking strengths and elongations prescribed in Table 5.

7.2 Transverse strength of Types I and IA strapping shall be a minimum of 2.75 lbf (12.2 N) and 2.25 lbf (10 N) respectively, when tested in accordance with 12.4.

7.3 *Joint Strength* (see 12.3).

7.3.1 Type I and Type IA Grade 1 and Grade 2 joined strapping shall have a strength of not less than 45 % of the minimum breaking strength of the strapping grade and size listed in Tables 1 and 2.

7.3.2 Type IA Grade 3, Grade 4, and Grade 5 joined strapping shall have a strength of not less than 75 % of the minimum breaking strength of the strapping grade and size listed in Table 2.

TABLE 2 Breaking Strengths and Elongation[A] of Type IA Bonded or Woven Polyester Cord Strapping

Nominal Width of Strapping, in. (mm)	Grade	Minimum Breaking Strength, lbf (N)
¼ (6.4)	2	300 (1 335)
(6.0)	3	540 (2 400)
⅜ (9.5)	2	450 (2 000)
(9.0)	3	780 (3 480)
½ (12.7)	1	400 (1 780)
(12.7)	2	600 (2 670)
(13.0)	3	1 050 (4 670)
(13.0)	4	1 360 (6 050)
⅝ (15.9)	1	500 (2 225)
(15.9)	2	725 (3 225)
(16.0)	3	1 310 (5 830)
(16.0)	4	1 650 (7 340)
¾ (19.1)	2	900 (4 000)
(19.0)	3	1 585 (7 060)
(19.0)	4	1 830 (8 150)
(19.0)	5	2 600 (11 570)
1¼ (32.0)	2	1 830 (8 150)
(32.0)	3	2 105 (9 370)
(32.0)	4	3 285 (14 620)
(32.0)	5	4 200 (18 680)

[A] Average range of elongation at break is 9 to 13 %. Minimum transverse strength is 2.25 lbf (10 N).

TABLE 3 Breaking Strengths of Type II Strapping

Nominal Width of Strapping, in. (mm)	Nominal Thickness of Strapping, in. (mm)[A]	Minimum Breaking Strength lbf (N)[B]
¼ (6.4)	0.0135 (0.34)	134 (600)
0.236 (6.0)	0.0130 (0.33)	135 (600)
¼ (6.4)	0.015 (0.38)	180 (810)
¼ (6.4)	0.018 (0.46)	190 (850)
0.236 (6.0)	0.0173 (0.44)	190 (850)
¼ (6.4)	0.025 (0.64)	200 (900)
0.354 (9.0)	0.0142 (0.36)	225 (1 000)
0.354 (9.0)	0.0173 (0.44)	270 (1 200)
0.354 (9.0)	0.0181 (0.46)	270 (1 200)
⅜ (9.5)	0.018 (0.46)	250 (1 120)
⅜ (9.5)	0.015 (0.36)	290 (1 300)
0.413 (10.5)	0.0201 (0.51)	360 (1 600)
⅜ (9.5)	0.020 (0.51)	390 (1 750)
⅜ (9.5)	0.025 (0.64)	400 (1 790)
0.413 (10.5)	0.0248 (0.63)	460 (2 050)
0.413 (10.5)	0.0295 (0.75)	550 (2 450)I
7⁄16 (11.1)	0.014 (0.36)	300 (1 340)
7⁄16 (11.1)	0.019 (0.48)	350 (1 570)
0.472 (12.0)	0.0169 (0.43)	360 (1 600)
7⁄16 (11.1)	0.023 (0.58)	420 (1 880)
7⁄16 (11.1)	0.025 (0.64)	450 (2 020)
0.472 (12.0)	0.0212 (0.54)	460 (2 050)
0.472 (12.0)	0.0252 (0.64)	550 (2 450)
7⁄16 (11.1)	0.030 (0.76)	550 (2 460)
½ (12.7)	0.017 (0.43)	350 (1 570)
½ (12.7)	0.015 (0.38)	390 (1 750)
½ (12.7)	0.022 (0.56)	450 (2 020)
½ (12.7)	0.020 (0.51)	530 (2 370)
½ (12.7)	0.026 (0.66)	550 (2 460)
½ (12.7)	0.025 (0.64)	660 (2 960)
½ (12.7)	0.030 (0.76)	810 (3 630)
⅝ (15.9)	0.015 (0.38)	500 (2 240)
⅝ (15.9)	0.020 (0.51)	680 (3 050)
⅝ (15.9)	0.030 (0.76)	950 (4 260)
¾ (19.1)	0.020 (0.51)	725 (3 250)
1¼ (31.8)	0.035 (0.89)	2 200 (9 860)
1¼ (31.8)	0.050 (1.27)	3 100 (13 900)

[A] When specified (see 5.1.4), the strapping as measured by flat anvil micrometer shall have an embossed finish which yields an overall nominal thickness no greater than twice the nominal thickness of smooth-surfaced strapping of the same width and breaking strength.

[B] Average range of elongation at break is 15 to 30 %.

7.3.3 Types II, III, and IV joined strapping shall have a strength of not less than 45 % of the minimum breaking strength corresponding to the dimensions of the strap listed in Table 3, 4, or 5.

8. Dimensions and Permissible Variations

8.1 The minimum length per coil of strapping shall be as specified in the following tables according to type, grade, size, and coil width. The coil shall be an oscillating or ribbon wind.

8.1.1 Type I and Type IA strapping shall be furnished with the minimum yards per coil according to size and grade as prescribed in Table 6.

8.1.2 Type II, III, and IV strapping shall be furnished with the minimum yards per coil according to the size and coil width as prescribed in Table 7.

TABLE 4 Breaking Strengths of Type III Strapping

Nominal Width of Strapping, in. (mm)	Nominal Thickness of Strapping, in. (mm)	Minimum Breaking Strength, lbf (N)[A]
7/16 (11.1)	0.017 (0.43)	420 (1 870)
	0.023 (0.58)	560 (2 490)
	0.029 (0.74)	700 (3 110)
1/2 (12.7)	0.015 (0.38)	420 (1 870)
	0.020 (0.51)	560 (2 490)
	0.025 (0.64)	700 (3 110)
	0.030 (0.76)	900 (4 000)

[A] Average range of elongation at break is 12 to 20 %.

9. Workmanship, Finish and Appearance

9.1 Type I and Type IA strapping shall be straight, clean, and free of cracks and other defects that may affect the serviceability.

9.2 Type II, III, and IV strapping shall be straight, clean, and free of kinks, edge curvature, cracks, and other defects that may affect the serviceability.

9.3 Splices may be made if they do not affect the serviceability of the strapping in strapping equipment and have a break strength equal to at least 50 % of the strap break strength. No more than one splice per coil will be permitted.

10. Sampling Plan

10.1 Unless otherwise specified in contract or order, sampling for acceptance inspection and testing by attributes shall be performed in accordance with applicable provisions set forth in MIL-STD-105; Tables I, and II-A. A random sample of coils (or joining methods) shall be selected from each lot.

10.2 *Lot Size*—The lot size shall be expressed in units of coils or joining methods, or both, and shall consist of all products of one type, grade, and size offered for acceptance at one time.

10.3 *Unit Sample*—One coil or one joining method shall be considered a unit. A minimum of 10 ft (3.05 m) per coil of strapping is required to perform the tests.

10.4 *Rate of Sampling*—The rate of sampling shall be in accordance with MIL-STD-105, inspection level S-3. Similarly the acceptable quality level shall be 4 %.

TABLE 5 Breaking Strengths of Type IV Strapping

Nominal Width of Strapping, in. (mm)	Nominal Thickness of Strapping, in. (mm)	Minimum Breaking Strength, lbf (N)[A]
3/8 (9.5)	0.015 (0.38)	310 (1 380)
0.354 (9.0)	0.0205 (0.52)	400 (1 780)
3/8 (9.5)	0.020 (0.51)	420 (1 870)
0.413 (10.5)	0.0205 (0.52)	460 (2 050)
0.413 (10.5)	0.024 (0.61)	560 (2 490)
7/16 (11.1)	0.016 (0.41)	360 (1 600)
7/16 (11.1)	0.020 (0.51)	460 (2 050)
0.472 (12.0)	0.0175 (0.44)	470 (2 090)
7/16 (11.1)	0.022 (0.56)	500 (2 220)
0.472 (12.0)	0.0205 (0.52)	560 (2 490)
7/16 (11.1)	0.024 (0.61)	560 (2 490)
0.472 (12.0)	0.0280 (0.71)	750 (3 340)
1/2 (12.7)	0.015 (0.38)	420 (1 870)
1/2 (12.7)	0.017 (0.43)	470 (2 090)
1/2 (12.7)	0.020 (0.51)	560 (2 490)
1/2 (12.7)	0.025 (0.64)	700 (3 110)
1/2 (12.7)	0.028 (0.71)	750 (3 340)
1/2 (12.7)	0.030 (0.76)	850 (3 780)
5/8 (15.9)	0.020 (0.51)	700 (3 110)
5/8 (15.9)	0.025 (0.64)	870 (3 870)
5/8 (15.9)	0.030 (0.76)	1 060 (4 720)

[A] Average range of elongation at break is 5 to 15 %.

TABLE 6 Minimum Yards Per Coil for Type I and Type IA Strapping

Nominal[A] Width of Strapping, in. (mm)	Grade	Minimum Yards (Metres) Per 5½-in. (140 mm) Wide Coil[B]
1/4 (6.4)	2	2600 (2380)
(6.0)	3	2750 (2520)
3/8 (9.5)	1 and 2	1650 (1510)
(9.0)	3	1650 (1510)
1/2 (12.7)	1 and 2	1300 (1190)
(13.0)	3	1390 (1270)
(13.0)	4	1100 (1010)
5/8 (15.9)	1 and 2	1000 (910)
(16.0)	3	1100 (1010)
(16.0)	4	830 (760)
3/4 (19.1)	2	700 (640)
(19.0)	3	830 (760)
(19.0)	4	555 (500)
1¼ (32.0)	2, 3 and 4	110 (100)
(32.0)	5	150 (140)

[A] The width tolerance shall be +1/32 in. (+0.79 mm) and −3/32 in. (−2.37 mm).
[B] 1¼ in. (32 mm) strapping is ribbon-wound 1¼ in. (32 mm) wide coil.

11. Number of Tests

11.1 *Tests*—For the determination of break strength, elongation, transverse strength, and sealed joint strength, an average of three specimens per coil shall be considered a complete test.

11.2 *Retests*—When testing for elongation, if the specimen fails outside of the center third of the gage length or within 1 in. of the jaws, a retest shall be made. If the percentage of elongation or the break strength is less than the specified minimum, a retest is permitted. In this retest, three consecutive specimens must meet the minimum requirements.

12. Test Methods

12.1 *Conditioning*—Condition test samples in a standard atmosphere as described in Practice D 4332.

12.1.1 Condition Type I and Type IA strapping to equilibrium when the change in weight during a 1-h or longer period of conditioning does not exceed 0.02 % of the sample weight at the end of the period.

12.1.2 Condition Type II, III, and IV strapping for 24 h prior to testing.

12.2 *Breaking Strength and Elongation*—Test breaking strength and elongation in conditions as described in 12.1.

12.2.1 *Breaking Strength*—Measure the breaking strength of strapping on a tension tester capable of load accuracies to within ±1 %. Select the load range in which the point of maximum force will fall between 30 to 90 % of full scale. Use suitable clamping arrangements to prevent slippage or breakage within the clamps. The crosshead speed shall be 2 in. (51 mm)/min and the free length between clamps shall be 5 in. (127 mm). Insert the sample of strap of a length at least the free length distance plus twice the length of the clamps, in the clamps, adjust manually to remove slack, and activate the tester. Read breaking strength directly from the recording device to the nearest 5 lbf (22 N) (or 10 lbf (44 N) for

TABLE 7 Minimum Yards (Metres) Per Coil; Types II, III and IV Strapping

Nominal Width[B] of Strapping, in. (mm)	Nominal Thickness[C] of Strapping, in. (mm)	Min Yards (Metres) Per 3-in. Wide Coil[A]	Min Yards (Metres) Per 5¾-in. Wide Coil[A]	Min Yards (Metres) Per 6-in. Wide Coil[A]	Min Yards (Metres) Per 7½-in. Wide Coil[A]
¼ (6.4)	0.0135 (0.34)		6 500[D] (5 940)		
0.236 (6.0)	0.0130 (0.33)		6 665 (6 095)		7 330 (6 705)
0.236 (6.0)	0.0130 (0.33)		830 (760)		
¼ (6.4)	0.015 (0.38)	3 000 (2 740)		6 000 (5 490)	
¼ (6.4)	0.018 (0.46)		5 000[E] (4 570)		
0.236 (6.0)	0.0173 (0.44)		5 000 (4 570)		5 330 (4 875)
¼ (6.4)	0.025 (0.64)				4 300 (3 930)
0.354 (9.0)	0.0142 (0.36)				4 165 (3 810)
0.354 (9.0)	0.0173 (0.44)		3 330 (3 045)		
0.354 (9.0)	0.0181 (0.46)		4 500 (4 115)		3 750 (3 430)
0.354 (9.0)	0.0205 (0.52)		4 165 (3 810)		
⅜ (9.5)	0.018 (0.46)		3 000 (2 740)	4 300 (3 930)	3 000 (2 740)
⅜ (9.5)	0.015 (0.38)	2 165 (1 980)		4 300 (3 930)	3 000 (2 740)
0.413 (10.5)	0.0201 (0.51)		3 500 (3 200)		2 750 (2 515)
0.413 (10.5)	0.0205 (0.52)		3 500 (3 200)		
0.413 (10.5)	0.0240 (0.61)		3 000 (2 740)		
⅜ (9.5)	0.020 (0.51)	1 600 (1 460)		2 800 (2 560)	
⅜ (9.5)	0.025 (0.64)			2 800 (2 560)	2 000 (1 830)
0.413 (10.5)	0.0248 (0.63)		2 665 (2 435)		2 230 (2 040)
0.413 (10.5)	0.0295 (0.75)		2 330 (2 130)		1 865 (1 705)
7/16 (11.1)	0.014 (0.36)			2 800 (2 560)	
7/16 (11.1)	0.019 (0.48)		3 175 (2 900)		
0.472 (12.0)	0.0169 (0.43)		3 330 (3 045)		
0.472 (12.0)	0.0169 (0.43)		830 (760)		
0.472 (12.0)	0.0175 (0.44)		3 500 (3 200)		
0.472 (12.0)	0.0205 (0.52)		3 000 (2 740)		
7/16 (11.1)	0.020 (0.51)			2 500 (2 290)	
7/16 (11.1)	0.022 (0.56)			3 500 (3 200)	
7/16 (11.1)	0.023 (0.58)			2 350 (2 150)	
7/16 (11.1)	0.024 (0.61)			2 250 (2 060)	
7/16 (11.1)	0.025 (0.64)		2 400 (2 200)		
0.472 (12.0)	0.0210 (0.54)		2 030 (1 860)		
0.472 (12.0)	0.0252 (0.64)		2 330 (2 130)		
7/16 (11.1)	0.030 (0.76)		2 150 (1 970)		
0.472 (12.0)	0.0280 (0.71)		2 165 (1 980)		
½ (12.7)	0.017 (0.43)	1 500 (1 370)	3 175 (2 900)		
½ (12.7)	0.015 (0.38)	1 500 (1 370)		2 950 (2 700)	2 330 (2 130)
½ (12.7)	0.022 (0.56)		2 600 (2 380)		
½ (12.7)	0.020 (0.51)	1 200 (1 100)		2 350 (2 150)	1 900 (1 740)
½ (12.7)	0.026 (0.66)		2 150 (1 970)		
½ (12.7)	0.025 (0.64)	950 (870)		1 850 (1 690)	
½ (12.7)	0.030 (0.76)	800 (730)		1 575 (1 440)	
½ (12.7)	0.028 (0.71)		2 100 (1 920)		
⅝ (15.9)	0.015 (0.38)	1 166 (1 070)		2 300 (2 100)	
⅝ (15.9)	0.020 (0.51)	866 (790)		1 700 (1 550)	
⅝ (15.9)	0.030 (0.76)	600 (550)		1 176 (1 080)	
⅝ (15.9)	0.025 (0.64)	733 (670)			
¾ (19.1)	0.020 (0.51)			1 700 (1 550)	
1¼ (31.8)	0.035 (0.89)	165 (150)		300 (270)	
1¼ (31.8)	0.050 (1.27)	165 (150)		300 (270)	

[A] Coil widths may vary ±⅛ in.

[B] Width tolerances:

Type II, ±0.030 in. (±0.76 mm)
Type III, ±0.015 in. (±0.38 mm)
Type IV, ±0.030 in. (±0.76 mm)

[C] Thickness tolerances: (Embossed strap should be measured for thickness with a flat-face micrometer or similar measuring device so as to cover as much area as possible.)

Type II and III—for strapping 0.025 in. thick or less, ±0.0025 in. (±0.06 mm).
Type II and III—for strapping greater than 0.025 in. thick, ±0.003 in. (±0.08 mm).
Type IV—for all thickness, ±0.0025 in. (±0.06 mm).

[D] 2 300 yd (2 100 m) in self-contained dispensing box.

[E] 1 800 yd (1 670 m) in self-contained dispensing box.

breaking strengths in excess of 1000 lbf (4448 N)).

NOTE 1—For Type I and Type IA strapping split-barrel type clamps, 2 to 4 in. (50 to 100 mm) in diameter are recommended.

NOTE 2—For Type II, III and IV strapping with a minimum breaking strength greater than 1000 lbf (4448 N), wedge action grips equipped with flat, serrated 25 teeth per inch (1 tooth per millimetre) faces are recommended.

12.2.2 *Elongation*—Measure percent elongation at break at the same time as breaking strength using the same procedure (see 12.2.1) except use a chart recorder to measure crosshead displacement. If greater accuracy is desired, use an extensometer to measure the extension as errors due to slippage or slack removal can be introduced if crosshead separation is used.

12.3 *Joint Strength:*

12.3.1 Determine joint strength by the same procedure

used for breaking strength (see 12.2.1). Read the maximum force directly from the chart to the nearest 5 lbf (22 N).

12.3.2 *Percent of Breaking Strength*—Determine the percent of breaking strength according to the type, grade, and dimensions tested (see 7.1).

12.4 *Transverse Strength*—Measure the transverse strength of Type I and Type IA strapping on a tension tester with a full-scale load of 10 lbf (44 N). Use flat-faced upper and lower clamps and a crosshead speed of 6 in. (152 mm)/min. Insert a sample of strap at least the width of the clamps in the clamps so that they each grip one half of the width of the strap. Activate the tester. Read the maximum strength directly from the recording device in pounds-force (or newtons).

12.5 *Precision and Bias*—Precision and bias measurements for these test methods are currently under study by the Strapping Task Group of ASTM Subcommittee D10.14.

13. Rejection and Rehearing

13.1 Material that fails to conform to the requirements of this specification may be rejected. Rejection should be reported to the producer or supplier promptly and in writing. In case of dissatisfaction with the results of the test, the producer or supplier may request retesting.

14. Certification

14.1. When specified in the purchase order or contract, a producer's or supplier's certification shall be furnished to the purchaser.

15. Preparation for Delivery

15.1 Unless otherwise specified in a contract or order, strapping and joining devices shall be packaged and packed in accordance with Practice D 3951. Such packaging shall assure safe arrival at destination in satisfactory condition and shall be acceptable to the carrier used at lowest rates.

16. Keywords

16.1 closing; joining methods; nonmetallic strapping; palletizing; reinforcing; strapping; unitizing

SUPPLEMENTARY REQUIREMENTS

The following supplementary requirements shall apply only when specified by the purchaser in the inquiry, contract, or order for Agencies of the U.S. Government.

S1. Scope

S1.1 These Supplementary Requirements describe the Department of Defense requirements for the packaging, packing, and marking of nonmetallic strapping and joining devices.

S2. Referenced Documents

S2.1 The following documents[3] form, in the parts referenced, a part of this document in the issue in effect at the time of placement of the order or contract:

Federal Specifications:
L-P-378 Plastic Sheet and Strip, Thin Gauge, Polyolefin.
PPP-B-636 Boxes, Shipping, Fiberboard
PPP-B-640 Boxes, Fiberboard, Corrugated, Triple-Wall
Military Specification:
MIL-P-116 Preservation—Packaging; Methods of
Military Standards:
MIL-STD-129 Marking for Shipment and Storage
MIL-STD-147 Palletizing Unit Loads

S3. Preparation for Delivery

S3.1 Packaging shall be Level A or C as specified in the purchase order or contract.

S3.1.1 *Level A*—Nonmetallic strapping and joining devices shall be unit protected in accordance with Method III of MIL-P-116.

S3.1.1.1 *Strapping*—Nonmetallic strapping shall be wound on a core of suitable material to support load and prevent collapse during dispensing. Core diameter and core width shall be of the proper size to be compatible for mounting in the designated dispenser. Secure each coil of strapping against unwinding and place in a close-fitting box conforming to PPP-B-636, class weather-resistant style RSC.

S3.1.1.2 *Joining Devices*—Unit Packaging—Unless otherwise specified in the procurement document, 100 of like type and size of joining devices shall be placed in bags constructed of material conforming to L-P-378 2 mil (0.05 mm) minimum or commercially equivalent polyethylene material. Bag closure shall be accomplished by heat sealing.

Intermediate Containers—Unless otherwise specified by the procuring agency, 10 unit packs of joining devices of a like type and size shall be placed in a fiberboard box conforming to PPP-B-636, class weather-resistant style RSC.

S3.1.2 *Level C*—Nonmetallic strapping and joining devices shall be packaged in accordance with Practice D 3951. Unless otherwise specified in the procurement document, quantity per unit pack requirements of Level A packaging shall apply.

S3.2 *Packing*—Unit packing need not be overpacked when paragraph S3.3 is specified. When the quantity procured does not lend itself to palletization, packing shall be specified as Level A, B, or C in the purchase order or contract.

S3.2.1 *Level A*—When nonmetallic material is specified in the procurement document, nonmetallic strapping and joining devices of like grade and type shall be packed in containers conforming to PPP-B-601, Overseas Type or PPP-B-621, Class 2, Style 4.

S3.2.2 *Level B*—Unit containers shall be packed for shipment as specified for Level A (S3.2.1) requirements except that containers shall conform to PPP-B-636, Class weather-resistant as a minimum or PPP-B-640, class 1.

S3.2.3 *Level C*—Unit containers of nonmetallic strapping and joining devices packaged as specified in S3.1 shall be packed for shipment in accordance with Practice D 3951.

S3.3 *Palletization*—Unless otherwise specified, all unit containers of material of the same type and grade shall be palletized in accordance with load type I requirements of MIL-STD-147. To meet Level A protection requirements, waterproof protection shall be provided by waterproof wrapping (storage aid 9) in accordance with MIL-STD-147. For Level C protection, the above waterproof requirements are not necessary. Pallets shall be expendable and nonreturnable. Gross weight of a palletized load shall not exceed 3000 lb (1360 kg).

S3.4 *Marking*—In addition to any special marking required by the contract or order, coils, packages, and shipping containers shall be marked in accordance with MIL-STD-129.

Designation: D 3951 – 90

Standard Practice for Commercial Packaging[1]

This standard is issued under the fixed designation D 3951; the number immediately following the designation indicates the year of original adoption or, in the case of revision, the year of last revision. A number in parentheses indicates the year of last reapproval. A superscript epsilon (ε) indicates an editorial change since the last revision or reapproval.

1. Scope

1.1 This practice establishes minimum requirements for packaging of supplies and equipment, exclusive of ammunition, explosives, or hazardous materials, as covered in Title 49 of the Code of Federal Regulations.

1.2 *This standard does not purport to address the safety problems associated with its use. It is the responsibility of the user of this standard to establish appropriate safety and health practices and determine the applicability of regulatory limitations prior to use.*

2. Referenced Documents

2.1 *ASTM Standards:*

D 996 Terminology of Packaging and Distribution Environments[2]

D 4169 Practice for Performance Testing of Shipping Containers and Systems[2]

2.2 *Code of Federal Regulations:*

Title 49 Transportation (Parts 100 to 199)[3]

2.3 *Military Standard:*

MIL-STD-129 Marking for Shipment and Storage[4]

2.4 *Federal Standards:*

Fed-Std-123 Marking for Shipment (Civil Agencies)[3]

3. Terminology

3.1 Terminology found in Terminology D 996 shall apply.

3.2 *Description of Term Specific to this Standard:*

3.2.1 *safe delivery*—delivery of a shipment to its destination with minimal damage to the package and no damage to the contents.

4. Significance and Use

4.1 This practice covers the requirements for the commercial preservation, packaging, packing unitization and marking for supplies and equipment. It includes the minimum requirements for physical and mechanical protection and provides for multiple handling, shipment by any mode, and limited storage periods of up to 1 year in enclosed facilities without damage to the product. It also provides for package quantities suitable for redistribution without additional repackaging or marking.

5. Requirements

5.1 *General*—The following requirements, unless specifically changed by the contract or order, shall apply. This practice neither requires nor precludes the use of federal or military packaging methods or materials. Suppliers are encouraged to use advanced packaging technology.

5.1.1 *Packaging*—Preservation, packaging, packing, and marking furnished by the suppliers shall meet or exceed the following minimum requirements:

5.1.1.1 *Cleanliness*—Items shall be free of dirt and other contaminants which would contribute to deterioration of the item or which would require cleaning by the customer prior to use. Coatings and preservatives applied to the item for protection are not considered contaminants.

5.1.1.2 *Preservation*—Items susceptible to corrosion or deterioration shall be provided protection such as preservative coatings, volatile corrosion inhibitors, or desiccated unit packs.

5.1.1.3 *Cushioning*—Items requiring protection from physical and mechanical damage or which are fragile in nature shall be protected by wrapping, cushioning, pack compartmentization, cartonizing, or other means to mitigate shock and vibration during handling and shipment.

5.1.2 *Unit Package:*

5.1.2.1 *Unit Package*—A unit package shall be so designed and constructed that it will contain the contents with no damage to them, and with minimal damage to the unit pack during shipment and storage in the shipping container, and will allow subsequent handling.

5.1.2.2 *Unit Package Quantity*—Unless otherwise specified, the unit package quantity shall be one each part, set, or assembly. Excepted are small lightweight items, for example, industrial hardware, which will be unit packaged by dozen, fifty, hundred, gross, or other quantities that are standard in the trade and suitable for retail sales. Bulk packaging is specifically excluded unless specified in the contract or order.

5.1.3 *Intermediate Package:*

5.1.3.1 The use of intermediate packaging is encouraged particularly when such use enhances handling and inventorying. Unless otherwise specified for shipments to the Department of Defense, intermediate packaging is required to facilitate handling and inventory whenever the quantity is over 1 gross and the size of the unit package is 64 in.^{3} (0.00104 m^3) or less.

5.1.4 *Packing:*

5.1.4.1 Unit packages and intermediate packages not meeting the requirements for a shipping container shall be packed in shipping containers.

5.1.4.2 *Shipping Containers*—The shipping container (including any necessary blocking, bracing, cushioning, or waterproofing) shall comply with the regulations of the

---

[1] This practice is under the jurisdiction of ASTM Committee D-10 on Packaging and is the direct responsibility of Subcommittee D10.14 on Closure and Reinforcement.

Current edition approved July 27, 1990. Published September 1990. Originally published as D 3951 – 80. Last previous edition D 3951 – 88.

[2] *Annual Book of ASTM Standards*, Vol 15.09.

[3] Available from Superintendent of Documents, U.S. Government Printing Office, Washington, DC 20402.

[4] Available from Naval Publication and Forms Center, 5801 Tabor Ave., Philadelphia, PA 19120.

carrier used and shall provide safe delivery to the destination at the lowest tariff cost. It shall be capable of multiple handling and storage under favorable conditions for a minimum of 1 year.

5.1.5 *Unitization*—Shipments should be considered for unitization where appropriate. Unitization encompasses, but is not limited to, consolidation in a container or placement on a pallet. Unless otherwise specified, shipment to the Department of Defense shall be unitized into a single load in such a manner that the load can be handled as a unit throughout the distribution system.

5.1.6 *Marking:*

5.1.6.1 *Unit and Intermediate Marking*—Marking for the unit and intermediate container shall mean those used in the commercial distribution or over-the-counter retail sales. The markings shall be sufficient to clearly and visibly identify the contents of the package.

5.1.6.2 *Shipping Container Markings*—Minimum markings shall include the name and address of consignee and consignor and any applicable precautionary markings. Other desired marking shall be as specified. All markings shall be legible and applied with a waterproof ink.

5.1.6.3 *Government Marking Requirements*—Unless otherwise specified, shipments to the Department of Defense shall be marked in accordance with MIL-STD-129; unless otherwise specified, shipments to other government agencies shall be marked in accordance with Fed-Std-123.

6. Options

6.1 *Export Shipments*—While not intended for export shipments, this practice can be used when favorable conditions are provided, such as containerization media and closed areas of receiving.

7. Performance Testing

7.1 The use of Practice D 4169 is recommended for determining the performance level of a shipping unit.

7.2 For shipments to the Department of Defense, when specified, the supplier is responsible for requirements specified herein. The purchaser reserves the right to perform any or all of the tests.

7.2.1 Packaging used shall successfully pass the test levels of Practice D 4169, Distribution Cycle 1. Export shipments shall also pass the test levels of Distribution Cycles 15, 16, or 17 as appropriate.

7.2.1.1 Assurance Level II shall be used. Acceptance shall be criterion I, no product damage.

8. Keywords

8.1 commercial packaging; intermediate; packaging; preservation; unit package

Designation: D 3953 – 91

Standard Specification for Strapping, Flat Steel and Seals[1]

This standard is issued under the fixed designation D 3953; the number immediately following the designation indicates the year of original adoption or, in the case of revision, the year of last revision. A number in parentheses indicates the year of last reapproval. A superscript epsilon (ε) indicates an editorial change since the last revision or reapproval.

1. Scope

1.1 This specification covers flat steel strapping and seals intended for use in closing, reinforcing, bundling articles for shipment, unitizing, palletizing, and bracing for carloading and truckloading.

1.2 This specification includes strapping and seals made from carbon steel and applies only to the types and sizes as cataloged by strapping suppliers to be a standard, commercially available product. It does not include "specialty"-type strapping as developed for special applications or strapping produced from alloy metals.

1.3 The values stated in inch-pound units are to be regarded as standard.

1.4 The following safety hazards caveat applies only to the test method portion, Sections 11, 12, and 13, of this specification: *This standard does not purport to address all of the safety problems, if any, associated with its use. It is the responsibility of the user of this standard to establish appropriate safety and health practices and determine the applicability of regulatory limitations prior to use.*

2. Referenced Documents

2.1 *ASTM Standards:*

A 90 Test Method for Weight of Coating on Zinc-Coated (Galvanized) Iron or Steel Articles[2]

A 109 Specification for Steel, Strip, Carbon, Cold-Rolled[3]

A 239 Test Method for Locating the Thinnest Spot in a Zinc (Galvanized) Coating on Iron or Steel Articles by the Preece Test (Copper Sulfate Dip)[2]

B 504 Method for Measurement of Thickness of Metallic Coatings by the Coulometric Method[4]

D 996 Terminology of Packaging and Distribution Environments[5]

D 3951 Practice for Commercial Packaging[5]

D 4332 Practice for Conditioning Containers Package, or Package Components for Testing[5]

2.2 *Military Standard:*

MIL-STD-105 Sampling Procedures and Tables for Inspection by Attributes[6]

2.3 *AAR General Rules 15*

AAR Rule Governing the Loading of Commodities on Open Top Cars[7]

3. Terminology

3.1 General definitions for packaging and distribution environments are found in Terminology D 996.

4. Strapping Classifications

4.1 Strapping shall be of the following types, finish and sizes:

Type 1, Flat Strapping: Power Machine and Hand Application:

Regular-duty

Heavy-duty

Heavy-duty, high elongation

Type 2: Nail-On

4.1.1 *Strapping Finish, Lubricated or Dry* (see 8.3):

4.1.1.1 *Finish A—Organic (Pigmented or Unpigmented Paint)*

4.1.1.2 *Finish B—Galvanized (Hot Dipped or Electro-Deposit Galvanized)* (see 8.10):

Grade 1—Heavy Coating

Grade 2—Moderate Coating

Grade 3—Commercial Coating

4.1.1.3 *Finish C—Organic Painted Metallic Filled (Aluminum or Zinc)*

4.1.1.4 *Finish D—Uncoated*

4.1.2 *Strapping Sizes*—Strapping shall have the widths and thicknesses in accordance with sizes shown in Tables 1 and 2.

5. Seal Classifications

5.1 Seals shall be of the following classes, finishes, types, and styles:

5.1.1 *Seal Classes:*

R—Regular duty

H—Heavy duty

5.1.2 *Seal Finishes:*

Finish A—Organic (painted)

Finish B—Galvanized (hot dipped or electro-deposit galvanized)

Grade 1—Heavy coating

Grade 2—Moderate coating

Grade 3—Manufacturer's standard commercial practice

Finish C—Uncoated

5.1.3 *Seal Styles:*

I—Snap-on or open

---

[1] This specification is under the jurisdiction of ASTM Committee D-10 on Packaging and is the direct responsibility of Subcommittee D10.14 on Closure and Reinforcement.

Current edition approved Jan. 25, 1991. Published May 1991. Originally published as D 3953 – 83. Last previous edition D 3953 – 87.$^{\epsilon 1}$

[2] *Annual Book of ASTM Standards*, Vol 01.06.

[3] *Annual Book of ASTM Standards*, Vol 01.03.

[4] *Annual Book of ASTM Standards*, Vol 02.05.

[5] *Annual Book of ASTM Standards*, Vol 15.09.

[6] Available from Standardization Documents Order Desk, Bldg. 4 Section D, 700 Robbins Ave., Philadelphia, PA 19111-5094, Attn: NPODS.

[7] AAR Rule Governing the Loading of Commodities on Open Top Cars, Section No. 1 General Rules—Rule 15, available from the Association of American Railroads, Mechanical Division 50 F St., NW, Washington, DC 20001.

TABLE 1 Type 1 Regular-Duty Strapping Widths, Thickness, and Breaking Strengths

Width, in. (mm)	Thickness, in. (mm)							
	0.010 (0.25)	0.012 (0.30)	0.015 (0.38)	0.018 (0.46)	0.020 (0.51)	0.023 (0.58)	0.028 (0.71)	0.035 (0.89)
	Breaking Strength, min, lbf (N)							
3/8 (9.52)	300 (1335)	360 (1600)	450 (2000)	540 (2400)	600 (2670)	690 (3070)	. . .	. . .
1/2 (12.70)	400 (1780)	480 (2135)	600 (2670)	720 (3200)	800 (3560)	920 (4090)	. . .	. . .
5/8 (15.87)	500 (2225)	600 (2670)	750 (3335)	900 (4000)	1000 (4450)	1150 (5115)	. . .	. . .
3/4 (19.05)	. . .	. . .	900 (4000)	. . .	1200 (5340)	1380 (6140)	1680 (7470)	2100 (9340)

TABLE 2 Type 1 Heavy-Duty, Heavy-Duty/High Elongation Strapping Widths, Thickness, and Breaking Strengths

Width, in. (mm)	Thickness, in. (mm)										
	0.020 (0.51)	0.022 (0.56)	0.023 (0.58)	0.025 (0.64)	0.028 (0.71)	0.029 (0.75)	0.031 (0.79)	0.035 (0.89)	0.044 (1.12)	0.050 (1.27)	0.057 (1.45)
	Breaking Strength, min, lbf (N)										
1/2 (12.70)	1 280 (5 690)	. . .	1 470 (6 540)	. . .	. . .	. . .	1 950 (8 675)	. . .	. . .	. . .	. . .
5/8 (15.87)	1 600 (7 120)	. . .	1 840 (8 185)	. . .	. . .	. . .	. . .	. . .	. . .	. . .	. . .
3/4 (19.05)	2 000 (8 896)	2 280 (10 140)	2 280 (10140)	2 280 (10 140)	2 280 (10 140)	2 850 (12 675)	2 850 (12 675)	2 850 (12 675)	4 050 (18 015)	4 050 (18 015)	. . .
1 (25.4)	. . .	. . .		. . .	. . .	3 700 (16 455)	3 700 (16 455)	. .	. . .	. . .	. . .
1¼ (31.75)	3 200 (14 240)	. . .	. . .	3 680 (16 370)	. . .	4 750 (21 130)	4 750 (21 130)	4 750[A] (21 130)	6 750 (30 025)	6 750 (30 025)	7 700 (34 250)
2 (50.80)	. . .	. . .	. . .	. . .	. . .	. . .	. . .	. . .	10 600 (47 150)	10 600 (47 150)	. . .

[A] Heavy-duty, or heavy-duty/high elongation.

TABLE 3 Number of Feet Per Pound (Metres per Kilogram) of Type 1 Strapping[A]

NOTE—Number of feet per pound (metres per kilogram) is based on nominal dimensions and will vary slightly with the width and thickness of the strapping within the tolerances as described in 8.2.

Width, in. (mm)	Thickness, in. (mm)														
	0.010 (0.25)	0.012 (0.311)	0.015 (0.38)	0.018 (0.46)	0.020 (0.51)	0.022 (0.56)	0.023 (0.58)	0.025 (0.64)	0.028 (0.71)	0.029 (0.75)	0.031 (0.79)	0.035 (0.89)	0.044 (1.12)	0.050 (1.27)	0.057 (1.45)
	Length per Unit Mass, ft/lb (m/kg), approximate														
3/8 (9.52)	78.5 (52.7)	65.4 (43.9)	52.4 (35.2)	43.6 (29.3)	39.3 (26.4)		34.1 (22.9)								
1/2 (12.70)	58.9 (39.6)	49.1 (33.0)	39.3 (26.4)	32.7 (22.0)	29.4 (19.7)		25.6 (17.2)				19.0 (12.7)				
5/8 (15.87)	47.1 (31.6)	39.3 (26.4)	31.4 (21.1)	26.2 (17.6)	23.6 (15.8)		20.5 (13.7)								
3/4 (19.05)			26.2 (17.6)		19.6 (13.1)	17.8 (11.9)	17.1 (11.5)	15.7 (10.5)	14.0 (9.4)	13.5 (8.95)	12.7 (8.5)	11.2 (7.5)			
1 (25.4)										10.1 (6.7)	9.5 (6.3)				
1¼ (31.75)					11.8 (7.9)			9.4 (6.3)		8.1 (5.5)	7.6 (5.1)	6.6 (4.4)	5.3 (3.5)	4.7 (3.1)	4.1 (2.8)
2 (50.80)													3.3 (2.2)	2.9 (1.9)	

[A] Commercial tolerances apply.

II—Thread-on or closed
III—Push-type or overlap
IV—Magazine-feed
V—Intersectional or cross-tie

5.2 *Seal Sizes*—Seals shall have widths in accordance with strapping width shown in Tables 1 and 2.

6. Ordering Information

6.1 The inquiry and order shall indicate the following:

6.1.1 Strapping type, duty, finish, grade, lubrication (when applicable), dimension required, and quantity of each size (see Section 4).

6.1.2 Coil of strapping dimensions, weight and feet per pound, oscillated, ribbon wound, or cut-to-length (see Section 9, Tables 3 and 4, and Figs. 1 and 2).

6.1.3 Seals (see Section 5) class, finish, grade (when applicable), type, style, and quantity by width (or nomenclature).

6.1.4 Make or model of strapping equipment that the strapping and seals must work in, if applicable.

6.1.5 Title, designation, and date of this specification.

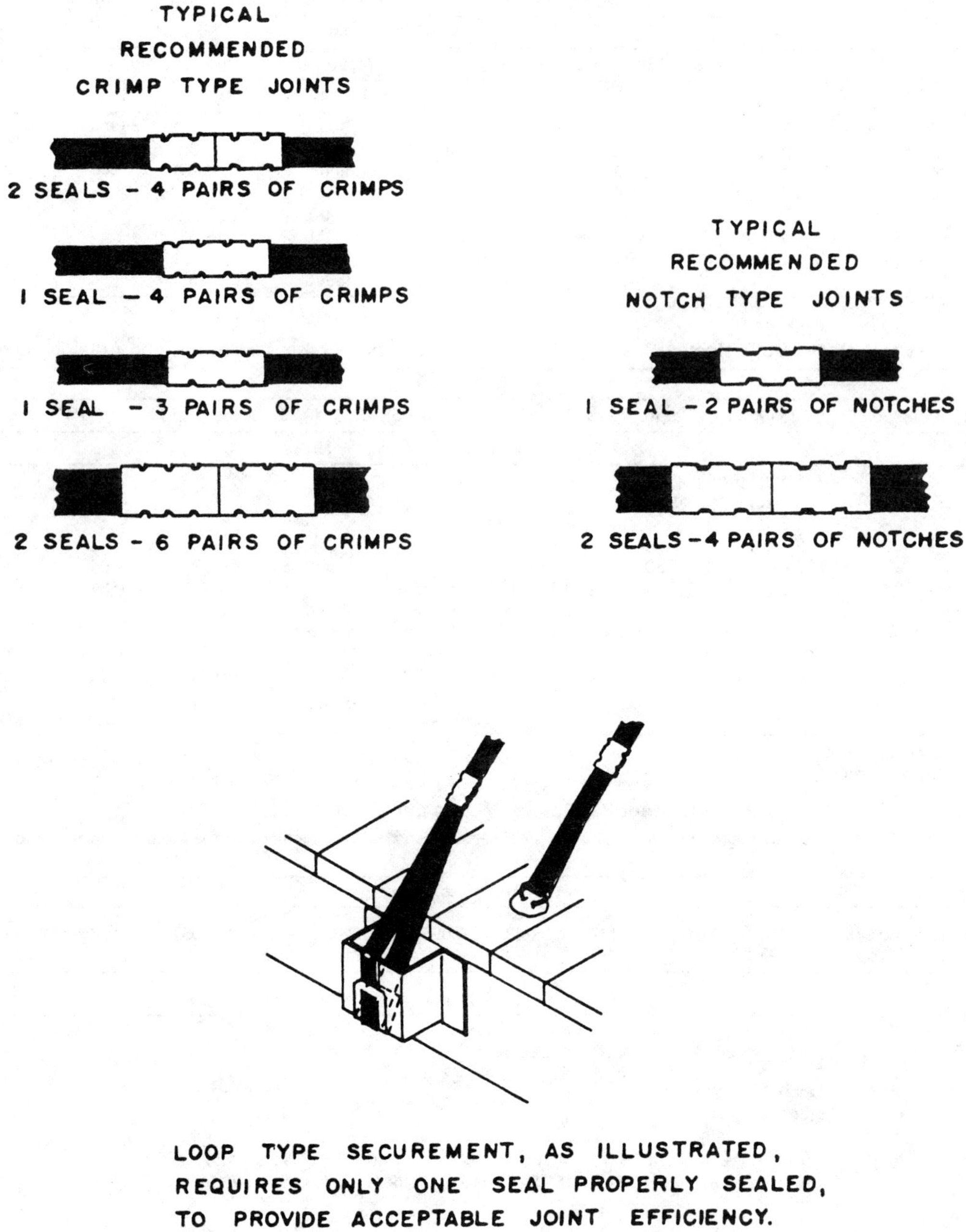

FIG. 1 Recommended Seal Joints

6.1.6 For Department of Defense, indicate selection of preservation, packing, marking, and palletization, if applicable (see Section 16).

6.2 Sizes or physical properties, or both, other than those covered by this specification shall be as specified in the contract or order.

7. Materials and Manufacture

7.1 Materials shall be cold-rolled carbon steel of the quality normally used for this purpose in commercial practice, and the physical mechanical property requirements, as determined by tests, shall be within the allowable limits.

7.1.1 *Type 1 Regular-Duty Strapping*—Regular-duty strapping in Table 1 shall be low-carbon cold-rolled steel (see 6.1.1).

7.1.2 *Type 1 Heavy-Duty Strapping*—Heavy-duty strapping in Table 2 shall be cold-rolled, high-carbon, or medium carbon high-manganese steel which has been fully heat-treated; for example, austenitized and quenched.

7.1.3 *Type 1 Heavy-Duty, High-Elongation Strapping*—Heavy-duty, high-elongation strapping in Table 2 shall be high-carbon steel which has been fully heat-treated; for example, austenitized and quenched to provide the elongations in 8.1.2.

7.1.4 *Type 2 Nail-On Strapping*—Unless otherwise specified, Type 2 nail-on strapping shall be strapping made from Type 1 regular-duty or Type 1 heavy-duty strapping, avail-

TABLE 4 Size of Coils and Package

Type Coils	Size of Strapping, in. (mm)	Inside Diameter, in (mm) (Commercial Tolerance)	Outside Diameter, max, in. (mm)	Coils Per Package, max
Oscillated	3/8 by 0.010 (9.52 by 0.25) through 3/4 by 0.035 (19.05 by 0.89)	16 (406)	23½ (597)	1
Ribbon	3/4 by 0.022 (19.05 by 0.56) through 3/4 by 0.035 (19.05 by 0.89)	16 (406)	27 (686)	2
	1 by 0.029 (25.4 by 0.75) through 1 by 0.031 (25.4 by 0.79)	16 (406)	27 (686)	1
	1¼ by 0.020 (31.75 by 0.51) through 1¼ by 0.057 (31.75 by 1.45)	16 (406)	27 (686)	1
	2 by 0.044 (50.80 by 1.12) through 2 by 0.050 (50.80 by 1.27)	16 (406)	25⅞ (657)	1

able in sizes 5/8 in. (15.87 mm) in width or greater with punched nail holes. Hole diameters and hole spacing can vary depending upon strap size and duty, which shall be in accordance with manufacturer's standard practice. However, nail holes shall be spaced at intervals no more than 3 in. (76 mm) apart.

7.1.5 *Seals for Type 1 Strapping*—Unless otherwise specified, seals for Type 1 strapping shall be made of steel in accordance with the manufacturer's commercial practice.

8. Physical and Mechanical Properties

8.1 *Breaking Strength and Elongation, Type 1 Regular-Duty and Heavy-Duty Strapping:*

8.1.1 *Breaking Strength*—The breaking strength of the strapping shall be as shown in Table 1 and Table 2 when tested as specified in 13.3.

8.1.2 *Elongation*—The elongation in 6 in. (152 mm) for Type 1 strapping, when tested as specified in 13.4, shall be as follows:

Regular-duty (see 7.1.1)	0.5 to 4.0 %
Heavy-duty (see 7.1.2)	
Thickness less than 0.031 in. (0.79 mm)	5.0 to 12 %
Thickness 0.031 in. (0.79 mm) and greater	6.5 to 12 %
Heavy-duty high elongation (see 7.1.3)	
Thickness 0.035 in. (0.889 mm) and greater	9.0 to 16 %

8.2 *Tolerances, Type 1 Strapping, Widths*—The widths of the strapping shall not vary from that shown in Tables 1 and 2 by more than ±0.005 in. (0.13 mm) for all cold-rolled strapping.

8.2.1 *Type 1 Strapping, Thickness*—The thickness of strapping sizes listed in Tables 1 and 2 shall not vary from nominal thickness by more than the following:

Cold-rolled strapping 0.031 in. (0.79 mm) or less in thickness +0.002 in. (0.05 mm), −0.001 (0.03 mm).

Cold-rolled strapping greater than 0.031 in. (0.79 mm) in thickness up to 0.057 in. (1.45 mm) ±0.002 in. (0.05 mm).

8.3 *Lubrication*—Unless otherwise specified, Type 1 strapping shall have a coating of wax or wax-like lubricant applied to the surfaces that will render the strapping suitable for application with tension tools of the feed wheel (friction) type (see 4.1.2 and 6.1.4). The lubrication of the strapping is intended to facilitate the use of strapping tools in which the two bands of the strapping are held in contact with each other under pressure by frictional grips, and sliding of the faces of the strapping is necessary during the tensioning operation. Test in accordance with 13.9.

8.4 *Type 1 Strapping, Coils*—Each coil shall consist of one continuous piece of strapping. Coils with lengths of strapping welded together shall be considered as one-piece coil. Type 1 strapping shall have no more than two welded joints in any coil.

8.4.1 *Dimensional Requirements at the Weld*—The thickness at the weld shall be no more than one and one half the

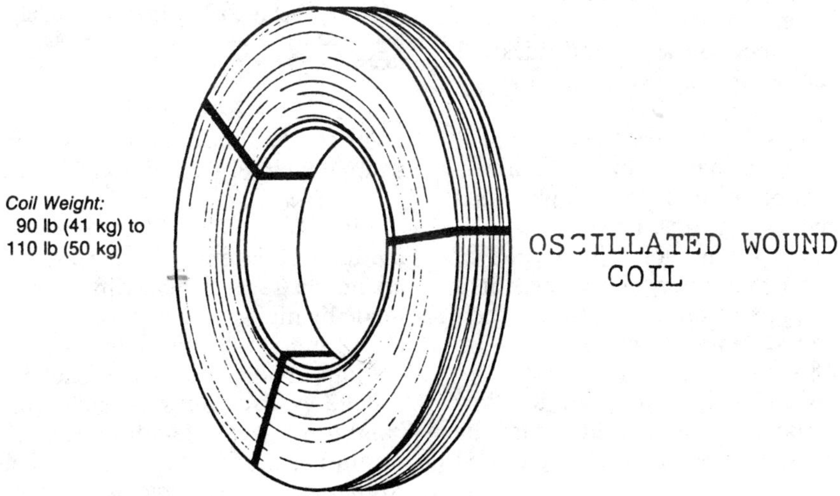

FIG. 2 Oscillated Wound Coil

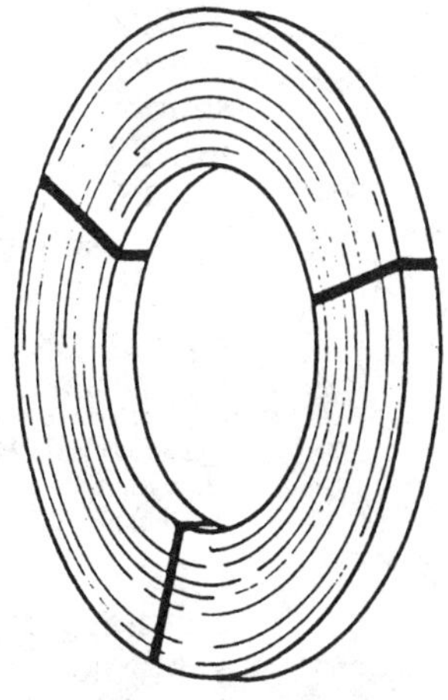

FIG. 3 Ribbon-wound Coil

thickness of the strap. The width of strapping at the weld shall be subject to the same dimensional requirements as the remainder of the strapping. Metal splattering or extruded points causing increased thickness are not permissible. Only machine welds shall be acceptable.

8.4.2 *Weld Efficiency*—Strength of manufacturer's process welded joints in coils of strapping shall have a joint efficiency of not less than 75 % based on the breaking strength shown in Tables 1 and 2 when tested as specified in 13.6.

8.5 *Identification Marking*—Each coil of Type 1 heavy-duty, 1¼-in. (31.75-mm) by 0.029 in. (0.75 mm) in thickness and greater and 2-in. wide (50.8-mm) strapping shall be marked in accordance with the Association of American Railroads General Rule 15. The marking shall consist of the letters "AAR," the manufacturer's or distributor's name, or abbreviated name, or registered trademark, or symbol, or AAR code consisting of two digits. Marking shall be spaced not more than 5 ft (1.52 m) apart.

8.6 *Type 2 Strapping:*

8.6.1 *Breaking Strength*—The breaking strength for Type 2 strapping shall be the same as shown in Table 1 or 2 for the applicable strapping size and type, but shall be based on the cross-sectional metal area of the strapping taken across the hole. Breaking strength will be reduced proportionally in accordance with nail hole diameter.

8.6.2 *Tolerances*—The tolerances on width and thickness for Type 2 strapping shall be the same as for Type 1 strapping (see 8.2).

8.7 *Seals, Type 1 Strapping*—Seals intended for Type 1 strapping shall be made of steel to manufacturer's commercial practice. Seal lengths shall be adequate for a single-notch joint (one pair of cut notches), a double-notch joint (two pairs of cut notches), a double-crimp joint (two pairs of crimps), a triple-crimp joint (three pairs of crimps), or a quadruple-crimp joint (four pairs of crimps). Unless otherwise specified, seals shall be suitable in size or design to satisfy the requirements of 8.9.2 and 8.9.3.

8.7.1 *Seal Width*—Seal width shall be adequate to allow the seal to slip on the applicable strapping and shall be compatible with the sealing tool used to provide the required joint strength.

8.7.2 *Seals, Regular- and Heavy-Duty*—Seals shall be in thickness and hardness suitable for use with regular- or heavy-duty strapping. Regular-duty seals shall be used with regular-duty strapping and heavy-duty seals with heavy-duty strapping.

8.8 *Application of Seals:*

8.8.1 *Style I*—Style I seals shall have one or both sides open so that the seal shall slip in the overlapping bands of strapping either during or after tensioning the strapping.

8.8.2 *Style II*—Style II seals shall have sides partially closed at the bottom of the seal so that this seal shall be prethreaded with the overlapping bands of strapping before the tensioning tool is applied.

8.8.3 *Style III*—Style III seals shall have the sides completely closed by overlapping at the bottom of the seal so that prethreading with the overlapping bands of strapping and bending of the bottom strap under the seal can be accomplished. The seal shall permit the tensioning tool to tension the strapping as the tool butts against the seal.

8.8.4 *Style IV*—Style IV seals shall have sides opened to allow nesting and stacking of the seals for hand or power tools with magazine feed. Seals shall fit and form joints with the seal manufacturer's tool model as specified (see 6.1.4). Means provided on Style IV seals to facilitate stacking are acceptable.

NOTE 1—Style IV seals are commonly designed for use with the seal manufacturer's tool and are not adaptable for other manufacturer's tools.

8.8.5 *Style V*—Style V seals shall have one or both sides open so that the seal shall slip on the crossing bands of strapping either during or after tensioning of the straps.

8.9 *Sealed Joints:*

8.9.1 *Fastening Flat Steel Strapping*—The various methods of fastening flat steel strapping in use commercially are notched seal joints, crimped seal joints, and sealless joints made by die cutting or spot welding the overlapping ends of the strapping. These methods are acceptable provided that the joint made meets the efficiency requirements, as follows:

8.9.2 *Notch Joints*—Notch-type seal joints made with hand tools, power hand tools, or power strapping machines recommended by the supplier, shall be made on applicable strapping, without cracking and tearing of the seal metal. Notched seal joints are usually intended for use with

TABLE 5 Mass Loss of Zinc Coating

Strapping Width		Minimum Mass Loss, mg/6-in. (152-mm) Specimen		
in.	(mm)	Grade 1	Grade 2	Grade 3
3/8	(9.52)	263	158	53
1/2	(12.70)	350	210	70
5/8	(15.87)	438	262	87
3/4	(19.05)	525	315	105
1	(25.40)	700	420	140
1 1/4	(31.75)	875	525	175
2	(50.8)	1400	840	280

lubricated (waxed) strapping. Notched seal joints when used with Type 1 regular-duty strapping, or Type 1 heavy-duty strapping must have a minimum joint strength of 75 % for double-notch joints and 45 % for single-notch joints. Joint strength is based on the minimum breaking strength of the applicable strapping size as shown in Tables 1 and 2 when tested as specified in 13.5. Examples of notch-type seal joints are shown in Fig. 3.

8.9.3 *Crimp Joints*—Crimp-type seal joints, made with hand tools or power hand tools, shall be made on applicable strapping without cracking and tearing of the seal metal. Crimped seal joints are usually intended for use with dry Type 1 heavy-duty strapping, primarily where severe impact or heavy stresses on the strap and joint may be encountered. Crimp joints may also be used with lubricated (waxed) strapping, provided the loss in friction is compensated for by adding sufficient additional seals and crimps, or by using "Grit"-type friction seals. Crimp seal joints when used with Type 1 heavy-duty (high-tension) strapping in sizes 3/4 by 0.031 in. (19.05 by 0.79 mm) or greater must have a minimum joint efficiency of 75 % based on the minimum breaking strength of the applicable strapping size as shown in Table 2 when tested as specified in 13.5. Three or more crimps will be required to attain or exceed the 75 % joint efficiency as determined by seal, strap, and strap finish combination. Increased joint strength can be obtained by the addition of more seals and crimps. Examples of crimp-type seal joints are shown in Fig. 1.

8.9.4 Seal joint efficiency tests should be performed periodically on both notch- and crimp-type joints to ensure that the seal and sealer combination provides the desired joint strength.

8.9.5 Die-cut sealless joints shall have a minimum joint efficiency of 75 % based on the minimum breaking strength of the applicable strapping size as shown in Table 1 or Table 2, when tested as specified in 13.5. For exceptions, see AAR General Rules 15.

8.9.6 Spot-welded sealless joints shall have a minimum joint efficiency of 75 % based on the breaking strength of the applicable strapping size as shown in Table 1 or Table 2, when tested as specified in 13.5.

8.10 *Galvanized Strapping*—Finish B galvanized strapping shall have the following minimum coating weights or thicknesses, or both, when tested as specified in 13.7 (see 10.2.3 and Table 5).

Grade 1—0.30 oz/ft^2 (92 g/m^2) of surface or 0.0005 in. (0.013 mm) thick.

Grade 2—0.18 oz/ft^2 (55 g/m^2) of surface or 0.0003 in. (0.008 mm) thick.

Grade 3—In accordance with manufacturer's commercial practice but no less than 0.06 oz/ft^2 (18 g/m^2) of surface or 0.0001 in. (0.003 mm) thick.

8.10.1 Edges of Grades 1, 2, and 3 shall not show signs of copper sulfate when tested in accordance with 13.10.

8.11 *Ductility of Strapping Coating*—Finish A, painted finish, Finish B, galvanized, Finish C, metallic-filled strapping shall not crack or flake on the outside of the bent portion, except that cracks on the strapping shall be disregarded within a distance from the edge of the strapping equal to one fourth of the width when tested as specified in 13.8.

8.12 *Base Metal Ductility (Bend Test)*—Strapping of various thicknesses shall withstand the minimum number of bends in one area as shown in Table 6. The test procedure is described under 13.2, and shall be conducted on the final finished product.

8.13 *Straightness—Type 1 Strapping:*

8.13.1 *Camber*—Camber in the strapping shall not be more than 1/2 in. (12 mm) in any 8-ft (2.4-m) length, when measured in accordance with Specification A 109, Table 10.

8.13.2 *Coil Set (Bundle Curve)*—Strap ends shall not rise from horizontal flat plain more than 2 in. (50.8 mm) in any 8-ft (2.4-m) long section of strap.

8.13.3 *Spiral Twist (Cork Screw Effect)*—Twist must not exceed 20° in any 8 ft (2.4 m) long section of strap, when laid on horizontal flat surface.

TABLE 6 Metal Bend Test Requirements for Strapping

Thickness		Number of Bends
in.	(mm)	
0.010	(0.25)	15
0.012	(0.31)	12
0.015	(0.38)	12
0.018	(0.46)	10
0.020	(0.51)	8
0.022	(0.56)	6
0.023	(0.58)	6
0.025	(0.64)	6
0.028	(0.71)	5
0.029	(0.75)	5
0.031	(0.79)	5
0.035	(0.89)	5
0.044	(1.12)	3
0.050	(1.27)	3
0.057	(1.45)	3

9. Coil Configuration and Permissible Weight Variations

9.1 *Type 1 Coils*—Type 1 and Type 2 strapping shall be oscillated (mill)-wound coils weighing not less than 90 lb (41 kg) or more than 110 lb (50 kg) (see Fig. 2) and ribbon-wound coils weighing not less than 55 lb (25 kg) nor more than 110 lb (50 kg) when put up as specified in Table 4 (also see Fig. 3). Oscillated- and ribbon-wound coils of strapping are used in different types of dispensers.

9.2 *Type 1 Core Liners*—Type 1 strapping in oscillated-wound coils shall be wound on core liners. Strapping in ribbon-wound coils shall not require core liners.

9.3 *Type 2 Coils*—Type 2 strapping shall be in small or large coils. Small coils shall be in one piece, not less than 200 ft (61 m) or more than 400 ft (122 m) in length and supplied in individual dispensing cartons. Large coils, oscillated (mill) wound, shall weigh no less than 90 lb (41 kg) nor more than 110 lb (50 kg), and large coils, ribbon wound, shall weigh not

TABLE 7 Single Sampling Plan for Normal Inspection

Properties & Attributes[A]	Rate of Sampling	Acceptable Quality Level
8.1 through 8.13.3	S-2	6.5 %
9.1 through 9.3	Level I	4.0 %
10.1 through 10.3.3	Level I	4.0 %

[A] References are made to paragraphs in this specification.

less than 50 lb (23 kg) nor more than 125 lb (57 kg) when put up as specified in Table 4.

10. Workmanship, Finish, and Appearance

10.1 *Type 1 Strapping and Seals*—Strapping shall be straight, smooth, and clean and shall be free of kinks, grooves, waves, edge curvature, indentations, excessive oil, and cracks. Seals shall be free of dents, burrs, and slivers, and not be deformed.

10.1.1 *Style of Edges*—Edges of Type 1 strapping shall be as cut, milled, or formed in accordance with current commercial practice and shall be free of sharp edges, burrs, and slivers prior to the finish application.

10.2 *Strapping Finish*—The finished coat of the strapping shall be adherent, free of scratches, runs, embedded foreign matter, uncoated areas, rust areas, or loose oxide scale.

10.2.1 *Finish A*—Finish A strapping shall have surfaces and edges coated with an organic coating that shall yield a film as specified in 10.2. On Types 1 and 2, cut-to-length strapping, cuts or nail holes or slots, or combination thereof, may be made after the application of the finish coating.

10.2.2 *Finish B*—Finish B strapping shall have surfaces and edges coated with zinc by either the hot-dipped or electro-deposit process (see 8.10 and 10.2.3).

10.2.3 *Finish B Galvanized Strapping*—The choice between Grades 1, 2, or 3 Finish B, galvanized strapping should be guided by the service expected, considering that the protection afforded by the coating is proportional to the weight of the coating. Grades 1 and 2 Finish B zinc-coated strappings are commonly referred to as government-grade galvanized coating, and Grade 3 having a lighter zinc coating is referred to as commercial-grade galvanized strapping (see 8.10).

10.2.4 *Finish D (Uncoated Strapping)*—Finish D strapping shall be finished a natural black or blued, browned, or bright, as resulting from the manufacturing process. Surfaces shall be lubricated as indicated in 8.3 unless otherwise specified.

10.3 *Seal Finish*—The finished coat of the seal shall be adherent, free of scratches, runs, embedded foreign matter, uncoated areas, rust areas, or loose oxide scale. Edges on the seals are bright unfinished metal as a result of the manufacturing process.

10.3.1 *Finish A*—Unless otherwise specified (see 5.1.2), seals shall be finished in accordance with the manufacturer's commercial practice. The outer surface of the seal shall be coated with an organic coating that shall yield a film as specified in 10.2.

10.3.2 *Finish B*—Galvanized seals, Grades 1 and 2, shall be coated as specified in 8.10. Grade 3 galvanized seals shall be coated in accordance with manufacturer's standard commercial practice.

10.3.3 *Finish C*—Finish C seals shall be finished a natural black or blued, browned or bright, as resulting from the manufacturing process.

11. Sampling Plan

11.1 Unless otherwise specified in contract or order, sampling for acceptance inspection and testing by attributes shall be performed in accordance with applicable provisions set forth in MIL-STD-105, Tables I and II-A. Inspection levels and acceptable quality level (AQL) as listed in Table 7 of this document apply.

11.2 *Lot Size*—The lot size shall consist of all product from a specific production run, or that amount offered for acceptance at one time as applicable. In the case of strapping, it shall be of one class, type, finish, grade, duty, and size. In the case of seals it shall be of one type, style, class, finish, and width.

11.3 *Unit Sample*—One coil or one seal shall be considered a unit. Examination of a coil for imperfections, as applicable, shall be made on eight consecutive feet (2.5 m) of the coil, provided that the condition of the remainder of the coil is representative of the length of strap inspected. The test specimen of strapping can be taken from either end of the coil. When the rest of the coil is not representative, examination for permissible number and strength of welds, as applicable, shall be made on the entire length of the coil regardless of the condition of the coil.

11.4 *Rate of Sampling*—The rate of sampling shall be in accordance with Table 7, excerpted from applicable portions of MIL-STD-105D. Similarly, the acceptable quality level (AQL) shall be as indicated in Table 7.

11.5 *Product Quality*—A supplier shall be responsible for providing product at a quality level that shall meet the requirements of this specification, or an equivalent as agreed upon between purchaser and seller as part of the order of contract. The purchaser does reserve the right to perform any inspections as deemed necessary to assure supplies and services conform to preestablished requirements.

12. Number of Tests and Retests

12.1 *Acceptance Tests*—For the determination of break strength, elongation, ductility, joint efficiency, and coating tests, a single specimen shall be considered a complete test of a unit sample.

12.2 *End-Use Testing*—Where field activities require testing to assure compliance with Tables 1, 2, 4, and 5, the average of test values from three strap (or sealed-joint) specimens of sufficient size are required.

12.3 *Retests*—When determining break strength, elongation, or joint efficiency, a retest shall be permitted if the location of strap failure is outside the 6-in. (152-mm) gage length, or within 1 in. of the holding jaws, or both.

13. Test Methods

13.1 *Dimensions and Weight*—Examine the end item to determine compliance with dimensional, weight, and tolerance requirements. Determine dimensions on flat strapping on base metal after the paint, wax, zinc, or oxide scale has been removed. Any nonconforming dimensions or weight shall be so classified.

13.2 *Metal Bend Test for Strapping*—Fasten a test specimen from a sample unit of strapping in the jaws of a vise and bend by hand over a radius of 1/8 ± 1/64 in. (3.18 ± 0.40 mm). Formed edges may be removed before testing. One bend shall consist of a 90° bend in one direction and return to the original position. Make successive bends in opposite directions. Disregard cracking or flaking of the coating during testing. Strapping of the various thicknesses shall withstand the minimum number of bends in one area, as shown in Table 6. The inability of the strapping to make the minimum bends before breaking into two pieces shall constitute failure of this test. Test with the final finished product.

13.3 *Breaking Strength of Strapping*—Measure breaking strength on a tension tester capable of load accuracies to within ±1 %. Select the load range in which the point of maximum force will fall between 10 and 95 % of full scale. The test specimen should be approximately 12 in. (305 mm) in overall length, but not less than 8 in. (203 mm). If elongation is also to be measured, a series of transverse scribe marks shall be placed along the length of specimen at 1-in. intervals in sufficient number to ensure breakage of specimen occurs within a scribed 6-in. (152 mm) length. Place the test specimen in the tension machine so that there is at least a 6-in. (152-mm) gage length centered between the jaws of the clamping device. Use suitable clamping arrangements to minimize slippage or breakage within the clamps. Make manual adjustments to remove slack before the tester is activated. Conduct the test using a maximum crosshead separation speed of 2½ in./min (64 mm). Read the maximum force required to break the strap. Discard all data when the break occurs within 1 in. (25 mm) of the clamp and retest in accordance with 12.3.

13.4 *Elongation of Strapping*—Measure percent elongation at break at the time as the breaking strength and using the same procedure (see 13.3). Elongation is the *permanent* increase in length which results from the breaking strength test. Place the test specimen in the tension machine so that a 6-in (152-mm) section between gage marks is centered between the clamps. After the break, the two parts of the specimen are removed from the clamps, placed on a flat surface, and carefully joined together at the break point. The distance between the gage marks is measured using an accurate scale or vernier. The distance between gage marks is measured using a scale marked in 0.01-in. (0.25-mm) increments, and read to the nearest 0.01 in. Percent elongation is defined as the length between gage marks after the test minus the original length (6 in.) divided by the length before the test, multiplied by 100. Discard all test results where the break occurs within 1 in. (25 mm) of a gage mark, and retest in accordance with 12.3.

13.5 *Joint Strength as Percent of Breaking Strength:*

13.5.1 *Method of Joining by Seals*—Make a seal joint with a sample seal with a single or double notch, or double or more crimps as applicable, joining the two pieces of Type 1 strapping. Check the joint to determine compliance with the joint-making requirements (see 8.9).

13.5.2 *Determination of Joint Strength*—Determine seal joint strength by the same procedure used for determining breaking strength (see 13.3). The sealed joint is to be positioned midway between the holding jaws. The maximum force can be read directly from the indicating device to the nearest scale division.

TABLE 8 Thickness of Steel Plate for Bend Test Fixture

Strapping Thickness		Plate Thickness			
		Finish A Strapping		Finish B Strapping	
in.	(mm)	in.	(mm)	in.	(mm)
0.028	(0.71)	3/8	(10)	3/16	(5)
0.029	(0.75)	3/8	(10)	3/16	(5)
0.031	(0.79)	3/8	(10)	3/16	(5)
0.035	(0.89)	3/8	(10)	3/16	(5)
0.044	(1.12)	1/2	(12)	3/16	(5)
0.050	(1.27)	1/2	(12)	3/16	(5)
0.057	(1.45)	1/2	(12)	3/16	(5)

13.5.3 *Determination of Percent of Breaking Strength*—Determine percent of breaking strength by dividing the seal joint strength value by the minimum breaking strength in accordance with strapping type and size tested and multiplying by 100 (see Tables 1 and 2).

13.6 *Determination of Welded Joint Strength*—Determine welded joint strength for manufactured process welds with Type 1 strapping by the same procedure used for determining breaking strength (13.3), except that joint efficiency shall be equal to the ratio of minimum breaking strength of the strapping multiplied by 100.

13.7 *Galvanized Coating:*

13.7.1 *Weight Loss Test*—Test a 6-in. (152-mm) test specimen from a sample unit of Finish B strapping in accordance with Test Methods A 90 for loss of coating weight. Loss of weight less than that shown in Table 5 applicable to strapping width shall constitute failure of the specimen to meet the zinc coating requirements of 8.10.

13.7.2 *Thickness Test*—Test a 6-in. (152-mm) test specimen from a sample unit of Finish B strapping for thickness of zinc coating in accordance with Method B 504, or the magnetic gage or eddy-current method shall be permitted with proper calibration.

13.8 *Ductility Test, Coating*—Test a 12-in. (304-mm) test specimen from a sample unit of strapping for ductility of coating to determine compliance with 8.11. Determination of cracking or flaking of coating shall be by visual examination without magnification. Perform the tests in accordance with Practice D 4332. Take care during the tests to avoid scratching or abrading the coating in the area of the bend. Perform the tests as follows:

13.8.1 Bend strapping 0.025 in. (0.64 mm) thick or less cold, once through 180°, over two thicknesses of the strapping material. Use vise with smooth jaws for this test.

13.8.2 Bend strapping 0.028 through 0.057 in. (0.71 through 1.45 mm) thick, once through 180°, over the rounded end of a steel plate of convenient size and a thickness within ±1/64 in. (0.40 mm) of that shown in Table 8. Bend by clamping the specimen against the plate in a vise, bending the specimen slowly by hand over the rounded end of the plate as far as possible, then completing the bend by inverting the plate and specimen and tightening the vise until the specimen is tight against the plate.

13.9 *Lubrication Test*—To determine an acceptable level of lubrication, apply the straps to a rigid work bench or fixture using the strapping manufacturer's manual or power-tensioning machine. The straps shall have a capability of being tensioned to the percent of rated breaking strength of the strapping as shown below, and should tension smoothly without binding (see 8.3):

50 %—All sizes of Type 1 regular-duty strapping (see Table 1).

40 %—Type 1 heavy-duty strapping up to and including ¾ by 0.028 in. (19.05 by 0.71 mm) (see Table 2).

30 %—Type 1 heavy-duty strapping ¾ by 0.029 in. (19.05 by 0.75 mm) (see Table 2). Larger sizes, for which feed wheel type tensioning tools may be unavailable, and strap finishes are ordered and supplied without lubrication coating, are exempt from the lubricity test.

13.10 *Edge Test*—Test a sample unit of Finish B galvanized-coated strapping in accordance with Test Method A 239 for the presence of zinc coating and absence of copper sulfate on the edge; use two Preece test dips for Grade 1 and one Preece test dip for Grades 2 and 3 (see 4.1.2.1 and 8.10). Nonconformance to the requirements shall constitute failure of the test.

14. Precision and Bias

14.1 *Precision and Bias*—Precision and bias measurements for applicable test methods are currently under study by the Strapping Task Group of ASTM Subcommittee D10.14.

15. Rejection and Retesting

15.1 Any coil of strapping (or specimen of seals) that fails to meet the test requirements of this specification may be rejected. Rejection should be reported to the producer or supplier promptly and in writing. In case of dissatisfaction with the results of the test, the producer or supplier may request retesting (see 12.3).

16. Certification

16.1 At the time of shipment when specified in the contract or order, the manufacturer shall certify that the end product complies with the requirements of this specification.

17. Preparation for Delivery

17.1 Preparation for delivery shall be either overseas/waterproof packaging or commercial packaging when specified in contract or order.

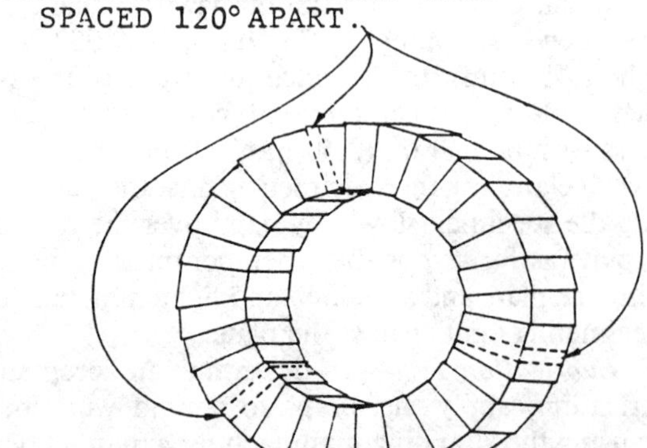

FIG. 4 Wrapping and Tie Straps

18. Overseas/Waterproof Packaging

18.1 *Strapping*—Type 1 and Type 2 strapping, oscillated- or ribbon-wound as specified (see 9.1) shall be wrapped or unit packed in sizes and quantities shown in Table 4.

18.1.1 *Wrapping*—Each coil or multiple of coils of Types 1 and 2 strapping, as specified in Table 4, shall be spiral-wrapped with paper by means of a coil-wrapping machine using tensioning and edge-folding device. Wrapping paper shall be either crepe or hard-sized noncreped paper at the discretion of the supplier (see Fig. 4). Alternatively, for protection instead of wrapping, Types 1 and 2 heavy-duty strapping, in ribbon-wound coils only (when stacked on a skid or platform), shall be capped with a waterproof reinforced paper as shown in Fig. 4.

18.1.1.1 *Creped Material*—Each coil that is wrapped in accordance with 18.1 shall be spiral-wrapped with 45/40/30 duplex reinforced waterproof kraft paper. The paper shall consist of two sheets of kraft paper, 45 lb (20.4 kg) and 30 lb (13.6 kg) basis weight per ream of 500 sheets 24 by 36 in. (6 ft^2 or 0.557 m^2 per sheet), respectively, laminated together with asphalt in which are embedded reinforcing cords or strands of fiber running in the machine direction of the paper, not more than ⅜ in. (9.52 mm) apart, having the 45-lb (20.4-kg) sheet saturated, coated, or infused with asphalt. Creping shall be in the machine direction. The minimum construction before creping shall be 45/40/30 when applying the paper. The tension shall be adjusted so that not over 8 % stretch will be left in the paper. The edge-folding device shall fold either one or both edges of the web of paper to give a rolled edge with a minimum of two thicknesses. Each wrap shall overlap each preceding layer of wrap at least 50 %. The saturated, coated, or infused 45-lb (20.4-kg) kraft shall be on the outside of the wrapped coil or coils. Wrapping shall be accomplished in such a manner as to provide a neat and compact unit pack (see Fig. 4).

18.1.1.2 *Hard-Sized Noncreped Materials*—Alternatively, each coil that is wrapped in accordance with 18.1 shall be spiral-wrapped using a hard-sized noncreped paper 2½ in. (63.5 mm) wide, having a basis weight per ream of 500 sheets 24 by 36 in. (6 ft^2 or 0.557 m^2) of 105 lb (47 kg) ±5 lb (2.3 kg). The paper may be of natural kraft color. The dry tensile strength shall be 105 lbf (467 N)/1 in. (25.4 mm) of width in the machine direction with a tolerance of ±5 lbf (22 N). The wet tensile strength shall be not less than 25 lbf (111 N)/1 in. of width in the machine direction when measured under standard laboratory conditions after 20-min immersion. In applying this material, the individual turns of the spiral wrapping shall be held under tension by spot gluing at intervals along the length of the spiral wrapping in such a way that each individual turn is firmly adhered to the preceding turn in several places along the circumference of each turn. Wrapping material shall be fed into the coil-wrapping machine in continuous strips 2½ in. (63.5 mm) wide. The strips, when glued in cross sections, shall resemble a modified letter *S*, with ¼ in. (6.3 mm) of material folded back on itself 180° along the entire length of each edge. The exposed edge of the wrap shall be folded under. The edges of the wrapping material shall be maximum of ¾ in. (19.05 mm) apart on the outer circumference of the coil. The use of three equally spaced steel straps on the outside of the strapping as specified in 18.2 does not apply.

18.2 *Coil-Tie Strapping*—Each individual coil of Types 1 and 2 regular- and heavy-duty strapping, either oscillated- or ribbon-wound, shall be strapped with a minimum of three equally spaced Type 1 regular-duty steel tie straps not less than ½ by 0.020 in. (12.7 by 0.51 mm). The straps may be applied either under or over the outer spiral wrap when used. When straps are applied under the spiral wrap, the end of the wrap shall be secured by either tucking the end under adjacent wrap or by adhesive, tape, or by the application of one Type 1 regular-duty strap not less than ½ by 0.020 in. (12.7 by 0.51 mm). See Fig. 4.

18.2.1 Alternatively, coils of Types 1 and 2 strapping, when ordered and unit packed in quantities as specified in Table 4, shall be packed in an end loading, full overlap slotted container constructed from V3C fiberboard or equivalent. When practical, containers shall be uniform shape and size, be of minimum cube, and tare consistent with protection required. Containers shall be closed, sealed, and reinforced in accordance with applicable container specifications or supplier's commercial practice.

18.3 *Seals*—Seals of one size, class, finish, type, and style (or nomenclature) unit packed in fiberboard boxes in accordance with supplier's commercial practice, shall be overpacked in a snug-fitting full overlap slotted container constructed from V3C fiberboard or equivalent. The box shall be reinforced with a liner constructed from V3C fiberboard or equivalent. The liner shall be scored to cover the end and side panel of the box. Flutes of the liner shall be perpendicular to the top of the box. Unless otherwise specified, the height of the liner shall be the full inside depth of the box. The liner shall fit snugly inside the intermediate box. This over-pack box shall be closed with not less than two ⅜ by 0.015 in. (9.52 by 0.38 mm) Type 1 regular-duty straps. An alternative method is to use eight overlapped corner sections forming an over-pack ("Adjusta-Pack" Method) constructed from V3C fiberboard or equivalent. A liner is not required with the "Adjusta-Pack". The "Adjusta-Pack" shall be closed with not less than four (2 by 2 pattern) ⅜ by 0.015 in. (9.52 by 0.38 mm) Type 1 regular-duty straps. Dimensions of the over-pack containers shall be determined by the supplier's commercial package and number of commercial packages to be fitted into the over-pack. The dimensions shall be as when specified in the order or as mutually agreed upon between supplier and purchaser. However, the maximum inside dimensions of the over-pack container shall not exceed 100 in. (2540 mm), and the gross weight of the over-pack shipping container shall not exceed 120 lb (54.4 kg).

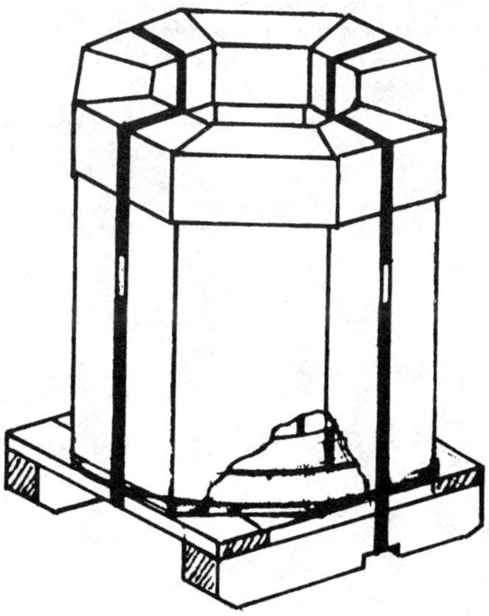

FIG. 5 Single Pile Capped with Corrugated Shroud

FIG. 6 Double Pile Capped

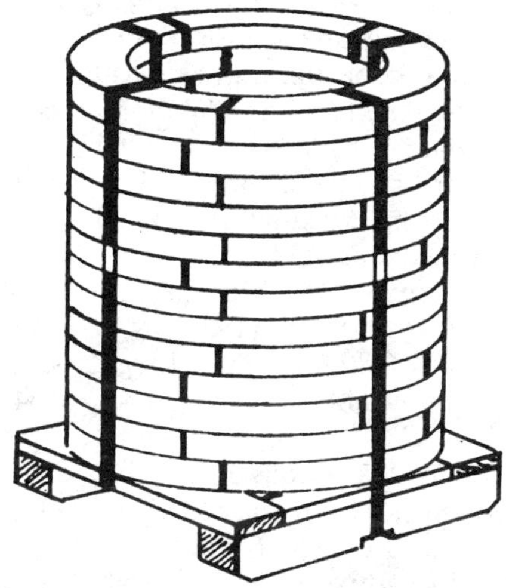

FIG. 7 Single Pile Bare

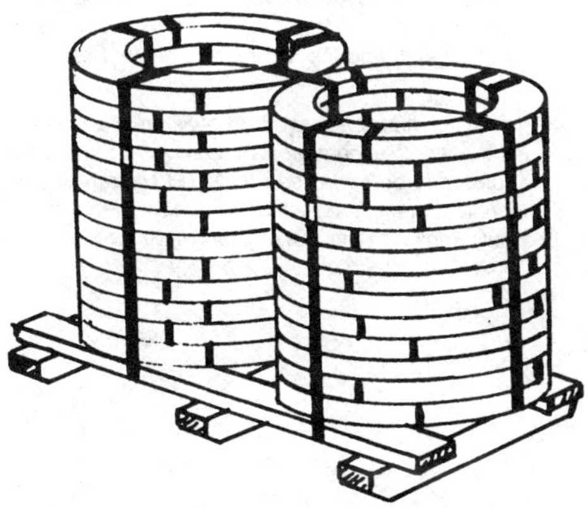

FIG. 8 Double Pile Bare

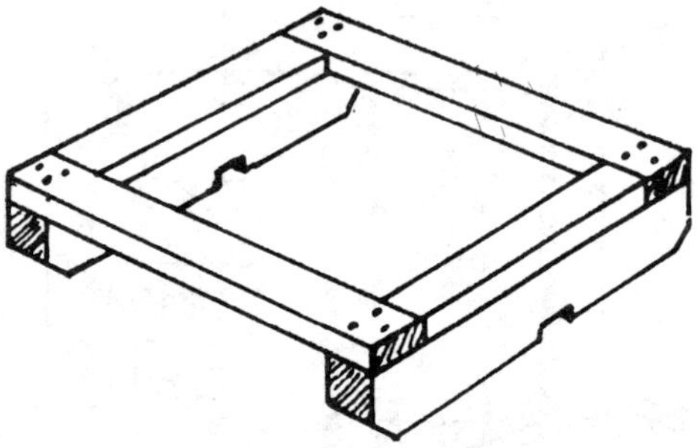

FIG. 9 Single Pile Platform

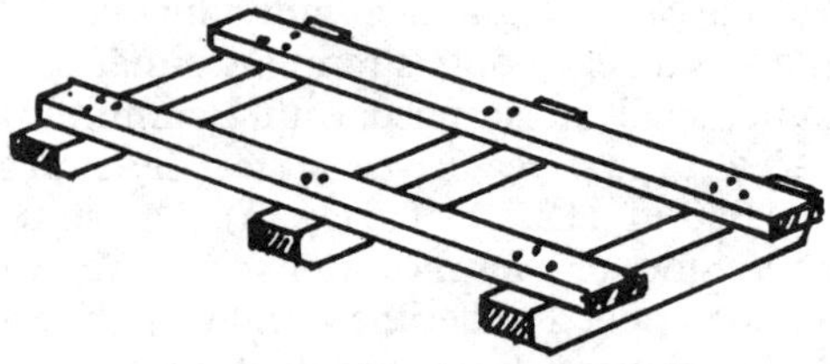

FIG. 10 Double Pile Platform

18.4 *Packing:*

18.4.1 *Strapping*—Coils of flat steel strapping unit packed as specified in 18.1 through 18.3 shall be packed as single or double pile.

18.4.1.1 *Single Pile*—Coils of strapping shall be evenly stacked in a single pile on a platform or skid (see Figs. 5 and 7) and secured to the platform or skid as follows:

(*1*) *Oscillated-Wound Coils*—With four equally spaced straps not less than ¾ by 0.028 in. (19.05 by 0.71 mm) Type 1 regular-duty or ¾ by 0.025 in. (19.05 by 0.64 mm) Type 1 heavy-duty strapping. Maximum stack height shall be 32 in. (813 mm) not including skid.

(*2*) *Ribbon-Wound Coils*—With four equally spaced straps not less than ¾ by 0.031 in (19.05 by 0.79 mm) or ¾ by 0.035 in. (19.05 by 0.89 mm) Type 1 regular-duty strapping. Maximum stack height shall be 34 in. (864 mm), not including skid.

18.4.1.2 *Double Pile*—Coils of strapping, either oscillated- or ribbon-wound, shall be stacked in two even piles in a platform or skid (see Figs. 6 and 8) and secured to the platform or skid with seven equally spaced straps not less than ¾ by 0.031 in. (19.05 by 0.79 mm) or ¾ by 0.035 in. (19.05 by 0.89 mm) Type 1 regular-duty strapping (see 18.4.1.1 for allowable stack heights). The two straps on each side of each pile shall encircle the inside and outside of the stacked coils and cross member of the skid. Each strap on the ends shall encircle the inside and outside of the stacked coils. The center strap shall be applied through the eyes of both piles and around the center runner of the skid. For basic design of platform or skid see Figs. 9 and 10.

18.4.1.3 *Exterior Container*—Coils of strapping unit packed as specified shall be packed in a close-fitting wood box. For shipments to DOD, containers shall conform to overseas type of PPP-B-601 or Class 2 of PPP-B-621. Closure and strapping (as applicable) shall be in accordance with container specifications.

18.4.2 *Seals*—Unit packs will be over-packed in a close-fitting wood box. For DOD shipments, containers shall conform to overseas type of PPP-B-601 or Class 2 of PPP-B-621. Closure and strapping (as applicable) shall be in accordance with container specifications.

19. Commercial Packaging

19.1 Commercial packaging shall be in accordance with Practice D 3951.

20. Keywords

20.1 baling; bundling; carloading; closing; palletizing; steel; strapping; unitizing

Designation: D 4003 – 86

Standard Methods of Controlled Horizontal Impact Test for Shipping Containers[1]

1. Scope

1.1 These test methods are intended to determine the ability of a package or product to withstand laboratory simulated horizontal impact forces.

1.2 The horizontal impacts used in the methods are programmed shock inputs that represent the hazards as they occur in the shipping and handling environments. The environmental hazards may include rail switching impacts, pallet marshalling impacts, and so forth. The following methods apply:

1.2.1 *Test Method A, Rail Car Switching Impacts*—suitable for tests of individual containers as they are shipped in less than carload of nonpalletized loads in rail cars. It may also be used to evaluate the effectiveness of pallet patterns to determine the effect of interaction between containers during rail switching operation impacts.

1.2.2 *Test Method B, Pallet Marshalling Impact Tests*—assess the ability of single containers and unit load quantities to withstand the forces encountered during pallet marshalling operations.

1.3 The test levels may be varied to represent the mode of shipping and handling used for the item under test.

1.4 *This standard does not purport to address all of the safety problems associated with its use. It is the responsibility of whoever uses this standard to consult and establish appropriate safety and health practices and determine the applicability of regulatory limitations prior to use.* Specific precautionary statements are given in Section 6.

2. Referenced Document

2.1 *ASTM Standard:*

D 880 Method of Incline Impact Test for Shipping Containers[2]

3. Terminology

3.1 *Definitions:*

3.1.1 *acceleration*—the rate of change of velocity of a body with respect to time measured in feet per second squared or metres per second squared (1 g = 32.2 ft/s^2 or 9.81 m/s^2).

3.1.2 *shock pulse*—a substantial disturbance characterized by a rise of acceleration from a constant value and decay of acceleration to the constant value in a short period of time.

3.1.3 *velocity change*—the sum of the impact velocity and rebound velocity (the area under the acceleration - time curve).

3.2 *Description of Terms Specific to This Standard:*

3.2.1 *backload*—a duplicate specimen similar to the test package or weights to simulate the other lading in the transport vehicle.

3.2.2 *shock pulse programmer*—a device used to control the parameters of the acceleration versus time-shock pulse generated by a shock test impact machine.

4. Significance and Use

4.1 These test methods provide a measure of a shipping container's ability to protect a product from failure due to horizontal impacts. These measures are based on controlled levels of shock input and may be used for arriving at the optimum design of a container to protect a product against a specified level of shipping environment hazard.

4.2 These methods also provide a measure of a packaged product's ability to withstand the various levels of shipping environment hazards. These measures may be used to prescribe a mode of shipping and handling that will not induce damage to the packaged product or to define the required levels of protection that must be provided by its packaging.

5. Apparatus

5.1 *Horizontal Impact Test Machine:*

5.1.1 The test machine shall consist of a guided test carriage with a flat test specimen mounting surface and an upright bulkhead that is at a 90° angle to the specimen mounting surface. The carriage should be of sufficient strength and rigidity so that the test specimen mounting surface and bulkhead remain rigid under the stresses developed during the test.

5.1.2 The test machine shall provide some means of moving the test carriage in a single guided direction of motion into a programmer impact surface. To permit wider use of this method, a 10° incline to the horizontal track may be used to set the test carriage in motion. Other systems include a carriage suspended on a pendulum device, or a horizontal mechanically driven test carriage. The motion of the carriage shall be controlled in such a manner that its velocity is known at the moment of impact.

5.1.3 The machine shall provide some means of programming shock pulses at the carriage bulkhead-impact surface interface. These devices may be hydraulic, springs, or cushioning materials.

5.1.4 To achieve control during the test, the impact surface should be of sufficient mass and rigidity to limit the foundation displacement during impact. The prescribed

---

[1] These test methods are under the jurisdiction of ASTM Committee D-10 on Packaging and are the direct responsibility of Subcommittee D10.22 on Handling and Transportation.

Current edition approved Feb. 28, 1986. Published April 1986. Originally published as D 4003 – 81. Last previous edition D 4003 – 81.

[2] *Annual Book of ASTM Standards*, Vol 15.09.

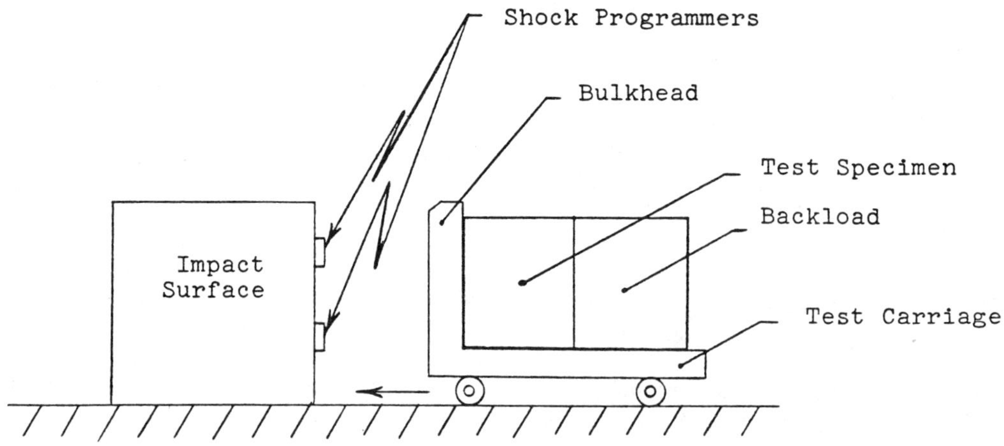

FIG. 1 Diagrammatic Sketch of a Test Configuration

shock pulse limits will provide the controlling factor as to the magnitude and location of the reaction mass required.

5.1.5 The machine and programmers shall be so designed as to arrest the motion of the carriage after the primary impact and prevent secondary impacts.

5.1.6 Provisions shall be made for determining the velocity change of the impact. This may be achieved by recalibration data based on impact velocity and the amount of rebound that occurs during the impact through the integration of the shock pulse. This pulse is measured on the carriage impacting surface, by using an electronic velocity transducer by integrating the acceleration versus time signal measured by the accelerometer instrument described in 5.3.1.

5.1.7 *Calibration and Machine Setting*—Since the test conditions are based on the response of the impacts in Test Methods A and B, pretest runs should be conducted with dummy weights prior to actual tests to establish the approximate calibration and machine equipment settings.

5.1.7.1 The control parameters that must be specified include:

5.1.7.2 The desired velocity change (impact plus rebound velocity of the test carriage).

5.1.7.3 The desired pulse, shape, duration, and acceleration levels.

5.1.7.4 The desired backload weight/friction relationships.

5.2 *Specimen Backload Equipment:*

5.2.1 During some horizontal impacts, the forces that the test unit encounter include both the shock forces of the deceleration as well as compressive forces resulting from other products impacting against the test unit. This will necessitate sufficient carriage strength and platform space to provide a location for the desired backload weights.

5.2.2 Specially adapted backloading fixtures may be used to provide an even loading of the backload weight, or additional product samples may be used to create the desired backload.

5.2.3 The backload weight and frictional characteristics must be specified for each test procedure and reported (see Fig. 1).

5.3 *Instrumentation:*

5.3.1 An accelerometer and signal conditioner, if required, and a recording apparatus are required to record the peak acceleration and duration of the impact shock pulse at the test carriage/product interface.

5.3.2 The monitoring equipment shall meet the following requirements:

5.3.2.1 *Frequency Response*—Response of five times the frequency being measured, minimum.

5.3.2.2 *Accuracy*—Readings shall be within ±2 % of the actual values as measured.

5.3.2.3 *Cross-Axis Sensitivity*—Maximum of 5 % of full scale.

NOTE 1—The test equipment described in Method D 880, Procedure C, may be used with these test methods.

6. Precautions

6.1 These methods may produce severe mechanical responses in the test specimen. Therefore, operating personnel must remain alert to the potential hazards and take necessary safety precautions. The test area should be cleared prior to each impact. The testing of hazardous materials or products may require special precautions that must be observed. Safety equipment may be required and its use must be understood before starting the test.

7. Test Specimen

7.1 The package and product as shipped or intended for shipment shall constitute the test specimen. Sensing devices may be applied to the package, product, or some component of the product to measure the response levels during impact. Dummy test loads of equal configuration, size, and weight distribution and packaging are acceptable if testing the actual product might be hazardous. Care must be taken to duplicate the load characteristics of the product.

8. Conditioning

8.1 Condition the test specimen prior to test or during test, or both, in accordance with the requirements of the applicable specifications. When no conditioning requirements are given, and container materials are moisture sensitive, a standard conditioning atmosphere of 23 ± 2° C

(73.4 ± 3.6° F) and 50 ± 2 % relative humidity for a minimum of 24 h is recommended.

9. Procedure

9.1 *Test Method A—Rail Car Switching Impact Test:*

9.1.1 Prior to initiating the test, write the test plan including the following information:

9.1.1.1 The number of impacts the unit will receive,

9.1.1.2 The velocity change for each of the impacts desired,

9.1.1.3 The pulse duration of the impact shock, and

9.1.1.4 The weight, configuration, and friction of the backload used.

NOTE 2—The number of impacts to which a product will be subjected in transit may range from 2 to 15. The velocity changes range between 1 and 10 mph with an average velocity of approximately 5 mph. The duration of the impact shocks is dependent on the draft gear of the rail cars used to transport the products. The duration normally ranges from 30 ms for standard draft gear to in excess of 300 ms for long travel gear and floating sill cushioning devices. The acceleration levels observed are normally a function of the velocity change and pulse duration rather than a controlling input parameter. The accelerations corresponding to the above durations are about 15 *g* and less than 1 *g*, respectively. It must be realized that rail car switching impacts normally occur many times during shipment. It is recommended that a test consists of a number of lower level impacts or an incremental series of increasing impact magnitude rather than a single large magnitude impact. This type of testing also provides better information by bracketing the failure between two impact levels.

NOTE 3—The backload weight/friction requirement is not well-defined due to a lack of environmental measurements of lading force levels. Through preliminary testing, backload pressures ranging from 200 to 700 kg/m^2 (0.3 to 1.0 psi) on the container impacting surface have created damage levels normally observed in distribution. These pressures are based on a coefficient of friction of 0.5 on a horizontal surface and the backload weight must be reduced to accommodate nonhorizontal surfaces if an inclined test apparatus is used. See appendix for further discussions.

9.1.2 After the test parameters have been established, place dummy weights equivalent to the weight of the product to be tested on the test carriage at the center position of the specimen mounting surface with the face or edge which is to receive the impact firmly positioned against the upright bulkhead. Use duplicate test specimens if available. Then backload the dummy weight with the backload weights to represent the backload conditions that are specified in the test plan. Impact the test carriage from various pullback distances into the selected programmers to produce the desired pulse durations (Note 4). Continue the pretesting until the pullback distance necessary for the desired range of velocity changes is obtained. This pretesting is not necessary if the levels of the major test parameters are known from previous experience.

NOTE 4—The type of programmers used shall be selected on the basis of the shock pulse, waveform, and duration desired.

9.1.3 Replace the dummy load with the test specimen and place it at the center position of the specimen mounting surface with the face or edge that is to receive the impact firmly positioned against the upright bulkhead. Backload the test specimen with an additional product or the dummy weight and pull the test carriage back to the distance necessary to achieve the desired velocity change.

9.1.4 Release the carriage to impact against the programmer for a single impact. Measure to determine the velocity change (impact plus rebound velocity) of the test carriage and the acceleration - time profile of the carriage bulkhead.

9.1.5 Inspection of the packaged product may be conducted between each test impact to examine the effect of the impact on the product and package.

9.1.6 The test container should be subjected to the desired number of impacts at various velocity changes and number of impacts specified in the test plan. Each axis of concern of the test package can be evaluated in a similar manner as described in 9.1.2 through 9.1.6.

9.2 *Test Method B—Pallet Marshalling Impact Test:*

9.2.1 As pallet loads of containers are handled with fork trucks, the pallets are subjected to impacts. These impacts may cause the containers thereon to become disoriented or crushed. The maximum impact levels that may be anticipated are 10 *g* acceleration − 50 ms and 40 *g* acceleration − 10 ms shocks.

9.2.2 A complete pallet load of containers with the pallet pattern and restraining devices including plastic overwrap, if used as is intended for shipment, should be used for this test.

9.2.3 Place the loaded pallet on the test carriage at the center position of the specimen mounting surface with the edge of the pallet or the load firmly positioned against the upright bulkhead. The test carriage should be impacted at the precalibrated test levels to impact at the acceleration and duration chosen. For a four-way entrance pallet the test pallet should be rotated 90° subjected to impacts on the second face.

10. Report

10.1 The report shall include the following:

10.1.1 Definition of the purpose of the test.

10.1.2 Complete identification of the product and package being tested or pallet load and configuration.

10.1.3 Method and orientation of item as it is positioned on the test carriage.

10.1.4 Recording of instrumented test input conditions used during the test and a complete description of the instrumentation.

10.1.5 Detailed description of type of damage resulting from the test. The criteria for damage to the package, product, or pallet load may be based on obvious failure as cracking or breaking of some structural part of the product or package or dislodging of packages from a pallet. The damage criteria also may be based on the physical dimensions or displacement of the product or package, or the relationship of the various packages in a multiple package test.

10.1.6 Recording of the conditioning used for the test.

11. Precision and Bias

11.1 The precision and bias of these test methods have not yet been determined.

APPENDIX

(Nonmandatory Information)

X1. GUIDE TO DETERMINATION OF BACKLOAD

X1.1 If one assumes a uniform package, the primary determinant of backload is package density. It would be expected that the force produced by decelerating a particular volume of lading would be proportional to the weight of that volume. In practice, under normal conditions, this relationship can be expressed as:

$$\text{Backload pressure, psi} = \text{package density, lb/in.}^3 \times F$$

F is an interaction factor, empirically determined to be 35 in. This is effectively a measure of a portion of the load exerting a lateral force on another package under normal impact conditions. This factor is dependent on the rail car draft gear, pulse duration, and the coefficient of friction of the car floor surface.

X1.2 In order to derive the total backload weight, one must multiply the backload by the area over which it is applied. This can be written as:

$$\text{Backload weight, lbf} = \text{backload pressure, psi} \times \text{area, in.}^2$$

By substituting in the original formula and canceling, this also can be expressed as:

$$\text{Backload weight, lbf} = (\text{package weight, lbf/package length, in.}) \times 35 \text{ in.}$$

$$\text{Density} = (\text{weight}/L \times W \times D)$$

Area = $W \times D$, and

Package length = dimension of package in direction parallel to impact direction.

This method of computing the backload provides a consistent method that has been shown to duplicate rail car switching impact damage empirically.

Standard Test Methods for Transmitted Shock Characteristics of Foam-in-Place Cushioning Materials[1]

This standard is issued under the fixed designation D 4168; the number immediately following the designation indicates the year of original adoption or, in the case of revision, the year of last revision. A number in parentheses indicates the year of last reapproval. A superscript epsilon (ε) indicates an editorial change since the last revision or reapproval.

1. Scope

1.1 These test methods determine the shock-absorbing characteristics of foam-in-place packaging materials.

1.2 *Test Method A* uses a free-fall package drop test apparatus.

1.3 *Test Method B* uses a shock-test apparatus.

1.4 *This standard does not purport to address the safety problems associated with its use. It is the responsibility of the user of this standard to establish appropriate safety and health practices and determine the applicability of regulatory limitations prior to use.*

2. Referenced Documents

2.1 *ASTM Standards:*

D 775 Test Method for Drop Test for Loaded Boxes[2]

D 3332 Test Methods for Mechanical-Shock Fragility of Products, Using Shock Machines[2]

3. Terminology

3.1 *Definitions:*

3.1.1 *acceleration*—the rate of change of velocity of a body with respect to time, measured in feet per second per second (or metres per second per second).

3.1.2 *velocity*—the rate of change of position of a body in a specified direction with respect to time, measured in feet per second (or metres per second).

3.2 *Descriptions of Terms Specific to This Standard:*

3.2.1 *equivalent free-fall height*—the height of free fall required for a falling body to attain a measured or given impact velocity.

3.2.2 *foam-in-place cushioning material*—a material that is formed by dispensing chemical components usually into a box or mold, such that the chemicals expand to produce a foam cushion.

3.2.3 *velocity change*—the sum of the impact velocity and any rebound velocity.

3.3 *Symbols:*

3.3.1 *g*—symbol for the acceleration due to the effects of the earth's gravitational pull. While somewhat variable, it is usually considered a constant of value 386 in./s^2 or 32.2 ft (9.8 m)/s^2.

3.3.2 *G*—symbol for the dimensionless ratio between an acceleration in length per time squared units and the acceleration of gravity in the same units.

4. Summary of Test Methods

4.1 The methods consist of using the cushioning material to be tested to support a weighted test block inside a package. The complete package is subjected to drops or controlled shocks and the accelerations transmitted to the test block are measured.

4.2 Either method A or B may be used. However, the two optional methods will not necessarily give the same result.

5. Significance and Use

5.1 Data obtained by these methods may be used to determine the transmitted shock cushioning characteristics of foam-in-place packaging materials. These data allow design of cushioning systems that can provide adequate and efficient use of foam for protection of goods during a distribution life-cycle.

5.2 These methods, in contrast to other methods that usually test only the cushioning foam, are designed to evaluate foam-in-place cushioning materials in a manner in which the foam-in-place packaging material is used. In particular, the method includes simultaneous use of a plastic film, the foam, and the box usually used in this method of packaging. See Fig. 1.

6. Apparatus

6.1 *Test Method A:*

6.1.1 The free-fall drop test apparatus shall conform to the requirements as described in Test Method D 775.

6.2 *Test Method B:*

6.2.1 The shock test apparatus shall conform to the requirements as described in Test Method D 3332.

6.3 *Instrumention and Shock Sensors*—Accelerometers, signal conditioners, and data storage apparatus are required to monitor acceleration versus time histories at various points on the test apparatus and test specimen. The instrumentation systems shall have the following minimum properties:

6.3.1 Frequency response range from 2 Hz or less to at least 1000 Hz.

6.3.2 Accuracy of reading to be within ±5 % of the actual value.

6.3.3 Cross axis sensitivity less than 5 % of full scale.

6.4 *Test Block:*

6.4.1 The test block shall have a rigid square bottom face of not less than 8 in. (203 mm) on a side. Its height shall be

---

[1] These test methods are under the jurisdiction of ASTM Committee D-10 on Packaging and are the direct responsibility of Subcommittee D10.13 on Interior Packaging.

Current edition approved Feb. 26, 1988. Published May 1988. Originally published as D 4168 – 82. Last previous edition D 4168 – 82.

[2] *Annual Book of ASTM Standards*, Vol 15.09.

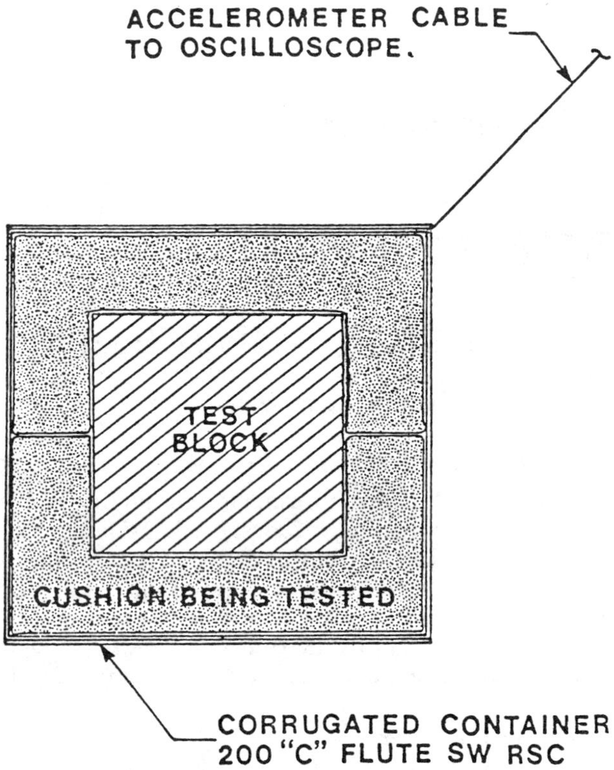

FIG. 1 Specimen Ready For Test

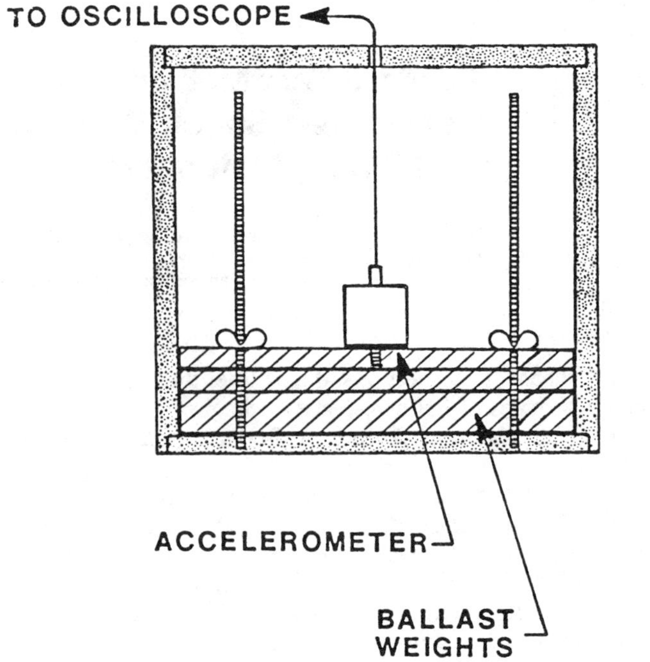

FIG. 2 Test Block

a minimum of half the bottom square dimension and a maximum of the full bottom square dimension.

6.4.2 The test block shall include provisions for firmly mounting ballast weight to adjust its total weight to a desired value.

6.4.3 An accelerometer mounting attachment shall be provided near the center of gravity of the block. The block shall be designed and constructed to be as rigid as possible and to minimize motion of the various components.

6.4.4 A recommended figuration is shown in Fig. 2. Also, it is suggested that the box weight be evenly distributed about the center of the bottom face of the test block.

6.4.5 Ballast weights are added or removed to achieve the desired static stress value on the cushioning material. The accelerometer shall be considered as a portion of the ballast weight.

6.5 *Outer Container*, shall be a regular, slotted container (RSC) corrugated box fabricated with 200 psi (1380 kPa) test "C" flute fiberboard.

7. Sampling

7.1 Because the user forms his own cushioning material, it is very important that an accurate record be prepared of all physical data pertaining to that material for later identification. In addition, the specific characteristics of the film used with the cushion shall be recorded. Care must be taken to ensure that samples are of representative quality.

7.2 *Specimens*—If any of the requirements of Sections 7 and 8 are not met, the container and cushioning material shall be discarded and the procedure repeated with new material. Figure 1 shows a typical specimen assembled and ready for test.

8. Preparation of Test Specimens

8.1 *Modified Free-Rise Bottom Cushion Fabrication* (*Option 1*):

8.1.1 Dispense a layer of the foam-in-place cushioning material into the bottom of the box. Place the barrier film over the foam; then position the test block such that the foam will rise to a depth equal to the thickness to be tested. Alternatively, a dummy test block jig referenced to the top of the box may be used to ensure the correct position of the test block. The sides of the foam cushion shall be 3 in. (76 mm) thick and should come up about half of the depth of test block. Care must be taken to dispense a proper amount of foam in order that the cushion samples are of representative quality (see 13.1.4), are not overly densified, and are free of large voids. See Fig. 3.

8.2 *Premolded Bottom Cushion Fabrication* (*Option 2*):

8.2.1 The bottom cushion may be premolded outside of the outer container by using a wooden mold having the inside dimensions of the outer container, and including an insert in the bottom that duplicates the size and shape of the test block. The sides of the foam cushion shall be 3 in. (76 mm) thick and should come up to half of the depth of the test block. Care must be taken to dispense a proper amount of foam in order that the cushions are of representative quality (see 13.2.4) and have the desired molded density. (See 12.3.2 and Fig. 4)

9. Test Machine

9.1 *Shock Test Apparatus Set-Up:*

9.1.1 Calculate the required carriage velocity change using the following equation:

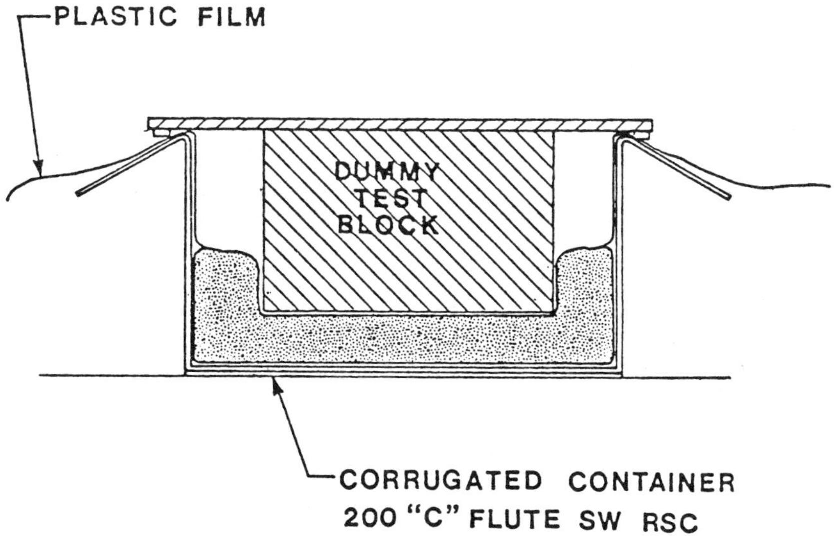

Option 1: FREE RISE

FIG. 3 Bottom Cushion Fabrication

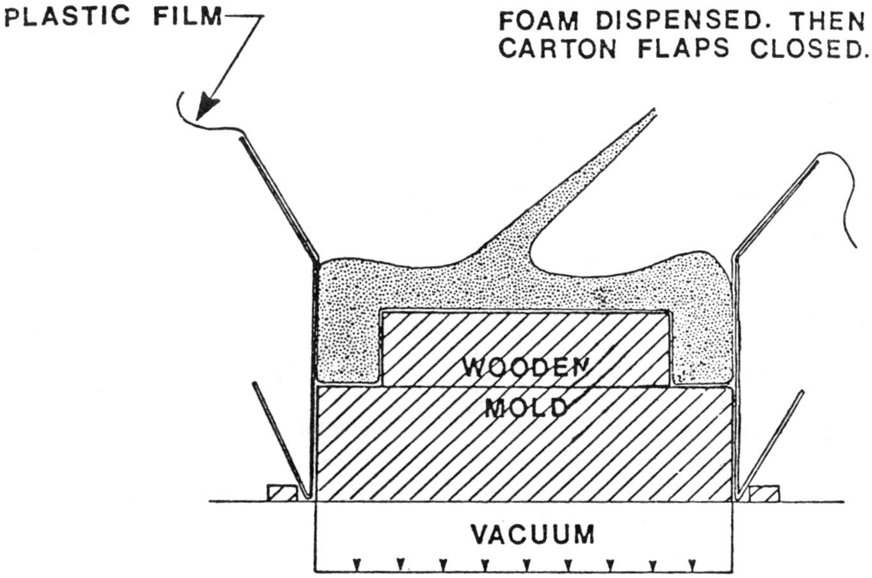

Option 2: PRE-MOLDED

FIG. 4 Bottom Cushion Fabrication

$$\Delta V = \sqrt{2gh}$$

where:

ΔV = required velocity change, in./s or (m/s),
g = acceleration of gravity, in./s^2 or (m/s^2), and
h = desired equivalent freefall drop height, in. (or m).

9.1.2 Following the recommendations of the shock machine manufacturer, adjust the machine to produce a shock pulse of not greater than 3 ms total duration and a velocity change equal to that calculated in 9.1.1.

9.1.3 Rigidly mount an accelerometer or other velocity-measuring instrumentation such as optical sensors on the carriage to measure this velocity change.

9.1.4 Secure the test specimen to the shock table carriage in such a way to ensure a flatface impact of the test specimen.

NOTE 1—Velocity change equals the area under the acceleration versus time pulse. This area may be obtained directly from an accurate graph or trace of the pulse by use of planimeter or equivalent technique. In cases where the pulse is described mathematically (that is, ½ sine, triangular, etc.) the area may be calculated. If it is required to "fair" or smooth the actual acceleration versus time history to obtain a smooth mathematical shape prior to calculation, take care to ensure that the area of the "faired" pulse is the same as the area of the actual pulse. See Annexes A1 and A2 of Test Method D 3332 for a more complete discussion.

NOTE 2—Equivalent drop height determined in 9.1.1 for the shock test machine will not be necessarily the same as free-fall drop height. For a more complete discussion of the relationship, refer to Test Method D 3332.

10. Conditioning

10.1 Condition the test material by exposure to standard or other fixed conditions of air temperature, relative humidity, or other anticipated service conditions for not less than 24 h prior to performance of the test. Perform the tests in the conditioned environment or immediately upon removal from that environment.

10.2 *Conditioning Apparatus*—In those cases where special atmospheric conditions may be required, provide adequate facilities for conditioning the test specimens, to the desired humidity and temperature prior to test.

11. Procedure

11.1 *Test Specimen Assembly:*

11.1.1 Assemble the test block with the weight adjusted to achieve the desired static load.

11.1.2 Insert the test cushion into the corrugated container.

11.1.3 Place the test block into the cavity of the test cushion with care.

11.1.4 Add the top cushion.

11.1.4.1 For option 1, modified free-rise bottom cushion, add a layer of plastic film, then dispense the foam-in-place cushioning material to form the top cushion. The sides of the top cushion should extend down about one half of the depth of the test block to mate with the sides of the bottom cushion.

11.1.4.2 For option 2, premolded bottom cushion, the top cushion shall also be premolded as discussed in 8.2.

11.1.5 Close the box securely. See Fig. 1.

11.1.6 Take proper care to ensure that the accelerometer leads are arranged so as not to interfere with a free fall.

11.2 *Test Method A, Free-Fall Drop Test:*

11.2.1 Set the release mechanism to support the test specimen at the desired drop height.

11.2.2 Place the test specimen on the release mechanism support. The center of gravity of the supported face shall be centered on the supporting mechanism.

11.2.3 Operate the machine in accordance with the manufacturer's instructions to produce a flat drop within ±2° as defined in Test Method D 775.

11.2.4 Repeat for a total of five drops at the chosen test conditions with the box positioned so that the weight rests on the test cushion as shown on Fig. 1 throughout the entire test.

11.2.4.1 There shall be a minimum of 1 min between the sequential drops.

11.2.4.2 The time interval between the drops shall not be excessive in order not to affect adversely the test specimen preconditioning.

11.2.5 The maximum number of drops to be conducted with one sample of the cushioning material is five. If more tests are to be run at different drop heights or static loadings, for example, use new samples.

11.3 *Test Method B, Shock Test:*

11.3.1 Securely fasten the properly prepared test specimen to the carriage of the shock test machine. Attach the input pulse-monitoring accelerometer to the carriage as closely as possible to the outer container. Properly connect the test block acceleration measuring channels.

11.3.2 Operate the machine in accordance with the manufacturer's instructions to produce the acceleration versus time, pulse, as determined in 9.1.

11.3.3 Monitor the carriage accelerometer output to verify that the proper pulse is being obtained.

11.3.4 Repeat for a total of five drops at the chosen test conditions with the box positioned so the weight rests on the test cushion, as shown in Fig. 1, throughout the entire test.

11.3.4.1 There shall be a minimum of 1 min between the sequential drops.

11.3.4.2 The time interval between the drops shall not be excessive in order not to affect adversely the test specimen preconditioning.

11.3.5 The maximum number of impacts to be conducted with one sample of the cushioning material is five. If more tests are to be run at different impact heights or static loadings, for example, use new material.

11.4 *Foam Specimen Weight:*

11.4.1 To determine the foam density when option 2 (premolded bottom cushion) is used (see 12.3.2), the weight of the bottom cushion is required. At the conclusion of the drop or impact sequence, carefully remove the barrier film from the test cushion and weigh the foam to within 1 % accuracy in pounds (or kilograms).

12. Calculation

12.1 *Drop Height:*

12.1.1 For Test Method A, measure drop height, in inches (or metres).

12.1.2 For Test Method B, calculate equivalent free-fall drop height as $h = \Delta V^2/2g$, in inches (or metres).

12.2 *Static Loading:*

12.2.1 Calculate the weight of the test block in pounds (or kilograms) divided by the surface area of the bottom face of the test block in square inches (or square metres).

12.3 *Density:*

12.3.1 For option 1, modified free rise foam-in-place bottom test cushion, determine the free rise core density of a representative foam sample as pounds per cubic foot (or kilograms per cubic metre).

12.3.2 For option 2, pre-molded bottom cushion:

12.3.2.1 Measure the inside dimensions of the mold used to fabricate the bottom cushion to a 1 % accuracy in inches (or millimetres).

12.3.2.2 Determine the inside mold volume in cubic feet (or cubic metres) using the inside mold dimensions as determined in section 12.3.2.1.

12.3.2.3 Calculate density in pounds per cubic feet (or kilograms per cubic metre) by dividing the weight determined in 11.4 by the volume determined in 12.3.2.2.

13. Report

13.1 The report shall include the following:

13.2 A description of cushioning material tested,

13.2.1 Name of manufacturer,

13.2.2 Generic name of foam,

13.2.3 Date of cushion fabrication, and

13.2.4 Any foam quality characteristics such as large voids, prominent knit lines, shrink, proper mixedness.

13.3 Description of film used, including nominal gage measured.

13.4 Number of specimens tested and date of test.

13.5 Bottom cushion preparation method: modified free rise (option 1) or pre-molded (option 2) (see Section 8).

13.6 Test apparatus and procedure used (Method A (free fall drop test) or Method B (shock test)).

13.7 Conditioning parameters.

13.8 Drop height in inches (or metres) as determined in 12.1.

13.9 Static loading in pounds per square inch (or kilograms per square metre) as calculated in 12.2.

13.10 Density in pounds per cubic foot (or kilograms per cubic metre) as determined in 12.3.

13.11 The peak G at each drop.

13.11.1 Peak duration in milliseconds (optional), and

13.11.2 The acceleration versus time curve (optional).

13.11.3 Filter frequency and filter technique used.

13.12 Detailed description of any deviations from the procedure as specified herein.

14. Precision and Bias

14.1 *Repeatability*—The findings of a limited round-robin test indicate the duplicate test results on uniform undamaged specimens agree in the range of ±20 % between laboratories.

14.2 *Reproducibility*—the degree of agreement of averages of triplicate specimens generally varies not more than in the range of ±20 %. More data are needed to establish a firm statement on precision.

14.3 If test results lack this degree of precision, the exact procedures should be reviewed to determine what possible differences exist. Special attention should be given to drop flatness, proper calibration and precision of the instrumentation, and consistency of the samples.

Standard Practice for Performance Testing of Shipping Containers and Systems[1]

This standard is issued under the fixed designation D 4169; the number immediately following the designation indicates the year of original adoption or, in the case of revision, the year of last revision. A number in parentheses indicates the year of last reapproval. A superscript epsilon (ε) indicates an editorial change since the last revision or reapproval.

This standard has been approved for use by agencies of the Department of Defense. Consult the DoD Index of Specifications and Standards for the specific year of issue which has been adopted by the Department of Defense.

1. Scope

1.1 This practice provides a uniform basis of evaluating, in a laboratory, the ability of shipping units to withstand the distribution environment. This is accomplished by subjecting them to a test plan consisting of a sequence of anticipated hazard elements encountered in various distribution cycles. This practice is not intended to supplant material specifications or existing preshipment test procedures. The suitability of this practice for use with hazardous materials has not been determined.

1.2 *This standard does not purport to address all of the safety problems associated with its use. It is the responsibility of the user of this standard to establish appropriate safety and health practices and determine the applicability of regulatory limitations prior to use.* Specific precautionary statements are given in 1.1.

2. Referenced Documents

2.1 *ASTM Standards:*

D 642 Method of Compression Test for Shipping Containers[2]

D 775 Test Method for Drop Test for Loaded Boxes[2]

D 951 Test Method for Water Resistance of Shipping Containers by Spray Method[2]

D 959 Method of Drop Test for Filled Bags[2]

D 996 Terminology of Packaging and Distribution Environments[2]

D 997 Test Method for Drop Test for Cylindrical Shipping Containers[2]

D 999 Methods for Vibration Testing of Shipping Containers[2]

D 1083 Test Methods for Mechanical Handling of Unitized Loads and Large Shipping Cases and Crates[2]

D 4003 Methods of Controlled Horizontal Impact Test for Shipping Containers[2]

D 4332 Practice for Conditioning Containers, Packages, or Packaging Components for Testing[2]

3. Terminology

3.1 *Descriptions of Terms Specific to This Standard:*

3.1.1 *acceptance criteria*—the acceptable quality level that must be met after the shipping unit has been subjected to the test plan. See Section 7.

3.1.2 *assurance level*—the level of test intensity based on its probability of occurring in a typical distribution cycle. Level I is a high level of test intensity and has a low probability of occurrence. Level III is a low level of test intensity, but has a correspondingly high probability of occurrence. Level II is between these extremes.

3.1.3 *coefficient of restitution*—the ratio of the rebound velocity to the impact velocity.

3.1.4 *distribution cycle (DC)*—the sequential listing of the elements expected to occur for a specific routing of a shipping unit from production to consumption. See Section 10.

3.1.5 *hazard element*—a specific event that occurs in a distribution cycle that may pose a hazard to a shipping unit. The element will usually be simulated by a single schedule. See Section 11.

3.1.6 *schedule(s)*—the specific procedure to be used, including the three assurance level intensities, and a reference to the test method that is the basis of the schedule. The purpose of the schedule is to simulate the forces occurring during any element of the distribution cycle. See Section 10.

3.1.7 *shipping unit*—the smallest complete unit that will be subjected to the distribution environment, for example, a shipping container and its contents.

3.1.8 *test plan*—a specific listing of the test sequence to be followed to simulate the hazards anticipated during the distribution cycle of a shipping unit. Included will be the test intensity and number of sequential tests to be conducted. See 9.3.5.

3.1.9 *total velocity change*, (ΔV)—the sum of the impact and rebound velocities.

3.2 *Abbreviations:*

3.2.1 *TOFC*—trailer on flatcar.

3.2.2 *COFC*—container on flatcar.

3.2.3 *TL*—truckload.

3.2.4 *CL*—carload.

3.2.5 *LTL*—less than truckload.

4. Significance and Use

4.1 This practice provides a guide for the evaluation of shipping units according to a uniform system, using established test methods at levels representative of those occurring in actual distribution. The recommended test levels are based on available information on the shipping and handling environment, and current industry practice and experience

[1] This practice is under the jurisdiction of ASTM Committee D-10 on Packaging and is the direct responsibility of Subcommittee D10.21 on Shipping Container Environment.

Current edition approved Dec. 28, 1990. Published February 1991. Originally published as D 4169 – 82. Last previous edition D 4169 – 89.

[2] *Annual Book of ASTM Standards*, Vol 15.09.

(1–12).[3] The tests should be performed sequentially on the same containers in the order given.

5. Test Specimen

5.1 Test specimens shall consist of representative samples of complete shipping units, including actual contents. Products with blemishes or minor defects may be used if the defective component is not to be studied by the test and if the defect is documented in the report. Dummy test loads are acceptable if testing the actual product might be hazardous. If a dummy load is used, it should be instrumented to determine if the fragility level of the actual product has been exceeded. Take care to duplicate the load characteristics of the actual product, and avoid unnecessary prehandling.

5.2 Care must be taken to ensure that no degradation has occurred to either the product or the package if the test packages have been shipped to the test site. If any doubt exists as to the condition of the package, repack the product in new packaging material before testing.

5.3 The number of test replications depends on the desired objectives of the testing and the availability of duplicate products and shipping containers. Replicate testing is recommended to improve the reliability of the test results.

6. Conditioning

6.1 If the distribution cycle contains climatic conditions that have an effect on the performance characteristics of the product, shipping container, or components such as cushioning, use one of the following procedures:

6.1.1 Conduct the test at standard conditions and compensate for the effects of any climatic condition. Condition the shipping units to a standard atmosphere of 73.4 ± 1.8°F (23 ± 1°C) and 50 ± 2 % relative humidity. Condition fiberboard containers in accordance with Practice D 4332. If testing cannot be conducted at the standard condition, conduct the tests as soon after removal from the conditioning chamber as practicable. Recondition the shipping units to the standard atmosphere as necessary during the test plan.

6.1.2 Condition the shipping units to the climatic conditions (salt spray, water immersion, humidity, or temperature) and conduct the tests at desired conditions as soon after removal from the conditioning chamber as practicable. Recondition the shipping units as necessary. If testing at desired conditions is impossible, conditions at time of test shall be recorded and reported.

6.1.3 In some circumstances, it may be necessary to conduct some or all of the tests at special climatic conditions, such as those given in Practice D 4332 or Test Method D 951.

7. Acceptance Criteria

7.1 Acceptance criteria must be established prior to testing and should consider the required condition of the product at receipt. The organizations conducting the test may choose any acceptance criteria suitable for their purpose. It is advisable to compare the type and quantity of damage that occurred to the test specimens with the damage that occurs during actual distribution and handling or with test results of similar containers whose shipping history is known.

7.2 In many cases, the acceptance criteria can be the following:

Criterion 1—Product is damage-free.

Criterion 2—Package is intact.

Often, this means that the shipping container and its contents are suitable for normal sale and use at the completion of the test cycle. Detailed acceptance criteria may allow for accepting specified damage to a product or its package. The form and content of acceptance criteria may vary widely, according to the particular situation. Methods may range from simple pass-fail judgments to highly quantitative scoring or analysis systems.

8. Procedure

8.1 Use the following steps to conduct a performance test:

8.1.1 *Define Shipping Unit*—Describe in terms of size, weight, and form of construction. See 3.6.

8.1.2 *Establish Assurance Level*—Specify a level of test intensity. The level should be one of three pre-established assurance levels. This must be pre-established based on the product value, the desired level of anticipated damage that can be tolerated, the number of units to be shipped, knowledge of the shipping environment, or other criteria. Assurance Level II is suggested unless conditions dictate otherwise. Assurance Level I provides a more severe test than II. Assurance Level III provides a less severe test than II. The assurance level may be varied between elements (see Section 11) if such variations are known to occur. The test levels used should be reported. See Section 12.

8.1.3 *Determine Acceptance Criteria*—Acceptance criteria are related to the desired condition of the product and package at the end of the distribution cycle. See Section 7 and 9.3.3.

8.2 *Select Distribution Cycle*, from the available standard distribution cycles. Use the DC that most closely correlates with the projected distribution. When the distribution is undefined, the general distribution cycle DC-1 should be selected. When the anticipated distribution is well understood, a special distribution cycle DC-2 may be specified. See Section 10. In using DC-2, the user selects elements from 11.1 and specifies the test sequence.

8.3 *Write Test Plan*—Prepare by using the sequence presented in Table 1 for the distribution cycle selected. Obtain the test intensities from the referenced schedules. The test plan intensity details must take into account the assurance levels selected as well as the physical description of the shipping unit. Table 1 thus leads to a detailed test plan consisting of the exact sequence in which the shipping unit will be subjected to the test inputs. The test schedules associated with each element reference the existing ASTM test methods for clarification of the equipment and techniques to be used to conduct the test.

8.4 *Select Samples for Test*—See Section 5.

8.5 *Condition Samples*—See Section 6.

8.6 *Perform Tests*, as directed in reference ASTM standards and as further modified in the special instructions for each test element.

[3] The **boldface** numbers in parentheses refer to the list of references at the end of this practice.

8.7 *Evaluate Results*, to determine if the shipping units meet the acceptance criteria. See Section 7.

8.8 *Document Test Results*, by reporting each step. See Section 12.

8.9 When possible, obtain feedback by monitoring shipments of the container that was tested to ensure that the type and quantity of damage obtained by the laboratory testing correlates with the damage that occurs in the distribution cycle. This information is very useful for the planning of subsequent tests of similar shipping containers.

9. Example

9.1 The following example will serve to illustrate the use of this practice:

9.2 *Example*—Test a packaged consumer product. The moderate value and volume of shipment is typical of other

TABLE 1 Performance Test Sequence

Sequence		1	2	3	4	5	6
Distribution cycle 1		General schedule—undefined distribution system					
Test type	Climate hazard	Handling	Vehicle stacking	Stacked vibration	Loose-load vibration	Rail switching	Handling
Element	1	A or B	D	E	F	H	A or B
Distribution cycle 2		Specially controlled environment, user specified					
Test type	Climate hazard			USER SPECIFIED			
Element	1	Select from 11.1					
Distribution cycle 3		Single package environment, up to 100 lb (45.4 kg)					
Test type	Climate hazard	Manual handling	Vehicle stacking	Loose-load vibration	Vehicle vibration	Manual handling	
Element	1	A	D	F	G	A	
Distribution cycle 4		Motor freight, single package over 100 lb (45.4 kg), or unitized					
Test type	Climate hazard	Handling	Vehicle stacking	Loose-load vibration	Vehicle vibration	Handling	
Element	1	B	D	F	G	B	
Distribution cycle 5		Motor freight, TL, not unitized					
Test type	Climate hazard	Handling	Vehicle stacking	Stacked vibration	Vehicle vibration	Handling	
Element	1	A or B	D	E	G	A or B	
Distribution cycle 6		Motor freight, TL, or LTL—unitized					
Test type	Climate hazard	Handling	Vehicle stacking	Stacked vibration	Handling	Warehouse stacking	
Element	1	B	D	E	B	C	
Distribution cycle 7		Rail only, bulk loaded					
Test type	Climate hazard	Manual handling	Vehicle stacking	Stacked vibration	Rail switching	Manual Handling	
Element	1	A	D	E	H	A	
Distribution cycle 8		Rail only, unitized					
Test type	Climate hazard	Handling	Vehicle stacking	Stacked vibration	Rail switching	Handling	Warehouse stacking
Element	1	B	D	E	H	B	C
Distribution cycle 9		Rail and motor freight, not unitized					
Test type	Climate hazard	Handling	Vehicle stacking	Vehicle vibration	Rail switching	Loose-load vibration	Handling
Element	1	A or B	D	G	H	F	A or B
Distribution cycle 10		Rail and motor freight, unitized					
Test type	Climate hazard	Handling	Vehicle stacking	Stacked vibration	Rail switching	Handling	Warehouse stacking
Element	1	B	D	E	H	B	C

TABLE 1 *Continued*

Sequence		1	2	3	4	5	6
Distribution cycle 11		Rail, TOFC and COFC					
Test type	Climate hazard	Handling	Vehicle stacking	Rail switching	Stacked vibration	Loose-load vibration	Handling
Element	1	A or B	D	H	E	F	A or B
Distribution cycle 12		Air (intercity) and motor freight (local), over 100 lb, unitized					
Test type	Climate hazard	Handling	Vehicle stacking	Stacked vibration	Vehicle vibration	Handling	
Element	1	A or B	D	E	G	A or B	
Distribution cycle 13		Air (intercity) and motor freight (local), single package up to 100 lb (45.4 kg)					
Test type	Climate hazard	Manual handling	Vehicle stacking	Loose-load vibration	Vehicle vibration	Manual handling	
Element	1	A	D	F	G	A	
Distribution cycle 14		Warehousing (partial cycle to be added to other cycles as needed)					
Test type	Climate hazard	Handling	Warehouse stacking				
Element	1	A or B	C				
Distribution cycle 15		Export/Import shipment for intermodal container or roll on/roll off trailer (partial cycle to be added to other cycles as needed)					
Test type	Climate hazard	Handling	Vehicle stacking	Handling			
Element	1	B	D	B			
Distribution cycle 16		Export/Import shipment for palletized cargo ship (partial cycle to be added to other cycles as needed)					
Test type	Climate hazard	Handling	Vehicle stacking	Handling			
Element	1	A or B	D	A or B			
Distribution cycle 17		Export/Import shipment for break bulk cargo ship (partial cycle to be added to other cycles as needed)					
Test type	Climate hazard	Handling	Vehicle stacking	Handling			
Element	1	A	D	A			

products in the shipper's line. No damage is acceptable and the package must be in good condition after the test. The fiberboard packaged product weighs 28 lb, is 9 in. high, and palletized 63 in. high for storage and LTL shipment. The product does not support any of the load.

9.3 *Procedure:*

9.3.1 *Step 1, Define Shipping Unit*—Shipping unit to be tested is a typical pallet load.

9.3.2 *Step 2, Establish Assurance Level*—Assurance Level II will be used, based on value and volume of shipment.

9.3.3 *Step 3, Determine Acceptance Criteria* at Assurance Level II:

Criterion 1—No product damage.

Criterion 2—All packages in good condition.

9.3.4 *Step 4, Select Distribution Cycle*—DC-6 will be used for this palletized, LTL shipment.

9.3.5 *Step 5, Write Test Plan:*

Sequence	Type Test	Method	Level
1	Handling	D 1083	One impact on two opposite base edges from 6 in.
2	Stacking	D 642	Compression to 756 lb (individual container[A]).
3	Vibration	D 999, Method C	Search 3 to 100 Hz at 0.5 *g* peak. Dwell 10 min at 0.5 *g* peak.
4	Handling	D 1083	One impact on two opposite base edges from 6 in.
5	Stacking	D 642	Compression to 756 lb (individual container[A]).

[A] Alternative full pallet load compression test, 529 lb per bottom tier container.

9.3.6 *Step 6, Select Samples for Test*—Select representative samples.

9.3.7 *Step 7, Condition Samples*—Condition to 23 ± 1°C, 50 ± 2 % relative humidity, in accordance with Practice D 4332.

9.3.8 *Step 8, Perform Tests*—Perform tests in accordance with the test plan in Step 5, as directed in the referenced ASTM standards and in the special instructions for each element.

9.3.9 *Step 9, Evaluate Results*—Examine products and packages to determine if the acceptance criteria have been met.

9.3.10 *Step 10, Document Test Results*—Write a report to cover all steps in detail, in accordance with Section 12.

10. Distribution Cycles

10.1 Distribution cycles are categorized as follows:

DC Number	Distribution Cycle	Table Section
1	General schedule—undefined distribution system.	1
2	Special—controlled environment, user specified.	2
3	Single package environment, up to 100 lb (45.4 kg).	3
4	Motor freight—single package over 100 lb (45.4 kg)	4
5	Motor freight—TL, not unitized.	5
6	Motor freight—TL or LTL unitized.	6
7	Rail only CL—bulk loaded.	7
8	Rail only CL—unitized.	8
9	Rail and motor freight—not unitized.	9
10	Rail and motor freight—unitized.	10
11	Rail TOFC and COFC.	11
12	Air (intercity) and motor freight (local)—over 100 lb, or unitized.	12
13	Air (intercity) and motor freight (local)—single package up to 100 lb (45.4 kg).	13
14	Warehousing, partial cycle.	14
15	Export/Import shipment by intermodal container or roll on/roll off trailer, partial cycle, to be added to other cycles as needed.	15
16	Export/Import shipment of unitized cargo by ship, partial cycle to be added to other cycles as needed.	16
17	Export/Import shipment by break bulk cargo ship, partial cycle to be added to other cycles as needed.	17

11. Hazard Elements

11.1 Elements are categorized as follows:

Element	Distribution Cycle Element	Test Simulation Hazard	Section
A	Manual handling up to 200 lb (90.7 kg).	drop	11.2
B	Mechanical handling over 100 lb (45.4 kg).	drop	11.3
C	Warehouse stacking.	compression	11.4
D	Vehicle stacking.	compression	11.4
E	Truck and rail transport stacked or unitized load.	vibration	11.5
F	Loose-load vibration.	repetitive shock	11.6
G	Vehicle vibration.	vibration	11.7
H	Rail switching.	longitudinal shock	11.8
I	Climate, atmospheric condition.	temperature, moisture, humidity	11.9

11.2 *Element A—Manual handling up to 200 lb (90.7 kg)*:

11.2.1 The test levels and the test method for this element of the distribution cycle are intended to determine the ability of the shipping unit to withstand the hazards occurring during manual handlings such as loading, unloading, stacking, sorting, or palletizing. The main hazards from these operations are the impacts caused by dropping or throwing. Size, weight, and shape of the shipping unit will affect the intensity of these hazards. Presented below are recommended drop heights, the number of drops, and the shipping unit orientation at impact:

Test methods—D 775, D 959, D 997.
Conditioning—See Section 6.
Special instructions:

Number of Drops At Specified Height[A]	Impact Surface: Box D 775	Bag D 959	Pail D 997
One	bottom	face	bottom
Two	adjacent bottom edges	two sides	two sides 90° apart
Two	diagonally opposite bottom corners	both ends	bottom edges 90° apart
One	top	opposite face	top

[A] On the last handling operation of a distribution cycle, the bottom drop shall be made at twice the specified height.

11.2.1.1 When a second series of tests is to be conducted, that is, a second handling operation is present in the distribution cycle, the specified drop heights should be conducted on faces, edges, or corners not previously tested, one drop per orientation, when possible. Choose critical orientations for first series of drops. In the case of UPS or Parcel Post shipments, the second series of tests should include impacts on all surfaces (faces) not previously tested.

11.2.1.2 For purposes of this procedure, the bottom of a small parcel is the surface on which the parcel rests in its most stable configuration.

11.2.1.3 Use Element B for packages over 48 in. (1.22 m) in length when the package width is less than half the package length.

11.2.2 Use the following test levels:

Shipping Weight, lb (kg)	Drop Height, in. (mm) Assurance Level I	II	III
0 to 20 (0 to 9.1)	24 (610)	15 (381)	9 (229)
20 to 40 (9.1 to 18.1)	21 (533)	13 (330)	8 (203)
40 to 60 (18.1 to 27.2)	18 (457)	12 (305)	7 (178)
60 to 80 (27.2 to 36.3)	15 (381)	10 (254)	6 (152)
80 to 100 (36.3 to 45.4)	12 (305)	9 (229)	5 (127)
100 to 200 (45.4 to 90.7)	10 (254)	7 (178)	4 (102)

11.3 *Element B—Mechanical handling over 100 lb (45.4 kg)*:

11.3.1 The test levels and the test method for this element of the distribution cycle are intended to determine the ability of large and heavy shipping units and unitized loads to withstand the mechanical handling hazards that occur during loading, unloading, sorting, or stacking. Different test methods are used for large shipping cases and crates versus unit loads. For various types of unit loads, test methods also vary, depending on the method of truck handling: fork, clamp, spade, or pull/pack.

Test Method—D 1083.
Conditioning—See Section 6.
Special Instructions—Use the following methods of Test Methods D 1083 as appropriate:

(*a*) *Large Shipping Cases and Crates*—Fork lift truck handling. One drop on each opposite base edge in accordance with 9.1.1 of Test Methods D 1083 and one drop on each of two opposite base corners in accordance with 9.1.2 of Test Methods D 1083.

Gross Weight, lb (kg)	Drop Height, in. (m) Assurance Level I	II	III
100 to 500 (45.4 to 226.8) or 48 in. (1.22 m) length option	12 (0.305)	9 (0.229)	6 (0.152)
Over 500 (226.8)	9 (0.229)	6 (0.152)	3 (0.076)

(*b*) *Large Shipping Cases and Crates*—Crane handling. One drop flat on bottom and one drop on base edge in accordance with 9.1.4 of Test Methods D 1083. Use the same drop heights versus shipping unit weight as in (*a*) above.

(*c*) *Large Cases and Crates*—Tip test. In accordance with Section 10 of Test Methods D 1083.

(*d*) *Large Cases and Crates*—Tipover test. In accordance with Section 11 of Test Methods D 1083.

(*e*) *Unitized Loads*—Fork lift truck handling.

(*1*) One drop on each opposite base edge in accordance with 8.1.2 of Test Methods D 1083. Use the same drop heights versus shipping unit weight as in (*a*) above.

(*2*) Pick up, transport around test course, and set down in accordance with 8.1.1 of Test Methods D 1083.

Assurance Level	Cycles (Round Trips)
I	8
II	5
III	3

(*f*) *Unitized Loads*—Spade or clamp truck handling. Pick up, transport around test course, and set down in accordance with 8.2 or 8.3 of Test Methods D 1083. Assurance levels and cycles are as specified in (*e*) (*2*) above.

(*g*) *Unitized Loads*—Pull pack handling.

(*1*) Test slip sheet tab durability in accordance with 8.4.2 of Test Methods D 1083. Only one test sequence is required per distribution cycle.

Assurance Level	Pull-Push Cycles
I	25
II	15
III	10

(*2*) Test stability of unitized load using the handling on test course procedure in accordance with 8.4.1 of Test Methods D 1083. Assurance levels and cycles are as specified in (*e*) (*2*) above.

11.4 *Element C—Warehouse stacking and Element D—Vehicle stacking*:

11.4.1 The test levels and the test methods for these elements of a distribution cycle are intended to determine the ability of the shipping unit to withstand the compressive loads that occur during vehicle transport or during warehousing. The required loading must consider, in addition to the overload, the effects of length of time in storage, vibration, the alignment or stacking pattern of the container, variability in container strength, moisture content, temperature, previous handling, and method of load support. The minimum required loads for typical shipping units which include the combined effects of the above factors are recommended below:

Test Method—D 642.

Conditioning—73.4 ± 1.8°F (23 ± 1°C), 50 ± 2 % relative humidity in accordance with Practice D 4332.

11.4.2 Use the following test levels:

Shipping Unit Construction	F Factors Assurance Level I	II	III
1. Corrugated, fiberboard, or plastic container that may or may not have stress-bearing interior packaging using these materials, and where the product does not support any of the load.	8.0	4.5	3.0
2. Corrugated, fiberboard, or plastic container that has stress-bearing interior packaging with rigid inserts such as wood.	4.5	3.0	2.0
3. Containers constructed of materials other than corrugated, fiberboard, or plastic that are not temperature or humidity sensitive or where the product supports the load directly, for example, compression package.	3.0	2.0	1.5
4. If the product supports a known portion of the load, the *F* factor is calculated in the following manner:			

$$F = P(F_p) + C(F_c)$$

where:

F_p = factor given above for compression package (construction type 3),

P = percentage of load supported by product,

F_c = factor given above for appropriate container construction, and

C = percentage of load supported by container.

If a full pallet load is tested, *F* factors may be reduced by 30 %.

11.4.3 Load the shipping unit to the maximum load value, as calculated below. Remove the load immediately after reaching the specified value.

$$L = W\left(\frac{H - h}{h}\right) \times F$$

where:

L = minimum required load, lbf or N,

W = weight of one shipping unit or individual container, lbf or N,

H = maximum height of stack in storage or transit vehicle, in. or m,

h = height of shipping unit or individual container, in. or m, and

F = a factor to account for the combined effect of the individual factors described above.

11.5 *Element E—Truck and rail transport stacked or unitized load*:

11.5.1 The test levels and test method for this element of the distribution cycle are intended to determine the ability of stacked or unitized shipping units to withstand the vertical vibration environment during transport. The test levels and method account for the magnitude, frequency range, duration, and direction of vibration.

Test Method—D 999, Method C.

Condition—See Section 6.

Special instructions—Perform the test along the vertical axis with the load in normal shipping orientation. Dwell time is for each noted product or package resonance up to four discrete resonances. If more than four resonances are noted, test at the four frequencies where the greatest response is noted. It is permissible to use a concentrated dead load to simulate an upper pallet load. In frequency sweeps it is advisable to consider the frequency ranges normally encountered in the type of transportation being considered. The resonant frequency(ies) may shift during test due to changing characteristics of the container system. It is suggested that the resonant frequency be varied slightly during the test to detect any shift and to continue testing at the frequency of maximum response.

11.5.2 Use the following test levels:

Assurance Level	Frequency Range, Hz	Amplitude (O-Peak), *g* Rail	Truck	Dwell Time, min
I	3 to 100	0.25	0.5	15
II	3 to 100	0.25	0.5	10
III	3 to 100	0.25	0.5	5

11.6 *Element F—Loose load vibration:*

11.6.1 The test levels and the test method for this element of the distribution cycle are intended to determine the ability of the shipping unit to withstand the repetitive shocks occurring during transportation of bulk or loose loads. The test levels and test method account for amplitude, direction, and duration of the repetitive shocks.

Test Method—D 999, Methods A1 and A2.

Conditioning—See Section 6.

Special instructions—Dwell time distributed 50 % along normal vertical shipping axis and remaining 50 % evenly along all other possible shipping orientations.

11.6.2 Use the following test levels:

Assurance Level	Dwell Time, min
I	60
II	40
III	30

11.7 *Element G—Vehicle vibration:*

11.7.1 The test levels and the test method for this element of the distribution cycle are intended to determine the ability of the shipping unit to withstand the vertical vibration environment during transportation. The test levels and test method account for the frequency range, amplitude, sweep rate, dwell time for resonances, and direction of vibration.

Test Method—D 999, Method B.

Conditioning—See Section 6.

Special instructions—Perform the test for each possible shipping orientation. Dwell time is for each noted package or product resonance up to four discrete resonances. Test at the four frequencies where the greatest response is noted. In frequency sweeps it is advisable to consider the frequency ranges normally encountered in the type of transportation being considered.

11.7.2 Use the following test levels:

Assurance Level	Frequency Range, Hz	Sweep and Dwell Amplitude (O-Peak), *g* Rail	Truck	Dwell Time, min
I	3 to 100	0.25	0.5	15
II	3 to 100	0.25	0.5	10
III	3 to 100	0.25	0.5	5

11.8 *Element H—Simulated rail switching:*

11.8.1. The test levels and test methods for this element are intended to determine the ability of the shipping unit to withstand the acceleration levels and compressive forces that might occur during rail switching operations.

Test method—D 4003, Test Method A.

Conditioning—See Section 6.

Special instructions—Three impacts shall be performed. For railcars with standard draft gear, shock durations of 40 ± 10 ms shall be used, as measured on the floor of the dolly. For railcars with long-travel draft gear, shock durations of 300 ± 50 ms shall be used.

For purposes of this test, the coefficient of restitution shall be considered 0.0, and the total velocity change shall be equivalent to the specifed impact velocity. Container impact surface should be the same as occurs in actual shipment. If more than one orientation is possible, impact that surface which is known to be most sensitive to damage.

Refer to Methods D 4003 for specific instructions on how to instrument and conduct the test.

11.8.2 *Procedure*—Load shipping unit on dolly against bulkhead. Total backload shall be equivalent to a minimum of 3 ft lineal (0.9 m) of cargo. The package used as backload in contact with the test package must be identical to the test package.

11.8.3 *Test Levels*—Allow the dolly to impact a cushioned barrier in accordance with the following assurance levels:

Assurance Level	Number of Impacts	Impact Velocity mph	(m/s)
I	2	8	(3.58)
	1	6	(2.68)
II	2	6	(2.68)
	1	4	(1.79)
III	1	6	(2.68)
	2	4	(1.79)

11.8.4 *Procedure Modification*—If more detailed information is available on backload or shock characteristics it is recommended that the above procedure be modified to use such information. The specified backload and shock characteristic are intended to simulate the conditions that occur on rail cars equipped with standard type draft gear.

11.9 *Element I—Climate, Atmospheric Condition:*

11.9.1 The levels of atmospheric conditioning for this element are intended to provide the minimum preconditioning necessary to expose the shipping unit to an atmospheric hazard that may affect performance in distribution. The user is cautioned to determine if climate hazards other than those given exist, or if additional conditioning is necessary between other elements of a distribution cycle. It should be noted that different atmospheric conditions are likely to exist between the origin and destination points of a distribution cycle, particularly for export/import cycles.

Test Method—Practice D 4332.

Special instructions—The user may select the Standard Conditioning Atmosphere of 23° C and 50 % relative humidity or any of the special standard atmospheres listed in Practice D 4332, as appropriate. The same atmospheric condition should be used for any assurance level. A conditioning period of 72 h, or sufficient time to reach equilibrium of all parts of the package and product is recommended.

Tests should be conducted in the conditioned atmosphere whenever possible. If not possible, test for one hazard element immediately after removal from the conditioned atmosphere and then immediately return to reconditioning before testing the next hazard element.

For atmospheres other than the Standard Conditioning Atmosphere, the user must determine the appropriate compressive load factor for Elements C and D, as the factors given there are based on testing under the Standard test atmosphere.

12. Report

12.1 Report fully all the steps taken. At a minimum, the report should include:

12.1.2 Description of product and shipping unit,

12.1.3 Distribution cycle (DC) and test plan,

12.1.4 Assurance levels and rationale,

12.1.5 Number of samples tested,

12.1.6 Conditioning used,

12.1.7 Acceptance criteria,

12.1.8 Variation from recommended procedures, and

12.1.9 Condition of specimens after test.

13. Precision and Bias

13.1 The precision and bias of this practice are dependent on those of the various test methods used, and cannot be expressly determined.

REFERENCES

(1) Ostrem, F. E., and Libovicz, B. A., "A Survey of Environmental Conditions Incident to the Transportation of Materials," General American Research Division, October 1971 PB 204 442.

(2) Henzi, A. N., "A Survey of Test Methods Currently Used for Simulating the Transportation Environment," General American Research Division, April 1971. PB 202 728. Prepared for Department of Transportation, Office of Hazardous Materials.

(3) Ostrem, F. E., and Rumerman, M. L., "Shock and Vibration Transportation Design Criteria Manual," General American Research Division, September 1965. N 66 34681. Prepared for National Aeronautics and Space Administration.

(4) National Safe Transit Association. "Preshipment Test Procedures," Project 1 and 1A. March 1977.

(5) EIA Standard RS-414-A, "Simulated Shipping Tests for Consumer Electronic Products and Electronic Components." Electronic Industries Assoc., February 1975.

(6) Preshipment Test Standard for Packaged Major Appliances, ANSI/AHAM MA-1-PS-1980.

(7) Federal Test Method Standard No. 101C. *Preservation, Packaging, and Packaging Materials*: Test Procedures, March 1980.

(8) "The Railroad Environment, A Guide for Shipper and Railroad Personnel," New York Central Railroad Co., 1966.

(9) Packaging for Parcel Post, U.S. Postal Service, November 1974.

(10) Ostrem, F. E., and Godshall, W. D., "An Assessment of the Common Carrier Shipping Environment." General Technical Report FPL 22. U.S. Forest Product Laboratory, 1979.

(11) "Transportability Criteria, Shock and Vibration," Department of Army Technical Bulletin TB 55-100 Hq., Department of Army, April 1964.

(12) *Environmental Test Methods*, Military Standard 810C, Department of Defense, March 1975.

Standard Test Methods for Water Vapor Transmission of Shipping Containers—Constant and Cycle Methods[1]

This standard is issued under the fixed designation D 4279; the number immediately following the designation indicates the year of original adoption or, in the case of revision, the year of last revision. A number in parentheses indicates the year of last reapproval. A superscript epsilon (ε) indicates an editorial change since the last revision or reapproval.

ε1 NOTE—Sections 3 and 4 were transposed editorially in July 1988.

1. Scope

1.1 These test methods cover the determination of water vapor transmission rates for bulk shipping containers, as follows:

1.1.1 Method A, for Reclosable Containers, and

1.1.2 Method B, for Containers Not Designed for Reclosing.

1.2 Within each procedure details are given for the constant and cycle methods of test atmosphere.

1.3 The test may be applied to the container as packed, or after one or more performance tests such as drum, vibration, drop, or actual shipping tests, as required.

1.4 For small shipping containers requiring greater accuracy in weighing, the water vapor transmission may be determined in accordance with Test Method D 895, or Test Method D 1251.

1.5 *This standard does not purport to address the safety problems associated with its use. It is the responsibility of the user of this standard to establish appropriate safety and health practices and determine the applicability of regulatory limitations prior to use.*

2. Referenced Documents

2.1 *ASTM Standards:*

D 895 Test Method for Water Vapor Permeability of Packages[2]

D 1251 Test Method for Water Vapor Permeability of Packages by Cycle Method[2]

3. Definitions

3.1 *water vapor transmission of a shipping container (constant atmosphere method)—for the purpose of this test method,* the rate at which water is transmitted into the container from the test atmosphere (normally of 90 ± 2 % relative humidity and a temperature of 100° ± 2°F (37.8° ± 1.1°C) surrounding it while a desiccant is sealed within.

3.2 *water vapor transmission of a container* (*cycle atmosphere method*)—for this test method, the amount of water transmitted into the container from the test atmosphere during one cycle while a desiccant is sealed within.

3.3 *cycle*—one series of test atmospheres to which the test specimens are exposed. Normally one cycle will consist of 1 week of exposure to a temperature of 0° ± 5°F (−17.8° ± 2.8°C) to be followed by 3 weeks of exposure at a temperature of 100° ± 2°F (37.8° ± 1.1°C) and a relative humidity of 90 ± 2 %.

4. Significance and Use

4.1 These test methods are normally used for the following purposes:

4.1.1 To evaluate materials and constructions for a specific type of container,

4.1.2 To compare performance of different types of containers,

4.1.3 To determine adequacy of protection for a specific product or application, and

4.1.4 To maintain quality control.

METHOD A—RECLOSABLE CONTAINERS

5. Apparatus

5.1 *Desiccant*—A desiccant shall be used that has a powerful affinity for water and a high drying efficiency, that is, a low vapor pressure after absorbing a large amount of water. The desiccant shall be in the form of small lumps that will pass a No. 8 (2.36-mm) sieve and be free from fines that will pass a No. 30 (600-μm) sieve. Anhydrous calcium chloride and anhydrous magnesium perchlorate have been found suitable. When the test is made to determine the suitability of a specific container for a particular product, that product may be used inside the test specimen instead of the desiccant, in which case the specimen shall be filled to normal capacity.

5.2 *Weighing Balance*—A weighing balance accurate to within 1 g shall be used. When the required amount of desiccant is greater than can be weighed on a balance of this sensitivity, two or more receptacles shall be used and weighed individually. When product tests are made, a regular laboratory balance and drying oven or other appropriate equipment are required for making standard moisture determinations peculiar to the product.

5.3 *Receptacle for Desiccant*—A nonmoisture-absorptive receptacle ¾ to 1½ in. (19 to 38 mm) deep shall be used for holding the desiccant within the container being tested. It should be equipped with a cover that will rest on the top rim of the receptacle to protect the desiccant from picking up

[1] These test methods are under the jurisdiction of ASTM Committee D-10 on Packaging and are the direct responsibility of Subcommittee D10.23 on Storage Test Methods.

Current edition approved Aug. 26, 1983. Published October 1983.

[2] *Annual Book of ASTM Standards*, Vol 15.09.

moisture while being weighed. The size of the weighing receptacle or receptacles shall be such that the exposed area of desiccant is not less than 10 % of the area of the test specimen. For testing large containers, this may require using two or more weighing receptacles.

5.4 *Test Chamber*—A test room or cabinet provided with conditioned air that is continuously circulated around the specimens under test. The conditions in the chamber shall be such that no condensation occurs on the specimens except during that portion of the cycle when the exposure is changed from a low temperature to a high temperature.

6. Test Specimens

6.1 Test specimens shall be representative of the containers being tested, and shall be closed and sealed in the normal manner.

6.2 The performance shall be based on tests of not less than two representative specimens of a given size and type of container.

7. Procedure

7.1 Locate the weighing receptacle or receptacles centrally within the test specimen, using a nonabsorptive support when required. Place the selected quantity of the desiccant in the receptacle sufficient to uniformly cover the area of the receptacle to a depth of not less than ½ in. (13 mm). Cover, and after weighing, immediately transfer into the test specimen. Uncover the receptacle and close and seal the specimen in the normal manner. The desiccant may require one or more replacements if it becomes noticeably moist during the test.

NOTE 1—When the test is conducted to determine the water vapor transmission of the shipping container for research, development, manufacturing control, specification acceptance, etc., a desiccant shall be used as the water absorbing medium. When the test is used to determine the suitability of the shipping container with respect to water vapor resistance for a particular product, the product shall be used in lieu of the desiccant.

7.2 Place the specimen inside the test room or cabinet in a position where free access of the conditioned circulating air is provided on all surfaces of the container according to the following:

7.2.1 *Constant Atmosphere Method*—Normally surrounding conditions are 90 ± 2 % relative humidity and a temperature of 100° ± 2°F (37.7° ± 1.1°C).

7.2.2 *Cycle Atmosphere Method*—Place the specimen in the low temperature test room or cabinet (normally maintained at 0°F (−17.8°C) for a period of 1 week. Follow immediately by 3 weeks of exposure to the high temperature and humidity (normally 100°F (37.8°C) and 90 % relative humidity).

7.3 Make successive weighings of the receptacle at suitable intervals and plot the mass gained against time using these schedules:

7.3.1 *Constant Method*—Accuracy of the test is adversely affected by too frequent weighings. For highly permeable containers a minimum weighing frequency of 3 days is recommended. For containers having a low rate of transmission, a weighing frequency of biweekly to monthly is recommended.

7.3.2 *Cycle Method*—Make successive weighings of the receptacle at the completion of each cycle. It is suggested that the test be carried on for a minimum of three complete cycles.

7.4 The weighing procedure, which should be conducted as rapidly as possible, shall be as follows: Remove the specimen from the test chamber, open, remove the receptacle, and place the cover on the receptacle. Temporarily close the specimen to prevent the entrance of moisture from the surrounding atmosphere. Weigh the receptacle, open the specimen, and return the receptacle. Uncover the receptacle, reclose the specimen, and return it without delay to the test room or cabinet. Continue until a constant rate of gain is attained as indicated by at least three successive points in a straight line. The slope of this portion of the curve will furnish a measure of water vapor transmission.

NOTE 2—In instances where tests are made in which specific products are used in place of the standard desiccant, a constant rate of gain may not be attainable. Therefore, for product tests, the container is filled to its normal capacity and the moisture content determined from a composite sample taken at the start of the test.

For the Constant Method the specimen is placed in the test room for a predetermined period (usually one month), removed, and the average moisture content determined from a composite sample taken at the end of this period).

For the Cycle Method the specimen shall be subjected to one exposure cycle for a predetermined period, removed, and the average moisture content determined from a composite sample taken at the end of this period.

In either method the test can be continued by immediately closing and resealing, after which the specimen is returned without delay to the test room or cabinet.

METHOD B—CONTAINERS NOT DESIGNED FOR RECLOSING

8. Apparatus

8.1 The apparatus shall consist of a desiccant, balance, and test chamber in accordance with 5.1, 5.2, 5.4 respectively, also desiccant receptacles in accordance with 5.3 if it is not practicable to weigh the entire container with the desiccant inside.

9. Test Specimens

9.1 Test specimens shall be representative of the containers being tested, and shall be closed and sealed in the normal manner. For containers that cannot be weighed to the accuracy prescribed in 5.2 with desiccant inside, an auxiliary opening large enough to permit inserting the desiccant receptacle shall be cut in one face. A means of sealing the opening during the test shall be provided that will permit repeated opening and reclosing during the test. Closure of this handhole can be accomplished with a piece of sheet metal, or other impervious sheeting somewhat larger than the opening, sealing it in place with wax (NOTE 3), or forming an equally impervious seal by other means.

NOTE 3—A satisfactory wax for this purpose consists of a mixture of 60 % microcystalline wax[3] and 40 % refined crystalline paraffin wax.

[3] Grade Numbers 2300, 2305, and 2310 of the Mobil Oil Corp., or their equivalent, have been found satisfactory for this purpose.

10. Procedure

10.1 For specimens that can be weighed accurately to the requirements prescribed in 5.2, place the selected quantity of desiccant or product inside the specimen, which shall then be closed and sealed in the normal manner. Place the specimen in the test atmosphere as described in 7.2 and make successive weighings at suitable intervals until a constant rate of gain is established as described in 7.3.

10.2 For specimens that are not practicable to weigh as described in 10.1 proceed exactly as in Section 7 except use the auxiliary opening described in Section 9 as the access opening. The regular closure is to remain sealed throughout the test.

REPORT AND PRECISION

11. Report

11.1 The report shall include the following:

11.1.1 Identification of container including data on closure, liners, etc., and whether Method A or B was used,

11.1.2 Desiccant used or identification of contents if a product is used. When a desiccant is used, the mass and area exposed shall be specified, and

11.1.3 Number of specimens tested.

11.2 If the Constant Method is used, record the following:

11.2.1 Temperature and relative humidity at which the test was conducted. If the tolerance of 2 % relative humidity or 2°F (±1.1°C) temperature is exceeded, the variation shall be specified, and

11.2.2 Water vapor transmission reported as grams of water per 30 days for the container as a unit.

11.3 If the Cycle Method is used record the following:

11.3.1 Temperature, relative humidity, and time for each test atmosphere of the cycle at which the test was conducted, and

11.3.2 Water vapor transmission reported as grams of water per cycle for the package as a unit.

11.4 The water vapor transmission rate calculation shall be made for the period of constant rate of gain. If a product is used instead of a desiccant, the moisture content at the start and completion of each cycle shall be given, together with the net weight of contents, from which data the actual amount of water pick up can be calculated.[3]

12. Precision and Bias

12.1 Duplicate determinations on uniform, undamaged specimens should agree within 10 % within a given laboratory. However, the reproducibility between different laboratories has been found to vary as much as 25 %. This must be recognized when making interlaboratory comparisons of data.

Standard Practice for Conditioning Containers, Packages, or Packaging Components for Testing[1]

This standard is issued under the fixed designation D 4332; the number immediately following the designation indicates the year of original adoption or, in the case of revision, the year of last revision. A number in parentheses indicates the year of last reapproval. A superscript epsilon (ε) indicates an editorial change since the last revision or reapproval.

1. Scope

1.1 This practice provides for special conditioning and testing atmospheres that may be used to simulate particular field conditions that a container, package, or packaging component may encounter during its life or testing cycle.

1.2 This practice describes procedures for conditioning these containers, packages, or packaging components so that they may reach equilibrium with the atmosphere to which they may be exposed.

1.3 *This standard does not purport to address the safety problems associated with its use. It is the responsibility of the user of this standard to establish appropriate safety and health practices and determine the applicability of regulatory limitations prior to use.*

2. Referenced Documents

2.1 *ASTM Standards:*

D 685 Method of Conditioning Paper and Paper Products for Testing[2]

D 996 Terminology of Packaging and Distribution Environments[2]

E 41 Definitions of Terms Relating to Conditioning[3]

E 171 Specification for Standard Atmospheres for Conditioning and Testing Materials[2]

E 337 Test Method for Measuring Humidity with a Psychrometer (the Measurement of Wet- and Dry-Bulb Temperatures)[4]

2.2 *ISO Standard:*

2233 Packaging—Complete, filled Transport Packages–a Part II: Conditioning for Testing[5]

3. Terminology

3.1 Terms and definitions used in this practice may be found in Terminology D 996, Definitions E 41, or Specification E 171.

4. Significance and Use

4.1 Many materials from which containers and packages are made, especially cellulosic materials, undergo changes in physical properties as the temperature and the relative humidity (RH) to which they are exposed are varied. Therefore, the package should be placed and kept in a specified atmosphere for a length of time such that subsequent measurements of physical properties will be meaningful and reproducible.

4.2 The conditions described in this practice are either historically accepted standard conditions or special laboratory conditions chosen to represent particular phases of the distribution environment. These special conditions do not necessarily duplicate actual field conditions, but tend to simulate them and have effects on packages and materials which may be related to their field performance.

5. Atmospheric Conditions

5.1 *Preconditioning Atmosphere*—20 to 40°C (68 to 104°F) and 10 to 35 % relative humidity.

5.2 *Standard Conditioning Atmosphere*—23 ± 1°C (73.4 ± 2°F) and 50 ± 2 % relative humidity.

NOTE 1—Average values must fall within these limits. Short-term fluctuations and measurement limitations may cause individual measurements to vary up to ±2°C (±3.6°F) and ±5 % relative humidity without significant impairment of test reproducibility.

5.3 *Special Atmospheres*—The special standard atmospheric conditions shown in Table 1 may be selected when appropriate.

NOTE 2—Temperature and humidity tolerances of ±1°C (±1.8°F) and ±2 % relative humidity are desirable but not always attainable.

NOTE 3—In the absence of a specific requirement for a particular atmospheric condition, use the conditioning atmosphere given in 5.2.

NOTE 4—Conditioning at the desert condition (Table 1) at a constant temperature of 60°C (140°F) may have effects on some materials that do not relate to effects of cyclical field conditions.

5.4 The measurement of temperature and relative humidity of the conditioning atmosphere shall be made as close to the specimen being exposed as is possible. (See Test Method E 337 for a detailed description of methods.) The

TABLE 1 Special Atmospheres[A]

Environment	Temperature	Relative Humidity
Cryogenic	−55 ± 3°C (−67 ± 6°F)	. . .
Frozen food storage	−18 ± 2°C (0 ± 4°F)	. . .
Refrigerated storage	5 ± 2°C (41 ± 4°F)	85 ± 5 %
Temperate high humidity	20 ± 2°C (68 ± 4°F)	85 ± 5 %
Tropical	40 ± 2°C (104 ± 4°F)	85 ± 5 %
Desert	60 ± 3°C (140 ± 6°F)	15 ± 2 %

[A] When compliance with ISO Standard 2233 is required, the relative humidities for the refrigerated storage environment, the temperate high humidity environment, and the tropical environment shall be maintained at 90 % ± 2 % relative humidity.

[1] This practice is under the jurisdiction of ASTM Committee D-10 on Packaging and is the direct responsibility of Subcommittee D10.21 on Shipping Container Environment.

Current edition approved Dec. 29, 1989. Published February 1990. Originally published as D 4332 – 84. Last previous edition D 4332 – 84.

[2] *Annual Book of ASTM Standards*, Vol 15.09.

[3] *Annual Book of ASTM Standards*, Vol 14.02.

[4] *Annual Book of ASTM Standards*, Vols 11.03 and 15.09.

[5] Available from the American National Standards Institute, 1430 Broadway, New York, NY 10018.

temperature and relative humidity indicated at the control point, may not be representative of conditions elsewhere in the conditioned space due to local effects or deficiency in air circulation. Tolerances at the controller usually must be smaller than those at the specimen.

6. Apparatus

6.1 *Room* (*or Cabinet*) of such size that sample containers or packages may be individually exposed to circulating air at the temperature and relative humidity chosen.

6.1.1 *Control Apparatus* capable of maintaining the room at the required atmospheric conditions within the tolerance limits.

6.2 *Hygrometer*—The instrument used to indicate the relative humidity should be accurate to ±2 % relative humidity. A psychrometer may be used either for direct measurement of relative humidity or for checking the hygrometer (see Test Method E 337).

6.3 *Thermometer*—Any temperature-measuring device may be used provided it can accurately indicate the temperature to within 0.1°C or 0.2°F. The dry-bulb thermometer of the psychrometer may be used either for direct measurement or for checking the temperature-indicating device.

7. Procedure

7.1 Obtain containers, packages, or packaging components in quantity in accordance with the standard governing the test to be performed. Fiberboard containers (and containers fabricated from other materials that are affected by changes in humidities greater than 40 %) may require preconditioning (see Method D 685).

7.2 Containers or packages that are sealed with aqueous adhesives should have all seals made sufficiently prior to preconditioning or conditioning so that the water in the adhesive will not be a factor in preconditioning or conditioning of the package.

7.3 Place the container, package, or packaging component within the conditioning chamber and expose it to the required conditions for an amount of time specified in the particular test procedure. If no time period is specified, a period of at least 72 h (hours) or that required to reach equilibrium is recommended. The container, package, or packaging component shall be supported in a manner such that all of the top and sides have free access to the conditioning atmosphere. Heavier packages may be placed on a pallet or other device that should allow for as much exposure of the surface bottom to the conditioning atmosphere as is possible so long as the device does not cause a permanent change in the box or affect the final test result.

7.4 Wherever possible, the desired test or measurement should be made at the specified conditioning atmosphere. When this is not possible, the test shall be performed as quickly as possible after the specimen is removed from the conditioning atmosphere.

7.5 Alternatively, the package may be enclosed in a plastic sheet prior to removal from the conditioning atmosphere, in a manner such that the plastic will not interfere with the particular test, so that the package will neither gain or lose moisture. Here too, the package shall be tested as quickly as possible after removal from the conditioning atmosphere.

8. Report

8.1 The report of the test on the conditioned package shall include the following:

8.1.1 Temperature, relative humidity, and time of exposure.

8.1.1.1 Preconditioning atmosphere and time of exposure when used.

8.1.2 Statement of compliance with this procedure.

8.1.3 Variations from this procedure, including the testing temperature and relative humidity if different from the conditioning atmosphere, the time elapsed, etc.

9. Precision and Bias

9.1 A statement of precision and bias is not applicable to this practice.

Standard Test Method for Coefficient of Static Friction of Corrugated and Solid Fiberboard[1]

This standard is issued under the fixed designation D 4521; the number immediately following the designation indicates the year of original adoption or, in the case of revision, the year of last revision. A number in parentheses indicates the year of last reapproval. A superscript epsilon (ε) indicates an editorial change since the last revision or reapproval.

1. Scope

1.1 This test method covers the determination of the coefficient of static friction of corrugated and solid fiberboard or of the materials used to make such board.

1.2 This test method contains the following two methods:

1.2.1 *Horizontal Plane Method*—The horizontal instrument requires some means of movement of the specimen in relation to the surface upon which it rests. The coefficient of friction is measured directly from the resistance to that force and the applied weight.

1.2.2 *Inclined Plane Method*—The incline plane is raised until sliding begins. The coefficient of friction is equal to the tangent of the angle at which sliding begins.

1.3 *This standard does not purport to address the safety problems associated with its use. It is the responsibility of whoever uses this standard to consult and establish appropriate safety and health practices and determine the applicability of regulatory limitations prior to use.*

2. Referenced Documents

2.1 *ASTM Standards:*

D 585 Method for Sampling and Accepting a Single Lot of Paper, Paperboard, Fiberboard, or Related Products[2]

D 2534 Test Method for Coefficient of Kinetic Friction for Wax Coatings[3]

D 4332 Practice for Conditioning Containers, Packages, or Packaging Components for Testing[2]

E 691 Practice for Conducting an Interlaboratory Study to Determine the Precision of a Test Method[4]

3. Terminology

3.1 *Descriptions of Terms Specific to This Standard:*

3.1.1 *coefficient of friction*—the ratio of the frictional force resisting movement of the surface being tested to the force applied normal to that surface (the weight of the material above that surface).

3.1.2 *kinetic coefficient*—the ratio of the force resisting motion of the surface once the motion is in progress.

NOTE 1—Kinetic coefficient is not determined in this test method (see Appendix X1).

3.1.3 *static coefficient*—the ratio of the force resisting initial motion of the surface.

4. Significance and Use

4.1 The coefficient of friction of corrugated and solid fiberboard indicates how containers made from that board will perform in many critical applications. A high coefficient of friction of one surface of board to itself means that containers having that surface will tend to resist sliding in unit loads. A low coefficient may indicate potential problems with the containers slipping from the load.

4.2 The coefficient of friction test is empirical. It describes the condition of that surface at the moment of test. This may or may not relate to the condition of the surface in use. Its condition will depend upon the treatment of the surface. Nevertheless, the test results are useful in determining the properties of the surfaces tested. Measurement is not made until the third slide. Experience has shown that variability is greatly reduced after the second slide, and friction measurements are essentially constant for repeated slides.

5. Test Specimens

5.1 From each test unit of a sample obtained in accordance with Method D 585, cut five specimens (see 5.3 for size). Carefully mark each specimen with its machine direction, on a place that will not be the actual test area. For corrugated board this will be perpendicular to the direction of the flutes.

5.2 Cut and orient the specimens as follows:

5.2.1 If the specimens are from a container, carefully cut them from a flap area of the container. It is usually desirable to test a flap area since this is the area that normally comes in contact with the next container in pallet loads.

5.2.2 Paper is anisotropic, and the coefficient of friction is particularly so. Care must be taken in making the test to orient the surfaces to each other just as they will be oriented in actual use. Therefore, if the layers of containers are to be alternated in palletization, the coefficient should be measured with one surface machine direction (MD) and the other cross-machine direction (CD).

5.2.3 If it is of interest to know how the container will perform on packaging machinery or be retained by conveyors, floors, shelves, etc., measure the coefficient between a sample of the paperboard and the material from which the conveyor or floor or shelf is made.

5.2.4 If there is no known relationship to use conditions, cut the test specimens so that the measurements will be made with both parts oriented in the CD direction.

5.3 Each specimen consists of two sheets. Cut the smaller sheet, and score if necessary, to fit the size of the sled

[1] This test method is under the jurisdiction of ASTM Committee D-10 on Packaging and is the direct responsibility of Subcommittee D10.17 on Paper and Paperboard Products.

Current edition approved Sept. 27, 1985. Published November 1985.

[2] *Annual Book of ASTM Standards*, Vol 15.09.

[3] *Annual Book of ASTM Standards*, Vol 05.02.

[4] *Annual Book of ASTM Standards*, Vols 06.03, 08.03, and 14.02.

allowing for excess material if it is to be clamped. Cut the larger sheet at least 1 in. (25 mm) larger in width than the test surface of the sled, and sufficient in length to fit the horizontal table, also allowing for material to be clamped. For some instruments it may be necessary to separate the outer facing of the board from the medium and inner facing or from the filler in order to clamp onto the sled or block.

NOTE 2—Take care not to touch the test area since skin oils, pencil marks, etc. can affect the test results.

6. Conditioning

6.1 Precondition, condition, and test the specimens in an atmosphere in accordance with Practice D 4332.

6.2 With a clean cloth or tissue, lightly wipe the test surface to remove any loose fibers or debris which may be present.

7. Test Methods

7.1 *Horizontal Plane Method* (Fig. 1):

7.1.1 *Apparatus* (see Fig. 1):

7.1.1.1 *Instrument Table*—A horizontal plane surface of a smooth incompressible material (metal, hard wood, plate glass, or plastic), having a width at least 1 in. (25 mm) wider than the test sled (7.1.1.2) with means of leveling it in two directions. The base serves as a support for the rest of the mechanism. When placed on a solid support, the instrument must be free of vibration.

7.1.1.2 *Sled*—A specimen block having a bottom area of at least 3 in. (20 cm^2) smoothly finished to exert a pressure of 0.50 ± 0.25 psi (3.44 ± 1.72 kPa) on the horizontal surface. A sled 2.5 by 2.5 in. (62.5 by 62.5 mm) weighing 3 lb (1.36 kg) has been found to be satisfactory. Means are provided for clamping the specimen to the sled and for fastening the sled to the force measuring device.

7.1.1.3 *Mechanical Power Unit*—Means for moving the sled horizontally along the instrument table or the horizontal table under the specimen block at a speed of 6.0 ± 1.0 in./min (2.54 ± 0.42 mm/s). This unit must not transmit vibration to the instrument table. The unit is connected to the sled or table with some kind of inelastic linkage.

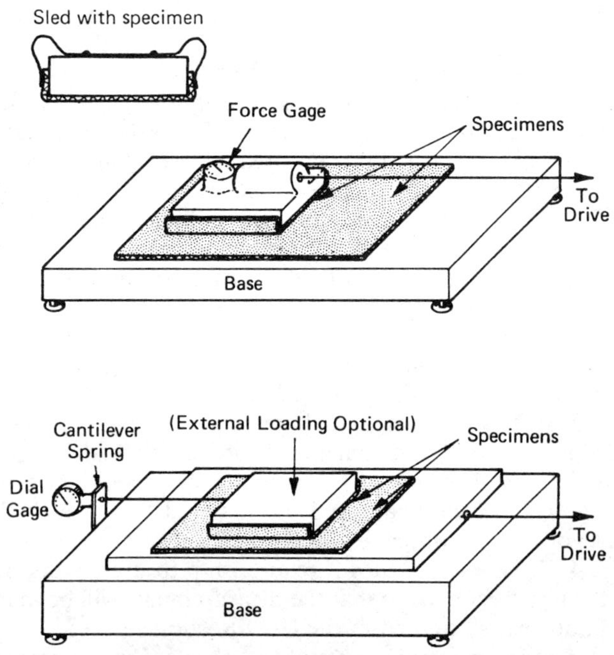

FIG. 1 Schematics for Two Horizontal Plane Instruments

7.1.1.4 *Measuring Device*—Means for measuring the force required to move or restrain the sled to the nearest 0.01 lb (0.044 N).

7.1.2 *Procedure:*

7.1.2.1 Place the instrument upon a solid and vibration-free base or table and level it.

7.1.2.2 If necessary, adjust the force gage to zero.

7.1.2.3 Place the larger sheet of the specimen on top of the slide base, and clamp the end of the sheet farthest from the measuring device in the sheet clamp.

7.1.2.4 Clamp the smaller sheet of the specimen in place on the sled. It is best to clamp one end of the sheet, and then draw the sheet uniformly tight before clamping the remaining end.

7.1.2.5 Place the sled with the specimen sheet in position on the other sheet, and fasten the cable on the block to the power unit or to the force gage.

7.1.2.6 Start the power unit, taking care to maintain the cable(s) taut as the drive takes up the load.

7.1.2.7 Continue the motion of the two specimen sheets in relation to each other unit until the entire test area has been traversed.

7.1.2.8 Repeat the steps in 7.1.2.4 to 7.1.2.6, and record the force to just begin the motion on the third slide.

7.1.2.9 Repeat the test on the other four specimens.

7.1.3 *Calculations*—Divide the force required to begin motion of the block by the weight of the block. This is the coefficient of static friction.

7.1.4 *Precision and Bias:*

7.1.4.1 *Repeatability*—The repeatability is ±6.4 % over the range exhibited by corrugated linerboard.

7.1.4.2 *Reproducibility*—The reproducibility is not known.

7.1.4.3 *Comparability*—The comparability is not known (in accordance with the definitions of these terms in Practice E 691).

7.2 *Inclined Plane Method:*

7.2.1 *Apparatus* (see Fig. 2):

7.2.1.1 *sled*—a rectangular metal block with a plane lower surface. A means for clamping the specimen to the sled is desirable but is not necessary if the lower surface is faced with soft rubber ⅛ in. (3 mm) thick. The size and weight of the metal block are not critical, but a 2 by 4 in. (50 by 100 mm) block weighing 1.65 lb (750 g) giving 0.2 ± 0.1 psi (1.38 ± 0.69 kPa) when horizontal has been found satisfactory.

7.2.1.2 *inclined plane*—a plane surface, hinged so that it can be tilted, with a smooth, incompressible top surface which may be made of metal, hard wood, plate glass, or plastic having a width at least 1 in. (25 mm) wider than the sled and a length sufficient to permit the sled to move at least ⅝ in. (15 mm). The plane is provided with a clamp at the upper end to fix the test specimen to the plane.

NOTE 3—When the plane is horizontal, means are provided to level the instrument in two directions.

7.2.1.3 Means to indicate the angular displacement of the plane to within 0.5°.

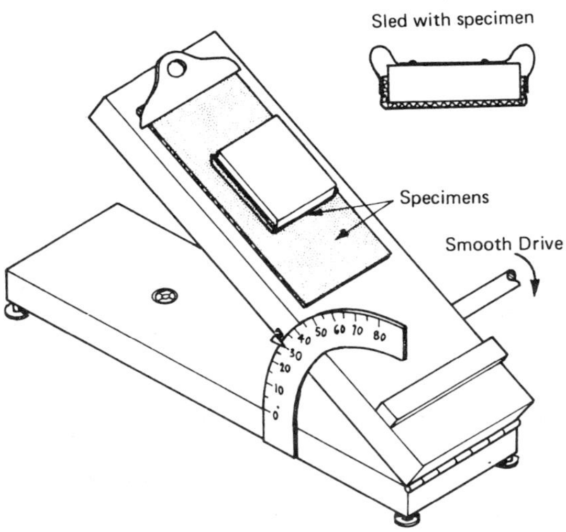

FIG. 2 Schematic of Inclined Plane Apparatus

7.2.1.4 Means to increase smoothly the inclination of the plane from the horizontal through an arc of at least 45° at a rate of 1.5 ± 0.5°/s.

7.2.2 *Procedure:*

7.2.2.1 Level the plane so that it is horizontal when the inclinometer indicates zero.

7.2.2.2 Mount the larger sheet on the plane with the surface to be tested upward. Clip the smaller sheet to the sled, and position it on top of the larger sheet with the surface to be tested facing downward.

7.2.2.3 Incline the plane at the specified rate. When the sled starts to move, stop the inclinator immediately and allow the sled to slide to the stop. Carefully lift the sled and attached sheet, and place the assembly in its original starting position with the plane in its horizontal position. Repeat this for a total of three slides, and read the angular displacement at the moment the sled starts to move on its third slide.

7.2.2.4 Repeat the procedures in 8.2.1 to 8.2.3 on the other four specimens.

7.2.3 *Calculation*—Calculate the tangent of the angle at which sliding begins on the third slide. This is the coefficient of static friction.

7.2.4 *Precision and Bias:*

7.2.4.1 *Repeatability*—The repeatability is ±0.026 (tangent of 1.5°). This is 5.6 to 8.1 % of the average in the range of values usually obtained for corrugated board.

7.2.4.2 *Reproducibility*—The reproducibility is not known.

7.2.4.3 *Comparability*—The comparability is not known (in accordance with the definitions of these terms in Practice E 691).

8. Report

8.1 The report for both methods shall include the following information:

8.1.1 Average coefficient of static friction of the five specimens with the maximum and minimum values for each combination of MD and CD directions tested, and

8.1.2 If any side other than the top side of the board was tested (see 7.1.2.3 and 7.1.2.4).

8.1.3 All test data, including statistical analysis and standard deviation.

8.1.4 Test conditions.

APPENDIX

(Nonmandatory Information)

X1. RELATION TO KINETIC COEFFICIENT

X1.1 Scope

X1.1.1 The kinetic coefficient is not measured in this test method. In the inclined plane method, the kinetic coefficient is not as easily determined to the same precision as the static coefficient. The static coefficient is usually of greater importance.

X1.2 Significance and Use

X1.2.1 For most corrugated and solid fiberboard and their components, the static and kinetic coefficients are usually related to each other (that is, a surface with high static coefficient will usually have a high (relatively) kinetic coefficient). Not all surfaces, particularly those with some surface treatment such as wax, retain this relationship. Therefore, if it is desirable to know the kinetic coefficient, it should be measured. Test Method D 2534 describes how to measure the kinetic coefficient for waxed coatings.

X1.3 Procedure for Testing Corrugated and Solid Fiber Containers

X1.3.1 With most instruments it is not possible to test actual containers unless they are unusually small; however, some inclined plane instruments are large enough to test whole containers. With such equipment, perform the test as follows:

X1.3.2 The instrument characteristics should conform to those stated above for the inclined plane instrument.

X1.3.3 Clamp one board specimen (top of one box) to the inclined plane, or if friction is to be measured against another surface, such as metal or a conveyor belt. Clamp that material to the inclined plane.

X1.3.4 Place a container on top of the test surface. Orientation should correspond to the orientation of the two surfaces in actual use.

X1.3.5 Raise the inclined plane just until sliding begins. Repeat for three slides.

X1.3.6 On the third slide, record the angle for which sliding just begins on the third slide.

X1.3.7 Report the coefficient of static friction that is the tangent of that angle.

Standard Test Method for Compression Resistance of a Container Under Constant Load[1]

This standard is issued under the fixed designation D 4577; the number immediately following the designation indicates the year of original adoption or, in the case of revision, the year of last revision. A number in parentheses indicates the year of last reapproval. A superscript epsilon (ε) indicates an editorial change since the last revision or reapproval.

1. Scope

1.1 This test method is designed to determine the resistance of a shipping container to a vertically applied constant load for either a specified time or to failure. The test method may also, include, for example, a package of a shipping container, palletized or unitized loads.

1.2 *This standard does not purport to address all of the safety problems associated with its use. It is the responsibility of whoever uses this standard to consult and establish appropriate safety and health practices and determine the applicability of regulatory limitations prior to use.* For specific precautionary statements, see Section 6.

2. Referenced Documents

2.1 *ASTM Standards:*

D 642 Method of Compression Test for Shipping Containers[2]

D 644 Test Method for Moisture Content of Paper and Paperboard by Oven Drying[2]

D 685 Method of Conditioning Paper and Paper Products for Testing[2]

D 996 Terminology of Packaging and Distribution Environments[2]

D 2016 Test Methods for Moisture Content of Wood[3]

3. Definitions

3.1 *load*—the force in newtons applied to a body.

3.2 *static load*—an imposed stationary force, constant in magnitude, direction, and sense.

3.3 *constant load*—a load that is invariable or unchanging.

3.4 For other definitions, refer to Terminology D 996.

4. Significance and Use

4.1 In the distribution system for many products there is a phase wherein the packaged product may be stored for a period of time in a manner such that one or more containers are superimposed one upon the other. The bottom package is thus continually stressed with a constant load.

4.2 This test method subjects a container, empty or filled, to a predetermined static load, and to specified atmospheric conditions, if required.

5. Apparatus

5.1 The testing apparatus shall be capable of imposing a constant load on the test specimen and may be hydraulically, pneumatically, or mechanically activated, similar to that shown in Figs. 1 and 2. It shall contain two platens, or suitable framework and fixturing, one stationary and one movable in the vertical direction. The one platen should be swiveled and should have proper mechanical, pneumatic, or hydraulic linkages to permit top-to-bottom loading. If the floor where the test is to be conducted is subject to severe vibration, it may be necessary to vibration-isolate the test apparatus. The test device should have a timer for measuring the period of time required to cause container failure and means such as a dial indicator to measure box deformation (inches or millimetres) while under load, or an autographic recording device that records load and deformation over a period of time.

5.2 *Closing Equipment for Fiberboard Boxes*—When empty boxes are to be tested, suitable closing facilities such as sealing boards and proper adhesive for closing the flaps of box specimens shall be used. See Method D 642.

5.3 *Conditioning Apparatus*—Adequate facilities shall be provided to maintain a conditioned atmosphere of temperature and humidity as required for the purpose of the test.

5.4 *Miscellaneous Equipment*—Drying oven, scales, knife, saws, etc., for use in determination of the moisture content or for making other supplementary tests of the materials from which the containers are made.

6. Safety Precautions

6.1 Performance of a test should never be considered without regard to safety. Some apparent precautions against injuries are:

6.1.1 Care and caution should be observed while placing the shipping container filled or unfilled on the testing apparatus.

6.1.2 The testing apparatus should have load arrestors or safety interlocks to prevent complete crushing of the container after initial failure.

7. Test Specimens and Number of Tests

7.1 The containers being tested shall be complete in all respects. Depending on the purpose of the test, interior packing may or may not be included. No related bracing material within the boxes that will give false results as to sample behavior shall be used. Tests may be made on containers with or without contents as prescribed. Packed containers should be closed and secured in the same manner

---

[1] This test method is under the jurisdiction of ASTM Committee D-10 on Packaging and is the direct responsibility of Subcommittee D10.23 on Storage Test Methods.

Current edition approved March 27, 1986. Published May 1986.

[2] *Annual Book of ASTM Standards*, Vol 15.09.

[3] *Annual Book of ASTM Standards*, Vol 04.09.

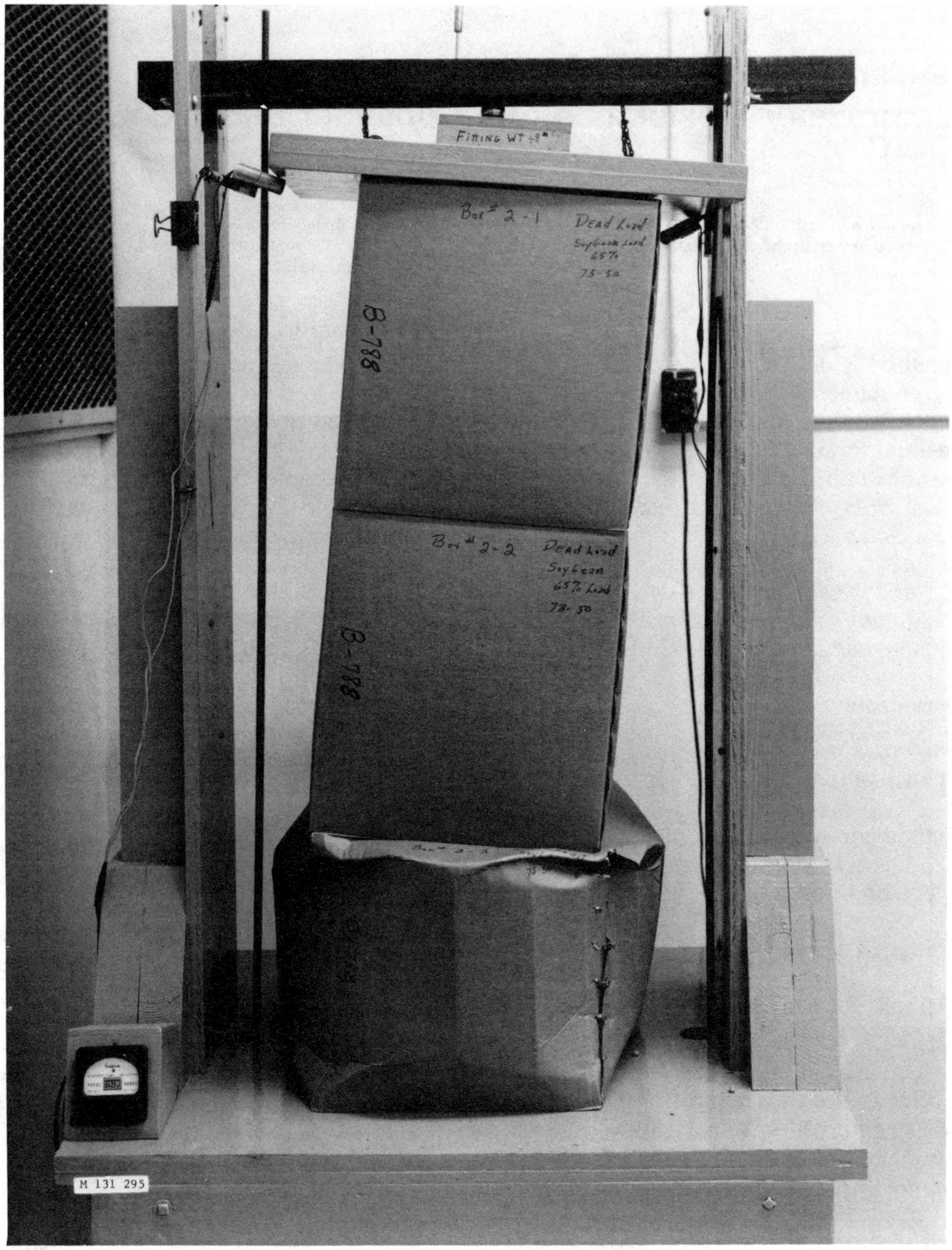

FIG. 1 Containers Under Constant Load

as will be used in preparing them for shipment (for example, tape, strapping).

7.2 Performance normally should be based on tests of not fewer than five representative specimens of a given size and type of container.

8. Closing Fiberboard Containers Using Adhesive

8.1 Close the box specimen so as to avoid distortions that may affect its loadbearing ability. The method of preparing the test specimen as described in the Annex of Method D 642 will accomplish this, but any method that will produce the same results may be used.

9. Conditioning

9.1 When required, the container should be conditioned for the static load test by exposure to fixed or controlled variable conditions of temperature and humidity.

9.2 The test container shall be preconditioned in the desired atmosphere for such a time as is necessary to bring the container into equilibrium with that atmosphere and using the techniques given in Method D 685.

10. Acceptance Criteria

10.1 Acceptance criteria must be established prior to testing and should consider the required condition of the product at receipt. The organizations conducting the test may choose any acceptance criteria suitable for their pur-

FIG. 2 Individual Containers Under Constant Load

pose. It is advisable to compare test results from proposed containers with the test results on similar containers whose shipping history is known.

10.2 In many cases, the acceptance criteria of a package that has been subjected to the test plan can be one of the following:

Criterion 1—Product is damage-free.
Criterion 2—Package is intact.

Other acceptance criteria, including provision for accepting minimal damage to the product or package, may be indicated. Acceptance criteria may include a provision for the condition of package. The form and content of acceptance criteria may vary widely, according to the particular situation. Criteria may range from the most rudimentary to highly quantitative scoring or analysis systems.

11. Procedure

11.1 Prior to test, select the constant load to which the container will be subjected. The selection may be based on:

11.1.1 The load to which a bottom container will be subjected to in storage, for example, warehouse.

11.1.2 A percentage of the value obtained by Method D 642 on a similar container.

NOTE 1—For corrugated containers, the work of Kellicutt and Landt[4] may be used as a guide to the selection of the percentage but at no time should the dead load exceed 85 % of the test value derived by Method D 642 testing.

11.2 Center the specimen on the bottom platen of the testing apparatus so as not to incur eccentric loading. Induce the test load on the specimen. The load shall be slowly applied at a near uniform rate until the container supports the entire load. When using dead weights (Figs. 1 and 2), this is accomplished by lowering the support jacks and completely freeing the upper platen. In a controlled fixture, set it to a constant load control mode. The timing device is then started.

11.3 An initial observation and record of zero deformation shall begin at the end of 60 s after full load to establish base-time reference. Additional observations will be made after 5 min, 10 min, 30 min, 60 min, and 2 h. Thereafter, observations and reading of deformation shall be made at predetermined intervals, until the specified test period is over, or the container fails.

11.4 *Moisture Content (When Specified):*

11.4.1 *Fiberboard Containers*—Determine the moisture content of fiberboard at the end of the test in accordance with Test Method D 644.

11.4.2 *Wood Container*—Determine the moisture content of wood at the end of the test in accordance with Methods A or B of Test Methods D 2016.

12. Calculation

12.1 This test produces time-to-failure data that does not conform to the normal or Gaussian distribution: The logarithms of the data, natural or common, do tend to be normally distributed. Before analysis, transform the data using the following equation:

$$C = \log D$$

where:
C = transformed datum, and
D = observed datum.

12.2 The arithmetic mean of the transformed data is the recommended measure of central tendency. The antilog of this average is a median estimate for the observed data.

12.3 The standard deviation of the transformed data is the recommended measure of variability or dispersion. Statistical calculations should use this standard deviation directly, without transforming it to observed units of time. The antilogs of upper and lower confidence limits will be in observed units of time and will not be symmetrical about the median estimate. (See Annex A1.)

NOTE 2—Calculations based on the Weibull distribution are an acceptable alternative to these procedures. (See Annex A2.)

13. Report

13.1 The report shall include the following:

13.1.1 Dimensions of the container under test; its complete structural specifications; kind of material; description and specification of blocking and cushioning if used; spacing, size, and kind of fasteners; method of closing; and net and gross mass.

13.1.2 Description of the contents of the container, if any.

13.1.3 Acceptance criteria that have been established.

13.1.4 A detailed report of the test on each container, including damage to the container and contents, together with any observations that may assist in correctly interpreting results, or aid in improving the design of the container.

13.1.5 A graph or table showing the deflection versus time for each test.

13.1.6 The method of conditioning the container including the temperature and relative humidity of the conditioning atmosphere, the moisture content of the material (where applicable), and the results of any supplementary test of the materials from which the container was made.

13.1.7 Report the number of containers tested, and the mean, and standard deviation in transformed units, and the median estimate in observed units of time.

13.1.8 The compression test results from Method D 642.

13.1.9 A statement to the effect that all tests were performed in full compliance with this test method noting any variations.

14. Precision and Bias

14.1 The precision and bias of this test method has not yet been determined. Data are being sought that will be suitable for use in developing precision statements.

[4] Kellicut, K. Q., and Landt, E. F., "Safe Stacking Life of Corrugated Boxes," *Fibre Containers*, Vol 36, No. 9, Sept. 1951.

ANNEXES

(Mandatory Information)

A1. EXAMPLE OF CALCULATIONS BASED ON LOGARITHMIC CONVERSION

A1.1 The following example demonstrates the method of calculation for time-to-failure data described in the test method. If ten identical containers were tested with the same load, some variation in the times-to-failure for the containers would be expected. For example, a set of ten data points might be: 5.26, 13.56, 6.74, 16.03, 25.23, 11.43, 1.84, 10.97, 19.18, and 8.08 days to failure. It is convenient to list the time to failure (Y) in one column and its logarithm (X) in another column as shown below:

Y	$X = \log(Y)$
5.26 days	0.72
13.56	1.13
6.74	0.83
16.03	1.20
25.23	1.40
11.43	1.06
1.84	0.26
10.97	1.4
13.18	1.12
8.08	0.91

Standard statistical calculations are conducted on the log data (X), which are assumed to be from a normally distributed population. For this example, the average (arithmetic mean) and the sample standard deviation of the log data (X) are calculated to be $\bar{X} = 0.966$ and $s = 0.314$, as estimates of the population parameters. The antilog of the average is 9.25 days and is reported as the central tendency. Other standard statistical calculations also would be based on the log data (X).

A2. EXAMPLE OF WEIBULL CALCULATIONS

A2.1 The Weibull distribution can be fit to the same data from the example in A1. Graphical as well as several computer aided techniques are available to estimate the population parameters. A computer analysis of this set of data yielded an Alpha Scale Parameter of 12.66 and a Beta Shape Parameter of 1.89. From these it would be estimated that 10 % of the population of containers would fail by 3.85 days and 50 % of the containers would fail by 10.43 days.

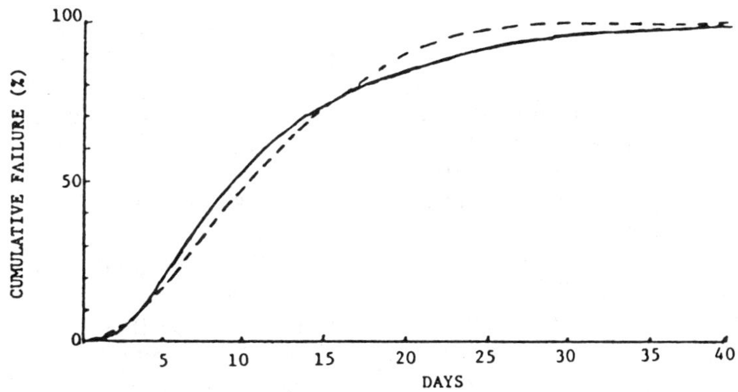

NOTE—The solid line represents the log-normal distribution

FIG. A2.1 Comparative Plots of the Cumulative Failure versus Time for the Example

Designation: D 4635 – 86$^{\epsilon 1}$

Standard Specification for Low-Density Polyethylene Films for General Use and Packaging Applications[1]

This standard is issued under the fixed designation D 4635; the number immediately following the designation indicates the year of original adoption or, in the case of revision, the year of last revision. A number in parentheses indicates the year of last reapproval. A superscript epsilon (ε) indicates an editorial change since the last revision or reapproval.

$^{\epsilon 1}$ NOTE—Table 1 was corrected editorially in February 1991.

1. Scope

1.1 This specification covers unpigmented, unsupported, tubular, low-density polyethylene films (hereafter referred to as film or films) having densities in the range from 910.0 to 925.0 kg/m^3 (0.910 to 0.925 g/cm^3) inclusive, as measured on molded plaques.

NOTE 1—Blends of ethylene/vinylacetate (EVA) with low-density polyethylene may have densities up to 0.929 g/cm^3.

1.2 This specification is applicable to homopolymer polyethylene, but is not restricted to it. It is applicable to films made from copolymer polyethylene commonly referred to in industry as low-pressure or linear low-density polyethylene, and also applicable to films made from blends of homopolymers and copolymers, including ethylene/vinylacetate copolymers.

1.3 The thickness of the films covered by this specification is 100 μm or less (0.004 in. or less) inclusive. The maximum width of the sheet or layflat is 3.00 m (120 in.).

1.4 This specification does not cover oriented heat-shrinkable films.

1.5 This specification defines the levels of the various physical properties from which specifications for specific films may be described. The levels of physical properties required by a film for a given application are selected from Section 6. However, Sections 2 through 5 relating to tolerances shall apply without change to all film falling within the scope indicated by the title and 1.1, 1.2, 1.3, and 1.4.

1.6 This specification covers dimensional tolerances, classifications, intrinsic quality requirements, and test methods. The dimensional tolerances include thickness, width, and length or yield. Classification defines types, classes, surfaces, and finishes. The intrinsic quality requirements include density, workmanship, tensile strength, heat sealability, and odor, as well as the classification properties for impact strength, coefficient of friction, optical properties, and surface treatment. A sampling method is included.

1.7 The values stated in SI units are to be regarded as the standard. The values in parentheses are given for information only.

1.8 The following precautionary caveat pertains only to the test methods portion, Section 10, of this specification: *This standard may involve hazardous materials, operations, and equipment. This standard does not purport to address all of the safety problems associated with its use. It is the responsibility of the user of this standard to establish appropriate safety and health practices and determine the applicability of regulatory limitations prior to use.*

2. Referenced Documents

2.1 *ASTM Standards:*

D 618 Methods of Conditioning Plastics and Electrical Insulating Materials for Testing[2]

D 882 Test Methods for Tensile Properties of Thin Plastic Sheeting[2]

D 883 Definitions of Terms Relating to Plastics[2]

D 1003 Test Method for Haze and Luminous Transmittance of Transparent Plastics[2]

D 1248 Specification for Polyethylene Plastics Molding and Extrusion Materials[2]

D 1505 Test Method for Density of Plastics by the Density-Gradient Technique[2]

D 1709 Test Method for Impact Resistance of Polyethylene Film by the Free Falling Dart Method[3]

D 1894 Test Method for Static and Kinetic Coefficients of Friction of Plastic Film and Sheeting[3]

D 1898 Practice for Sampling of Plastics[3]

D 2103 Specification for Polyethylene Film and Sheeting[3]

D 2457 Test Method for Specular Gloss of Plastic Films[3]

D 2578 Test Method for Wetting Tension of Polyethylene and Polypropylene Films[3]

D 4321 Test Method for Package Yield of Plastic Film[4]

E 462 Test Method for Odor and Taste Transfer from Packaging Film[5]

F 88 Test Methods for Seal Strength of Flexible Barrier Materials[6]

2.2 *Military Standard:*

MIL-STD 105 Sampling Procedures and Tables for Inspection by Attributes[7]

NOTE 2—Relevant government regulations also apply to this specification.

---

[1] This specification is under the jurisdiction of ASTM Committee D-20 on Plastics and is the direct responsibility of Subcommittee D20.19 on Sheeting.

Current edition approved Oct. 31, 1986. Published December 1986.

[2] *Annual Book of ASTM Standards*, Vol 08.01.

[3] *Annual Book of ASTM Standards*, Vol 08.02.

[4] *Annual Book of ASTM Standards*, Vol 08.03.

[5] *Annual Book of ASTM Standards*, Vol 15.07.

[6] *Annual Book of ASTM Standards*, Vol 15.09.

[7] Available from Naval Publications and Forms Center, 5801 Tabor Ave., Philadelphia, PA 19120. ATTN: NPFC 105.

NOTE 3—If this product is intended for packaging foods, medicines, drugs, and cosmetics it is subject to applicable regulations of the Food and Drug Administration or the Department of Agriculture and must comply with such regulations. If it is necessary to comply with regulations of other government agencies, such as the Consumer Product Safety Commission, Environmental Protection Agency, Department of Transportation, or others, such compliance should be arranged between the buyer and the seller prior to placing an order.

3. Terminology

3.1 *General*—Unless otherwise indicated the terminology used in this specification is in accordance with Definitions D 883.

4. Classification

4.1 The low-density polyethylene film is, by this specification, classified by Types 1, 2, and 3; Surfaces 1, 2, and 3; Classes 1, 2, 3, and 4; and Finishes 1, 2, 3, and 4. These classifications are described in detail in 6.1.

5. Materials

5.1 The film shall be made from an ethylene homopolymer, ethylene copolymers, or blends of homopolymers or copolymers, or homopolymer and copolymer, so that it meets the density and other film requirements listed herein.

5.2 The density of the film shall be between 910.0 and 925.0 kg/m^3 (0.910 and 0.925 g/cm^3) inclusive.

5.3 The film shall be natural in color (essentially colorless).

6. Physical Requirements

6.1 *Classification Properties:*

6.1.1 *Type*—The dart drop impact for all thickness of film shall be as specified in Table 4 for Types 1, 2, and 3.

6.1.2 *Surface*—The kinetic coefficient of friction shall be as specified in Table 5 for Surfaces 1, 2, and 3.

6.1.3 *Class*—The optical properties shall be as specified in Table 6 for Classes 1, 2, and 3. The optical properties of gloss and haze do not always correlate. The particular property of most importance for the specific application shall be established and the value for this property shall then govern, in case of any inconsistency.

6.1.4 *Finish*—The surface treatment level of the film shall be as specified in Table 7 for Finishes 1, 2, 3, and 4.

6.2 *Other Properties:*

6.2.1 *Tensile Properties*—The tensile strength and elongation at break for all thicknesses shall be as specified in Table 8.

6.2.2 *Heat Sealability*—The minimum ratio of heat-seal strength to the film strength in the two principal directions shall be as specified in Table 9.

6.2.3 *Odor*—The odor level of the film shall average no more than a 3.5 rating level.

7. Dimensions

7.1 *Size*—The nominal thickness, width, length per roll or roll diameter, and yield of the film shall be established by mutual agreement between customer and supplier.

7.2 *Thickness Tolerance*—The thickness variation across the film shall be within the tolerances given in Table 1.

7.3 *Width Tolerance*—The width shall be within the tolerances given in Table 2.

7.4 *Yield Tolerance*—The deviation of the actual yield from nominal yield shall be within the tolerances given in Table 3.

7.5 *Flatness*—The flatness of the film shall be within limits as mutually agreed upon between buyer and seller.

7.6 Dimension tolerances for "J-Sheeting" lip and gusset depth shall be established by mutual agreement between customer and supplier.

8. Workmanship, Finish, and Appearance

8.1 *Film*—The film shall have workmanship qualities conforming to good commercial practice. The quality of film with regard to gels, streaks, pinholes, particles of foreign matter, scratches, wrinkles, wind chatter, undispersed raw materials, holes, tears, and blisters shall be mutually established by customer and supplier.

8.2 *Roll Formation:*

8.2.1 The diameter of cores upon which film is wound shall be established by mutual agreement between customer and supplier. Cores upon which film is wound must not be recessed at either edge of the roll, but may extend up to 6 mm (¼ in.) beyond either edge of the roll. Rolls with cores that are crushed and are not able to be mounted on customers' equipment are to be considered rejects.

8.2.2 Ridges and soft spots which result in bagginess and looseness of the unwound film are unacceptable and may be rejected by the customer on a roll-to-roll basis if the conditions contribute to poor performance of the film in end-use application.

8.2.3 The edges of the roll must be free of nicks and cuts and the general condition of roll edges must not interfere with the unwinding of the rolls.

8.2.4 The type and number of splices, flaggings of splices, or breaks (if unspliced) in rolls of more than one piece shall be established by mutual agreement between the supplier and purchaser.

9. Sampling

9.1 Statistically based sampling plans which are appropriate for each particular product or quantity may be used to obtain samples for use in determining compliance with this specification. Refer to Practice D 1898 for guidance.

9.2 For the purposes of developing seller-buyer specifications a lot size generally refers to the number of rolls in a lot. Sampling units are those rolls selected by random numbers from the lot. A unit sample is the sample of film taken from the roll. Care must be exercised in taking unit samples. Unwind and then discard several turns from the roll and then take more than enough sample to run all specified tests. Keep the sample from becoming soiled. Ensure that the sample is not folded or creased excessively.

10. Test Methods

10.1 *Conditioning*—Condition the samples or test specimens, or both, at 23 ± 2°C (73.4 ± 3.6°F) before testing. In cases of disagreement condition the test specimens for not less than 40 h prior to test, in accordance with Procedure A of Methods D 618 for those tests where conditioning is required.

TABLE 1 Tolerance, Percent from Nominal Thickness[A]

Film Width		Nominal Thickness		Across Film[B] % Tolerance
mm	in.	µm	in.	
1250 or less	50 or less	25 to 65	0.001 to 0.004	± 20
>1250 to 3000	>50 to 120	25 to 65	0.001 to 0.004	± 25

[A] Use Table 3 as the controlling table for average gage in terms of yield. For minimum gage requirements order film specifying nominal gage greater than the required minimum by at least the percent tolerance set above.

[B] No single measurement shall differ from the nominal gage by more than the tolerance listed in Table 1.

TABLE 2 Width Tolerances[A]

Film Widths	Sheeting	Layflat Tubing
375 mm (15 in.) or less	−0, + 5 mm (3/16 in.)	± 5 mm (3/16 in.)
>375 to 750 mm (15 to 30 in.)	−0, + 6 mm (1/4 in.)	± 10 mm (3/8 in.)
>750 to 1.5 m (30 to 60 in.)	−0, + 10 mm (3/8 in.)	± 16 mm (5/8 in.)
>1.5 to 3 m (60 to 120 in.)	−0, + 13 mm (1/2 in.)	± 25 mm (1 in.)

[A] Across sheet or layflat tubing.

TABLE 3 Deviation,[A] Actual Yield from Nominal Yield

Quantity	Tolerances
Any one roll	± 10 %
500 kg (1000 lb) or less	± 10 %
>500 to 1000 kg (1000 to 2500 lb)	± 5 %
Over 1000 kg (2500 lb)	± 3 %

[A]Negative deviation generally infers that the average page is greater than nominal. Positive deviation generally infers that the average gage is less than nominal.

TABLE 4 Classification for Type

Film Thickness		Drop Dart (g, min)[A]		
µm	in.	Type 1	Type 2	Type 3
<25	not specified	not specified	not specified	not specified
25	0.001	40	75	105
38	0.0015	65	105	140
50	0.002	85	135	175
75	0.003	125	195	245
100	0.004	165	255	315

[A] Impact limits for thickness not covered in the table will be determined by linear interpolation between successive values in table.

TABLE 5 Classification for Surface

Surface	Coefficient of Friction
1	>0.5
2	>0.2 to 0.5
3	0.2 or less

10.2 *Test Conditions*—Conduct tests at the Standard Laboratory Temperature of 23 ± 2°C (73.4 ± 3.6°F).

10.3 *Width*—Measure width with a metal rule capable of measuring to an accuracy of ±1 mm (1/16 in.).

10.4 *Thickness*—Measure thickness in accordance with the General Test Methods section of Specification D 2103, with the modification that the specimen shall be taken as a 50-mm (2-in.) strip across the web and measured at intervals of not more than 50 mm (2 in.).

10.5 *Yield*—See Test Method D 4321.

10.6 *Flatness*—Measure flatness using a method mutually agreed upon between buyer and seller.

TABLE 6 Classification for Classes

Class	Gloss Units	Haze, %
1	30 or less	>25
2	>30 to 50	>10 to 25
3	>50 to 70	>5 to 10
4	>70	0 to 5

TABLE 7 Classification for Finish

Finish	Wetting Tension, mN/m (dynes/cm)
1	32, 33, 34
2	35, 36, 37
3	38, 39, 40
4	41 and over

TABLE 8 Tensile Properties

Property	Units	Machine Direction	Transverse Direction
Tensile Strength, min	MPa (psi)	11.7 (1700)	8.3 (1200)
Tensile Elongation, min	%	225	350

TABLE 9 Heat Sealability[A]

Finish of Contact Surfaces[B]	Heat Sealability, min
2 to 2	0.60
1 to 2	0.60
1 to 1	0.75

[A] Heat sealability is the ratio of the tensile strength of the heat-sealed specimen to the tensile strength of the original film specimen.

[B] Heat sealability is not applicable to films with finish greater than two; this does not infer that films with finishes greater than two cannot be sealed.

10.7 *Density*—Measure density of the film in accordance with Test Method D 1505.

10.8 *Coefficient of Friction*—The kinetic coefficient of friction shall be measured in accordance with Test Method D 1894. The test shall be conducted film to film, both in the machine direction, and transverse direction.

10.9 *Optical Properties:*

10.9.1 *Clarity*—Measure clarity of the film by visual or instrumented means, as agreed upon by the customer and supplier.

10.9.2 *Gloss*—Measure gloss of the film in accordance with Test Method D 2457, using a 45° gloss head.

10.9.3 *Haze*—Measure haze in accordance with Test Method D 1003.

10.10 *Wetting Tension*—Measure wetting tension in accordance with Test Method D 2578.

10.11 *Impact Resistance*—Measure impact resistance in accordance with Test Method D 1709, except that Test Method A shall be used for all gages of film.

10.12 *Tensile Properties*—Measure tensile strength and elongation at break in accordance with Test Methods D 882.

10.13 *Heat Sealability*—Measure heat sealability in accordance with Test Methods F 88, Test Method B, Dynamic Load Test.

10.14 *Odor*—Measure odor level in accordance with Test Method E 462, low to moderate scale.

11. Inspection

11.1 Inspection of the material shall be agreed upon by the customer and supplier as part of the purchase contract.

12. Rejection and Rehearing

12.1 Material that fails to conform to the requirements of this specification may be rejected. Rejection should be reported to the producer or supplier promptly and in writing. In case of dissatisfaction with the results of the test, the producer or supplier may make claim for a rehearing.

13. Certification

13.1 When specified in the purchase order or contract, a certified report of the results shall be furnished.

14. Packaging and Package Marking

14.1 *Labels and Literature*—So that purchasers may know that the film has been manufactured in compliance with this specification, it is suggested that the supplier include an appropriate statement to this effect on labels, invoices, or another place.

14.2 Identification of the film shall include:

14.2.1 *Manufacturer's Name,*

14.2.2 *Type,*

14.2.3 *Surface,*

14.2.4 *Class,*

14.2.5 *Finish,* (if treated, the treated side of the film shall be clearly identified), and

14.2.6 Reference to this specification number.

QUALITY ASSURANCE PROVISIONS FOR GOVERNMENT/MILITARY PROCUREMENT

These requirements apply *only* to Federal/Military procurement, not domestic sales or transfers.

S1. Sampling for inspection and testing shall be carried out in accordance with the Practice D 1898.

S2. Selection of Acceptable Quality Level (AQL) and of Inspection Level (IL) shall be made, with consideration of the specific use requirements. This is discussed in Practice D 1898, in the sections "Means and Standard Deviations" and "Comparison of Sampling Plans" with reference to MIL-STD 105.

S3. In the absence of contrary requirements, the following values shall apply:

	IL	AQL
Defects of appearance and workmanship	II	2.5
Defects of preparation of delivery	S-2	4.0
Testing (products)	S-1	1.5

Designation: D 4727 – 89

Standard Specification for Corrugated and Solid Fiberboard Sheet Stock (Container Grade) and Cut Shapes[1]

This standard is issued under the fixed designation D 4727; the number immediately following the designation indicates the year of original adoption or, in the case of revision, the year of last revision. A number in parentheses indicates the year of last reapproval. A superscript epsilon (ε) indicates an editorial change since the last revision or reapproval.

1. Scope

1.1 This specification covers fiberboard primarily used for the fabrication of boxes and interior details such as pads, sleeves, liners, partitions, die-cut sheets, etc.

1.2 The performance of fiberboard boxes is largely dependent on the paper components from which they are fabricated and, in the case of corrugated boxes, on the flute structure as well. Therefore, a variety of grades reflecting varied performance levels are specified.

1.3 The values stated in either inch-pound units or SI units are to be regarded separately as standard. Within the text, the SI units are shown in brackets. The values stated in each system are not exact equivalents; therefore, each system must be used independently of the other. Combining values from the two systems may result in non-conformance with the standard. See Practice E 380 for conversion of units.

1.4 The following safety hazards caveat pertains only to the test portion, Sections 8 and 9, of this specification: *This standard does not purport to address the safety problems associated with its use. It is the responsibility of the user of this standard to establish appropriate safety and health practices and determine the applicability of regulatory limitations prior to use.*

2. Referenced Documents

2.1 *ASTM Standards:*

D 585 Method for Sampling and Accepting a Single Lot of Paper, Paperboard, Fiberboard or Related Product[2]

D 685 Method for Conditioning Paper and Paper Products for Testing[2]

D 996 Terminology of Packaging and Distribution Environments[2]

D 3950 Specification for Strapping, Nonmetallic (and Joining Methods)[2]

D 3951 Practice for Commercial Packaging

D 3953 Specification for Strapping, Flat Steel and Seals[2]

E 162 Test Method for Surface Flammability of Materials Using a Radiant Heat Energy Source[3]

E 380 Practice for the Use of the International System of Units (SI) (The Modernized Metric System)[4]

E 662 Test Method for Specific Optical Density of Smoke Generated by Solid Materials[3]

2.2 *TAPPI Standards:*

T 410 Weight per Unit Area (Basis Weight or Substance)[5]

T 411 Thickness (Caliper) of Paper and Paperboard[5]

T 441 Water Absorptiveness of Sized (Non-Bibulous) Paper and Paperboard (Cobb Test)[5]

T 803 Puncture and Stiffness Test of Container Board[5]

T 810 Bursting Strength of Corrugated and Solid Fiberboard[5]

T 812 Ply Separation of Solid and Corrugated Fiberboard (Wet)[5]

2.3 *Federal Specifications:*

PPP-B-640 Boxes Fiberboard, Corrugated, Triplewall[6]

2.4 *Other Publications:*

National Motor Freight Classification[7]

Uniform Freight Classification[8]

3. Terminology

3.1 General definitions for packaging and distribution environments are found in Terminology D 996.

4. Classification

4.1 *Type*—Corrugated fiberboard (CF).

4.1.1 *Classes*—Domestic and domestic/fire retardant (D/FR).

4.1.1.1 *Variety*—Singlewall (SW).

(*a*) *Grades*—125, 150, 175, 200, 275, and 350.[9]

4.1.1.2 *Variety*—Doublewall (DW).

(*a*) *Grades*—200, 275, 350, 500 and 600.[9]

4.1.1.3 *Variety*—Triplewall (TW).

(*a*) *Grade*—1100.[10]

4.1.2 *Classes* —Weather-resistant and weather-resistant/fire retardant (WR/FR).

4.1.2.1 *Variety*—Singlewall (SW).

(*a*) *Grades*—V3c, W5c and W6c.

4.1.2.2 *Variety*—Doublewall (DW).

(*a*) *Grade*—V11c, V13c and V15c.

---

[1] This specification is under the jurisdiction of ASTM Committee D-10 on Packaging and is the direct responsibility of Subcommittee D10.17 on Paper and Paperboard Products.

Current edition approved Oct. 27, 1989. Published December 1989. Originally published as D 4727 – 87. Last previous edition D 4727 – 87.

[2] *Annual Book of ASTM Standards*, Vol 15.09.

[3] *Annual Book of ASTM Standards*, Vol 04.07.

[4] *Annual Book of ASTM Standards*, Vol 14.02.

[5] Available from the Technical Association of the Pulp and Paper Industry (TAPPI), P.O. Box 105113, Atlanta, GA 30348.

[6] Available from Naval Publications and Forms Center, 5801 Tabor Avenue, Philadelphia, PA 19120.

[7] National Motor Freight Classification is available from National Motor Freight Classification, Inc., 2200 Mill Road, Alexandria, VA 22314.

[8] Uniform Freight Classification is available from National Railroad Freight Committee, Suite 1120, 222 South Riverside Plaza, Chicago, IL 60606.

[9] Grade numbers for single and doublewall corrugated varieties and solid fiber types refer to the bursting strength in lbs./in.^{2} as determined by TAPPI T 810.

[10] Grade number for triplewall refers to the dry puncture resistance in units as determined by TAPPI T 803.

4.1.2.3 *Variety*—Triplewall.

(*a*) *Grade*—1100.[10]

4.1.3 *Classes*—Water and water vapor resistant (WWVR) and water and water vapor resistant/fire retardant (WWVR/FR).

4.1.3.1 *Variety*—Singlewall (SW).

(*a*) *Grades*—V3c WWVR, W5c WWVR.

4.1.3.2 *Variety*—Doublewall (DW).

(*a*) *Grades*—V11c WWVR and V13c WWVR.

4.2 *Type*—Solid Fiberboard (SF).

4.2.1 *Class*—Domestic (D).

4.2.1.1 *Grades*—125, 175, 200, 275, 350, 500 and 600.[9]

4.2.2 *Class*—Weather-resistant.

4.2.2.1 *Grades*—V2s, V3s, V4s, W5s and W6s.

5. Ordering Information

5.1 Purchasers should select the preferred options permitted herein, and include the following information in procurement documents:

5.1.1 Title, number and date of this specification,

5.1.2 Type, class, variety and grade (see Section 4 and 7.1),

5.1.3 Size of sheet, pad, etc. (see 7.3.1.4),

5.1.4 Center line of score to center line of score or sheet edge (see 7.3.1.4),

5.1.5 Flute design of type corrugated fiberboard; variety SW and DW (see 6.4 and 6.4.1),

5.1.6 Mill run or trimmed sheets,

5.1.7 Dimensional direction of flutes of type CF, fiberboard,

5.1.8 Packing and marking requirements (see S3 and S4), and

5.1.9 Palletization requirements (see S3.1.1.1).

6. Materials and Manufacture

Materials

6.1 *Paperboard Components:*

6.1.1 *Facings and Outer Plies*—The facings of corrugated fiberboard and the plies of solid fiberboard shall have bending qualities to satisfy the requirements of 7.6. The paperboard components of weather-resistant class material shall be treated with a suitable wet strength resin to make them water resistant. Similarly, the paperboard components of fire retardant class material shall be treated to make them fire retardant to meet the requirements of 7.7.1 and 7.7.2.

6.1.1.1 *Outer Facing of Corrugated Fiberboard, Water-and Water Vapor Resistant Class*—One outer facing of this material shall be a composite sheet comprised of one ply of sized, wet strength kraft linerboard laminated to a ply of linerboard conforming to 6.1.1 with a minimum of 6 lbs/1000 ft^2 (29 g/m^2) of polyethylene. The sized ply shall be on the exterior side of the facing, with the unsized linerboard next to the corrugated medium. At the supplier's option, both plies of linerboard may of a sized wet strength material.

6.1.2 *Corrugated Medium*—The corrugating medium of corrugated fiberboard shall be made from any suitable fibers.

6.2 *Adhesives:*

6.2.1 *Domestic and Domestic/Fire Retardant Classes*—The adhesive used in the construction of the domestic/fire retardant class of fiberboard should be that which is commercially used by the industry, and shall enable the end item to meet the strength requirements specified herein.

6.2.2 *Weather-Resistant and Weather-Resistant/Fire Retardant Classes*—The adhesive used in the construction of weather-resistant fiberboard shall be water-resistant to the extent that the end item will meet the requirements of this specification (see 7.4).

6.2.3 *Water and Water Vapor-Resistant and Water and Water Vapor-Resistant/Fire Retardant Classes*—The adhesives used shall enable the combined board to meet the requirements of 7.4. Paragraph 7.4 also applies to the polyethylene linerboard bonds in the laminated facings.

6.3 *Triplewall and Triplewall/Fire Retardant Corrugated Fiberboard Classes*—Triplewall corrugated fiberboard, domestic and domestic/fire retardant classes, non-weather-resistant and weather-resistant in both regular and fire-retardant classes shall conform to the requirements of Federal Specification PPP-B-640.

Construction

6.4 *Corrugated Fiberboard*—Corrugated fiberboard shall be made with two, three or four facings for single, doublewall or triplewall, respectively. Each facing shall be separated by and securely adhered to the corrugated medium.

6.4.1 *Flutes*—The number of flutes per unit length of fiberboard shall be as follows:

	Flutes/Ft	Flutes/M
A-Flute	33 to 39	108 to 128
B-Flute	47 to 53	154 to 174
C-Flute	39 to 45	128 to 148
E-Flute	90 to 98	295 to 321

6.5 *Solid Fiberboard:* Solid fiberboard shall be constructed by laminating plies securely and continuously together.

6.5.1 *Domestic:*

6.5.1.1 *Grades 125 and 175*—The fiberboard shall consist of not less than two plies of components.

6.5.2 *Grades 200 through 600*—The fiberboard shall consist of not less than three plies of components.

6.6 *Scored Sheets*—The scorelines on scored sheets shall show no visual continuous surface break greater than specified in 7.6 when tested as specified in 9.3.

6.7 *Pads and Cut Shapes*—Pads and cut shapes shall be made in accordance with specified dimensions or drawings.

6.7.1 Pads and cut shapes may be single sheets, or may be made by laminating two or more sheets of specified types and classes of material to the specified thickness.

7. Physical Requirements

7.1 *Corrugated Fiberboard:*

7.1.1 *Domestic and Domestic/Fire Retardant Classes, All Varieties (D/FR):*

7.1.1.1 *Facings*—The facings shall conform to the requirements in Table 1 when tested as specified in 9.1.2.

7.1.1.2 *Corrugating Medium*—The corrugating medium for the domestic class of corrugated fiberboard shall weigh not less than 26 lbs/1000 ft^2 (127 g/m^2) when tested in accordance with 9.1.2.

7.1.1.3 *End Item*—Domestic and domestic/fire retardant classes of corrugated fiberboard, constructed as specified in 6.4, shall conform to the requirements in Table 1 for bursting strength when tested as specified in 9.1.5, except

TABLE 1 Type CF (Corrugated Fiberboard), Domestic and Domestic/Fire Retardant Classes (D/FR)

Variety	Grade	Combined Weight Facings Only, min	Bursting Strength, Dry, min[A]	
		lb/1000 ft² (g/m²)	psi (kPa)	
SW	125	52 (254)	125 (862)	
SW	150	66 (322)	150 (1034)	
SW	175	75 (366)	175 (1207)	
SW	200	84 (410)	200 (1379)	
SW	275	138 (674)	275 (1896)	
SW	350	180 (879)	350 (2413)	
DW	200	92 (449)	200 (1379)	
DW	275	110 (537)	275 (1896)	
DW	350	126 (615)	350 (2413)	
DW	500	222 (1084)	500 (3447)	
DW	600	270 (1318)	600 (4137)	
TW	1100	264 (1289)	Puncture	
			in.·oz/inches of tear	(J)
			1100	(33)

[A] Only one burst may fall beneath the minimum required. Domestic board failing to pass this test will be accepted if, in a retest consisting of 24 bursts (12 from each side of the board), not more than 4 bursts fall below the minimum value required.

that tests shall consist of an initial six bursts (three from each side of the board) and, if required, a supplementary 24 bursts (twelve from each side of the board).

7.1.2 *Weather-Resistant and Weather-Resistant/Fire Retardant Classes*—The components and end items conform to the applicable thickness and bursting strengths specified in Table 2 when tested as specified in 9.1.1 and 9.1.5.

7.1.3 *Water and Water Vapor-Resistant and Water- and Water Vapor-Resistant/Fire Retardant Classes:*

7.1.3.1 *Components*—The components shall conform to the applicable requirements as specified in Table 2 and 6.1.1.1, when tested as specified in 9.1.1.

7.1.3.2 The end items shall conform to the applicable requirements of Table 2 when tested as specified in 9.1.5 and 9.2.1. In addition, the end item shall not absorb more than 1.1 g of water with wetting agent when tested as specified in 9.5.

7.2 *Type SF (Solid Fiberboard):*

7.2.1 *Class Domestic:*

7.2.1.1 *Facings and Filler Plies*—The combined weights of the facings and filler plies shall conform to the requirements in Table 3 when tested as specified in 9.1.2.

TABLE 2 Type CF (Corrugated Fiberboard), Weather-Resistant and Water and Water Vapor Resistant Classes (WWVR)

Variety	Grade[A,C]	Thickness, in. (mm)[B]		Bursting Strength, psi (kPa), min avg	
		Corrugating Medium	Outer Facings	Dry	Wet[D]
SW	V3c	0.010 (0.254)	0.023 (0.584)	400 (2758)	150 (1034)
SW	W5c	0.010 (0.254)	0.016 (0.406)	275 (1896)	100 (689)
SW	W6c	0.010 (0.254)	0.010 (0.254)	175 (1207)	50 (345)
DW	V11c	0.010 (0.254)	0.023 (0.584)	600 (4137)	300 (2068)
DW	V13c	0.010 (0.254)	0.016 (0.406)	400 (2758)	200 (1379)
DW	V15c	0.010 (0.254)	0.010 (0.254)	300 (2068)	100 (689)

[A] Includes WWVR grades.
[B] A −5 %, or unlimited plus tolerance shall be permitted.
[C] For doublewall fiberboard, the inner facing shall be the same thickness as the outer facing.
[D] After 24 h immersion (see 9.2.1).

TABLE 3 Type SF (Solid Fiberboard): Class Domestic, All Grades

Grade	Combined Weight Of Plies Before Lamination, lb/1000 ft² (g/m²), min	Bursting Strength, psi (kPa), min average
125	114 (557)	125 (862)
175	149 (727)	175 (1207)
200	190 (928)	200 (1379)
275	237 (1157)	275 (1896)
350	283 (1382)	350 (2413)
500	330 (1611)	500 (3447)
600	360 (1758)	600 (4137)

7.2.1.2 *End Item*—Domestic class of solid fiberboard constructed as specified in 6.5.1 shall conform to the bursting strength requirements in Table 3, when tested as specified in 9.1.5.

7.2.1.3 *Weather-Resistant Class, All Grades*—The weather-resistant solid fiberboard shall conform to the requirements in Table 4, when tested as specified in 9.1.1, 9.1.5 and 9.2.1.

7.3 *Dimensions*—Sheet and pad sizes, and shape dimensions shall be as specified in 7.3.1.4.

7.3.1 *Dimension Tolerance:*

7.3.1.1 *Mill Run (Untrimmed) Sheets*—The dimensional tolerance in the machine direction of untrimmed fiberboard sheets shall be −¼ in. (6 mm) and +2 in. (51 mm). The cross machine dimensions shall be not less than those specified in 5.1.3.

7.3.1.2 *Trimmed Sheets*—The dimensional tolerances for trimmed fiberboard sheets shall be ±⅛ in. (3 mm) for width and −⅛ in. (3 mm) and +½ in. (12 mm) for length.

7.3.1.3 *Pads and Shapes*—The dimensional tolerances for all other materials made from fiberboard shall be ±1⁄16 in. (2 mm) for dimensions under 18 in. (457 mm) and ±⅛ in. (3 mm) for dimensions 18 in. (457 mm) and above, unless otherwise specified.

7.3.1.4 *Scored Sheets*—The center line of score to center line of score dimension, and the center line of score to sheet edge dimension shall be as specified ±1⁄16 in. (2 mm), unless otherwise specified.

7.4 *Glue Bond Separation of Weather-Resistant and Water and Water-Vapor Resistant Classes*—The facings and corrugating mediums of corrugated fiberboard and the plies of solid fiberboard shall remain securely and continuously adhered to their contacting surfaces when tested as specified in 9.1.4. Edge separation shall not exceed ¼ in. (6 mm) in depth.

7.4.1 *Lamination of Pads and Cut Shapes*—The bonding agent used in the lamination process for fabricating pads and

TABLE 4 Type SF, Class Weather-Resistant, All Grades

Grade	Thickness, in. (mm)[A]	Bursting Strength, psi (kPa), min avg	
		Dry	Wet (After 24-h Immersion)
V2s	0.090 (2.29)	550 (3792)	500 (3447)
V3s	0.090 (2.29)	400 (2758)	150 (1034)
V4s	0.080 (2.03)	400 (2758)	150 (1034)
W5s	0.075 (1.91)	275 (1896)	100 (689)
W6s	0.060 (1.52)	175 (1207)	50 (345)

[A] A ±10 % tolerance shall be permitted.

cut shapes shall be equal to or greater than the requirements of 7.4 and shall pass the tests specified therein.

7.5 *Warp*—The amount of warp when received shall not exceed ½ in. when measured over a 2 ft span (12 mm/610 mm) when tested in accordance with 9.6.

7.6 *Bending Requirements*—Unless otherwise specified, fiberboard shall show no continuous visual surface break of the plies, or split completely through, when folded as specified in 9.3 and applicable subparagraphs.

7.7 *Fire Retardant Class*—When fire retardant class is specified, the requirements of 7.7.1 and 7.7.2 as tested by 9.1.6 and 9.1.7, respectively, shall be met.

7.7.1 *Flame Spread Index*—Tests shall be conducted in accordance with Test Method E 162. Test samples shall be exposed continuously to 95°F and 95 % relative humidity in a weathering test for seven days and subsequently conditioned in accordance with 8.2 to a constant weight prior to testing. A flame spread index of 20 or less is acceptable.

7.7.2 *Specific Optical Density*—Tests shall be conducted in accordance with Test Method E 662. Test samples shall be exposed continuously to 95°F and 95 % relative humidity in a weathering test for seven days and subsequently conditioned in accordance with 8.2 to a constant weight prior to testing. A specific optical density of 100 or less is acceptable.

8. Sampling

8.1 The fiberboard shall be sampled for all tests in accordance with Method D 585, using Plan II for all properties.

8.2 Samples shall be preconditioned and tested in an atmosphere maintained at 50 % relative humidity ±2 %, and 73.4°F ± 2°F (23.0°C) in accordance with Method D 685.

9. Test Methods

9.1 *ASTM and TAPPI Standards*—Conduct the tests in accordance with the following ASTM and TAPPI Standards:

9.1.1 *Thickness*—TAPPI T 411.

9.1.2 *Basis Weight*—TAPPI T 410.

9.1.3 *Puncture Resistance*—TAPPI T 803.

9.1.4 *Ply Separation*—TAPPI T 812.

9.1.5 *Bursting Strength*—TAPPI T 810.

9.1.6 *Flame Spread*—Test Method E 162.

9.1.7 *Specific Optical Density*—Test Method E 662.

9.2 *Wet Burst Test:*

9.2.1 For the wet burst test, the specimen shall be immersed in water as specified in the applicable paragraphs of TAPPI T 812 prior to performing the bursting strength test in accordance with TAPPI T 810.

9.3 *Bending:*

9.3.1 *Apparatus:*

9.3.1.1 *Scoring Devices*, on commercial production equipment having the proper profiles for the type, class, variety and grade of fiberboard to be tested.

9.3.2 *Test Specimen*—Specimen shall be 12 in. (305 mm) by 12 in. (305 mm).

9.3.3 *Test Procedure:*

9.3.3.1 Score the specimen using the proper scoring device both along and across the machine direction of the fiberboard. Cracks and ruptures occurring at intersection of the scorelines shall be disregarded.

9.3.3.2 *Degree of Bending:*

(*a*) Singlewall corrugated fiberboard shall be folded 180° toward the inner facing along scorelines.

(*b*) Doublewall corrugated fiberboard shall be folded 180° toward the inner facing on scores parallel to the corrugations. For scores perpendicular to the corrugations, the fiberboard shall be folded first 90° toward the outer facing, returned to the flat position, and then folded 90° toward the inner facing.

(*c*) Solid fiberboard shall be folded 180° inward (male score) on scores parallel to the machine direction. Scores parallel to the cross direction shall be folded 90° outward (female), returned to the unfolded position, and then folded 90° inward.

9.4 *Determination of the Amount of Polyethylene:*

9.4.1 *Test Specimen*—Specimen shall be 6 in. (150 mm) by 4 in. (100 mm) with each dimension having a ±1/32 in. (1.0 mm) tolerance.

9.4.2 *Test Procedure*—Soak the sample in a 2 % water solution of a wetting agent (Triton X-114 or equivalent). Peel the fibers away from the film with a firm rubbing action until only those fibers remain which are strongly adhered to the polyethylene film. To remove the remaining fiber, place the film in a shallow pan of Cupriethylendiamine, or hot (160 to 180°F (57 to 68°C)) zinc chloride solution (three parts zinc chloride to one part water). When the remaining fibers soften, they can be removed from the film by scraping it with a wooden spatula. When the film becomes clear, it shall be thoroughly washed in cold water and subsequently dried to constant weight. The weight of the film shall be determined accurately to the nearest 0.01 g. The weight of the film in lbs/1000 ft^2 is determined by multiplying the weight of the film in grams by 13.2. The weight of the film in g/m^2 is determined by multiplying the weight of the film in grams by 64.6

9.5 *Wet Water Absorption of Weather-Resistant, Water Vapor-Resistant Fiberboard (TAPPI T 441):*

9.5.1 *Test Procedure*—The test shall be performed in accordance with TAPPI T 441 using a 2 % water solution of a wetting agent (Triton X-114 or equivalent).

9.5.2 *Test Period*—The test shall be performed for a period of 24 h ± 1 h.

9.6 *Warp*—Measure warp by placing the blank fiberboard on a flat surface so that the bow rises between the ends of the blank. The maximum distance from the bottom of the sample to the flat surface shall then be measured.

9.7 *Precision and Bias*—The precision and bias of the ASTM and TAPPI Test Methods are as stated by the individual methods. No justifiable statement can be made on the precision and bias of 9.3, 9.4, and 9.6.

10. Preparation for Delivery

10.1 Unless otherwise specified in the contract or purchase order, packing and marking shall be in accordance with Practice D 3951.

10.2 Unless otherwise specified in the contract or purchase order, fiberboard sheet stock may be strapped in accordance with Specifications D 3950 or D 3953.

10.3 For shipment to the Department of Defense, when specified, packing and marking shall be in accordance with Supplementary Requirements.

SUPPLEMENTARY REQUIREMENTS

S1. Scope

S1.1 The following supplementary requirements shall apply only when specified by the purchaser in solicitation, contract, or order for agencies of the U.S. government.

S2. Applicable Documents

S2.1 The following documents[6] form, in the parts referenced, a part of this document in the issue in effect at the time of placement of the order or contract:

S2.1.1 *Federal Specifications:*

PPP-B-636 Box, Shipping, Fiberboard

S2.1.2 *Military Specifications:*

MIL-C-52950 Crates, Wood, Open and Covered

S2.1.3 *Federal Standards:*

Fed. Std. No. 123 Marking for Shipment (Civil Agencies)

S2.1.4 *Military Standards:*

MIL-STD-129 Marking for Shipment and Storage

MIL-STD-731 Quality of Wood Members for Containers and Pallets

S3. Preparation for Delivery

S3.1 *Unitization:*

S3.1.1 *Mill Run Sheets*—Unless otherwise specified, a quantity of fiberboard sheets of one type, class, variety, grade, and size, not exceeding a net weight of 800 lbs (365 Kg) nor a height of 24 in. (610 mm) shall be packed in a wood crate conforming to MIL-C-52950, Style A, Type II. A top and bottom waster sheet, of the same board packed, shall be provided. When specified, uncrated fiberboard sheets shall be palletized. A load, including a pallet (constructed as described in S3.1.1.1 and shown in Fig. 1) shall not exceed an overall height of 54 in. (1370 mm) and a gross weight of 2,000 lbs (910 Kg). A top waster sheet of the same board packed shall be used. The stack of fiberboard sheets shall be positioned on the pallet so that they do not protrude beyond any pallet edge. Each stack of sheets shall be secured to its pallet with flat steel or plastic strapping conforming to Specification D 3953, Type I, Grade N or II, or Specification D 3950, Type I or III. The size strapping used is dependent upon the gross weight of the pallet load and number of straps used; Table S1.1 shows the minimum allowable strapping sizes. One strap per stringer shall encircle the load in the direction of the pallet length with each strap being inserted through the stringer strap slot. The load shall also be strapped parallel with the pallet width with each strap passing under the top deck but under the stringers. The number of widthwise straps shall be one less than the number of top deck boards, with one being positioned midway between each adjacent pair of deck boards. All straps shall be tensioned and sealed tightly. Suitable edge protectors shall be used under the strapping at the top of the load to prevent the straps from cutting or creasing the fiberboard sheets.

S3.1.1.1 *Construction of Pallet:*

(*a*) The pallet shall have overall length and width dimensions not less than the dimensions of the fiberboard sheets being packed, but not greater than ½ in. (13 mm) more than

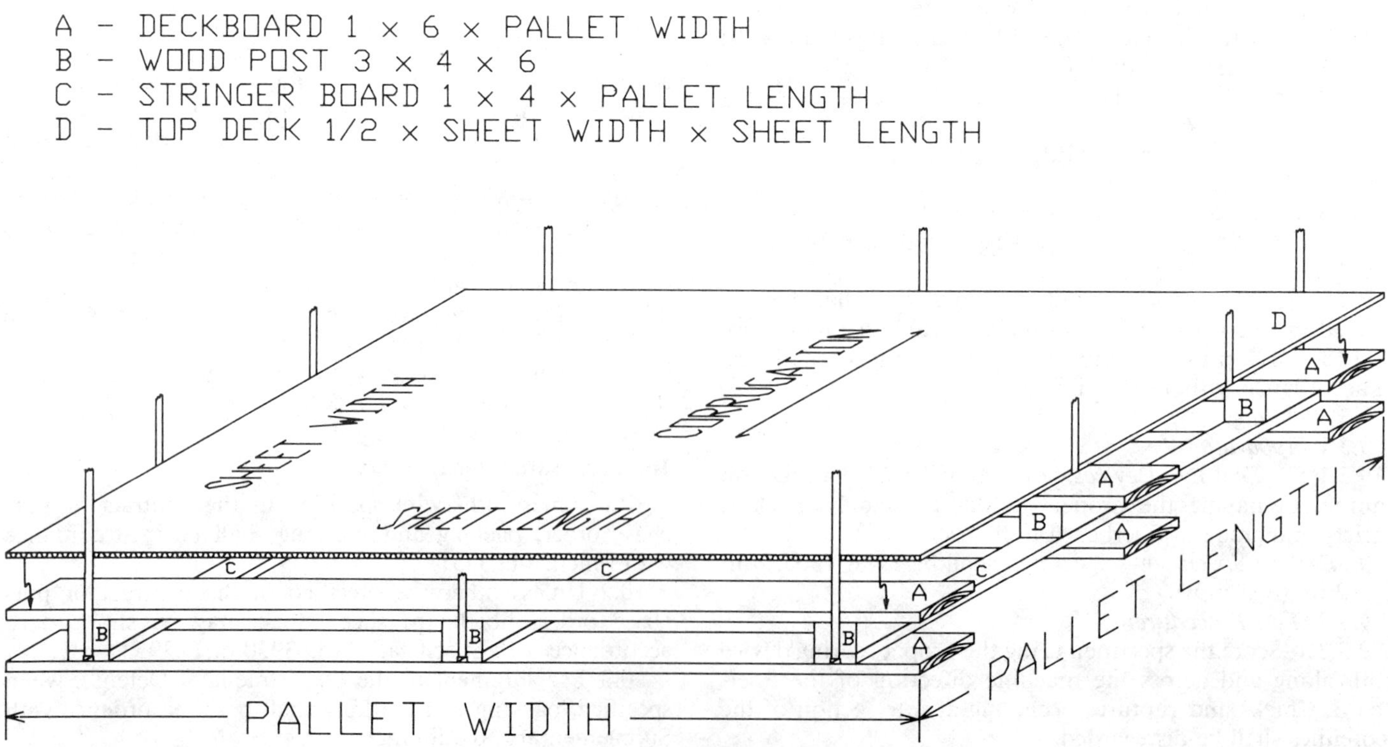

FIG. 1 Shipping Pallet Construction

TABLE S1.1 Minimum Allowable Strapping Size

Steel Strap		Plastic Type II		Plastic Type III		Maximum Gross Load Per Strap	
in.	(mm)	in.	(mm)	in.	(mm)	lb	(kg)
5/8 by 0.020	(16 by 0.5)	5/8 by 0.020	(16 by 0.5)	7/16 by 0.023	(11 by 0.6)	300	(136)
...	...	1/2 by 0.025	(13 by 0.6)	1/2 by 0.020	(13 by 0.5)	300	(136)
5/8 by 0.023	(16 by 0.6)	1/2 by 0.030	(13 by 0.8)	7/16 by 0.029	(11 by 0.7)	400	(182)
3/4 by 0.020	(19 by 0.5)	5/8 by 0.025	(16 by 0.6)	1/2 by 0.025	(13 by 0.6)	400	(182)
3/4 by 0.023	(19 by 0.6)	5/8 by 0.030	(16 by 0.8)	1/2 by 0.030	(13 by 0.8)	500	(227)

the fiberboard dimensions. The wood used shall be softwood conforming to MIL-STD-731, Group 1, Class 2. All boards and posts shall be finished top and bottom to a uniform thickness. The pallet shall consist of the following members: (*a*) nominal 1 by 6 in. (25 by 150 mm) upper and lower deck boards; (*b*) nominal 1 by 4 in. (25 by 100 mm) lower wood stringers; (*c*) nominal 3 by 4 by 6 in. (75 by 100 by 150 mm) wood posts; and (*d*) a top deck of corrugated fiberboard conforming to paragraph 4.1.2.3 or two full sized sheets of V11c laminated together with a water-resistant adhesive.

(*b*) The outside deck boards shall be flush with the ends of the stringers and all deck boards shall be evenly spaced across the length of the pallet. The number of deck boards used is dependent upon the length of the pallet; spacing between adjacent deck boards shall not exceed 17 in. (430 mm). The outside stringers shall be set 6 in. (150 mm) ± 1/4 in. (6 mm) from the ends of the deck boards; the inside stringers shall be evenly spaced over the distance between the outside stringers. Not less than three stringers shall be used in pallets up to 88 in. (2235 mm) wide, and not less than four stringers in wider pallets. Each stringer shall be slotted so that the strapping required (see Table S1.1) can be threaded between the stringer and bottom deck boards. The stringers, deck boards, and posts shall be nailed together, at the joint, with six (three through the top deck boards and three through the bottom deck boards) cement coated or chemically etched, box or T-head 8d nails, 2½ in. (64 mm) long with a 0.97 in. (2 mm) shank. The nails shall not protrude into the strap slots of the stringers. The corrugated fiberboard top deck shall be secured to top deck boards with water-resistant adhesive. Fig. 1 illustrates a typical example of the pallet described in this paragraph.

S3.2 *Pads, Scored Sheets and Cut Shapes*—Pads, scored sheets and cut shapes of one class, type, variety, grade, and size shall be packed in a snug fitting fiberboard shipping container conforming to PPP-B-636, Type CF or SF, Class weather-resistant. Each shipping container shall be closed in accordance with the appendix of PPP-B-636. The weight of contents of each shipping container shall not exceed 65 lbs (30 kg). When specified, pads, scored sheets and cut shapes too large to box, may be unitized in accordance with S3.1.

S4. Marking

S4.1 *Compliance marking, when specified*—Each sheet, pad, or cut shape shall be marked with identification of the ASTM specification, class, grade, and date of manufacture along one edge, with characters not less than 3/16 in. (4 mm) in height.

S4.2 *Civil Agencies*—In addition to any special marking required by the contract or order, shipments shall be marked in accordance with Fed. Std. No. 123.

S4.3 *Military Agencies*—In addition to any special marking required by the contract or order, shipments shall be marked in accordance with MIL-STD-129.

S5. Ordering Data

S5.1 Purchasers should select the preferred options permitted herein, and include the following information in procurement documents:

S5.1.1 Number, title, and date of the specification,

S5.1.2 Type, class, variety, and grade (see Section 4),

S5.1.3 Flute design (see 6.4.1),

S5.1.4 Mill run or trim sheets,

S5.1.5 Dimensional direction of flute of type CF, when applicable,

S5.1.6 When compliance symbol is not required, (see Section 4),

S5.1.7 Size of sheet, pad, shape, etc., and

S5.1.8 Unitization required (see S3.1).

Designation: D 4728 – 87

Standard Test Method for Random Vibration Testing of Shipping Containers[1]

This standard is issued under the fixed designation D 4728; the number immediately following the designation indicates the year of original adoption or, in the case of revision, the year of last revision. A number in parentheses indicates the year of last reapproval. A superscript epsilon (ε) indicates an editorial change since the last revision or reapproval.

1. Scope

1.1 This test method covers the random vibration testing of filled shipping units. Such tests may be used to assess the performance of a container with its interior packing and means of closure in terms of its ruggedness and the protection that it provides the contents when subjected to random vibrational inputs.

1.2 This test method provides guidance in the development and use of vibration data in the testing of shipping containers.

1.3 Three alternative vibration control techniques are described.

Method A Closed Loop—Automatic Equalization (5.2.1, 10.1.1)
Method B Closed Loop—Manual Equalization (5.2.2, 10.1.2)
Method C Open Loop—Magnetic Tape (5.2.3, 10.1.3)

NOTE—Sources of supplementary information are listed in the Reference section **(1–6)**.[2]

1.4 *This standard does not purport to address all of the safety problems associated with its use. It is the responsibility of the user of this standard to establish appropriate safety and health practices and determine the applicability of regulatory limitations prior to use.* Specific safety hazard statements are given in Section 6.

2. Referenced Documents

2.1 *ASTM Standards:*

D 996 Terminology of Packaging and Distribution Environments[3]

D 999 Methods for Vibration Testing of Shipping Containers[3]

D 3580 Test Method of Vibration (Vertical Sinusoidal Motion) Test of Products[3]

D 4332 Practice for Conditioning Containers, Packages or Package Components for Testing[3]

3. Terminology

3.1 *vibration:*

3.1.1 *complex vibration*—vibration whose components are sinusoids not harmonically related to one another.

3.1.2 *deterministic vibration*—vibration whose instantaneous value at any future time can be predicted by an exact mathematical expression.

3.1.3 *periodic vibration*—(see also *deterministic vibration*) an oscillation whose waveform repeats at equal increments of time.

3.1.4 *random vibration*—random vibration may be broad-band covering a wide and continuous range of frequencies, or narrow-band, covering a relatively narrow range of frequencies. Random vibration contains no periodic or quasiperiodic constituent. If random vibration has instantaneous magnitudes that occur according to the Gaussian distribution, it is called "Gaussian random vibration."

3.1.4.1 *broad-band random vibration*—random vibration that covers a wide and continuous range of frequencies. Any frequencies that exceed the narrow-band limitations are considered broad-band.

3.1.4.2 *narrow-band random vibration*—random vibration having frequency components only within a narrow band. It has the appearance of a sine wave whose amplitude varies in an unpredictable manner. A narrow band should be ±10 % or ±3 Hz, whichever is greater, of the center frequency of interest.

3.1.4.3 *white random vibration*—that form of broad-band random vibration in which the PSD level is constant over a wide frequency range.

3.2 *vibration control:*

3.2.1 *attenuate*—to reduce the amplitude of an electronic control signal or vibration output.

3.2.2 *closed-loop*—a condition of control where the input may be modified over time by the effect of the output or response of the system.

3.2.3 *equalization*—adjustment or correction of the amplitude characteristics of an electronic control signal throughout a desired frequency range to maintain a desired vibration output spectrum and level.

3.2.4 *equalizer*—instrumentation used to conduct equalization.

3.2.5 *open loop*—a condition of control where the input of a system is preestablished and is not affected by the output or response of the system.

3.2.6 *transfer function*—the dynamic relationship between output and input. In terms of a vibration system, it is the ratio of output response to a constant input over a defined frequency bandwidth.

3.3 *vibration frequency:*

3.3.1 *discrete frequency, sinusoidal*—a periodic function having a sinusoidal waveform of only one frequency.

---

[1] This test method is under the jurisdiction of ASTM committee D-10 on Packaging and is the direct responsibility of Subcommittee D10.22 on Handling and Transportation.

Current edition approved Aug. 28, 1987. Published October 1987. Previously published as D-10 Proposal P 186.

[2] The boldface numbers in parentheses refer to the list of references at the end of this test method.

[3] *Annual Book of ASTM Standards*, Vol 15.09.

3.3.2 *filter*—a device capable of passing certain frequencies with little loss (pass band) but of causing large losses so that other frequencies (stop band) are attenuated. Filters may be classified as low-pass (high stop), high-pass (low stop) or band-pass (stopping frequencies below and above the pass band).

3.3.3 *filtered signal*—signal is said to be filtered when components have been removed by passing it through a filter.

3.3.4 *spectrum*—a definition of the magnitude of the frequency components within a specified band width.

3.4 *vibration quantities:*

3.4.1 *mean-square*—the time average of the square of the function.

3.4.2 *overall* g *rms*—the square root of the integral of *power spectral density* over the total frequency range.

3.4.3 *power spectral density (PSD)*—a term used to quantify the intensity of random vibration in terms of mean-square acceleration per unit of frequency. The units are g^2/Hz (g^2/cycles per second). Power spectral density is the limiting mean square value per unit bandwidth, that is, the limit of the mean square value in a given rectangular bandwidth divided by the bandwidth, as the bandwidth approaches zero.

3.4.4 *root-mean-square (rms)*—the square root of the mean-square value. In the exclusive case of a sine wave, the rms value is 0.707 times peak value.

4. Significance and Use

4.1 Shipping containers are exposed to complex dynamic stresses in the distribution environment. Approximating the actual damage, or lack of damage, experienced in real life may require subjecting the container and its contents to random vibration tests. In this way, many product and container resonances are simultaneously excited.

4.2 Resonance buildups during random vibration tests are less intense than during sinusoidal resonance dwell or sweep tests. Therefore, unrealistic fatigue damage is minimized.

4.3 Random vibration tests should be based on representative field data. When possible, confidence level may be improved by comparing laboratory test results with actual field shipment effects. (See Appendix X1 for related information.)

4.4 There is no direct equivalence between random vibration tests and sinusoidal vibration tests. Equivalent tests between sine and random, in a general sense, are difficult to establish due to nonlinearities, damping and product response characteristics.

4.5 Vibration exposure affects the shipping container, its interior packing, means of closure, and contents. This test allows analysis of the interaction between these components. Design modification to one or all of these components may be used to achieve optimum performance in the shipping environment.

4.6 Random vibration tests may be simultaneously performed with transient or periodic data to simulate known stresses of this type, that is, rail joints, pot holes, etc.

4.7 Random vibration may be conducted in any axis (vertical or horizontal) or in any package orientation. However, different test levels must be utilized for each size or package orientation depending on the field environment that is to be duplicated.

5. Apparatus

5.1 *Vibration Test Machine*—The vibration test machine (shaker) shall have a vibration table of sufficient strength and rigidity so that the applied vibrations are essentially uniform over the entire test surface when loaded with the test specimen. The vibration table shall be supported by a mechanism capable of producing single axis vibration inputs at controlled levels of continuously variable amplitude throughout the desired range of frequencies. Suitable fixtures and guides to restrict undesired movement of the test specimens shall be provided.

5.2 *Electronic Controls*—Controls shall provide the capability of generating vibration command inputs necessary to produce the desired test levels at the table surface adjacent to the test specimen. The following methods provide this capability with varying degrees of accuracy and versatility.

5.2.1 *Method A, Closed Loop-Automatic Equalization*—Automatically equalized closed loop control systems allow the operator to enter PSD data by the input of desired test levels and frequency break points via a control panel. The controller automatically generates equalized vibration test machine drive signals to achieve the desired PSD thus maintaining closed loop control. The equalized drive signals automatically compensate for specimen and vibration test machine characteristics. Typical systems include an analog to digital converter for conditioning drive and feedback signals, digital processor with real time analysis capability, graphics display terminal, hard copy printer, flexible disk storage unit, random vibration control software programs, and a digital to analog converter for output signals.

5.2.2 *Method B, Closed Loop-Manual Equalization*—Manually equalized closed loop control systems allow predetermined PSD data to be generated and controlled. Typical systems include a white random noise generator covering the desired test frequency range, a manual equalizer to shape the signal from the random noise generator, and a spectrum analyzer to verify that the desired test levels are produced at the vibration table. The human operator must manually adjust and continuously monitor the test so as to close the vibration control loop.

5.2.3 *Method C, Open Loop-Magnetic Tape*—Magnetic tape open loop control systems enable prerecorded equalized PSD data to be produced as long as the combined transfer function of the specific tape player, vibration system and load configuration was incorporated in the tape's original preparation. Typical systems include a tape player, prerecorded magnetic tapes, and a means of calibrating the shaker's drive signal. The tape player must have adequate frequency response to prevent attenuation of the output vibration spectrum. Ideally, magnetic tapes should be equalized for each specific specimen to ensure that the dynamic response of the actual specimen(s) does not cause significant differences between the actual PSD test levels and the desired test levels. On the other hand, equalization with a very stiff ballast weight equivalent in mass to the test item is sometimes employed as a more practical approach. Caution must be exercised, however, since the tape player system will not compensate for actual specimen responses.

5.3 *Instrumentation*—Accelerometers, signal conditioners, analyzers, data display, storage devices, and the control techniques described in 5.2 are required to measure and control the PSD levels at the table surface. Instrumentation may also be desirable for monitoring the response of the test specimen(s). The instrumentation system shall have an accuracy of ±5 % throughout the frequency range specified for the test.

6. Safety Precautions

6.1 This method may produce severe mechanical responses of the test specimen(s). Therefore, fences, barricades, and other restraints must have sufficient strength and must be adequately secured. Operating personnel must remain alert to the potential hazards and take necessary precautions for their safety. Stop the test immediately if a dangerous condition should develop.

7. Test Specimens

7.1 The test specimen shall consist of the container as intended for shipment, loaded with the interior packaging and the actual contents for which it was designed. Blemished or rejected products may be used if the defect will not affect test results and is recorded prior to the test. Dummy test items may be used for developmental testing when necessary, but may not be used for final acceptance testing. Sensors and transducers should be applied with minimum possible alteration of the test specimen to obtain data on the container or packaged item. When it is necessary to observe the contents during the test, holes may be cut in noncritical areas of the container.

7.2 Whenever sufficient containers and contents are available, it is highly desirable that replicate tests be conducted to improve the statistical reliability of the data obtained.

8. Calibration and Standardization

8.1 The accuracy of instrumentation and test equipment used to control or monitor the test parameters should be verified prior to conducting each test to ensure that desired test levels and tolerances are maintained.

8.2 The specified PSD data and resulting RMS acceleration level should be derived from actual field measurements or published PSD data made on typical transport vehicles under representative conditions of speed, load, terrain, road surfaces, etc. Field measurements must be accurately recorded with equipment having adequate frequency response to prevent attenuation of the acceleration energy levels. Multiple independent field measurements must be sampled to assure representative test levels. This data must then be reduced to PSD format and equalized for proper control of the shaker.

8.3 Shaker table input levels to the test specimen provide the only common benchmark for repeatability between various test systems. Therefore, control analysis based on monitoring table motion rather than actual package response is recommended. This table feedback signal is generated by an accelerometer mounted directly to the table. Accelerometer mounting location should be next to the test specimen or directly below it on the underside of the table.

8.4 The power spectral density of the random vibration test profile shall not deviate from the specified requirements by more than ±3 dB except that deviations as large as ±6 dB shall be allowed over a cumulative bandwidth of 80 Hz between 2 and 200 Hz.

8.4.1 Maintenance of this tolerance requires close attention to equalizer bandwidths. Test frequencies below 5 Hz typically require at least 1 Hz control bandwidths whereas ⅓ and ½ octave bandwidths usually provide adequate control at higher frequencies.

8.5 The shaker's drive signal must be equalized as described in 5.2 to compensate for payload dynamics, the shaker's transfer function, and the control system's transfer function.

9. Conditioning

9.1 Test specimens shall be conditioned prior to test or during test, or both, in accordance with Practice D 4332. When no specific conditioning requirements are given, and container materials are moisture sensitive, a standard conditioning atmosphere of 23 ± 2°C (73.4 ± 3.6°F) and 50 ± 2 % relative humidity is recommended, for a minimum of 24 h prior to performance of test(s).

10. Procedure

10.1 *Specific Procedures:*

10.1.1 *Method A, Closed Loop–Automatic Equalization*—Place the test package on the vibration table and securely fix or restrain the package to the vibration table (see 10.2.1). Initiate the tests at least 6 dB below full test level and increment in one or more subsequent steps to full test level. This enables the closed loop control system to complete its equalization at lower test levels and provides the operator adequate opportunity to visually verify that the test package and fixture are receiving a realistic test, prior to full test level exposure. This is important since random vibration will produce relatively large, low frequency displacements in an unpredictable sequence.

10.1.2 *Method B, Closed Loop–Manual Equalization*—Place the test package on the vibration table and securely fix or restrain the package to the vibration table (see 10.2.1). Initiate the test at least 6 dB below full test level and make adjustments to the equalizer as required to bring the PSD within specification. Increment in one or more subsequent steps to full test level repeating the manual equalization adjustments as required. Upon reaching full test level, periodically view the PSD analyzer monitor and make equalization corrections, as necessary. Any changes to the test configuration will require re-equalization.

10.1.3 *Method C, Open Loop–Magnetic Tape*—Place the test package on the vibration table and securely fix or restrain the package to the vibration table (see 10.2.1). The test should be gradually brought up to full test levels as determined by the system's calibration method.

10.2 *General Procedures:*

10.2.1 Place the unit(s) to be tested in its normal shipping orientation so that the desired vibration condition (vertical or horizontal) is transmitted to the outer container. The specimen may be either securely fastened to the vibration table or allowed to vibrate freely. In the latter case, attach restraining devices to the vibration table to prevent excessive rocking and movement off the vibration table. Adjust the restraining devices to permit free movement of the specimen of approximately 10 mm (0.4 in.) in any horizontal direction from its centered position. Unit loads, stacked columns, or single units may be tested in this manner.

10.2.2 Conduct the random vibration test for the length of time stated in the applicable specification, if any, or for a predetermined period, or until a predetermined amount of damage may be detected. The test may be stopped momentarily to inspect for damage.

10.2.3 Test levels are often increased over the actual field data to shorten the test time. Any attempt to do so should be done with caution. Use of "equivalance" techniques of this type assume linearity of specimen response to test input which is, in fact, not likely.

10.2.4 When shipping information becomes available for the test item, the test duration should be modified based on observed damage levels. For example, if the laboratory test does not produce a realistic level of damage, then test duration should be adjusted.

11. Report

11.1 The report shall include the following:

11.1.1 Identification and description of the test specimens, including the container, the interior packaging, the product (give size, weight, and any other pertinent details), and photographs (before and after) of the test items, where possible,

11.1.2 If unitized loads are tested, description of the unitized load, the height of the stack, the unitizing method employed, and photographs (before and after) of the test items, where possible,

11.1.3 Purpose of the test and the applicable performance specification, if any,

11.1.4 Rationale for the random vibration PSD levels' pertinance including a detailed description of the measurement and analysis techniques utilized,

11.1.5 Details of the test method, test levels and durations used,

11.1.6 Verification of compliance with the test method or descriptions of any deviations,

11.1.7 Number of replications of each test,

11.1.8 Atmospheric conditions to which the specimens were subjected, both prior to test and during test,

11.1.9 Any other tests the specimens were subjected to prior to this test,

11.1.10 Description of the apparatus and instrumentation used,

11.1.11 Results of the tests, and a comparison between damage levels observed as a result of the test versus actual damage observed in transportation, if historical data exists,

11.1.12 Descriptions and photographs of any damage or deterioration to the containers or their contents as a result of the tests,

11.1.13 All significant resonant responses and any observations that may assist in correct interpretation of results or lead to improvements in design of container, interior packaging or product, and

11.1.14 Statement of whether or not the specimens complied with the requirements of the applicable specification.

12. Precision and Bias

12.1 No statement is made about either the precision or bias of this method since the results merely indicate the presence or absence of damaging responses from the random vibration input.

APPENDIX

(Nonmandatory Information)

X1. SAMPLE PSD DATA

X1.1 Sample PSD data is provided in Figs. X1.1 and X1.2 for *informational purposes only.* It does not purport to accurately describe a specific transportation mode or distribution environment. The user of random vibration must verify accuracy and applicability of any data of this type prior to its use.

X1.2 Figure X1.1 illustrates that there are relative differences in vibration intensity and frequency content for various types of commercial transportation.

X1.2.1 Figure X1.1 has evolved from a compilation of field measurements made by several organizations over a period of time.

X1.2.2 Whereas the data in Fig. X1.1 averages vibration intensities measured under various loading conditions, suspension types, road conditions, weather conditions, travel speeds, etc., it does not represent the environment that exists in any specific transportation environment. Exact situations must be verified by the user of this method.

X1.2.3 Transients caused by pot holes, rail joints, takeoff/landing, etc. are not reflected in Fig. X1.1.

X1.3 Figure X1.2 illustrates the various vibration intensities attributed to different road conditions by a single truck/payload combination. Keep in mind that this is one sample; numerous samples must be taken to develop statistically valid random vibration PSD test levels.

X1.3.1 In this particular study, a single truck was instrumented and driven over various routings. Each condition is reduced to PSD format and plotted in Fig. X1.2.

X1.3.2 Specific conditions and criteria associated with this study are as follows.

X1.3.2.1 *Type of Truck*—A standard tractor trailer truck with steel leaf spring suspension was used for all routes analyzed.

X1.3.2.2 *Type of Equipment*—A tractor-trailer was instrumented with peizo-resistive accelerometers. The output was recorded on a FM reel-to-reel magnetic tape recorder. The data recorded on magnetic tape was interpreted on a Fast Fourier Transform (FFT) spectrum analyzer to produce a PSD plot.

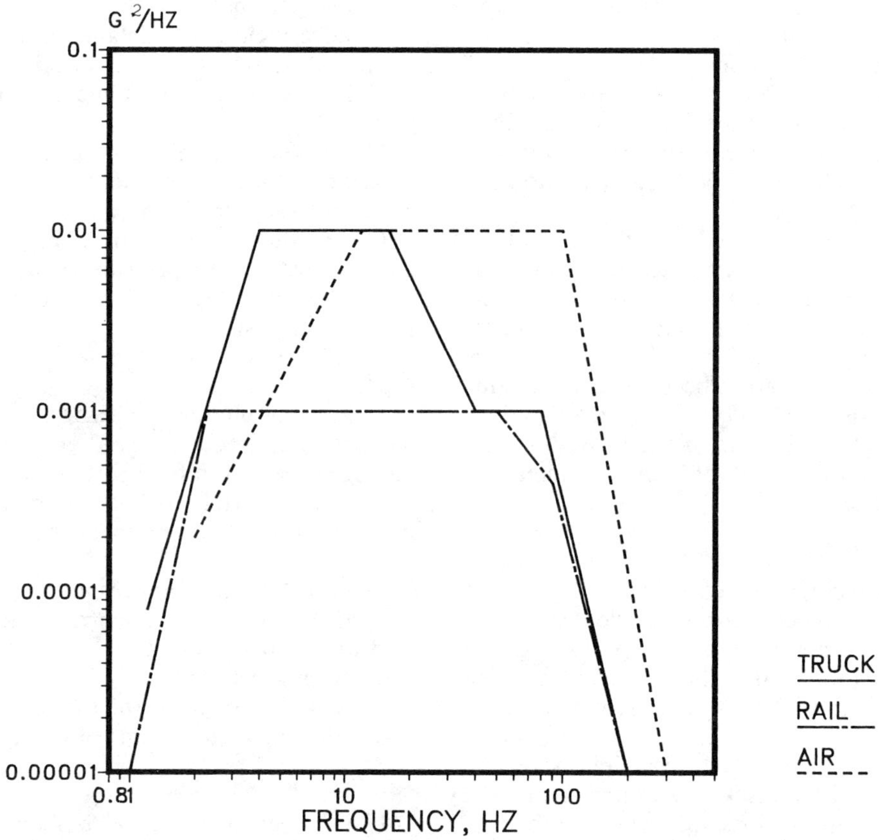

FIG. X1.1 Commercial Transport Random Vibration Spectra Summary

X1.3.2.3 *Location of Accelerometer*—One accelerometer was mounted on the wood floor above the rear wheels.

X1.3.2.4 *Type of Load*—The tractor-trailer contained a stretch wrapped pallet of plastic film with a 1500-lb steel load on top (simulating a pallet on a pallet, about 3000 lb).

X1.3.2.5 *Location of the Load*—The pallet was positioned over the rear wheels.

X1.3.2.6 *Routes and Surface Conditions: Production Area*—Slow speeds (<20 mph), rough roads.

City Roads—Slow speeds (10 to 15 mph), extremely rough, cracks, and railroad tracks.

Expressway—Concrete highway, indicative of Interstate Highway, high speeds (≥55 mph).

Suburban Roads—Slow speeds (<20 mph), sharp turns, and numerous stops.

Main Suburban Roads—Asphalt road with many pitched areas, railroad tracks, numerous stops, average speeds (35 to 45 mph).

Secondary Suburban Roads—Two lane, asphalt, lots of bumps, slow speed (<30 mph).

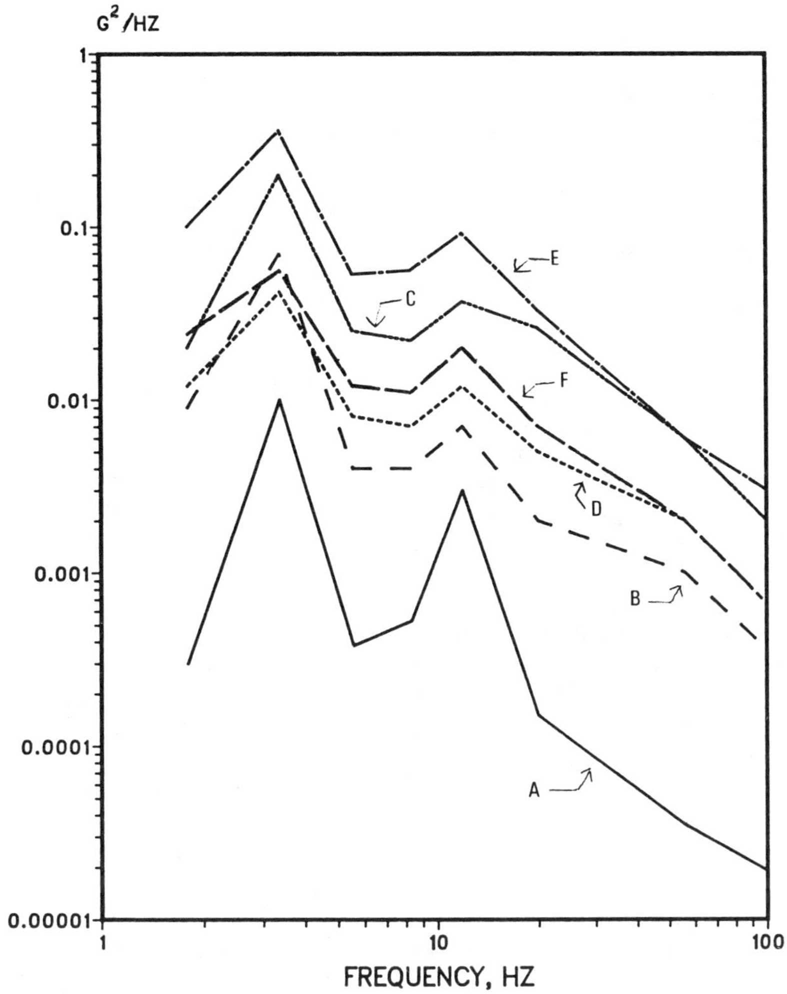

NOTE—Refer to X1.3.2.6 for descriptions of the following routes and surface conditions: A—Production Area, B—City Roads, C—Expressway, D—Suburban Roads, E—Main Suburban Roads, and F—Secondary Suburban Roads.

FIG. X1.2 Route Comparison Leaf Spring Truck Vertical Vibration

REFERENCES

(1) Harris and Crede, *Shock and Vibration Handbook*, McGraw-Hill Co., New York NY, 1976.

(2) "Method 514.3 Vibration," *Environmental Test Methods and Engineering Guidelines MIL-STD-810D,* Department of Defense, July 1983.

(3) "Procedure for the Generation and Application of Random Vibration Utilizing a Cassette Tape Deck," *Navy Manufacturing Screening Program NAVMAT P-9492,* Department of the Navy, May 1979.

(4) Ostrem, F. E., and Godshall, W. D., "An Assessment of the Common Carrier Shipping Environment," *General Technical Report FPL 22,* U.S. Forest Product Laboratory, 1979.

(5) Andress, E. A., "Generation of a Transportation Vibration Test Specification," *DRC-009 11/81,* Scientific Atlanta, Spectral Dynamics Division, 1981.

(6) Tustin and Mercado, *Random Vibration in Perspective,* Tustin Institute of Technology, Inc., Santa Barbara, CA, 1984.

Standard Specification for Testing of Hazardous Materials Packagings[1]

This standard is issued under the fixed designation D 4919; the number immediately following the designation indicates the year of original adoption or, in the case of revision, the year of last revision. A number in parentheses indicates the year of last reapproval. A superscript epsilon (ε) indicates an editorial change since the last revision or reapproval.

ε1 NOTE—The approval date was editorially corrected in December 1989.

1. Scope

1.1 This standard covers the testing of packagings to United Nations standards intended for transportation of hazardous materials, excepting packagings for radioactive substances, cylinders and other receptacles for gases, packagings for net masses exceeding 400 kg (880 lb) or capacities exceeding 450 L (120 gallons).

1.2 The following safety hazards caveat pertains only to the test method portions, Sections 8–14, of this specification: *This standard does not purport to address the safety problems associated with its use. It is the responsibility of the user of this standard to establish appropriate safety and health practices and determine the applicability of regulatory limitations prior to use.* See also A1.2.

2. Referenced Documents

2.1 *ASTM Standards:*

D 685 Method of Conditioning Paper and Paper Products for Testing[2]

D 775 Test Method for Drop Test for Loaded Boxes[2]

D 959 Method of Drop Test for Filled Bags[2]

D 996 Terminology of Packaging and Distribution Environments[2]

D 997 Test Method for Drop Test for Loaded Cylindrical Containers[2]

D 999 Methods for Vibration Testing of Shipping Containers[2]

D 4169 Practice for Performance Testing of Shipping Containers and Systems[2]

D 4332 Practice for Conditioning Containers, Packages, or Packaging Components for Testing[2]

D 4577 Test Method for Compression Resistance of a Container Under Constant Load[2]

E 4 Practices for Load Verification of Testing Machines[3]

2.2 *ISO Standard:*

ISO 535 Determination of Water Absorption of Paper and Board (Cobb Method)[4]

2.3 *ANSI/ASQC Document:*

ANSI/ASQC Z-1.15 Generic Guidelines for Quality Systems[4]

2.4 *TAPPI Standard:*

TAPPI T 441 Water Absorptiveness of Sized (Non-bibulous) Paper and Paperboard (Cobb Test)[5]

2.5 *United Nations Document:*

ST/SG/AC.10/1 Recommendations on the Transport of Dangerous Goods[6]

3. Terminology

3.1 *Descriptions of Terms Specific to This Standard:*

3.1.1 *bags*—flexible packagings made of paper, plastic film, textiles, woven materials or other suitable materials.

3.1.2 *barrels*—receptacles of circular cross-section, with bulging walls. Wooden barrels are constructed with staves and ends of natural wood, and are held together with hoops of metal or wood.

3.1.3 *boxes*—packagings with complete rectangular or polygonal faces, made of metal, wood, plywood, reconstituted wood, fiberboard, plastic, or other suitable material.

3.1.4 *closures*—devices which close an opening in a receptacle.

3.1.5 *combination packaging*—a combination of packagings for transport purposes, consisting of one or more inner packagings secured in an outer packaging.

3.1.6 *composite packagings*—packagings consisting of an outer packaging and an inner receptacle so constructed that the inner receptacle and the outer packaging form an integral packaging. Once assembled it remains, thereafter, an integrated single unit; it is filled, stored, transported and emptied as such.

3.1.7 *drums*—flat-ended or convex-ended cylindrical packagings made of metal, fiberboard, plastic, plywood, or other suitable materials.

Discussion—This definition also includes packagings of other shapes made of metal or plastic, for example, round taper-necked packagings or pail-shaped packagings. *Jerricans* are not covered by this definition.

3.1.8 *inner packagings*—packagings for which an outer packaging is required for transport.

3.1.9 *inner receptacles*—receptacles which require an outer packaging in order to perform their containment function.

---

[1] This specification is under the jurisdiction of ASTM Committee D-10 on Packaging and is the direct responsibility of Subcommittee D10.21 on Shipping Container Environment.

Current edition approved Nov. 10, 1989. Published December 1989.

[2] *Annual Book of ASTM Standards*, Vol 15.09.

[3] *Annual Book of ASTM Standards*, Vols 03.01, 04.02, and 08.03.

[4] Available from the American National Standards Institute, 1430 Broadway, New York, NY 10018.

[5] Available from the Technical Association of the Pulp and Paper Industry, P.O. Box 105113, Atlanta, GA 30348.

[6] Available from United Nations, 866 United Nations Plaza, New York, NY 10017.

3.1.10 *jerricans*—metal or plastic packagings of rectangular or polygonal cross-section.

3.1.11 *maximum capacity*—the maximum inner volume of packagings.

3.1.12 *outer packaging*—the outer protection of a composite or combination packaging together with any absorbent materials, cushioning, and any other components necessary to contain and protect receptacles or inner packagings.

3.1.13 *packages*—the complete product of the packing operation, consisting of the packaging and its contents prepared for transport.

3.1.14 *packagings*—receptacles and any other components or materials necessary for the receptacle to perform its containment function.

3.1.15 *receptacles*—containment vessels for receiving and holding substances or articles, including any means of closing.

Discussion—The following explanations and examples are meant to assist in clarifying the above definitions:

a) The *inners* of *combination packagings* are always termed *inner packagings* not *inner receptacles*. A glass bottle is an example of such an *inner packaging*.

b) the *inners* of *composite packagings*, are normally termed *inner receptacles*. For example, the *inner* of a plastic receptacle with outer steel drum composite packaging (plastic material) is such an *inner receptacle* since it is normally not designed to perform a containment function without its *outer packaging* and is not therefore an *inner packaging*.

4. Significance and Use

4.1 Packagings successfully tested to this specification meet only the performance standards established for international transportation of hazardous materials, based on recommendations of the United Nations Committee of Experts on the Transport of Dangerous Goods, as endorsed by the United Nations Economic and Social Council.

4.2 Packages successfully tested to this specification may not meet national regulatory requirements nor withstand the North American distribution environment. It is strongly recommended that tests required by national regulations and additional sequential tests, as detailed in Practice D 4169 for Assurance Level 1 for the planned system of distribution, be carried out to further establish suitability of the package.

4.3 Tests prescribed are of varying degrees of severity, depending on the degree of hazard presented by the proposed contents, and are grouped as follows:

Packing Group I	Substances presenting great danger
Packing Group II	Substances presenting medium danger
Packing Group III	Substances presenting minor danger

Substances and articles which are hazardous are assigned to a specific packing group in UN ST/SG/AC.10/1 or in international or national regulations.

5. Test Specimens

5.1 Test specimens shall consist of packagings prepared for transport including inner packagings of combination packagings. Test specimens of packagings which are in production shall be representative samples taken at random.

6. Filling and Closure Procedures

6.1 *Liquids:*

6.1.1 Packagings intended for the containment of liquids shall be filled to 98 % of their maximum capacity with water at 23 ± 2°C (73 ± 4°F) for subsequent testing at normal and high temperatures unless otherwise specified.

6.1.2 Plastic packagings, with the exception of bags and expanded polystyrene boxes, intended for the containment of liquids shall be filled with a solution which remains liquid at −18°C (0°F) and has a minimum density of 1 g per cm^3 (specific gravity at least 1) for testing at low temperatures.

6.2 *Solids:*

6.2.1 Packagings intended for the containment of solids shall, unless otherwise specified, be filled to not less than 95 % of their maximum capacity with the intended contents or with other substances with similar physical properties (mass, particle size, etc.). Mass as tested shall be not less than the shipping mass. It is permissible to use additives such as bags of lead shot to achieve the required total package mass, as long as they are placed so that the results are not invalidated.

6.3 Each packaging shall be closed for testing in the same manner as it would be closed for actual shipment. All closures shall be installed using the techniques or torques specified by the closure manufacturer, container manufacturer or shipper.

7. Conditioning

7.1 Standard conditioning shall be at 23 ± 2°C (73 ± 4°F). For packagings fabricated from paper, paperboard, or fiberboard, conditioning shall be in accordance with Method D 685.

7.2 High temperature conditioning shall be at 40 ± 2°C (104 ± 4°F) in accordance with Practice D 4332.

7.3 Low temperature conditioning shall be at −20 ± 2°C, (−4 ± 4°F), in accordance with Practice D 4332.

7.4 The packaging, prepared as for testing, shall be conditioned for at least 7 min/L (qt) or/dm^3 (200 min/ft^3) of maximum capacity or 24 h whichever is longer. Longer times shall be used if necessary to ensure that packagings and contents reach equilibrium with the conditioning atmosphere and are maintained in that condition for a minimum of four hours.

8. Compatibility Test

8.1 Unless otherwise specified, the compatibility test shall be performed on plastic packagings prior to initial use of the packaging for each hazardous material to be carried. Evidence of satisfactory user experience may be accepted in lieu of performance of this test.

8.2 Perform the compatibility test in accordance with Annex A1. The storage temperature and the duration of storage are set out in this test method.

8.3 Following the specified storage period and weighing, remove the hazardous material, fill and reclose the packaging as specified in Section 6; perform the drop test appropriate to the type of packaging and the intended contents, as specified in Section 9; perform the stacking test as specified in Section 10; perform the hydrostatic test for packagings intended for the containment of liquids as specified in Section 11; and perform the leak resistance test as specified in Section 12.

8.4 *Criteria for Passing Test*—The packages shall show no evidence of stress cracking or crazing, oxidation, embrittlement, vapor pressure build-up, collapse of walls, seepage, or other defect likely to cause or indicate premature failure. The packaging shall pass the tests specified in 8.3.

9. Drop Test

9.1 The drop test shall be performed on all types of packagings prior to initial use and, where specified, at periodic intervals.

9.2 Tests shall conform to Test Method D 775, Method D 959 (procedure A without beam hazard), or Test Method D 997 (without hazard), as applicable.

9.3 *Test Preparation*—The packagings shall be filled and closed with the procedures specified in Section 6.

9.4 *Drop Height:*

9.4.1 For solids and liquids, if the test is performed with the solid or liquid to be carried, or with another substance having essentially the same physical characteristics, or for liquids if the test is performed with water and the intended liquid contents have a density not exceeding 1.2 g/cm^3 (specific gravity not exceeding 1.2), the drop height shall be as specified below:

	Packing Group I	Packing Group II	Packing Group III
Drop Height	1.8 m (70. 9 in.)	1.2 m (47.2 in.)	0.8 m (31.5 in.)

9.4.2 Where packagings are filled with water and the density of the intended liquid contents is more than 1.2 g/cm^3 (specific gravity more than 1.2), to obtain the drop height in meters, multiply the relative density (specific gravity) by 1.5 for Packing Group I products, by 1.0 for Packing Group II products or by 0.67 for Packing Group III products. Round up the drop height to the first decimal. See below:

	Packing Group I	Packing Group II	Packing Group III
Drop Height	dx1.5 m (59.1 in.)	dx1.0 m (39.4 in.)	dx0.67 m (26.4 in.)

9.5 *Test Procedure:*

9.5.1 Drop the packages from the heights specified in 9.4, in the conditioning atmosphere or immediately following removal from the conditioning atmosphere. Following the drop, examine the exterior of each package for evidence of leakage, spillage, and damage.

9.5.2 *Steel Drums, Aluminum Drums, Plywood Drums, Wooden Barrels and Steel Jerricans*—Test six packagings. No conditioning is required. Drop test three packages, one drop each, diagonally on the chime or, if the packaging has no chime, on a circumferential seam or edge. Drop test three additional packages, one drop each, on the weakest part not tested by the first drop, for example a closure or, for some cylindrical drums, the welded longitudinal seam of the drum body. Perform drop tests in accordance with Test Method D 997 without the beam hazard.

9.5.3 *Fiberboard Drums*—Test six drums, conditioned in accordance with 7.1. Drop test three packages, one drop each, diagonally on the chime or, if the packaging has no chime, on a circumferential seam or edge. Drop test three additional packages, one drop each, on the weakest part not tested by the first drop, for example a closure. Perform drop tests in accordance with Test Method D 997 without the beam hazard.

9.5.4 *Plastic Drums, Plastic Jerricans, Plastic Composite Packagings*—Test six packages conditioned in accordance with 7.3. Drop test three packages, one drop each, diagonally on the chime or, if the packaging has no chime, on a circumferential seam or edge. Drop test three additional packages, one drop each, on the weakest part not tested by the first drop, for example a closure. Perform drop tests in accordance with Test Method D 997 without beam hazard if cylindrical in shape, or Test Method D 775 if rectangular in shape.

9.5.5 *Wooden Boxes, Plywood Boxes, Reconstituted Wood Boxes, Steel Boxes and Expanded Plastic Boxes*—Test five boxes, one box for each drop. No conditioning is required. Drop individual boxes flat on bottom, top, one long side, one short side. Drop box diagonally on bottom corner. Perform drop tests in accordance with Test Method D 775.

9.5.6 *Fiberboard Boxes*—Test five boxes, one for each drop, conditioned in accordance with 7.1. Drop individual boxes flat on bottom, top, one long side, one short side. Drop box diagonally on the manufacturer's joint bottom corner (5-2-3 corner). Perform drop tests in accordance with Test Method D 775.

9.5.7 *Textile Bags*—Test three bags, two drops per bag. No conditioning is required. First drop bags flat on face, followed by drop on end (valve end if applicable). Perform drop tests in accordance with Method D 959, without beam hazard.

9.5.8 *Paper Bags*—Test three bags, conditioned in accordance with 7.1, two drops per bag. First drop bags flat on face, followed by drop on end (valve end if applicable). Perform drop tests in accordance with Method D 959, without beam hazard.

9.5.9 *Plastic Fabric Bags and Plastic Film Bags*—Test three bags, three drops per bag. No conditioning is required. First drop bags flat on face, followed by flat drop on side, followed by drop on end. Perform drop tests in accordance with Method D 959, without beam hazard.

9.5.10 *Composite Packagings, Plastic Receptacle with Outer Plywood Box*—Test five packages conditioned in accordance with 7.3. Perform tests specified in 9.5.5.

9.5.11 *Composite Packagings Plastic Receptacle with Outer Fiber Drum*—Test six packages, one drop per package. Perform tests specified in 9.5.3 except that conditioning shall be in accordance with 7.3.

9.5.12 *Composite Packagings Plastic Receptacle with Outer Fiber Box*—Test five packages, one drop per package. Perform tests specified in 9.5.6, except that conditioning shall be in accordance with 7.3.

9.5.13 *Composite Packagings Plastic Receptacle with Outer Steel Drum*—Test six packages, one drop per package. Perform tests specified in 9.5.2 except that conditioning shall be in accordance with 7.3.

9.5.14 *Combination Packagings*—Test packages as specified for the outer packaging, except that combination packagings with inner plastic packagings other than bags and expanded polystyrene boxes shall be conditioned in accordance with 7.3.

9.6 *Criteria for Passing Test:*

9.6.1 When equilibrium has been reached between the internal and external pressures, and the package is rotated through all planes, there shall be no leakage of the filling substance. A slight discharge from the closure(s) upon impact shall not be considered to be a failure provided that no further leakage occurs.

9.6.2 There shall be no release of any of the inner packages from the outer packaging, but minor exposure of the inner packages which does not permit their withdrawal is acceptable. Inner packages shall be examined for leakage. There shall be no leakage of the filling substance from inner packages.

10. Stacking Test

10.1 The stacking test shall be performed on all types of packagings, other than bags, prior to initial use and, where specified, at periodic intervals.

10.2 The stacking test shall conform to Test Method D 4577, or be performed with the actual stacked load. If a compression testing machine is used, it shall be capable of applying a constant force to an accuracy of ±1 % when determined in accordance with Practices E 4.

10.3 Subject three separate packages to the stacking test by applying a force equivalent to the total actual weight of identical packages which might be stacked on a package during transport. The force should be calculated on the basis of the actual mass of the packages shipped. (See X4.) The force shall be applied to the top of the package in the same manner as applied by the actual loaded package. Unless otherwise specified a stacking height of at least three meters (118 in.) shall be used in calculating the load (see 10.5.1).

10.4 *Test Preparation:*

10.4.1 Fill and close the packaging with the procedures specified in Section 6.

10.4.2 Condition plastic drums, plastic jerricans, and plastic boxes intended for liquids and composite packagings consisting of plastic receptacles for liquids contained in outer plastic drums or plastic boxes in accordance with Practice D 4332 at 40 ± 2°C (104 ± 4°F) and test under those conditions.

10.4.3 Condition packages fabricated from fiberboard, paperboard, or paper including composite packagings with outer fiberboard packagings to 23 ± 1°C (73.4 ± 2°F) and 50 ± 2 % relative humidity, in accordance with Method D 685 or Practice D 4332 as applicable (see Section 7) and test under those conditions.

10.4.4 Packagings other than those covered in 10.4.2 and 10.4.3 require no conditioning prior to the stacking test.

10.5 *Test Procedures:*

10.5.1 The test sample shall be subjected to a force applied evenly on the top surface of the test specimen. The force to be applied shall be not less than as calculated using the following formula:

$$F = \frac{9.8\ m\ (3000\text{-}h)}{h}$$

where:

m = gross mass of package (as shipped), in kilograms,
h = height of package, in millimeters, and
F = force, in newtons.

or:

$$F = \frac{w\ (118\text{-}h)}{h}$$

where:

w = gross weight of package (as shipped), in pounds,
h = height of package, in inches, and
F = force, in pounds.

Where a constant load is used, the mass of the constant load shall be not less than as calculated using the following formula:

$$M = \frac{m\ (3000\text{-}h)}{h}$$

where:

m = gross mass of package (as shipped), in kilograms,
h = height of package, in millimeters, and
M = constant load mass, in kilograms.

or:

$$W = \frac{w\ (118\text{-}h)}{h}$$

where:

w = gross weight of package (as shipped), in pounds,
h = height of package, in inches, and
W = constant load weight, in pounds.

10.5.2 For packagings other than plastic drums and plastic jerricans intended for liquids and composite packagings consisting of plastic receptacles for liquids contained in outer plastic drums, the force or constant load shall be applied for a period of 24 h.

10.5.3 For plastic drums and plastic jerricans intended for liquids and composite packagings consisting of plastic receptacles for liquids contained in outer plastic drums or solid plastic boxes, the force or constant load shall be applied for a period of 28 days at 40 ± 2°C (104 ± 4°F). Upon removal of the force or constant load, examine the package for evidence of leakage, spillage, and damage.

NOTE 1—It may be desirable to measure the deflection during the test and prior to removal of the load.

10.6 *Criteria for Passing Test*—No test sample shall leak. There shall be no leakage of the filling substance from the inner receptacle of composite packagings or from inner packages of combination packagings. The packages shall show no deformation likely to reduce their strength or integrity or to cause instability in 3 m (118 in) stacks of packages. Sufficient stacking balance has been obtained when, after the stacking test—in the case of plastic packagings, after cooling to ambient temperature—two filled packages of the same type placed on the filled test sample maintain their position for a minimum of 1 h.

11. Hydrostatic Test

11.1 The hydrostatic test shall be performed on all types of packagings intended for the transportation of liquids, except inner packagings of combination packagings which are not required to be tested unless otherwise specified, prior to initial use and, where specified, at periodic intervals.

NOTE 2—Special requirements for air transport may not be covered in their entirety by this test.

11.2 Perform the hydrostatic test in accordance with Annex A2. Three samples per design type and manufacture shall be tested.

11.3 Metal packagings and composite packagings (glass, porcelain, or stoneware) including their closures shall be subjected to the test pressure for 5 min.

11.4 Plastic packagings and composite packagings (plastic material) including their closures shall be subjected to the test pressure for 30 min.

11.5 The hydraulic pressure (gage), as determined by any one of the following methods, shall be:

(a) not less than the total gage pressure measured in the packaging (that is, the vapor pressure of the filling substance and the partial pressure of the air or other inert gases, minus 100 kPa) (14.5 psi) at 55°C (131°F)), multiplied by a safety factor of 1.5. This total gage pressure shall be determined on the basis of a maximum degree of filling such that the packaging shall not become liquid full at 55°C (131°F) and a filling temperature of 15°C (50°F);

(b) not less than 1.75 times the vapor pressure at 50°C (122°F) of the substance to be transported, minus 100 kPa (14.5 psi) with a minimum test pressure of 100 kPa (14.5 psi);

(c) not less than 1.5 times the vapor pressure at 55°C (131°F) of the substance to be transported, minus 100 kPa (14.5 psi) with a minimum test pressure of 100 kPa (14.5 psi).

11.6 In addition, packagings intended to contain substances of Packing Group I shall be tested at least to a minimum test pressure of 250 kPa (36.3 psi) (gage) for a test period of 5 or 30 min depending upon the material of construction of the packaging, in accordance with 11.3 or 11.4.

11.7 Alternative means of pressure testing, such as a vacuum chamber, may be used where applicable.

11.8 *Criteria for Passing the Test*—There shall be no leakage as evidenced by visible liquid on any external surface of the packaging or on the paper that was placed on the floor.

12. Leak Resistance Test

12.1 The leak resistance test shall be performed as a pre-production test on all types of packagings intended to contain liquids prior to production for use. For the pre-production test three samples per design type and manufacture shall be tested.

12.2 The leak resistance test shall be performed on each production packagings intended to contain liquids prior to first use and, after reconditioning, prior to reuse. For the production leak resistance test, the packagings need not have their own closures fitted.

12.3 Unless otherwise specified, the leakage test is not required for the inner packagings of combination packagings.

12.4 The leak resistance test shall be conducted in accordance with Annex A3.

12.5 The air pressure (gage) to be applied shall be:

Packing Group I	Packing Group II	Packing Group III
Not less than 30 kPa (4.4 psi)	Not less than 20 kPa (2.9 psi)	Not less than 20 kPa (2.9 psi)

12.6 *Criteria for Passing Test*—There shall be no evidence of leakage which is indicated by formation of air bubbles.

13. Water Resistance Test

13.1 The water resistance test shall be performed on all types of outer fiberboard boxes prior to initial use and, where specified, at periodic intervals.

13.2 Condition the fiberboard in accordance with Method D 685.

13.3 Perform the test on the specimens as specified in ISO 535 or TAPPI T 441.

13.3.1 The flutes in the test specimen may be crushed with a heavy roller or other suitable means to flatten them so a good ring seal can be achieved.

13.3.2 Expose the test material to the water for 30 min.

13.4 *Criteria for Passing*—The water absorption shall not exceed 155 g/m^2 (0.508 oz/ft^2).

14. Cooperage Test

14.1 The cooperage test shall be performed on bung type wooden barrels prior to initial use and, where specified, at periodic intervals.

14.2 Perform the cooperage test in accordance with Annex A4.

14.3 The increase in the cross section diameter, expressed as a percentage of the original diameter, shall not exceed 10 %.

15. Recommendations

15.1 It is recommended that for transportation of hazardous materials in packages in Canada and the United States that the additional tests as described in Appendix X1 be carried out.

16. Report

16.1 The report shall include the following:

16.1.1 Identification of the packagings using exploded view drawings or cross-sectional drawings, including detailed data on material construction of inner packagings and cushionings of combination packagings, inner receptacles of composite packagings, and data on closures.

16.1.2 Identification of contents as tested, including net and gross masses.

16.1.3 The degree of severity of the test: Packing Group I, Packing Group II, or Packing Group III.

16.1.4 Method of conditioning, if any, prior to each test, and whether the conditioning atmosphere was maintained during the test.

16.1.5 For compatibility test, the specific hazardous material tested, including concentration, the temperature and the duration of the test, the condition of the packaging prior to performance of the drop tests, the stacking test, and where applicable, the leak resistance pre-production and hydrostatic tests, and the results of these tests.

16.1.6 For drop tests, the orientation of each drop, the height of each drop, details of observed damage, and whether the packaging met the criteria for passing the test.

16.1.7 For the stacking test, the duration of the test, the force in newtons, whether a compression testing machine or a constant load tester was used, and whether the criteria for passing the test were met. When determined, the deflection shall be reported.

16.1.8 For the hydrostatic test, whether a vented closure was replaced or sealed, the test pressure, time of maintenance of the test pressure, the test pressure at failure where applicable, and whether the packaging met the criteria for passing the test. For Packing Group I packaging, whether the actual closure was used in performing the test required in 11.4.6.

16.1.9 For the leak resistance pre-production test, whether a vented closure was replaced or sealed, the test pressure and whether the packaging met the criteria for passing the test.

16.1.10 For the leak resistance production test no test report is required.

16.1.11 For the water resistance test, the water absorption measured in grams per square meter (ounces per square foot).

16.1.12 For the cooperage test, the age of the barrel since manufacture in days, the diameter of the cross section of the barrel at the location of the head before the test, and after the test, the time to measurement after removal of the last hoop, the diameter measured after the test.

16.1.13 A statement to the effect that all tests were made in full compliance with the requirements of this standard, or noting any variations and detailing them.

16.1.14 The name and address of the testing agency, the date, and the signature of a responsible officer of the testing agency.

17. Precision and Bias

17.1 The precision and bias of this test method is dependent on those of the various test methods used and have not been expressly determined.

ANNEX

(Mandatory Information)

A1. Compatibility Test

A1.1 *Scope:*

A1.1.1 This test method determines the compatibility of a receptacle to the intended contents.

A1.1.2 *This test method involves hazardous materials. The method does not purport to address all of the safety problems associated with its use. It is the responsibility of the user of this test method to establish appropriate safety and health practices and determine the applicability of regulatory limitations prior to use.*

A1.2 *Significance and Use*—This test method is used to determine permeability and compatibility of receptacles by filling packagings to rated capacity with the specific hazardous material and storing them at room temperature for at least 180 days, or as defined in A1.4.6.

A1.3 *Apparatus:*

A1.3.1 *A suitable storage area*, capable of maintaining the required minimum temperature, is required.

A1.3.2 The storage area must be suitable for storage of the particular hazardous materials involved.

A1.4 *Procedure:*

A1.4.1 Fill sufficient test packagings to perform the drop tests, stacking tests, leak resistance tests and, where applicable, the hydrostatic tests.

A1.4.2 Fill packagings with each of the actual substances and at the actual concentrations to be carried, and close with the procedures in Section 6.

A1.4.3 Weigh each package and record the initial gross mass.

A1.4.4 Inner packages must be placed in outer packagings.

A1.4.5 The packaging shall be subjected to a superimposed load as specified in Section 10.

A1.4.6 Store the test packages at room temperature, not less than 18°C (64°F) for at least 180 days, or at not less than 50°C (122°F) for 28 days, or at not less than 60°C (140°F) for 14 days. Any substance likely to become unstable at elevated temperatures shall be stored at not over 18°C (64°F) for 180 days. Appropriate precautions must be taken to ensure that filled packages are stored in a safe manner.

A1.4.7 At the end of the test period, weigh each package and record the gross mass after storage. Calculate the percentage gross mass loss or gain as a percentage of the original gross mass.

A1.4.8 Empty and rinse each package and examine the exterior and interior of each package for evidence of leakage or damage.

A1.5 *Precision and Bias*—The precision and bias of this test method have not been determined.

A2. Hydrostatic Test

A2.1 *Scope*—This test method determines the ability of a packaging to withstand internal positive pressure.

A2.2 *Significance and Use:*

A2.2.1 This test method is used to determine the ability of a packaging to withstand a positive internal pressure by subjecting the packaging to a hydrostatic pressure and observing for any leakage while under pressure.

A2.2.2 This test method is conducted on all packagings intended to contain liquids, except inner packagings of combination packagings which do not require testing unless otherwise specified. This test method consists of pressurizing three packagings hydrostatically to the specified pressure. While at this pressure, the packagings are inspected for leakage.

A2.3 *Apparatus:*

A2.3.1 *A pressure gage*, of laboratory precision quality (± 0.25 % full scale accuracy) with graduations not exceeding 5 kPa (0.5 psi).

A2.3.2 *Equipment*, to provide regulated hydrostatic pressure.

A2.3.3 *Fittings and plumbing*, required to hydrostatically pressurize the packagings.

A2.3.4 *Light-Colored Absorbent Paper.*

A2.3.5 *Water*, or other suitable liquid as the hydrostatic test liquid.

A2.4 *Conditioning*—This test method is performed while the packagings are at ambient temperature, except that packagings fabricated in part of fiberboard shall be prepared in accordance with A2.5, then conditioned as specified in 7.1 and tested without delay.

A2.5 *Test Preparation:*

A2.5.1 Vented closures shall be replaced with non-vented closures, or the vents shall be sealed.

A2.5.2 Install an appropriate fitting in the packagings in such a manner as not to affect the performance of the packaging.

A2.5.3 Completely fill the packaging, eliminating all air pockets, with water.

A2.5.4 Close the packagings as specified in Section 6 and connect the hydrostatic pressure supply to the test fitting in the packaging, making sure the system is completely filled with water.

A2.5.5 Completely dry all external surfaces of the packaging.

A2.5.6 Position the packaging in the center of a sheet of absorbent paper previously placed on the floor. The paper should be dry and extend at least 300 mm (12 in.) in all directions beyond the base of the package.

A2.6 *Test Procedure:*

A2.6.1 The manner in which the packagings are supported shall not invalidate the test.

A2.6.2 Hydrostatically pressurize the packagings at a rate of 25 kPa/min (3.6 psi/min) to the specified pressure and maintain for the specified time.

A2.6.3 At the end of the specified time and before the pressure is released, carefully inspect the packaging for evidence of leakage.

A2.6.4 Release the pressure and carefully inspect the paper and the bottom of the packaging for evidence of leakage.

A2.7 *Precision and Bias*—The precision and bias of this test method have not been determined.

A3. Leak Resistance Test

A3.1 *Scope*—This test method determines the resistance to leakage of a receptacle.

A3.2 *Significance and Use*—This test method is used to determine the leak resistance of a receptacle by subjecting it to a low positive internal air pressure and, by immersion or other means, detecting any leaks. It is used as a pre-production test and as a production test on all production packagings intended to contain liquids prior to first use and, after reconditioning, prior to reuse. Unless otherwise specified, this test is not required for the inner packagings of combination packagings.

A3.3 *Apparatus:*

A3.3.1 *A pressure gage*, of laboratory precision quality (± 0.25 % full-scale accuracy) with graduations not exceeding 5 kPa (0.5 psi).

A3.3.2 *Equipment*, to provide regulated air pressure.

A3.3.3 *A tank*, large enough to immerse the receptacle under test and which permits examination of all surfaces and seams of the receptacle while it is under constant air pressure to detect leakage by formation of bubbles.

A3.3.4 Any other apparatus capable of detecting the same threshold rate of leakage as is detected by the immersion method may be used.

A3.4 *Procedure (Pre-Production Tests):*

A3.4.1 Vented closure shall be replaced with non-vented closures or the vents shall be sealed.

A3.4.2 The packagings including their closures shall be restrained under water while the specified internal air pressure is applied. The method of restraint shall not affect the results of the test. Other methods at least equally effective may be used.

A3.4.3 The packagings shall be carefully examined for leakage as evidenced by a stream of bubbles.

A3.5 *Procedure (Production Tests):*

A3.5.1 Vented closures shall be replaced with non-vented closures or the vents shall be sealed.

A3.5.2 For the production leak resistance test the packagings need not have their own closures fitted.

A3.5.3 The packagings including their closures shall be restrained under water while the specified internal air pressure is applied.

A3.6 *Precision and Bias*—The precision and bias of this test method have not been determined.

A4. Cooperage Test

A4.1 *Scope*—This test method determines the stability of the cooperage of a wooden barrel.

A4.2 *Significance and Use*—This test method is used to determine the stability of the cooperage of a bung type wooden barrel by removing the hoops of the barrel above the bilge and measuring the tendency of the staves to straighten.

A4.3 *Apparatus—A flexible tape measure* or *ruler.*

A4.4 *Procedure:*

A4.4.1 Store the barrel for at least 48 h at ambient temperature and humidity prior to making initial measurements.

A4.4.2 Measure and record the diameter of the cross section of the barrel at the location of the head, to the nearest millimeter (0.04 in.).

A4.4.3 Remove all hoops above the bilge, starting with the hoop closest to the bilge, and progressing towards the head.

A4.4.4 Measure and record the diameter of the cross section of the barrel at the location of the head to the nearest millimeter (0.04 in.), at least 5 but not more than 10 min after removal of the last hoop.

A4.5 *Precision and Bias*—The precision and bias of this test method have not been determined.

APPENDIX

(Nonmandatory Information)

X1. Other Test Procedures

X1.1 The procedures described in the text of the standard are to be considered a minimum requirement for adequate testing of packaging for hazardous materials. Additional test procedures should be conducted to evaluate the packaging as is suggested by shipping experience, improved packaging technology, and good engineering practice.

X1.2 A vibration test using an appropriate method in accordance with Methods D 999 is strongly recommended.

X2. Quality Assurance

X2.1 The test procedures described in this document are intended to be part of a more comprehensive quality assurance program for the packaging of hazardous substances. It is recommended that ANSI/ASQC Z-1.15 be consulted for basic guidance. This includes the consideration of subjects such as the inspection and control of raw materials and of packaging components, the periodic inspection and test of completed packaging based on sound statistical practice, and the monitoring of field performance.

X2.2 When this standard is referenced as part of a contract or purchase order or is referenced by a regulatory agency, quality assurance provisions such as an acceptance sampling plan or more detailed material and component descriptions may be included. Other provisions for a package to retain a "qualified" status may also be considered.

X3. Stacking Height

X3.1 The forces calculated for the stacking test are a minimum, based on top loading of similar packages. The package may be top loaded with heavier cargo. Consideration should be given to using a stacking force based on actual shipping experience.

Standard Test Method for Leakage Testing of Empty Containers by Vacuum Method[1]

This standard is issued under the fixed designation D 4991; the number immediately following the designation indicates the year of original adoption or, in the case of revision, the year of last revision. A number in parentheses indicates the year of last reapproval. A superscript epsilon (ε) indicates an editorial change since the last revision or reapproval.

1. Scope

1.1 This test method covers the testing of empty containers for resistance to leakage under differential pressure conditions such as those which can occur during air transport. It is suitable for testing containers intended for the transportation of some hazardous liquids in accordance with the United Nations Recommendations On The Transport Of Dangerous Goods (UN TDG) and the International Civil Aviation Organization Technical Instructions For The Safe Transport Of Dangerous Goods By Air (ICAO TIs).

1.2 *This standard does not purport to address the safety problems associated with its use. It is the responsibility of the user of this standard to establish appropriate safety and health practices and determine the applicability of regulatory limitations prior to use.*

2. Referenced Documents

2.1 *ASTM Standards:*

D 996 Terminology of Packaging and Distribution Environments[2]

D 3078 Test Method for Leaks in Heat-Sealed Flexible Packages[2]

2.2 *United Nations Document:*

ST/SG/AC.10/1 Recommendations On The Transport Of Dangerous Goods[3]

2.3 *International Civil Aviation Organization:*

DOC 9284-AN/905 Technical Instructions For The Safe Transport Of Dangerous Goods By Air[4]

3. Terminology

3.1 Terms and definitions used in this standard may be found in Terminology D 996.

4. Summary of Test Method

4.1 An empty container to be tested is pressurized by immersion in an ethylene glycol-water solution in a transparent test chamber and subjected to a gradually increasing partial vacuum. The container is observed for signs of leakage, as evidenced by escaping air bubbles during depressurization or fluid in the container after re-pressurization.

[1] This test method is under the jurisdiction of ASTM Committee D-10 on Packaging and is the direct responsibility of Subcommittee D10.21 on Shipping Container Environment.

Current edition approved Oct. 27, 1989. Published December 1989.

[2] *Annual Book of ASTM Standards*, Vol 15.09.

[3] Available from United Nations Sales Section, United Nations, New York, NY.

[4] Available from Intereg Group, Inc., 5724 N. Pulaski Rd., Chicago, IL 60646.

5. Significance and Use

5.1 Containers may be pressurized in accordance with this test method without modification to the closure or to the body of the container. This method may be used for testing containers intended for the transportation of some liquids by air in accordance with the ICAO TIs or in accordance with the UN TDG.

5.2 This test method establishes the point at which leakage commences, with a limit of approximately 95 kPa (13.8 psi) differential. For flexible containers, the pressure differential at leak cannot be accurately determined using this test method. This test method, rather, establishes resistance to leakage of a flexible container of specified volume at normal pressure when the surrounding atmospheric pressure is subsequently reduced. See Test Method D 3078.

5.3 This test method may not be suitable for some packages, such as paper laminated pouches or packages with paper cap seals, where the test fluid may rapidly deteriorate the packaging.

6. Apparatus

6.1 *Transparent Vessel*, large enough to permit the test specimen(s) to be immersed in the test fluid with a minimum headspace of 20 %, capable of withstanding at least 1½ atm pressure differential, fitted with a flat vacuum-tight cover is required.

6.2 *Vacuum Gauge, Inlet Tube from a Source of Vacuum, and Outlet Tube to the atmosphere*, shall be sealed into the cover. The inlet and outlet tubes shall be equipped with hand operated valves. The vacuum gage shall be laboratory quality with a full scale range of 0 to 100 kPa (0 to 14.5 psi) with minimum graduations no greater than 2 kPa (0.3 psi) and accuracy to within 1 %. A suitable moisture trap, to prevent back-flow of liquid, should be fitted between the vacuum gauge and the source of vacuum.

6.3 *Solution of Ethylene Glycol in Water, 50 % by Volume* for use as a test fluid, at ambient temperature unless otherwise specified, is required. Where a test temperature other than ambient is specified, the solution shall be at the specified test temperature. Other solutions may be required for high and low temperatures.

NOTE—If ethylene glycol anti-freeze solution is used, it should be clarified by filtering through activated charcoal.

6.4 A suitable means to hold the test specimen(s) with the closure(s) not more than 25 mm (1 in.) under the surface of the test fluid is required. The method of restraint should not affect the results of the test.

7. Test Specimens

7.1 Test specimens shall consist of containers, including

their closures. Test specimens of containers which are in production shall be representative samples, taken at random.

7.2 Containers shall be empty. Empty containers which previously contained any substance and were cleaned of the contents prior to testing. New containers and closures, where available, should be used.

7.3 Each container shall be closed for testing in the same manner as it would be closed for actual shipment. All closures shall be installed using the techniques or torques specified by the closure manufacturer, container manufacturer or shipper. Close containers at ambient pressure and temperature.

7.4 Where a test temperature other than ambient is specified, the sample containers shall be conditioned at the test temperature for at least 24 h, and until they reach equilibrium of temperature with the conditioning atmosphere.

7.5 Unless otherwise specified, test at least three specimens.

8. Preparation of Apparatus

8.1 Assemble the apparatus as shown in Fig. 1.

8.2 Prior to testing, draw a vacuum to at least 95 kPa (13.8 psi) differential and maintain this vacuum for 10 min in order to remove air dissolved in the test fluid.

9. Procedure

9.1 Immerse the test specimens in the ethylene glycol-water solution inside the transparent test chamber. Remove as many bubbles as possible from the outer surface of the test specimens by agitation or other means.

9.1.1 More than one specimen can be tested at one time provided that there is sufficient space in the test chamber to allow clear observation of each specimen. If this is done, affix the test specimens so closures are sufficiently separated to permit individual observation of each, to detect leakage.

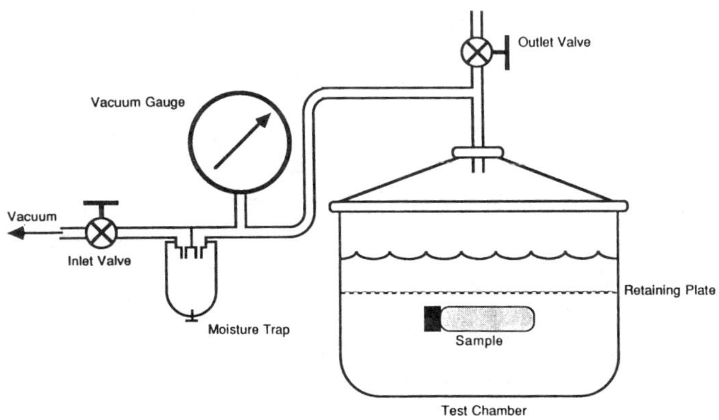

FIG. 1 Typical Test Setup for Leakage Testing of Containers by Vacuum Method

9.2 Seal the lid, open the hand valve on the inlet tube and close the hand valve on the outlet tube to the atmosphere. Turn on the vacuum so that the gage rises slowly (from 30 s to 1 min) to a pressure differential of 95 kPa (13.8 psi), or other specified pressure differential. The vacuum can be controlled by cracking open the hand valve on the outlet tube. During the rise in vacuum, observe the test specimen(s) for leakage as evidenced by a continuous stream or recurring succession of small bubbles. Isolated bubbles caused by trapped air are not considered evidence of leakage. Maintain the vacuum for a minimum of 10 min. For test specimens which are single packagings or combination packagings in whole or part of plastic material, maintain the vacuum for 30 min. Then turn the vacuum off and release it slowly by opening the hand valve on the outlet tube. Keep the test specimen(s) immersed for at least 30 s at ambient pressure. Remove the test specimen(s) and examine the inside for fluid.

9.3 Leakage is indicated by a continuous stream or recurring succession of bubbles. Fluid within the test specimen is also evidence of leakage.

10. Report

10.1 The report shall include the following:

10.1.1 Identification of the containers and closures, using manufacturer's part numbers and descriptions or using exploded view drawings or cross-sectional drawings, including detailed data on material construction,

10.1.2 Total number of specimens tested,

10.1.3 Number of specimens tested at one time,

10.1.4 Test pressure requirement, where applicable,

10.1.5 Whether test was performed at ambient temperature and if not, the test temperature. If a test solution other than 50 % ethylene glycol was used, identify the solution,

10.1.6 Whether test specimens passed or failed the test, where applicable,

10.1.7 Whether each test specimen leaked and at what pressure differential,

10.1.8 Whether test fluid was apparent in each test specimen after testing,

10.1.9 A statement to the effect that all tests were made in full compliance with the requirements of this test method, or noting any variations and detailing them, and

10.1.10 The name and address of the testing agency, the date, and the signature of a responsible representative of the testing agency.

11. Precision and Bias

11.1 No statement is made about either precision or bias of this test method for measuring leakage since the result merely states whether there is conformance to the criteria for success specified in the procedure.

APPENDIX

(Nonmandatory Information)

TABLE X1 Pressure Conversion Table—Pressure Differential[A]

kiloPascals (kPa)	pounds per square inch (psi)	inches of mercury in Hg (20°C)	bar
10	1.45	2.96	0.10
20	2.90	5.93	0.20
30	4.35	8.89	0.30
40	5.80	11.85	0.40
50	7.25	14.82	0.50
60	8.70	17.78	0.60
70	10.15	20.75	0.70
75	10.88	22.23	0.75
80	11.60	23.71	0.80
85	12.33	25.19	0.85
90	13.05	26.67	0.90
95	13.78	28.16	0.95
100	14.50	29.64	1.00
101.325	14.70	30.03	1.01325

[A] psi × 6.894757 = kPa
in Hg × 3.37411 = kPa
bar × 100 = kPa

Standard Terminology Relating to Electrostatic Discharge (ESD) Packaging Materials[1]

This standard is issued under the fixed designation D 5077; the number immediately following the designation indicates the year of original adoption or, in the case of revision, the year of last revision. A number in parentheses indicates the year of last reapproval. A superscript epsilon (ϵ) indicates an editorial change since the last revision or reapproval.

1. Scope

1.1 The terms and definitions in this standard are related to Electrostatic Discharge (ESD) Packaging Materials and ESD Protective Materials.

2. Referenced Documents

2.1 *ASTM Standard:*

D 996 Terminology Relating to Packaging and Distribution Environments[2]

2.2 *EIA Standard:*

EIA-541 Packaging Material Standards for ESD, Sensitive Items[3]

3. Terminology

antistatic agent—a chemical compound which, when impregnated or formulated into or topically applied to a primary material or substrate, gives the primary material antistatic properties. See **antistatic property**.

antistatic property—the prevention of triboelectric charge generation by effectively minimizing the production of a static charge when materials are separated from another surface.

DISCUSSION—The forword of EIA-541 states, "'Antistatic' no longer refers to a resistivity range ... 'Antistatic' refers to a material's ability to resist triboelectric charge generation. A material's antistatic propensity depends upon the nature of the material itself and the material with which it is in contact along with the means of surface separation. The antistatic property is not a dependent function of material resistivity. Material resistivity is an intrinsic property used to define its degree of conductivity without regard to other materials."

conductor—a substance or body that allows a flow of electric current to pass continuously along it or through it when a sufficient voltage is applied across any two points.

dielectric breakdown—a threshold effect in a dielectric medium where, at some electric field strength across the medium, bound electrons become unbound and travel through the medium as a current. In solid media, the region of the current path is permanently damaged. The unit of measurement is usually volts per unit of thickness.

electrically continuous surface—a surface that is electrically conductive in that current can be passed at an applied voltage between any two points of its physical surface.

electrical overstress (EOS)—overstress which may be due to ESD or the operation of items beyond their electrical specifications.

electromagnetic shield—a screen or other housing placed around a device or circuit to reduce the effects on them from both electric and magnetic fields.

electrostatic discharge—the transfer of electrostatic charge between bodies at different electrostatic potentials.

electrostatic discharge (ESD) protective—a property of materials capable of one or more of the following:

(*1*) preventing the generation of static electricity.

(*2*) dissipating electrostatic changes over its surface or volume.

(*3*) providing shielding from ESD or electrostatic fields.

electrostatic discharge sensitive (ESDS)—a property of items in which they are inherently sensitive (ESDS) susceptible to either catastrophic failure or latent damage when exposed to sources of ESD. Items are often categorized as to their levels of sensitivity but in all cases require some means of ESD protective packaging and handling.

electrostatic shield—a barrier or enclosure that prevents the penetration of an electrostatic field.

DISCUSSION—An electrostatic shield may not offer much protection against the effects of an electromagnetic field. Electromagnetic shields, however, are good electrostatic shields.

ground—a metallic connection with the earth to establish zero potential.

DISCUSSION—The ground is the voltage reference point in a circuit. There may or may not be an actual connection to earth, but it is understood that a point in the circuit said to be at ground potential could be connected to earth without disturbing the operation of the circuit in any way. Grounds that can be used for static control work stations include metal water pipes, any power ground, or any large metal structural member of a building, vessel hull, etc. (See **grounding**.)

grounding—connecting to ground or to a conductor that is grounded.

insulator (electric)—a device having high electrical resistance and used for supporting or separating conductors to prevent undesired flow of current from them to other objects. (Also known as electrical insulator.)

ionization—the process by which neutral atoms or molecules, such as air, acquire a positive or negative charge.

packaging—See Terminology D 996.

DISCUSSION—The term packaging as used in this context should not be confused with the term as it is used in the electronics industry where it describes the assembly, black box enclosure, or physical encasing of microelectronic devices.

surface resistivity, ρ_s—the surface resistivity of a material is the ratio of the potential gradient parallel to the current along its surface to the current per unit width of the

[1] This terminology is under the jurisdiction of ASTM Committee D-10 on Packaging and is the direct responsibility of Subcommittee D10.13 on Interior Packaging.

Current edition approved June 29, 1990. Published August 1990.

[2] *Annual Book of ASTM Standards*, Vol 15.09.

[3] Available from the Electronic Industries Association, 2001 Eye St., N.W., Washington, DC 20006.

surface.

Discussion—Surface resistivity of a material is numerically equal to the surface resistance between two electrodes forming opposite sides of a square. The size of the square is immaterial.

volume resistivity, ρ_v—the volume resistivity of a material is the ratio of the potential gradient parallel to the current in the material to the current density.

Discussion—In the metric system, volume resistivity of an electrical insulating material in ohm-cm is numerically equal to the volume resistance in ohms between opposite faces of a 1-cm cube of the material. (Volume resistivity in Ω·m has a value of 1/100 of the alue in Ω·cm.)

This standard is subject to revision at any time by the responsible technical committee and must be reviewed every five years and if not revised, either reapproved or withdrawn. Your comments are invited either for revision of this standard or for additional standards and should be addressed to ASTM Headquarters. Your comments will receive careful consideration at a meeting of the responsible technical committee, which you may attend. If you feel that your comments have not received a fair hearing you should make your views known to the ASTM Committee on Standards, 1916 Race St., Philadelphia, PA 19103.

Designation: D 5094 – 90

Standard Test Methods for Gross Leakage of Liquids from Containers with Threaded or Lug-Style Closures[1]

This standard is issued under the fixed designation D 5094; the number immediately following the designation indicates the year of original adoption or, in the case of revision, the year of last revision. A number in parentheses indicates the year of last reapproval. A superscript epsilon (ε) indicates an editorial change since the last revision or reapproval.

1. Scope

1.1 These test methods cover the determination of gross leaks in rigid and semi-rigid containers (up to 4 L (1.06 gal)) with threaded or lug-style closures. Such tests may be used to indicate the ability of a liquid container to survive the distribution environment without leaking. These test methods may not be suitable to determine the leak resistance of containers intended for transport of hazardous materials.

1.2 *Test Methods:*

1.2.1 *Test Method A, Shipping Container Vibration and Storage Test*, covers the ability of a shipping container and its interior packaging to protect the contents from leakage after transportation induced vibration and high-temperature storage.

1.2.2 *Test Method B, Shipping Container Vibration and Vacuum Chamber Test*, is suitable for individual containers and is usually less severe than Test Method A. The advantage of Test Method B is the shortness of the test.

1.3 *This standard does not purport to address the safety problems associated with its use. It is the responsibility of the user of this standard to establish appropriate safety and health practices and determine the applicability of regulatory limitations prior to use.*

2. Referenced Documents

2.1 *ASTM Standards:*

D 996 Terminology of Packaging and Distribution Environments[2]

D 999 Methods for Vibration Testing of Shipping Containers[2]

D 3198 Test Method for Application and Removal Torque of Threaded or Lug-Style Closures[2]

D 3474 Practice for the Calibration and Use of Torque Meters Used in Packaging Applications[2]

D 4169 Practice for Performance Testing of Shipping Containers and Systems[2]

D 4332 Practice for Conditioning Containers, Packages, or Packaging Components for Testing[2]

3. Terminology

3.1 General definitions for packaging are found in Terminology D 996.

3.2 *Descriptions of Terms Specific to This Standard:*

3.2.1 *gross leak*—any opening in the container or closure seal that produces visible leakage such that the container would not be suitable for transport or subsequent distribution.

3.2.2 *leak*—any opening in a container which, contrary to intention, either lets contents escape or permits substances to enter.

3.2.3 *leakage*—that which passes through a leak.

4. Summary of Test Methods

4.1 *Method A, Shipping Container Vibration and Storage Test*—Test specimens are filled to their expected fill capacity with product or liquid simulating product and the closure is applied to the container. Closures should be applied with the same torque as would be encountered in production. The specimens are packed into shipping containers and vibrated. The specimens, stored on their sides, are subjected to 40°C (104°F) (or other conditions as appropriate) for four weeks. Each specimen is examined for leakage.

4.2 *Method B, Container Vibration and Vacuum Chamber Test*—Test specimens are partially filled with product or liquid simulating product and the closure is applied to the container. Closures should be applied with the same torque as would be encountered in production. The specimens are packed into shipping containers and vibrated. Following vibration, the individual specimens are removed from the shipping container and subjected to 10 in. Hg (33.7 kPa) vacuum for 10 min and examined for leakage.

NOTE 1—Other levels of vacuum, or time, or both may be appropriate.

5. Significance and Use

5.1 These test methods are used to indicate the integrity of the container and closure system in the distribution environment.

5.2 These test methods measure the ability of the container and closure system to prevent leakage when stored or transported upright, inverted, or on the side.

5.3 These test methods allow for comparison of container/closure designs of threaded and lug-style closures of similar or different materials, or different manufacturers. These tests are suitable for packaging development, engineering and tooling evaluations.

6. Apparatus

6.1 *Method A, Shipping Container Vibration and Storage Test:*

6.1.1 *Drying Oven*, constant-temperature, equipped with

---

[1] This test method is under the jurisdiction of ASTM Committee D-10 on Packaging and is the direct responsibility of Subcommittee D10.32 on Consumer Packages.

Current edition approved June 29, 1990. Published August 1990.

[2] *Annual Book of ASTM Standards*, Vol 15.09.

means for ensuring adequate temperature control, 40 ± 1°C (104 ± 1.8°F) and air circulation, ambient humidity, or other conditions as appropriate.

6.1.2 *Torque Meter*, with an appropriate scale which accurately measures within the expected torque range.

6.1.3 *Vibration Test Machine*, the apparatus described in the Apparatus section of Methods D 999 may be used.

6.2 *Method B, Shipping Container Vibration and Vacuum Test:*

6.2.1 *Vacuum Chamber*, any suitable chamber capable of withstanding approximately one atmosphere pressure differential, fitted with a vacuum-tight seal. A vacuum gauge, an inlet tube from a source of vacuum, and an outlet tube to the atmosphere shall be sealed to the chamber. The inlet and outlet tubes shall be equipped with shut-off valves. The vacuum gauge shall be laboratory quality with a full scale range of 0 to 30 in. Hg (0 to 100 kPa).

6.2.2 *Torque Meter*, with an appropriate scale which accurately measures within the expected torque range.

6.2.3 *Vibration Test Machine*, the apparatus described in the Apparatus section of Methods D 999 may be used.

7. Sampling

7.1 The complete test unit shall consist of the shipping container, with interior packaging and the actual test specimens (container(s) and closure(s)).

7.2 A minimum of three specimens representing each variable, such as cap or bottle mold from each cavity but not less than ten specimens are to be selected. The total number of specimens may need to be increased to completely fill the shipping container. More than one shipping container may be required.

7.3 Select unused closures and unused containers that are within specifications.

8. Test Specimens

8.1 *Test Method A*—A specimen shall be one container filled to expected fill capacity with product or liquid simulating product. The closure is secured in accordance with 11.1.

8.2 *Test Method B*—A specimen shall be one container one-third filled with product or liquid simulating product. The closure is secured in accordance with 11.1.

8.3 Colorant should be added to the contents when necessary to give permanent indication where there is leakage.

9. Calibration

9.1 Calibrate the torque meter in accordance with Practice D 3474.

10. Conditioning

10.1 Store the shipping container components at ambient laboratory conditions for at least 4 h before testing, or other conditions in accordance with Practice D 4332.

10.2 Store the torqued container/closure test specimens upright at least 24 h at ambient conditions before commencing the test.

11. Procedure

11.1 Apply closures to the containers according to the Application Torque Measurement section of the procedure in Test Method D 3198.

11.2 *Test Method A, Shipping Container Vibration and Storage Test:*

11.2.1 Place the filled specimens in their shipping container or in an appropriate carton so that they will remain firmly in place during vibration.

11.2.2 Proceed as directed according to Methods D 999 (Method A1 or A2) 20 min each on bottom, top, and one side, or run an entire shipping cycle in accordance with Practice D 4169.

11.2.3 Remove the specimens from the shipping container or carton. Examine all specimens for visible leakage without removing or disturbing the closure. Note the location(s) of leakage.

11.2.4 Place the test specimens on their sides (liquid in contact with finish/closure) in a container capable of holding the liquid should a leak develop. It is recommended that absorbent blotting paper be placed beneath the samples to more easily detect leakage. Store the test specimens at 40° ± 1°C (104° ± 1.8°F), or other appropriate temperature, ambient humidity and inspect for liquid leakage at 24 h and at one-week intervals for four weeks. Note location(s) of leakage.

11.2.5 At the conclusion of the test, remove the closure and inspect the threads for signs of leakage or crusting from partial leakage.

11.3 *Test Method B, Shipping Container Vibration and Vacuum Chamber Test:*

11.3.1 Place the partially filled specimens in their shipping container or in an appropriate carton so that they will remain firmly in place during vibration.

11.3.2 Proceed as directed according to Methods D 999 (Method A1 or A2) 20 min each on bottom, top, and one side, or run an entire shipping cycle in accordance with Practice D 4169.

11.3.3 Remove the specimens from the shipping containers or carton. Examine all specimens for visible leakage without removing or disturbing the closure. Note location(s) of leakage.

11.3.4 Place the partially filled specimens inverted in a beaker or other receptable to contain leakage.

NOTE 2—Other orientations of the specimens may be appropriate. Orientation of the container should be such that suspected leak areas are covered with the liquid inside the container.

11.3.5 Place specimens in the vacuum chamber and close the vacuum chamber.

11.3.6 Close the chamber inlet valve.

11.3.7 Open the chamber outlet valve and turn on vacuum source so that the gauge rises slowly (from 30 s to 1 min) to 10 in. Hg (33.7 kPa) and close the outlet valve.

11.3.8 Maintain the vacuum for 10 min.

11.3.9 Partially open the chamber inlet valve and release the vacuum slowly (30 s to 1 min).

11.3.10 Open the chamber.

11.3.11 Remove the specimens and examine for leakage, cracks, or other visible damage. Record the results of this examination. Note location(s) of leakage.

12. Interpretation of Results

12.1 If there is leakage to the outside of the specimen, the package fails the test.

12.2 If there is no leakage to the outside of the specimen the package passes the test.

12.3 If there is leakage to the threads but not to the outside, the specimen may or may not be acceptable. This effect may be used as a comparative evaluation between closures and mold cavities.

13. Report

13.1 Report the following information:

13.1.1 Statement that testing was done in accordance with ASTM Test Method D 5094, Test Method A or Test Method B, or that the testing was done with indicated changes.

13.1.2 Materials of construction of the closure, liner and container including finish and manufacturer's designation.

13.1.3 Description of the shipping container and interior packaging, the test frequencies, intensities, and durations used.

13.1.4 Product or substitute used and fill level.

13.1.5 Statement whether or not leakage occurred and, if possible, a report of the location of each leak.

13.1.6 Statement of the number of specimens included in the test and the number of failures, if any.

13.1.7 Conditioning time, temperature, and relative humidity.

13.1.8 Temperature and relative humidity of the test area.

13.1.9 Time between application and removal of closures.

13.1.10 Compilation of measured torques.

13.1.11 If Test Method B is used, a statement of vacuum and duration.

13.1.12 Statement of observations that may lead to improvements.

14. Precision and Bias

14.1 No statement is made about either the precision or bias of this test method since the result merely states whether there is conformance to the criteria for success specified in the procedure.

15. Keywords

15.1 closure leakage; container leakage; leak; leakage; liquid leakage; vacuum

Standard Test Method for Vibration (Horizontal Linear Sinusoidal Motion) Test of Products[1]

This standard is issued under the fixed designation D 5112; the number immediately following the designation indicates the year of original adoption or, in the case of revision, the year of last revision. A number in parentheses indicates the year of last reapproval. A superscript epsilon (ε) indicates an editorial change since the last revision or reapproval.

1. Scope

1.1 This test method covers the determination of resonances of unpackaged products and components by means of horizontal linear sinusoidal motion applied at the surface on which the product is mounted. For vertical vibration testing of products see Method D 3580.

1.2 This information may be used for the design of a container or interior package that will minimize transportation vibration inputs at the critical frequencies. Since vibration damage is most likely to occur at product resonant frequencies, these may be thought of as potential product fragility points.

1.3 Information obtained from this test method may be used for product modification. This may become necessary if a product's response would require design of an impractical or excessively costly shipping container.

1.4 This test method does not simulate vibration effects the product will encounter in operating or end-use environments. Other more suitable test procedures are referenced in Section 2.

1.5 *This standard does not purport to address all of the safety problems associated with its use. It is the responsibility of the user of this standard to establish appropriate safety and health practices and determine the applicability of regulatory limitations prior to use.* For specific precautionary statements, see Section 6.

2. Referenced Documents

2.1 *ASTM Standards:*

D 996 Terminology of Packaging and Distribution Environments.[2]

D 3580 Method for Vibration (Vertical Sinusoidal Motion) Test of Products[2]

D 4332 Practice for Conditioning Containers, Packages, or Packaging Components for Testing[2]

E 122 Practice for Choice of Sample Size to Estimate a Measure of Quality for a Lot or Process[3]

2.2 *Military Standard:*

MIL STD 810C, Method 514, Vibration[4]

2.3 *International Electromechanical Commission Document:*

68-2-6, Part 2, Test F Vibration, Basic Environmental Testing Procedures for Electronic Components and Electrical Equipment[5]

3. Terminology

3.1 *Definitions*—For definitions of terms used in this test method, see Terminology D 996.

3.2 *Descriptions of Terms Specific to This Standard:*

3.2.1 *amplification*—the ratio of measured acceleration level of the component to the measured input acceleration to the product.

3.2.2 *amplitude*—the maximum value of sinusoidal quantity (zero to peak).

3.2.3 *octave*—the interval of two frequencies having a basic ratio of 2 (1 octave/min = 0.301 decade/min).

3.2.4 *resonance*—for a system undergoing forced vibration, the frequency at which any change of the exciting frequency causes a decrease in the response of the system.

4. Significance and Use

4.1 Products are exposed to complex dynamic stresses in the transportation environment. The determination of the resonant frequencies of the product, either horizontal, vertical or both, aids the package designer in determining the proper packaging system to provide adequate protection of the product, as well as providing an understanding of the complex interactions between the components of the product as they relate to expected transportation vibration inputs.

5. Apparatus

5.1 *Vibration Test Machine,* consisting of a flat horizontal test surface of sufficient strength and rigidity such that the applied vibrations are essentially uniform over the entire test surface. The test surface shall be driven to move only horizontally along a single axis in sinusoidal motion (rotary motion is not acceptable). The frequency and amplitude of the motion shall be variable, under control, to cover the range specified in 9.3.

5.2 *Specimen-Mounting Devices,* of sufficient strength and rigidity to attach the product securely to the test surface. The resonant frequency of the mounting devices shall be above the intended test range for the product. The device(s) shall support the product in a manner similar to the way in which it will be supported in its shipping container. Relative

---

[1] This test method is under the jurisdiction of ASTM Committee D-10 on Packaging and is the direct responsibility of Subcommittee D10.15 on Fragility Assessment.

Current edition approved Aug. 31, 1990. Published November 1990.

[2] *Annual Book of ASTM Standards*, Vol 15.09.

[3] *Annual Book of ASTM Standards*, Vol 14.02.

[4] Available from the Naval Publications and Forms Center, 5801 Tabor Ave., Philadelphia, PA 19120.

[5] Available from American National Standards Institute, Inc., 1430 Broadway, New York, NY 10018.

motion between the test surface and the product shall not be permitted.

5.3 *Instrumentation:*

5.3.1 Sensors, signal conditioners, filters, and data acquisition apparatus are required to monitor or record, or both, the accelerations and frequencies at the test surface of the apparatus and at points of interest in the product. The instrumentation system shall have a response accurate to within ±5 % over the test range.

NOTE 1—Strain gage type accelerometers may be required to monitor the product, control the test system, or both.

5.3.2 For visual examination of the specimen under test, a stroboscope or video system may be beneficial.

6. Safety Precautions

6.1 **Caution:** This test method may produce severe mechanical response in the product being tested. Therefore, the means used to fasten the product to the test surface must be of sufficient strength to keep it adequately secured. Operating personnel must remain alert to potential hazards and take necessary precautions for their safety. Stop the test immediately if a dangerous condition should develop.

7. Sampling, Test Specimens, and Test Units

7.1 Test specimens and the number of samples shall be chosen to permit an adequate determination of representative performance. Whenever sufficient products are available, five or more replicate samples should be tested to improve the statistical reliability of the data obtained (see Practice E 122).

7.2 The product as intended for packaging shall constitute the test specimen. Sensor(s) may be applied as appropriate to measure data at points of interest with the minimum possible alteration of the test specimen. Parts and surfaces of the specimen may be marked for identification and reference. When necessary to observe the interior components of the product during testing, holes may be cut in noncritical areas, or noncritical panels may be removed.

8. Conditioning

8.1 Condition test specimens prior to testing and maintain them in accordance with any specific requirements applicable to the item being tested. In the absence of other specific requirements, conditioning in accordance with Practice D 4332 is recommended (standard conditioning atmosphere of 23 ± 2°C (73.4 ± 3.6°F) and 50 ± 2 % relative humidity).

9. Preparation of Apparatus

9.1 Perform the tests in the conditioned environment or immediately upon removal from that environment.

9.2 Attach the test specimen to the test surface of the apparatus in a manner that will prevent the specimen from leaving or moving across the test surface during vibration. Caution is necessary to avoid mounting methods that cause excessive stress or strain that could alter the response of the product.

9.3 Test intensities shall be sufficient to vibrate the specimen at acceleration and frequency levels that determine if product resonances exist in the expected frequency range of the transportation environment. Typical products may exhibit resonant frequencies ranging from 1.0 to above 100 Hz. Acceleration levels sufficient to excite resonance normally range from 0.1 to 0.5 g.

10. Procedure

10.1 Sweep the frequency range from the lower frequency limit to the upper frequency limit and return using either manual or automatic sweep. Starting at the lower limit, vary the frequency of the vibration at a continuous logarithmic rate of 0.5 to 1.0 octaves/min to the upper limit and back to the lower limit. Maintain an approximately constant acceleration throughout the frequency sweep, as required, to determine the product resonances. Repeat the cycle if necessary, recording all resonant responses of the specimen.

NOTE 2—Low frequency vibration requires a long stroke. It may be necessary to reduce the acceleration level inputs at low frequencies in order to stay within the stroke capabilities of the test equipment. This can be accomplished by dividing the test into two or more frequency ranges with different input acceleration levels. Response levels from different input acceleration levels may result in different transmissibilities.

NOTE 3—Frequencies above and below the frequency range from 1.0 to 100 Hz may be necessary or desirable for some products.

10.2 Monitor the acceleration and frequency data sensed on the test surface to ensure that the desired test conditions are produced. Mount the accelerometer, in the direction of motion, as close as possible to the test specimen or in a protected location on the test surface which will provide a representative response of the input to the test specimen. Shielding of the accelerometer may be desirable to minimize the potential for damage.

10.3 Monitor the test specimen and its components for any resonant vibrations. Use a stroboscope, sensor, visual, auditory or other means, as applicable, to determine these resonances. If necessary, the frequency sweep may be interrupted for short time periods in order to properly identify a resonating component.

10.4 Record the frequencies of any resonances and identify the product components that are resonating.

10.5 Repeat the test in each of the potential shipping orientations of concern.

11. Report

11.1 Report the following information:

11.1.1 Description of the test specimen, in sufficient detail for proper identification.

11.1.2 Identification of the purpose of the test.

11.1.3 Descriptions of the test sequence, the input acceleration level(s), frequency range(s) swept and the sweep rate.

11.1.4 Descriptions of any deviations from the specified test method.

11.1.5 A statement of the number of test replications, if any.

11.1.6 Identification of apparatus and instrumentation used, including date of last calibration, manufacturers' names, model numbers, and serial numbers. Details of any known modifications thereto shall be included.

11.1.7 Conditioning methodology and levels.

11.1.8 Results of any prior tests of this product.

11.1.9 Components that displayed resonant vibration and their corresponding frequencies. Any other significant data including measurements and observations shall be included.

11.1.10 If applicable, identification of components that

could be redesigned to eliminate excessive resonant vibration or to change the resonant frequency.

11.1.11 Resonant frequencies and corresponding amplifications or acceleration magnitude at resonance, when measured, shall be recorded.

11.1.12 If applicable, a description of shipping container characteristics desirable for vibration protection of the product.

12. Precision and Bias

12.1 The precision and bias of this test method have not yet been determined. A round-robin test is currently underway to provide data to determine the precision and bias of this test method.

Standard Practice for Fabrication of Fiberboard Shipping Boxes[1]

This standard is issued under the fixed designation D 5118/D 5118M; the number immediately following the designation indicates the year of original adoption or, in the case of revision, the year of last revision. A number in parentheses indicates the year of last reapproval. A superscript epsilon (ϵ) indicates an editorial change since the last revision or reapproval.

1. Scope

1.1 This practice covers the fabrication of new fiberboard boxes, liners and sleeves.

1.2 This practice points out the factors and components that must be controlled in the manufacture of corrugated and solid fiberboard boxes, liners and sleeves.

1.3 This practice does not cover the adequacy of fiberboard containers under all conditions of exposure to atmosphere, handling, shipping and storage.

1.4 The values stated in either inch-pound units or SI units are to be regarded separately as standard. Within the text, the SI units are shown in brackets. The values stated in each system are not exact equivalents; therefore, each system must be used independently of the other. Combining values from the two systems may result in non-conformance with the standard.

1.5 *This standard does not purport to address the safety problems associated with its use. It is the responsibility of the user of this standard to establish appropriate safety and health practices and determine the applicability of regulatory limitations prior to use.*

2. Referenced Documents

2.1 *ASTM Standards:*

D 685 Method of Conditioning Paper and Paper Products for Testing[2]

D 996 Terminology of Packaging and Distribution Environments[2]

D 3950 Specification for Strapping, Plastic (and Seals)[2]

D 3951 Practice for Commercial Packaging[2]

D 3953 Specification for Strapping, Flat Steel and Seals[2]

D 4727 Specification for Corrugated and Solid Fiberboard Sheet Stock (Container Grade) and Cut Shapes[2]

2.2 *Federal Specifications:*[3]

MMM-A-250 Adhesives, Water-Resistant (For Closure of Fiberboard Boxes)

PPP-B-638 Packing of Boxes, Caps, Liners, and Sleeves

FED-STD-123 Marking for Shipment (Civil Agencies)

MIL-STD-129 Marking for Shipment and Storage

2.3 *Code of Federal Regulations:*[3]

Title 49

2.4 *Other Publications:*

National Motor Freight Classification[4]

Uniform Freight Classification[5]

Federal Food, Drug and Cosmetic Act[6]

3. Terminology

3.1 Definitions of terms relating to packaging are found in Definitions D 996.

4. Significance and Use

4.1 Corrugated and solid fiberboard boxes, sleeves and liners are used to unitize products into packages of size and shape suitable for manual or mechanical handling and to protect the contents against environmental, handling, shipping, and storage conditions.

4.2 This practice covers some of the basic constructions and styles of commercially available fiberboard packaging used to unitize and protect contents.

4.3 *Use of Other Specifications*—Nothing in this practice shall be construed to prohibit the use of boxes of special design or of fiberboard packages identified by package number in the current Uniform Freight Classification and National Motor Freight Classification when in the experience and judgement of the purchaser, the nature of the articles or material to be shipped justifies such boxes or packages. Some commodities may require less protection while other commodities may require better boxes than are specified herein. Containers for explosives and dangerous articles must also comply with regulations for the Transport of Hazardous Materials (CFR Title 49).

5. Styles

5.1 Fiberboard boxes covered by this practice may be of the following styles:

5.1.1 *RSC/0201*[7]—Regular slotted box (Fig. 1).

5.1.2 *OSC/0202*[7]—Overlap slotted box (Fig. 2).

5.1.3 *FOL/0203*[7]—Full overlap slotted box (Fig. 3).

5.1.4 *SFF/0206*[7]—Special full flap slotted box (Fig. 4).

5.1.5 *CSSC/0204*[7]—Center special slotted box (Fig. 5).

5.1.6 *CSOSC/0205*[7]—Center special overlap slotted box (Fig. 5).

5.1.7 *HSCC/0312*[7]—Half slotted box with cover (Fig. 6).

---

[1] This practice is under the jurisdiction of ASTM Committee D-10 on Packaging and is the direct responsibility of Subcommittee D10.17 on Paper and Paperboard Products.

Current edition approved Dec. 5, 1990. Published February 1991.

[2] *Annual Book of ASTM Standards*, Vol 15.09.

[3] Available from Standardization Documents, Order Desk, Bldg. 4, Section D, 700 Robbins Ave., Philadelphia, PA 19111-5094, attn: NPODS.

[4] Available from National Motor Freight Classification, Inc., 2200 Mill Road, Alexandria, VA 22314.

[5] Available from National Railroad Freight Committee, Suite 1120, 222 Riverside Plaza, Chicago, IL 60606.

[6] Available from European Federation of Manufacturers of Corrugated Board (FEFCO), 37 Rue d'Amsterdam, 75008, Paris, France. Also known as the International Box Code System, that is, in RSC/0201, RSC stands for regular slotted container and 0201 is the international box code number for the RSC.

[7] Available from the Superintendent of Documents, US Government Printing Office, Washington, DC 20402.

5.1.8 *DBLCC/0310*[7]—Double cover box (Fig. 7).
5.1.9 *IC/0325*[7]—Interlocking double cover box (Fig. 8).
5.1.10 *FTC/0301*[7]—Full telescope box (Fig. 9).
5.1.11 *FTHS/0320*[7]—Full telescope half slotted box (Fig. 10).
5.1.12 *OPF/0401*[7]—One piece folder (Fig. 11).
5.1.13 *FPF/0410*[7]—Five panel folder (Fig. 12).
5.1.14 *TS*—Triple slide box (type CF only) (Fig. 13).
5.1.15 *TSC*—Tongue and slot closure box (Fig. 14).
5.2 Other styles may also be applicable (see 8.1.3).

6. Requirements

6.1 *Materials:*

6.1.1 Fiberboard shall conform to type, class, variety, and grade of Specification D 4727 unless otherwise specified. Flute structure shall be as specified in Specification D 4727 (see 8.1.2).

6.1.2 Manufacturer's joints shall be secured by one of the following materials: (The joint is that seam of a box where the ends of the box blank are joined.)

6.1.2.1 *Metal fasteners* shall be commercially preformed staples or staples formed from commercial steel stitching wire. All metal fasteners shall have commercially applied coating of zinc or copper wash.

6.1.2.2 *Manufacturer's joint tapes* shall be of the following constructions:

(*1*) For boxes not exceeding 65 lb [30 kg] gross weight sealing strips must not be less than 2 in. [51 mm] wide, and must comply with one of the following requirements:

(*a*) Two thicknesses of sulphate paper, total basis weight not less than 80 lb [130 g/m^2] combined with a water-resistant compound and reinforced with not less than double strand nylon fiber not less than 210 denier forming a pattern with strands not more than ½ in. [13 mm] apart.

(*b*) Cloth having crosswise (filler) threads having a minimum tearing strength of 40 Elemendorf units. Inside and outside strips of sulphate paper not less than 2 in. [51 mm] wide; outside strip having a basis weight of not less than 60 lb [100 g/m^2] having a bursting strength of not less than 60 psi [414 kPa]; inside strip having a basis weight not less than 40 lb [65 g/m^2], and a bursting strength of not less than 40 psi [275 kPa].

(*2*) For boxes exceeding 65 lb [30 kg] gross weight, sealing strips must be not less than 3 in. [76 mm] wide, unless otherwise provided, and must comply with one of the following requirements:

(*a*) One thickness of sulphate paper, the fibers of which are integrally bonded and reinforced with rubber, basis weight not less than 175 lb [285 g/m^2], embossed and varnished. Two thicknesses of sulphate paper, total basis weight not less than 80 lb [130 g/m^2], reinforced in the cross direction with alternating bands of 840 denier and 420 denier nylon, not less than 4 bands every 2 in. [51 mm]. One thickness of sulphate paper having a basis weight not less than 70 lb [114 g/m^2] reinforced in the cross direction with alternating bands of 840 denier and 420 denier nylon, not less than 4 bands every 2 in. [51 mm], nylon bands firmly adhered to the surface of the paper by means of a water-resistant compound.

NOTE 1—Kraft tape basis weights are calculated as a ream weight of 500 sheets of 24 in. [610 mm] by 36 in. [915 mm] paper. The base stocks are 90 lb [41 kg], 120 lb [55 kg] or 140 lb [64 kg] kraft ream weight. These convert to 30 lb/1000 ft^2 [150 g/m^2], 40 lb/1000 ft^2 [195 g/m^2], and 47 lb/1000 ft^2 [230 g/m^2] kraft liner board equivalents.

(*b*) Two thicknesses of sulphate paper, total basis weight not less than 80 lb [130 g/m^2] combined with a water-resistant compound and with reinforcing as follows:

—Asphalt reinforced with sisal fibers.

—With glass, rayon, or glass and rayon fibers combined, running crosswise of tape not more than ⅜ in. [10 mm] apart, the rayon fibers to be not less than 1100 denier.

—With glass fibers in a diamond pattern the sides of which, parallel to each other, are formed by a cluster of yarn not less than two per inch as measured in the machine direction. Each cluster shall be formed of at least two 150-1/0 threads or the equivalent in weight of 75-1/0 yarn.

(*c*) Two thicknesses of sulphate paper, one 30 lb [50 g/m^2] basis weight and the other 60 lb [98 g/m^2] basis weight reinforced with cotton or linen fibers.

(*d*) Two or more thicknesses of sulphate paper, total basis weight not less than 150 lb [244 g/m^2] and having a bursting strength of not less than 150 psi [1034 kPa], all plies firmly glued together not less than ¼ in. [6 mm] wide along both edges. Cloth having crosswise (filled) threads having a minimum tearing strength of 70 Elemendorf units. Inside and outside strips of sulphate paper not less than 2 in. [50.0 mm] wide, each having a basis weight not less than 90 lb [147 g/m^2], and having a bursting strength of not less than 90 psi [621 kPa].

6.1.2.3 *Cold adhesive* shall conform to MMM-A-250 and as specified herein. When boxes are used for packaging or packing food and the adhesive used may contact or be in proximity to the food, the adhesive shall comply with the Federal Food, Drug and Cosmetic Act and regulations promulgated thereunder.

6.1.2.4 *Hot melt adhesives* can be used when gross weight does not exceed 65 lb [30 kg], by overlapping the sides of box forming the joint not less than 1¼ in. [32 mm] and by firmly gluing this joint with hot melt adhesive consisting of 100 % solids contents of thermal-plastic materials, which will maintain bond at temperatures ranging from −20°F [−29°C] to +165°F [+74°C] above zero.

6.2 Boxes shall be designed for type, class, variety, and grade as set forth in Specification D 4727 Tables 1 and 2 and styles specified herein (see 5.1, 5.2 and 6.2.1).

6.2.1 *Style:*

6.2.1.1 The style of box or folder shall be as specified (see 8.1.3). The location of the openings shall be determined by the specified dimensions, which shall always be furnished in the order of length, width, and depth (see 6.3 and Figs. 1 and 15).

6.2.1.2 *Regular Slotted (RSC/0201)*—This box shall meet the requirements of Fig. 1. The box shall be scored and slotted to form a body piece having four flaps for closing each of two opposite faces. The flaps along the longer edge of the box openings are the outer flaps and those along the shorter edge are the inner flaps. Flaps shall not project beyond an edge of the box. All flaps shall be of equal length with the outer flaps meeting in the center of the width panel but shall not overlap. The gap not to exceed ¼ in. [6 mm] will be permitted unless otherwise specified.

6.2.1.3 *Overlap Slotted (OSC/0202)*—This box shall meet

TABLE 1 Size and Weight Limitations for Types CF[A] and SF[A] Class Domestic Fiberboard Boxes[B]

Type CF Variety		Type SF	Max Weight of Boxes and Contents		Max Inside Dimensions Length + Width + Depth	
SW[A]	DW[A]					
Grade	Grade	Grade	lb	[kg]	in.	[mm]
125	. . .	125	20	[9.1]	40	[1016]
150	. . .	. . .	30	[13.6]	50	[1270]
175	. . .	175	40	[18.1]	60	[1524]
200	200	200	65	[29.5]	75	[1905]
275	275	275	90	[40.8]	90	[2286]
350	350	350	120	[54.4]	100	[2540]
	500	500	140	[63.5]	110	[2794]
	600	600	160	[72.6]	120	[3048]

[A] Explanation of abbreviations in Table 1:
CF—Corrugated Fiberboard
SF—Solid Fiberboard
SW—Singlewall Fiberboard
DW—Doublewall Fiberboard
[B] Size and weight limitations may be modified or waived by the ordering activity, as necessary, when boxes are used for non-shipper or special purposes.

TABLE 2 Size and Weight Limitations for Class Weather-Resistant [WR] and Water and Water Vapor Resistant [WWVR] Fiberboard Boxes Used as Exterior Containers[A]

NOTE— The gross weight and size limit expressed in the circular or rectangular boxmaker's certificate shall conform to the requirements of the Uniform Freight Classsification or National Motor Freight Classification Rules, as applicable, and may not necessarily be the same as those stipulated in Table 2.

Grade[B] [Compliance Symbol]	Max Weight of Boxes and Contents		Max Inside Dimensions Length + Width + Depth	
	lb	[kg]	in.	[mm]
V2s	120	[54.4]	100	[2540]
V3s, V4s, and V3c	90	[40.8]	90	[2286]
W5s and W5c	65	[29.5]	75	[1905]
W6s and W6c	30	[13.6]	30	[762]
V11c	160[C]	[72.6]	120	[3048]
V13c	120	[54.4]	100	[2540]
V15c	90	[40.8]	90	[2286]

[A] Not applicable to interior boxes.
[B] Reference Specification D 4727 for specific details of construction.
[C] Maximum weight may be increased to 225 lb [102 kg] provided the manufacturer's body joint is fastened with metal fasteners spaced not more than 1 in. [25 mm] apart [see 8.1.8].

the requirements of Fig. 2. This box shall be scored and slotted to form a body piece having four flaps for closing each of two opposite faces. When closed, the inner flaps shall not overlap and the outer flaps shall overlap the distance specified (see 8.1.9). Inner flaps shall be the same length as the outer flaps, except where the relation of width to length would cause the inner flaps to overlap, in which case, the inner flaps shall be cut so that, when in closed position, they shall meet.

6.2.1.4 *Full Overlap Slotted (FOL/0203)*—This box shall meet the requirements of Fig. 3. The box shall be constructed in accordance with 6.2.1.3, except that the length of the outer flaps shall be the full width of the box and shall not extend beyond the edge of the box (see 6.6).

6.2.1.5 *Special Full Flap Slotted (SFF/0206)*—This box shall meet the requirements of Fig. 4. This box shall be constructed in accordance with 6.2.1.3, except that the length of the inner flaps in the closed position shall be such that they meet in the center of the box but do not overlap. A gap not to exceed ¼ in. [6 mm] will be permitted.

6.2.1.6 *Center Special Slotted (CSSC/0204)*—This box shall meet the requirements of Fig. 5. The box shall be constructed in accordance with 6.2.1.2, except that the length of the inner and outer flaps shall be such that they meet in the center of the box but do not overlap. A gap not to exceed ¼ in. [6 mm] will be permitted.

6.2.1.7 *Center Special Overlap Slotted (CSOCS/0205)*—(See Fig. 5.) This box shall be as specified herein. When closed, the inner flaps shall meet in the middle of the face with a gap not to exceed ¼ in. [6.4 mm] when in the closed position. Outer flaps shall be the same length as the inner flaps and may overlap (no flap cutting is required). Dimension of the box shall be such that the outer flaps do not extend beyond the configuration of the box.

6.2.1.8 *Half Slotted With Cover (HSCC/0312)*—This box shall meet the requirements of Fig. 6. The box consists of a box body and a cover. The body shall be scored, slotted and secured to form a tube having four flaps, of equal length on the bottom. The outer flaps shall meet when closed with a gap not to exceed ¼ in. [6.4 mm]. Unless otherwise specified, the cover shall be a Type I (see Fig. 16) stitch locked cover. When specified (see 8.1.10), the cover shall be Type II (see Fig. 16). Unless otherwise specified, (see 8.1.11 and 8.1.12) the cover depth shall be 3 in. [76 mm] and shipped unassembled. The body may be used as an HSC alone.

6.2.1.9 *Double Cover (DBLCC/0310)*—This box shall meet the requirements of Fig. 7. The box consists of a body tube and two covers. The body consists of fiberboard, scored and secured to form a tube having parallel ends. Unless otherwise specified, the covers shall be Type I (see Fig. 16) stitch lock covers, 3 in. [76 mm] deep. When specified (see 8.1.10 and 8.1.11), the cover shall be Type II (see Fig. 16), and the cover or covers of any type may be other than 3 in. [76 mm] in depth. Unless otherwise specified (see 8.1.12), the cover shall be unassembled.

6.2.1.10 *Interlocking Double Cover (IC/0325)*—This box shall meet the requirements of Fig. 8. The box consists of a body tube with top and bottom flanges and two interlocking covers. The box body shall be fiberboard (SW or DW) scored, slotted, and secured to form a tube having double scored short flanges (flaps) which form a lock with the flanges of the cover. The top and bottom covers shall be Type III (see Fig. 16) flange interlock covers and shall be secured by means of horizontal straps. Unless otherwise specified, (see 8.1.3) the flanges shall be 3 in. [76 mm] long for boxes made of variety SW fiberboard and 4 in. [102 mm] long for boxes made of variety DW fiberboard.

6.2.1.11 *Full Telescope (FTC/0301)*—This box shall meet the requirements of Fig. 9. The box consists of a body and a cover, each constructed of one piece of fiberboard, scored and slotted. The box dimensions shall be the inside measurements of the assembled box body. The cover shall be a snug fit on the body. When specified, flaps shall be positioned inside the side panels of the body and inside the end panels of the cover or inside the side panels of the body and outside the end panels of the cover (see 8.1.15). When set up, the flaps shall not overlap but shall be of sufficient length to allow them to be securely fastened to the adjoining walls by one of the following means:

(*1*) With not less than five staples applied per flap illustrated in Fig. 9. The edge flaps on the body and cover (see Fig. 9) shall not protrude more than ¼ in. [6 mm] beyond the edge of the member to which it is secured.

Fastening along the free edges shall be not more than 4 in. [102 mm] apart.

(*2*) The flaps, when specified (see 8.1.14) shall be securely glued together with adhesive specified in 6.1.2.3 covering a minimum of 50 % of the area.

(*3*) The flaps and adjoining walls may be die cut to allow them to be fastened securely to the adjoining wall with a snap-together tongue or tuck lock corner without causing the fiberboard to rupture.

(*4*) Unless otherwise specified (see 8.1.15), the flaps on the body and cover shall be positioned in one of the following combinations, at the option of the supplier:

(*a*) Flaps outside the side panels of the body and inside the end panels of the cover.

(*b*) Flaps outside the end panels of the body and inside the side panels of the cover.

(*c*) For snap-together or tuck lock corner, the flaps shall be outside the end panels of the body and outside the side panel of the cover.

When specified (see 8.1.12), flaps shall be inside the side panels of the body and outside the end panels of the cover. Unless otherwise specified (see 8.1.12), the body and cover shall be shipped unassembled.

6.2.1.12 *Full Telescope Half Slotted Box (FTHS/0320)*—This box shall meet the requirements of Fig. 10. The box consists of a body and a telescoping cover, each constructed of one piece of scored and slotted fiberboard. The box dimensions shall be the inside measurements of the assembled box body. The cover shall be a snug fit on the body. The flaps along the longer edge of the box openings are the outer flaps and those along the shorter edge are the inner flaps. Flaps shall not project beyond an edge of the box. All flaps shall be of equal length with the outer flaps meeting in the center of the box but shall not overlap. A gap not to exceed ¼ in. [6 mm] will be permitted unless otherwise specified.

6.2.1.13 *One Piece Folder (OPF/0401)*—This folder shall meet the requirements of Fig. 11. The folder shall be constructed of one piece of fiberboard, scored and slotted as indicated in the referenced figure. When closed, outer flaps shall meet. A gap not to exceed ¼ in. [6 mm] will be permitted. Unless otherwise specified, the inner flaps shall not be less than 2 in. [51 mm] long for folders under 18 in. [457 mm] in width and not less than 3 in. [76 mm] long for folders 18 in. [457 mm] and over in width.

6.2.1.14 *Five-Panel Folder (FPF/0410)*—This folder shall meet the requirements of Fig. 12. The folder shall be scored and slotted as indicated in the referenced figure. When set up, outer flaps shall overlap (full overlap). The tuck flap length (see Fig. 12, dimension "D") shall be equal to the length of the side panel (see Fig. 12, dimension "D") less ¼ in. [6 mm] ± ⅛ in. [3 mm].

6.2.1.15 *Triple Slide (TS)*—This box shall meet the requirements of Fig. 13. The box shall be constructed of three pieces of fiberboard scored to provide three tubes which, when assembled, completely cover six faces of the box. The inner tube shall be left open as illustrated; the middle tube shall be taped at the body joint and shall be a sliding fit on the assembled inner tube; the outer tube shall be taped at the body joint and shall be a sliding fit on the assembled inner and middle tubes. The box dimensions shall be the inside dimensions of the inner tube in the sequence of length, width, and depth and shall be as illustrated in the referenced figure. Corrugations shall be at right angles to the scoreline in each tube.

6.2.1.16 *Tongue and Slot Closure (TSC)*—This box shall meet the requirements of Fig. 14. The box shall be constructed of one piece of fiberboard, scored and slotted as indicated in the referenced figure.

6.2.2 *Reinforcements:*

6.2.2.1 *Sleeves*—When specified (see 8.5) sleeves shall be constructed from fiberboard as specified in 6.4.1.2. If butt joint, it may be taped, or the overlap joint may be stapled, stitched or glued (see Fig. 17). Alternatively, the location of the body joint shall be in the center of the top or bottom panel, providing the joint does not interfere with the required marking. The sleeve shall fit closely over the top, bottom and end panels of the box for which it is intended. Space between the sleeve and box shall not exceed 3/16 in. [5 mm] when opposite surfaces of sleeve and box are in direct contact with each other.

6.2.2.2 *Liners*—When specified (see 8.1.6), liners shall be constructed from fiberboard as specified in 6.4.1.3 and 6.4.2.2 (see Fig. 18). The liner shall be scored to cover the end and side panels of the box for which it is intended. The flutes of the liner shall be perpendicular to the top of the box. Unless otherwise specified (see 8.1.16), the height of the liner shall be the full inside depth of the box for which intended, and the ends of the liner shall abut (gap not to exceed ⅛ in. [3 mm]) in the center of a side panel of the box.

6.3 *Dimensions*—Box dimensions shall be as specified in the contract or order (see 8.1.17). The dimensions shall be the inside measurements and shall be cited in the sequence of length, width, and depth (see Figs. 1 and 15). The length dimension shall be the larger of the two dimensions of the open face of the box or folder; the width dimension shall be the lesser dimension of the open face of the box or folder; the depth dimension shall be the distance between the innermost surfaces of the box or folder measured perpendicular to the length and width. Unless otherwise specified, a tolerance of ±⅛ in. [3 mm] shall be permitted in each overall dimension of the box, when conditioned according to Method D 685. When boxes are ordered complete with sleeves, for example, RSC-SL, the fit shall be as specified in 6.2.2.1. When boxes are ordered complete with liners, for example, RSC-L, the liners shall fit snugly inside the box.

6.3.1 *Size and Weight Limits*—Unless otherwise specified, (see 8.1.4) size and weight limitations of boxes furnished under this specification shall be in accordance with Tables 1 and 2. Boxes, folders, sleeves, or liners shall be made from one piece of fiberboard, except that when the length of the box blank (two lengths plus two widths plus manufacturer's joint) exceeds the largest fiberboard sheet length available to the manufacturer, two pieces of fiberboard may be used provided that the manufacturer's joints are placed at diagonally opposite corners.

6.4 *Construction of Boxes, Sleeves, and Liners:*

6.4.1 *Type CF and SF, Class Weather-Resistant and WWVR Boxes*—Unless otherwise specified (see 8.1.7), type CF or SF, class weather-resistant and WWVR boxes shall be of style (see 6.2.1), material (see 6.1.1), dimensions (see 6.3), and grade of fiberboard (specified in Table 2). Variety SW fiberboard used to fabricate type CF boxes shall be either A,

B, or C flute, as specified (see 8.1.2). Variety DW fiberboard used to fabricate type CF boxes may be any combination of A, B, C, or E flutes, except BB, EE or BE. Unless otherwise specified (see 8.1.18), conventional slotted, type CF, variety SW or DW boxes shall have the flutes running perpendicularly to the scoreline of the box opening. When specified (see 8.1.18), end loading, conventional slotted, type CF, variety SW or DW boxes shall have the flutes running parallel to the scoreline of the box opening. The flutes in DBLCC (see 6.2.1.9) and IC (see 6.2.1.10) boxes shall run the depth of the box perpendicular to the openings.

6.4.1.1 *Manufacturer's Joint, Class Weather-Resistant and WWVR Boxes (see Figs. 19 and 20)*—The joint of type CF and SF boxes shall be a fiberboard overlap not less than 1½ in. [38 mm] wide extending the full inside depth of the box. The joint tab may be an extension of either the end or side panel of the box and shall not extend beyond the top and bottom scorelines of the adjoining panel. The joint shall be fastened either inside or outside the adjoining panel, and the top and bottom edges of the joint tab shall be not more than 3/16 in. [5 mm] below the top scoreline or 3/16 in. [5 mm] above the bottom scoreline of this panel. When specified, the inside joint tab may extend into the flap area and be secured (see 8.1.19). Except as specified in 6.4.1.1 (*1*), the overlapped joint shall be secured with metal fasteners specified in 6.1.2.1. The metal fasteners shall be spaced not more than 2 in. [51 mm] apart, center to center, and the distance between the ends of the joint and the nearer end of the fastener shall not exceed 1 in. [25 mm]. An additional metal fastener (tie stitch) shall be placed ¼ in. [6 mm] to ¾ in. [19 mm] from each end of the joint. Tie stitches will not be required provided the joint is fastened with the same total number of fasteners required (including tie stitches) equally spaced on not more than 2 in. [50.0 mm] centers with the distance between the outer fasteners and the ends of the joint not exceeding 1 in. [25 mm].

(*1*) *Glued Manufacturer's Joint*—Alternatively, the joint for grades W5c, W5s, W6c, W6s and V3c boxes may be secured with adhesive specified in 6.1.2.3 and 6.1.2.4 except the toxicity requirements may be waived when packing items other than food that may contact the adhesive. The adhesive shall cover 85 % of the area between the joint tab and adjoining panel with a minimum of 75 % fiber tear of the total overlap area.

(*2*) *Butted Manufacturer's Joint*—When specified (see 8.1.20), the joint of W5c or W6c boxes shall be made by butting the panels closely together and securing them with tape. The tape shall be a minimum of 2 in. [50.8 mm] wide, and affixed to the outside surface of the joint. The tape shall be centered ±¼ in. [6 mm] on the joint, extend full depth, or within ⅜ in. [10 mm] of full depth, and shall adhere over not less than 90 % of area of contact.

6.4.1.2 *Sleeves of Class Weather-Resistant and WWVR Boxes*—Sleeves for class weather-resistant and WWVR boxes shall be the design specified in 6.2.2.1 and fabricated from class weather-resistant and WWVR fiberboard of the same type and grade as the box for which it is intended. Sleeves fabricated from type CF fiberboard shall have the flutes parallel to the scores. Overlap joints shall be formed in the same manner as in 6.2.2.1.

6.4.1.3 *Liners for Class Weather-Resistant and WWVR Boxes*—Liners for class weather-resistant and WWVR boxes shall be of the design specified in 6.2.2.2, unless otherwise specified, (see 8.1.22) shall be fabricated from material conforming to Specification D 4727, grade V15c. When specified (see 8.1.6), material conforming to Specification D 4727 grades W5c, W6c, V3c or V13c shall be used to fabricate liners. Unless otherwise specified, (see 8.1.21) the abutting ends shall be taped, at least three quarters the length of the joint, with tape at least 2 in. [51 mm] wide conforming to 6.1.2.2 (*1*). The tape shall adhere over not less than 90 % of the area of contact and shall be centered on the joint. Liners fabricated from type CF, variety SW fiberboard shall be A or C flute, and liners fabricated from type CF, variety DW fiberboard shall be any combination of A, B, C, or E flutes, except that BB, EE or BE flutes shall not be used. When boxes are ordered complete with liner, the dimensions as stated in the contract or order will be the dimensions of the box (see 6.3). The dimensions of the liner shall be such that the liner shall fit snugly (see 6.2.2.2) inside the box. When specified (see 8.1.21), the liner joint shall be affected by overlapping and fastening with adhesive, as specified for type CF box joints in 6.4.2.1 (*1*) or 6.1.2.4.

6.4.2 *Type CF and SF, Class Domestic Boxes*—Unless otherwise specified, (see 8.1.7) type CF and SF, class domestic boxes shall be of the style (see 6.2.1), material (see 6.1.1 and 8.1.2), dimensions (see 6.3) and grade of fiberboard specified in Table 1. Variety SW fiberboard used to fabricate Type CF boxes shall be A, B, C, or E flute. Variety DW fiberboard used to fabricate type CF boxes may be a combination of A, B, C, or E flutes, except that they shall not be BB, EE, or BE flute. Type CF boxes shall have flute direction as specified in 6.4.1. The manufacturer's joint of the boxes shall be formed and secured in accordance with 6.4.2.1.

6.4.2.1 *Manufacturer's Joint, Class Domestic Boxes, (see 6.1.2)*—The joint of the type CF, variety SW box shall be overlapped or butted. The joint of the type SF, shall be overlapped. The joint of the type CF, variety DW box, shall be overlapped or butted. The formed joint shall be secured as specified in 6.4.2.1 (*1*) or 6.4.2.1 (*2*), as applicable. Large boxes (see 6.3.1) may be fabricated with two joints positioned at diagonally opposite corners at the option of the supplier.

(*1*) *Overlapped Joint (Joint Tab)*—The joint shall be made with a fiberboard joint tab overlap not less than 1¼ in. [32 mm] wide with the length of the overlap equal to the inside depth of the box. The joint tab may be an extension of either the end or side panel of the box. When specified, (see 8.1.19) the joint tab may extend into the flap area and be secured. The joint tab shall be fastened either inside or outside the adjoining panel, and the top and bottom edges of the overlap shall be no more than 3/16 in. [5 mm] below the top scoreline or 3/16 in. [5 mm] above the bottom scoreline of this panel. The overlapped joint of type CF boxes shall be fastened with adhesive as specified in 6.1.2.3 or 6.1.2.4, except the toxicity requirement may be waived when packing items other than food that may contact the adhesive, or with metal fasteners. When adhesive is used it shall be applied so as to cover the full area between the joint tab and the adjoining panel. The adhesive shall substantially extend to all edges of the overlap. The overlapped joint of type SF boxes shall be fastened with metal fasteners as specified in 6.1.2.1. Metal fasteners for the type CF and type SF boxes having a depth dimension of 18 in. [457 mm] or less shall be spaced not more than 3 in. [76 mm] apart, center to center. Metal fasteners for the type SF box having a depth dimension greater than 18 in. [457 mm] shall be spaced not more than 2½ in. [64 mm] apart, center to center. The distance between the ends of the joint and the nearer end of the nearest fastener shall not exceed 1 in. [25 mm]. Metal fasteners may be applied diagonally, vertically, or horizontally at the option of the supplier.

(*2*) *Butted Joint (Type CF) Boxes Only*—The butted joint shall be made by fitting the edges of the panels to be joined closely together and securing them with gummed tape. Tape used to secure the body joint of boxes having a gross weight of 40 lb [18 kg] or less (grade 125 to 175) shall be that normally used by the industry for this purpose (see 6.1.2.2 (*1*)). Tape used to secure the joints of boxes having a gross weight of more than 40 lb [18 kg] (grades over 175) shall be reinforced with sisal, cloth, glass, rayon or double strand nylon fibers (see 6.1.2.2 (*2*)). The tape shall be not less than 2 in. [51 mm] in width for boxes having a gross weight of 65 lb [30 g] or less (grade 200 and below) and not less than 3 in. [76 mm] in width for boxes having a gross weight over 65 lb [30 kg] (grade above 200). The tape shall be centered on the joint and shall extend its full length, or within ⅜ in. [10 mm] of full length and shall adhere over not less than 90 % to the entire area of contact with the fiberboard.

6.4.2.2 *Liners for Class Domestic Boxes*—Unless otherwise specified (see 8.1.22), liners shall be constructed of fiberboard material conforming to Specification D 4727, type CF, class domestic, variety DW, grade 275. The liners shall be designed, as specified in 6.4.1.3. Liners for class domestic/fire retardant boxes shall be class domestic/fire retardant.

6.5 *Compliance Marking:*

6.5.1 *Types CF and SF, Class Weather-Resistant and WWVR Boxes*—Types CF and SF, class weather-resistant and WWVR boxes made to comply with this specification shall be imprinted with the following data which will legibly occupy a total area of not more than 36 in.^{2} [232 cm^2] nor less than 16 in.^{2} [103 cm^2] (see Fig. 21):

6.5.1.1 Boxmaker's name or boxmaker's certificate (see Note 2).

6.5.1.2 Month and year of manufacturer (for example, "6-87").

6.5.1.3 Individual grade or identification symbols.

6.5.1.4 Specification compliance data, specification number, and minimum average bursting strength of _____ psi (_____ kPa). The figure to be inserted should be that corresponding to the dry mullen requirements in Specification D 4727 for the particular grade of board involved (see Note 2).

6.5.1.5 For shipments to Government agencies the national stock number (NSN), inside dimensions, and outside cube shall be marked below the specification data on all exterior type boxes procured as an item of supply. When filled boxes are procured for direct shipment to the Government, the supplemental requirements set forth in Section 9 shall apply.

6.5.1.6 For class weather resistant/fire retardant boxes, the words "Fire Retardant" shall be marked below the box maker's certification.

NOTE 2—The minimum bursting strength expressed in the circular or rectangular boxmaker's certificate shall conform to the requirements of the National Railroad Freight Classification rules and National Motor Freight Classification rules, and may not be the same as that stipulated in the line reading "minimum average bursting strength _____ psi (_____ kPa)."

6.5.2 The above markings shall be located as follows:

6.5.2.1 *Slotted Style Boxes*—On the outer flap in a corner as close to the scoreline as practicable.

6.5.2.2 *All Other Styles*—In a corner of the top panel as close to the scoreline as practicable. Unless otherwise specified (see 8.1.23), no other marking shall appear on the box. The boxmaker's certificate does not relieve the supplier of his responsibility for meeting the requirements of this practice.

6.5.2.3 When the box panel is not large enough to accommodate all of the markings, the markings may cross the scoreline (see 6.5.3).

6.5.3 *Types CF and SF, Class Domestic Boxes*—Each box shall be plainly marked with the appropriate boxmaker's certificate signifying compliance with the National Railroad Freight Classification rules, and the National Motor Freight Classification rules, as applicable. The certificate may be located on the box wherever it is customarily placed; however, the preferred location is on the bottom panel or bottom outer flap. The certificate on boxes having a length of less than 10 in. [254.0 mm] or width of less than 9 in. [228.6 mm] may be reduced in size but the outside diameter shall not be less than 2 in. [50 mm]. In the event the box panel is too small to accommodate even the reduced certificate of compliance, the certificate will be permitted to cross the scoreline and cover more than one panel. For class domestic/fire retardant boxes, the words FIRE RETARDANT shall be marked below the boxmaker's certificate. Unless otherwise specified (see 8.1.23), boxes will require no further markings. When specified (see 8.1.23), the specification number, dimensions, and cube shall be marked below the boxmaker's certificate.

6.6 *Workmanship*—The completed box shall be clean, free of frayed or torn edges, improperly aligned panels, improper scores and slots, and the marking shall be clear and legible. All dimensions of the boxmaker's blank shall be accurately cut, scored, and slotted so that the assembled box parts fit closely without binding. No flap shall project beyond an edge of the box by more than ⅛ in. [3 mm] when the box is set up and closed. All metal fasteners shall be well clinched, flush with or below the interior and exterior surfaces of the corrugated fiberboard joined, and shall be flush or slightly above the surfaces for solid fiberboard.

7. Assembly of Class Domestic Boxes

7.1 *Slotted Styles*—All slotted style boxes shall be assembled in accordance with the figures referenced herein.

7.1.1 *Slotted Styles With Covers*—All slotted style boxes with covers shall be assembled in accordance with the figures referenced herein.

7.1.1.1 *Type I Cover (Stitch Lock)*—The type I cover end flaps (see Figs. 6, 7 and 16) shall be securely fastened to the inside of the adjoining side flange by not less than three metal fasteners located not more than 1¼ in. [32 mm] nor less than 1/2 in. [13 mm] by not less than three stitches located not more than 0.47 in. [12 mm] from free edges and scorelines. Metal fasteners shall conform to 6.1.2.1. When specified the flaps may be joined with adhesives detailed in 6.1.2.3, 6.1.2.4 and 6.4.1.1 (*1*). See 8.1.24.

7.1.1.2 *Type II Cover (Friction Lock)*—The type II cover end flanges B and C (see Fig. 16) shall be slightly wider than the side flanges "A" of the cover in order to produce a friction lock. Metal fasteners shall not be used to assemble these covers.

7.1.1.3 *Type III Cover (Flange Interlock)*—All interlocking flange boxes shall be assembled in accordance with the referenced figures herein. The interlocked flanges shall be secured by means of flat steel strapping applied horizontally one band width above the center line of the fold (see Fig. 8). The strapping shall be not less than 3/8 in. [9.5 mm] by 0.015 in. [0.38 mm], finish A or B, grade 2, as specified in Specification D 3953 or Type IV polyester plastic strapping specified in Specification D 3950.

7.1.2 *Style FTC*—All body and cover flaps shall be fastened to the adjoining walls, as illustrated in Fig. 9.

7.2 *Assembly of Class Weather-Resistant and WWVR Boxes:*

7.2.1 *Slotted Styles with Covers*—Slotted style boxes shall be assembled in accordance with the figures referenced herein.

7.2.2 *Style FTC*—Body and cover flaps of style FTC boxes shall be securely fastened to the adjoining walls (see 6.2.1.11).

8. Ordering Information

8.1 Purchasers should select the preferred options permitted herein, and should include the following information in procurement documents:

8.1.1 Title, number and date of this practice.

8.1.2 Type, class, variety, and grade of fiberboard (see 6.1.1, 6.4.1 and 6.4.2).

8.1.3 Box style required (see 5.1, 5.2 and 6.2.1).

8.1.4 Size and weight limitations of boxes (see 6.3.1).

8.1.5 When sleeve is required (see 6.2.2.1 and 6.4.1.2).

8.1.6 When liner is required of SW or DW fiberboard and grade of fiberboard required (see 6.2.2.2 and 6.4.1.3).

8.1.7 When box is for special requirements (see 6.4.1 and 6.4.2).

8.1.8 When maximum requirements for V11c boxes may be increased (see Table 2).

8.1.9 Distance of overlap for OSC boxes (see 6.2.1.3).

8.1.10 When type II cover is required for HSC and DBLCC boxes (see 6.2.1.8 and 6.2.1.9).

8.1.11 Depth of cover other than 3 in. [76.2 mm] for HSC and DBLCC boxes (see 6.2.1.8 and 6.2.1.9).

8.1.12 When covers and bodies are to be shipped assembled (see 6.2.1.8, 6.2.1.9 and 6.2.1.11).

8.1.13 Length of flange for IC boxes, if other than specified (see 6.2.1.10).

8.1.14 When glued flaps of FTC are permitted (see 6.2.1.11).

8.1.15 When flaps of the FTC box are required to be arranged in a manner other than that specified (see 6.2.1.11).

8.1.16 When the ends of the liner shall abut in other than the center of the side panel of the box (see 6.2.2.2).

8.1.17 Inside dimensions of box, specified in inches, in order of length, width and depth (see 6.3).

8.1.18 When the direction of flutes is to run parallel to scoreline of box opening (see 6.4.1).

8.1.19 When the manufacturer's joint tab may extend into the flap area (see 6.4.1.1 and 6.4.2.1 (*1*)).

8.1.20 When taping of manufacturer's joint is permitted for W5 and W6 boxes (normally for interior boxes) (see 6.4.1.1 (*2*)).

8.1.21 When liner joints secured with other than tape (see 6.4.1.3).

8.1.22 When other variety or grade fiberboard is required for liners (see 6.4.1.3 and 6.4.2.2).

8.1.23 Printing and marking required in addition to certification (see 6.5.1, 6.5.3, and 9.3).

8.1.24 When other methods of sealing cover assembly joints are required (see 7.1.1.1).

9. Preparation for Delivery

9.1 Unless otherwise specified, packing and marking shall be in accordance with Practice D 3951.

9.2 For delivery to agencies of the U.S. Government, packing shall be level A, B, or C as specified in contract or order in accordance with PPP-B-638.

9.3 *Marking for Delivery to U.S. Government Agencies (See 8.1.23):*

9.3.1 *Civil Agencies*—In addition to any special marking required by the contract or order, shipping containers shall be marked in accordance with FED-STD-123.

9.3.2 *Military Requirements*—In addition to any special marking required by the contract or order, shipping containers shall be marked in accordance with MIL-STD-129.

10. Keywords

10.1 boxes; box fabrication; box styles; fiberboard

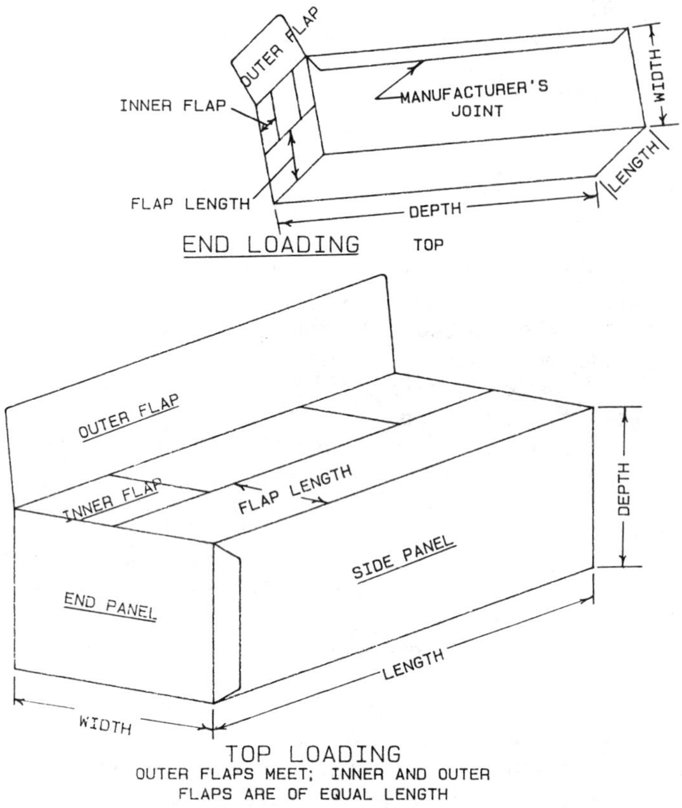

FIG. 1 Box, Fiberboard; RSC—Regular Slotted Box

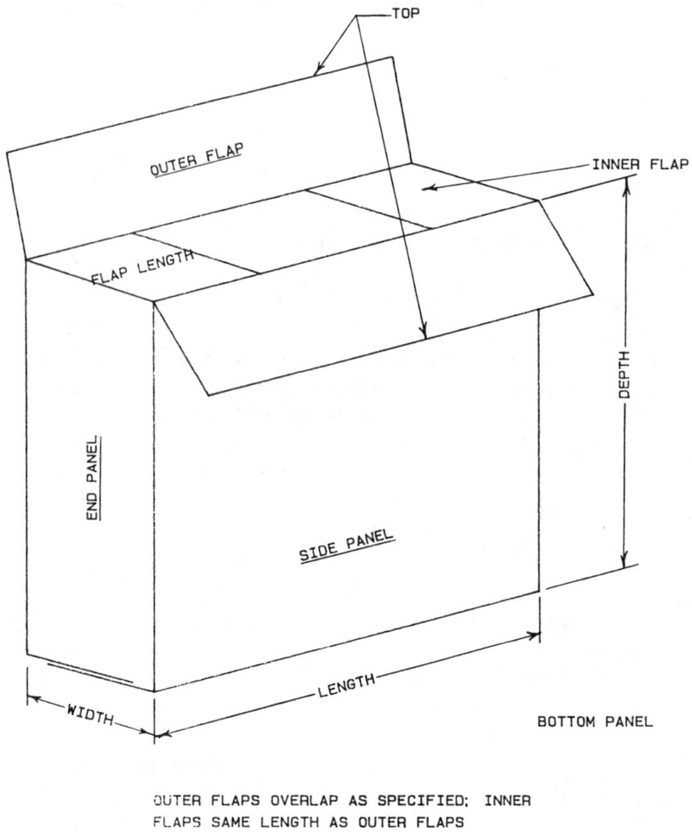

FIG. 2 Box, Fiberboard; OSC—Overlap Slotted Box

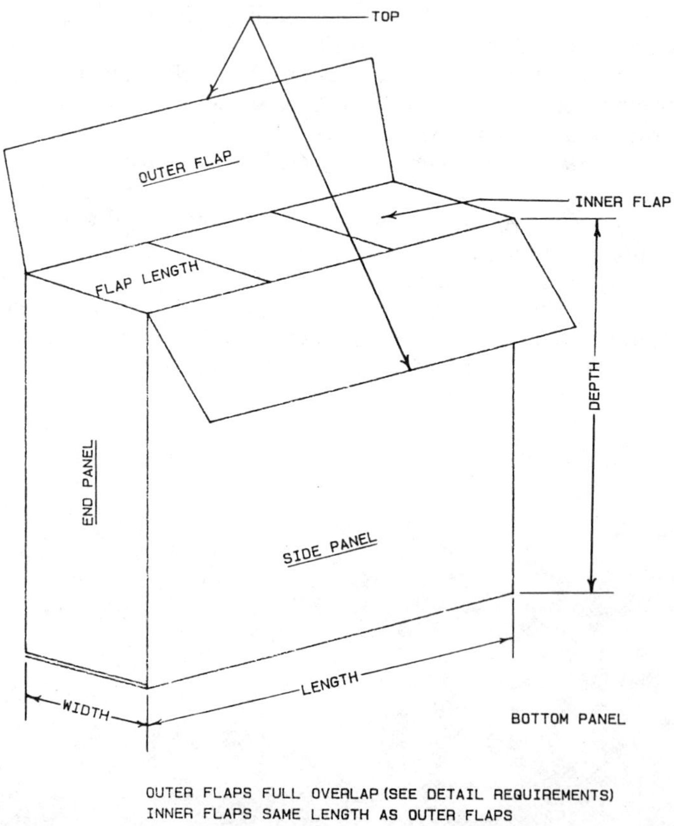

FIG. 3 Box, Fiberboard; FOL—Full Overlap Slotted Box

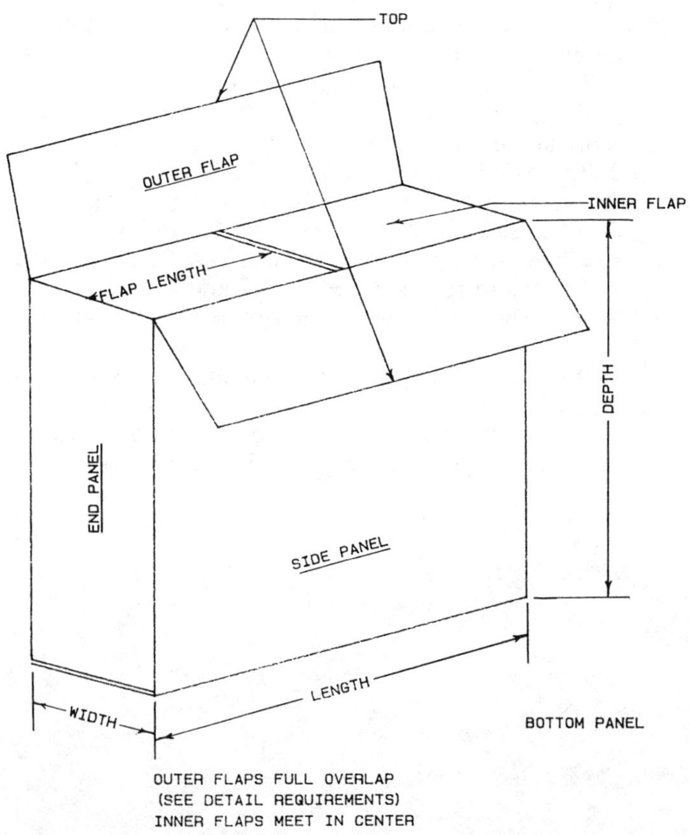

FIG. 4 Box, Fiberboard; SFF—Special Full Flap Slotted Box

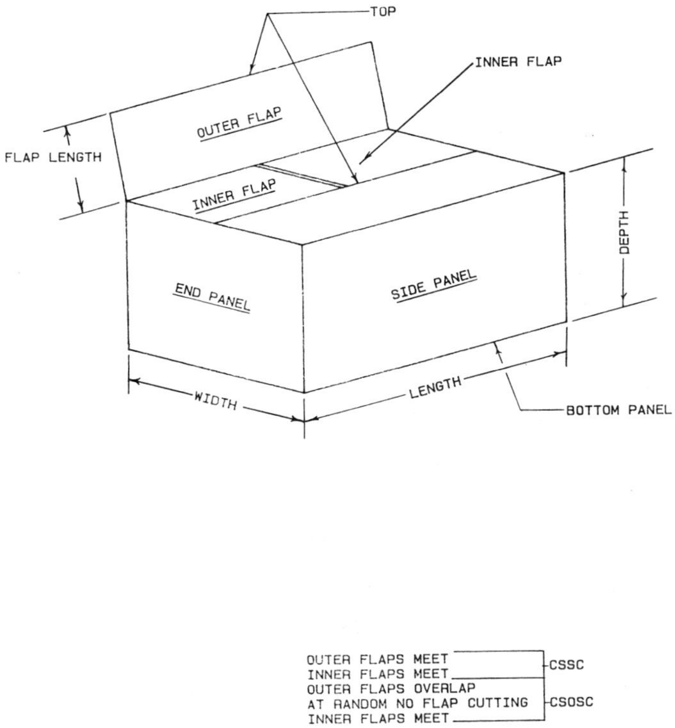

FIG. 5 Box, Fiberboard; CSSC—Center Special Slotted Box and CSOSC—Center Special Overlap Slotted Box

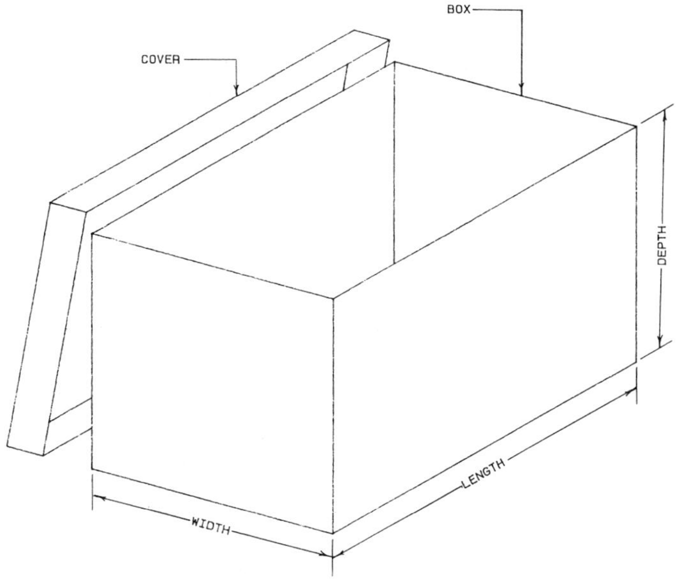

FIG. 6 Box, Fiberboard; HSCC—Half Slotted Box with Cover

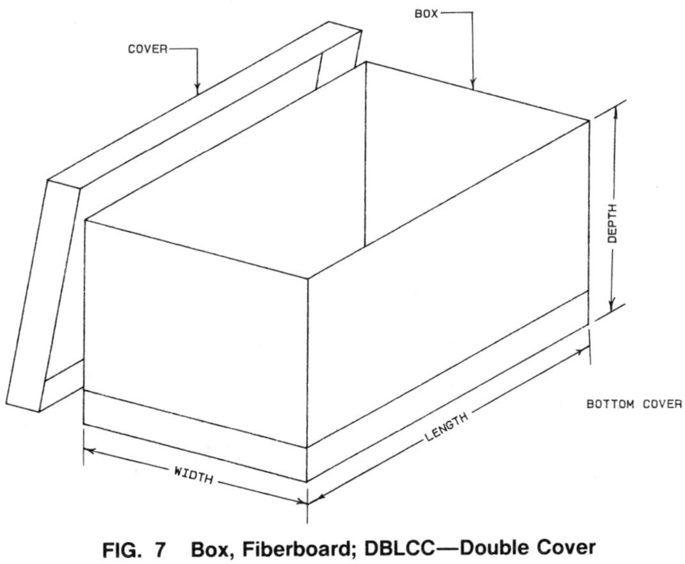

FIG. 7 Box, Fiberboard; DBLCC—Double Cover

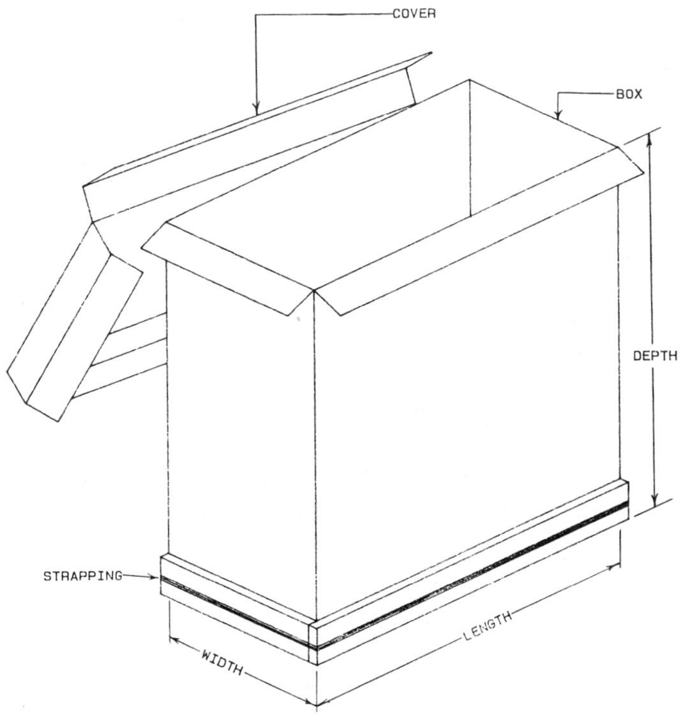

FIG. 8 Box, Fiberboard; IC—Interlocking Double Cover

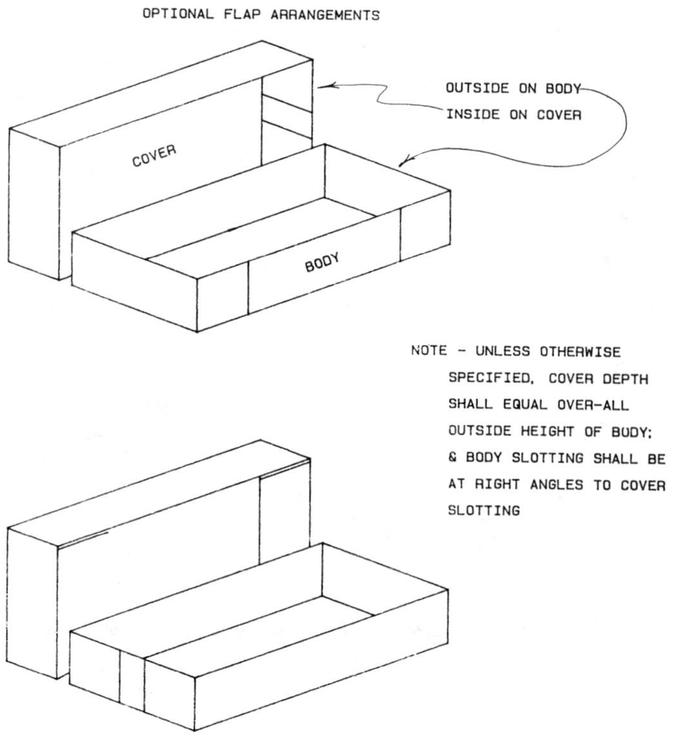

FIG. 9 Box, Fiberboard; FTC—Full Telescope

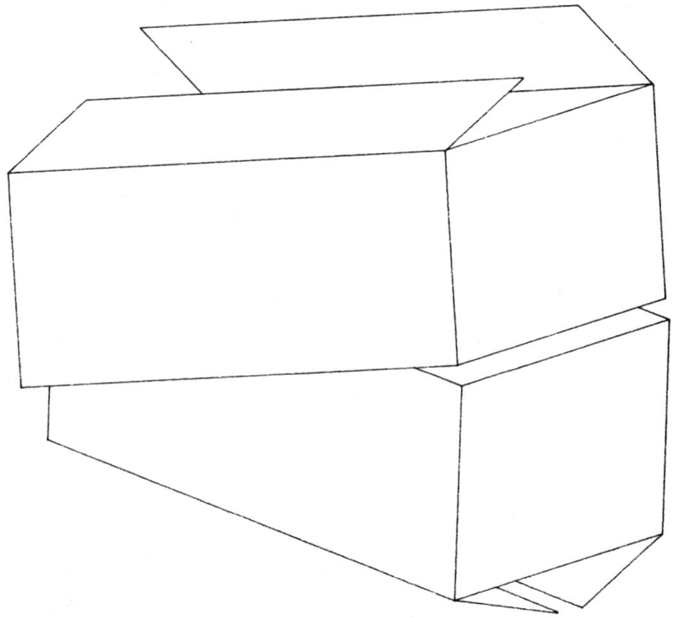

FIG. 10 Box, Fiberboard; FTHS—Two Piece Full Telescope Half Slotted Box

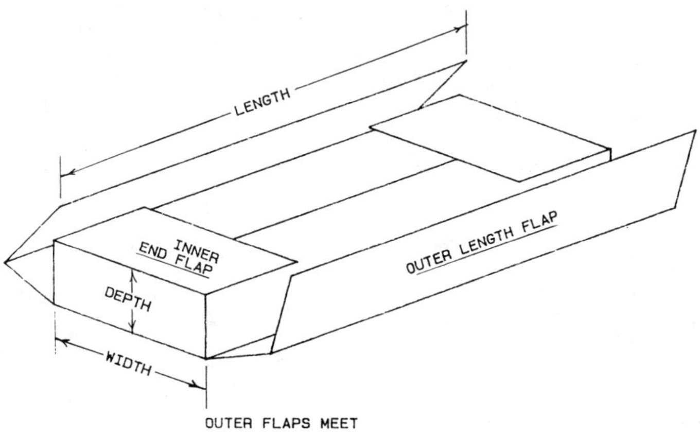

FIG. 11 Folder; Fiberboard; OPF—One Piece Folder

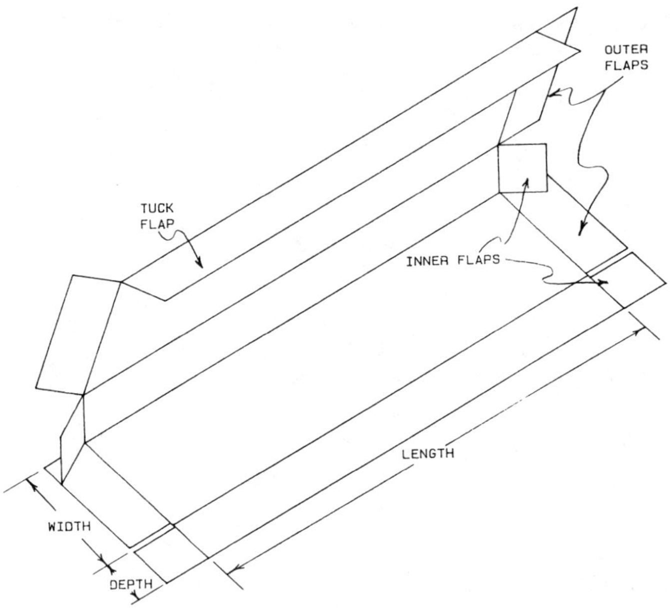

FIG. 12 Folder, Fiberboard; FPF—Five Panel Folder

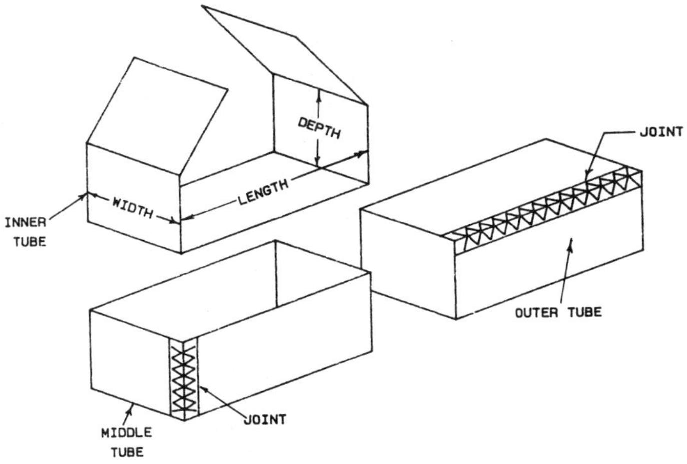

FIG. 13 Box, Fiberboard; TS—Triple Slide Box

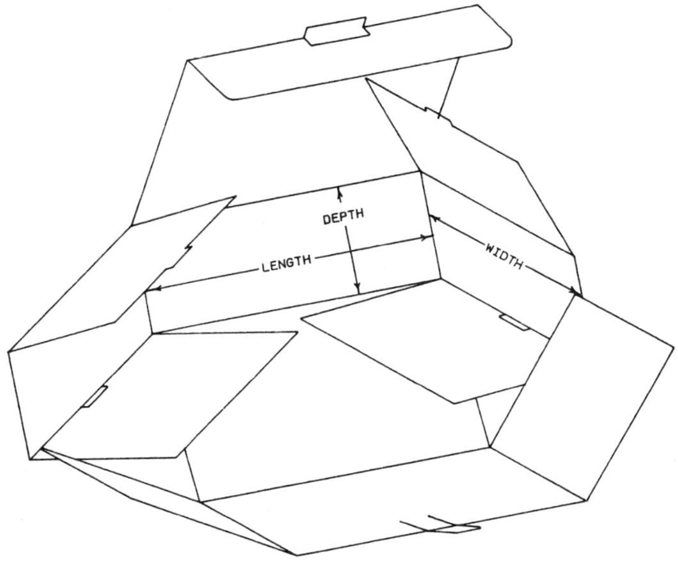

FIG. 14 Folder, Fiberboard; TSC—Tongue and Slot Closure

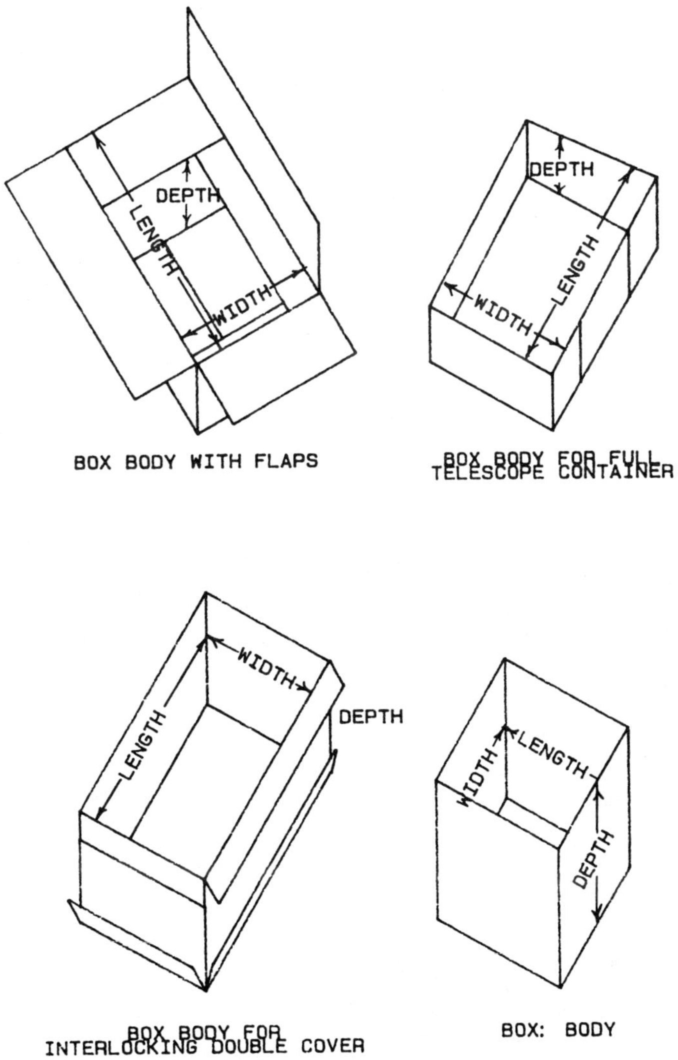

FIG. 15 Fiberboard Box Dimensioning

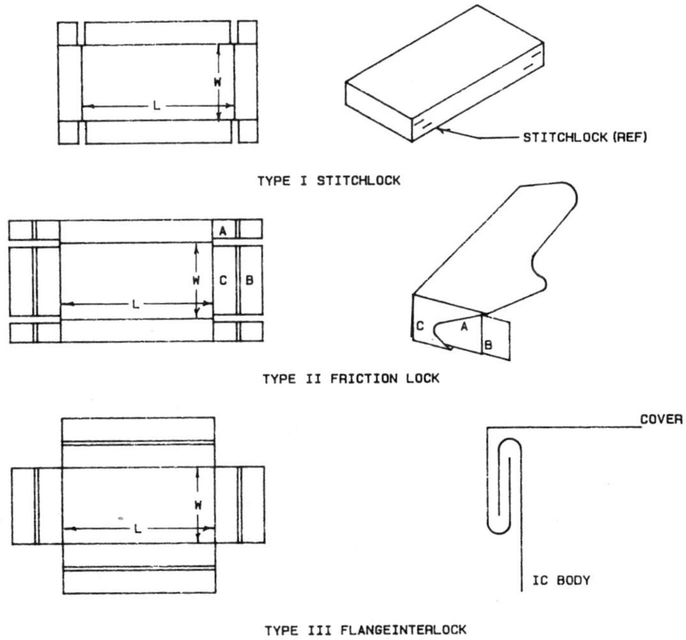

FIG. 16 Cover Assembly

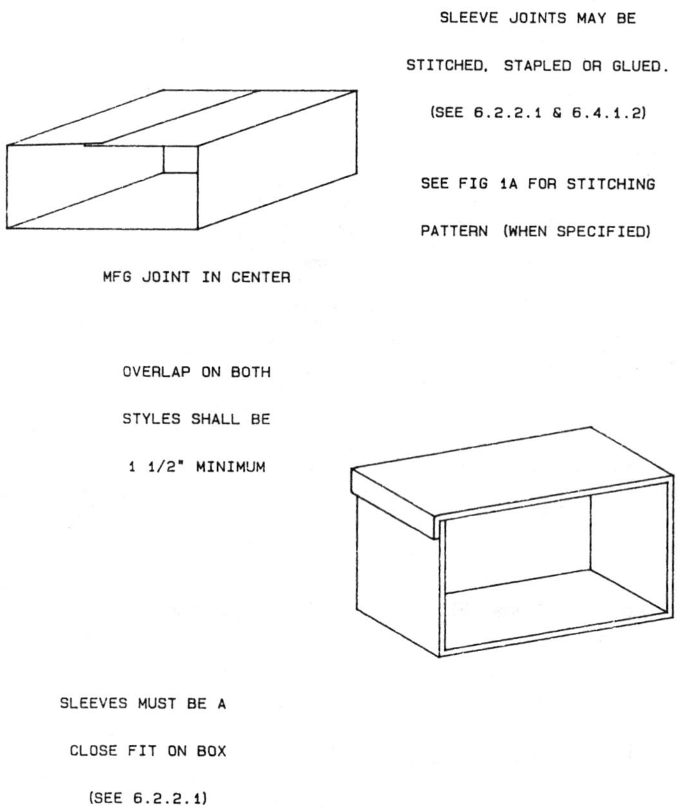

FIG. 17 Box, Fiberboard; SL—Sleeve

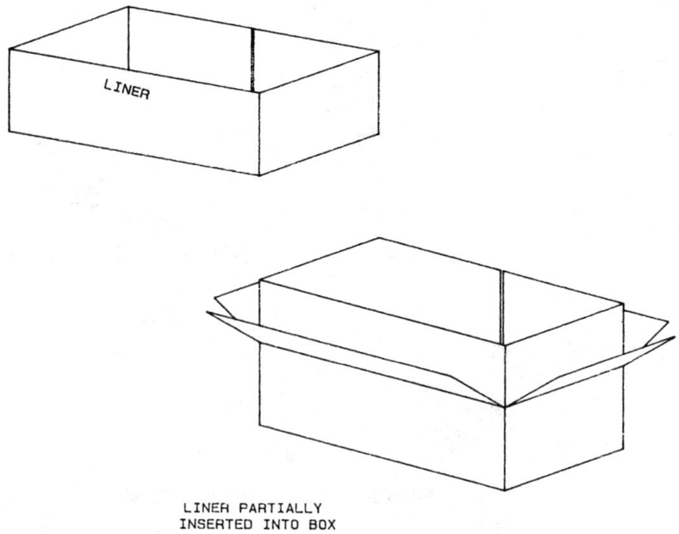

FIG. 18 Box, Fiberboard; L—Liner

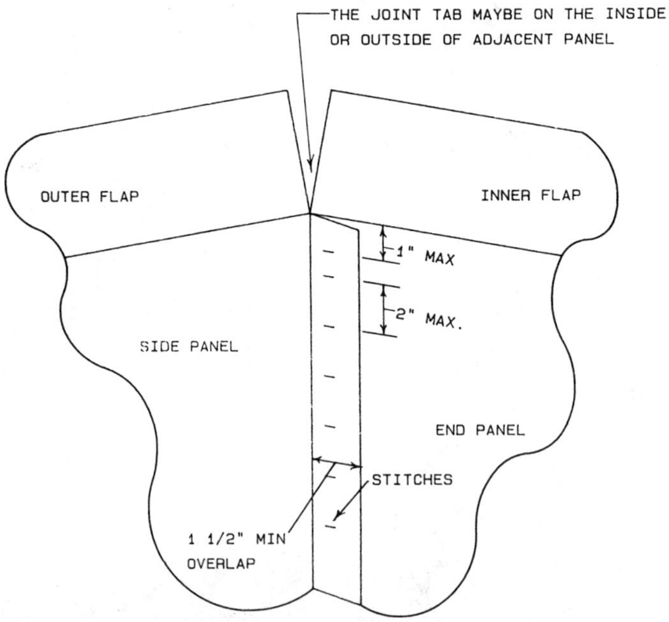

NOTE—The joint tab may be intergral with end or side panel; stitches may be horizontal or diagonal in accordance with standard manufacturing practice.

FIG. 19 Box, Fiberboard; Detail of Manufacturer's Joint Construction for Class Weather Resistant and WWVR Boxes

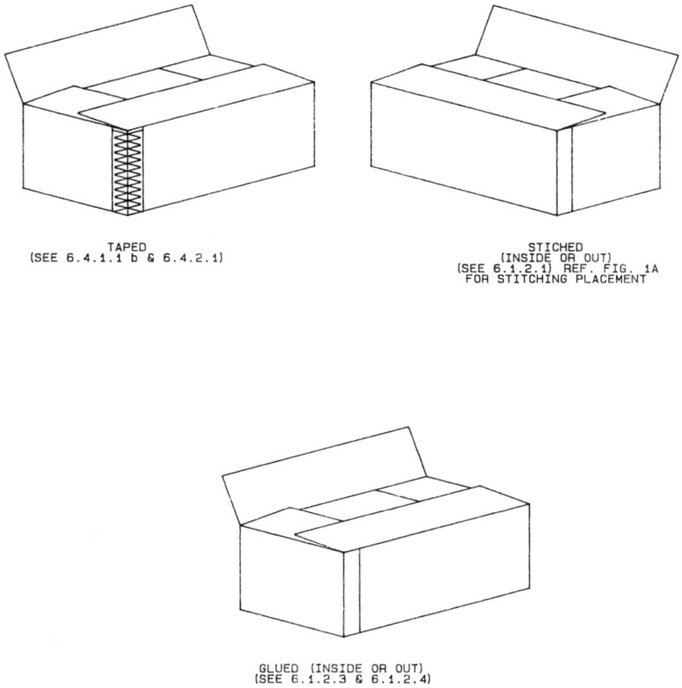

FIG. 20 Box, Fiberboard; Details of Manufacturers Joint Types

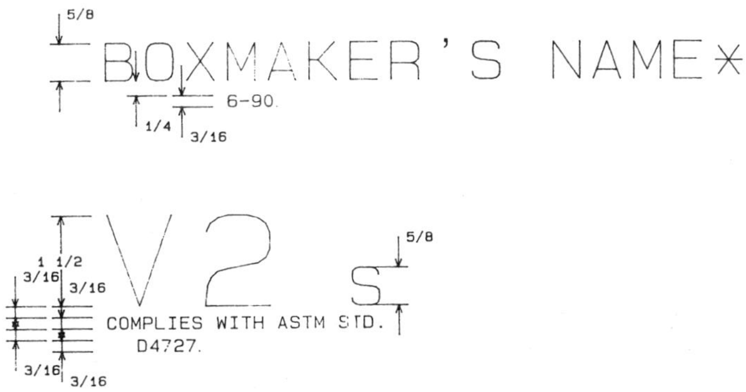

NOTE 1—Dimension and cube can be boxmaker's certificate, round or rectangular as applicable and may be located on the box where it is customarily placed.

NOTE 2—Dimensions and cube other than specified in MIL-STD-129 shall be printed only on government procured boxes (minimum bursting strength, psi).

FIG. 21 Compliance Marking for Class Weather-Resistant and WWVR Boxes

Standard Test Methods for Water Vapor Transmission of Materials[1]

This standard is issued under the fixed designation E 96; the number immediately following the designation indicates the year of original adoption or, in the case of revision, the year of last revision. A number in parentheses indicates the year of last reapproval. A superscript epsilon (ϵ) indicates an editorial change since the last revision or reapproval.

1. Scope

1.1 These test methods cover the determination of water vapor transmission (WVT) of materials through which the passage of water vapor may be of importance, such as paper, plastic films, other sheet materials, fiberboards, gypsum and plaster products, wood products, and plastics. The test methods are limited to specimens not over 1¼ in. (32 mm) in thickness except as provided in Section 9. Two basic methods, the Desiccant Method and the Water Method, are provided for the measurement of permeance, and two variations include service conditions with one side wetted and service conditions with low humidity on one side and high humidity on the other. Agreement should not be expected between results obtained by different methods. That method should be selected which more nearly approaches the conditions of use.

1.2 The values stated in inch-pound units are to be regarded as the standard. Metric inch-pound conversion factors for WVT, permeance, and permeability are stated in Table 1. All conversions of mm Hg to Pa are made at a temperature of 0°C.

1.3 *This standard does not purport to address the safety problems associated with its use. It is the responsibility of the user of this standard to establish appropriate safety and health practices and determine the applicability of regulatory limitations prior to use.*

2. Referenced Documents

2.1 *ASTM Standards:*

C 168 Terminology Relating to Thermal Insulating Materials[2]

D 449 Specification for Asphalt Used in Dampproofing and Waterproofing[3]

D 2301 Specification for Vinyl Chloride Plastic Pressure-Sensitive Electrical Insulating Tape[4]

E 691 Practice for Conducting an Interlaboratory Study to Determine the Precision of a Test Method[5]

3. Terminology

3.1 Definitions of terms used in this standard will be found in Terminology C 168, from which the following are quoted:

"*water vapor permeability*—the time rate of water vapor transmission through unit area of flat material of unit thickness induced by unit vapor pressure difference between two specific surfaces, under specified temperature and humidity conditions."

NOTE—Permeability is a property of a material, but the permeability of a body that performs like a material may be used. Permeability is the arithmetic product of permeance and thickness.

"*water vapor permeance*—the time rate of water vapor transmission through unit area of flat material or construction induced by unit vapor pressure difference between two specific surfaces, under specified temperature and humidity conditions."

NOTE—Permeance is a performance evaluation and not a property of a material.

"*water vapor transmission rate*—the steady water vapor flow in unit time through unit area of a body, normal to specific parallel surfaces, under specific conditions of temperature and humidity at each surface."

TABLE 1 Metric Units and Conversion Factors [A,B]

Multiply	by	To Obtain (for the same test condition)
	WVT	
g/h·m²	1.43	grains/h·ft²
grains/h·ft²	0.697	g/h·m²
	Permeance	
g/Pa·s·m²	1.75×10^{7}	1 Perm (inch-pound)
1 Perm (inch-pound)	5.72×10^{-8}	g/Pa·s·m²
	Permeability	
g/Pa·s·m	6.88×10^{10}	1 Perm inch
1 Perm inch	1.45×10^{-9}	g/Pa·s·m

[A] These units are used in the construction trade. Other units may be used in other standards.

[B] All conversions of mm Hg to Pa are made at a temperature of 0°C.

4. Summary of Test Methods

4.1 In the Desiccant Method the test specimen is sealed to the open mouth of a test dish containing a desiccant, and the assembly placed in a controlled atmosphere. Periodic weighings determine the rate of water vapor movement through the specimen into the desiccant.

4.2 In the Water Method, the dish contains distilled water, and the weighings determine the rate of vapor movement through the specimen from the water to the controlled atmosphere. The vapor pressure difference is nominally the same in both methods except in the variation, with extremes of humidity on opposite sides.

[1] These test methods are under the jurisdiction of ASTM Committee C-16 on Thermal Insulation and are the direct responsibility of Subcommittee C16.33 on Thermal Insulation Finishes and Vapor Transmission.

Current edition approved Feb. 9, 1990. Published April 1990. Originally published as E 96 – 53 T. Last previous edition E 96 – 80ϵ2.

[2] *Annual Book of ASTM Standards,* Vol 04.06.

[3] *Annual Book of ASTM Standards,* Vol 04.04.

[4] *Annual Book of ASTM Standards,* Vol 10.01.

[5] *Annual Book of ASTM Standard,* Vol 14.02.

5. Significance and Use

5.1 The purpose of these tests is to obtain, by means of simple apparatus, reliable values of water vapor transfer through permeable and semipermeable materials, expressed in suitable units. These values are for use in design, manufacture, and marketing. A permeance value obtained under one set of test conditions may not indicate the value under a different set of conditions. For this reason, the test conditions should be selected that most closely approach the conditions of use. While any set of conditions may be used and those conditions reported, standard conditions that have been useful are shown in Appendix X1.

6. Apparatus

6.1 *Test Dish*—The test dish shall be of any noncorroding material, impermeable to water or water vapor. It may be of any shape. Light weight is desirable. A large, shallow dish is preferred, but its size and weight are limited when an analytical balance is chosen to detect small weight changes. The mouth of the dish shall be as large as practical and at least 4.65 in.^{2} (3000 mm^2). The desiccant or water area shall be not less than the mouth area except if a grid is used, as provided in 12.1, its effective area shall not exceed 10 % of the mouth area. An external flange or ledge around the mouth, to which the specimen may be attached, is useful when shrinking or warping occurs. When the specimen area is larger than the mouth area, this overlay upon the ledge is a source of error, particularly for thick specimens. This overlay material should be masked as described in 10.1 so that the mouth area defines the test area. The overlay material results in a positive error, indicating excessive water vapor transmission. The magnitude of the error is a complex function of the thickness, ledge width, mouth area, and possibly the permeability. This error is discussed by Joy and Wilson.[6] This type of error should be limited to about 10 to 12 %. For a thick specimen the ledge should not exceed ¾ in. (19 mm) for a 10-in. (254-mm) or larger mouth (square or circular) or ⅛ in. (3 mm) for a 5-in. (127-mm) mouth (square or circular). For a 3-in. (76-mm) mouth (square or circular) the ledge should not exceed 0.11 in. (2.8 mm) wide. An allowable ledge may be interpolated for intermediate sizes or calculated according to Joy and Wilson.[6] A rim around the ledge (Fig. X2.1) may be useful. If a rim is provided, it shall be not more than ¼ in. (6 mm) higher than the specimen as attached. Different depths may be used for the Desiccant Method and Water Method, but a ¾-in. (19-mm) depth (below the mouth) is satisfactory for either method.

6.2 *Test Chamber*—The room or cabinet where the assembled test dishes are to be placed shall have a controlled temperature and relative humidity. The temperature chosen shall be between 70 and 90°F (21 and 32°C), and shall be maintained constant within 1°F (0.6°C). A temperature of 90°F (32°C) is recommended (Note 1). The relative humidity shall be maintained at 50 ± 2 %, except where extremes of humidities are desired, when the conditions shall be 100 ± 1°F (38 ± 0.6°C) and 90 ± 2 % relative humidity. Both temperature and relative humidity shall be measured frequently, or preferably recorded continuously. Air shall be continuously circulated throughout the chamber, with a velocity sufficient to maintain uniform conditions at all test locations. Its velocity over the specimen, in feet per minute, shall be (numerically) not less than ten times the permeance of the specimen, in perms, but at least 500 ft/min, or in centimetres per second not less than 7.7 times the permeance, in metric perms, but at least 250 cm/s.

NOTE 1—Simple temperature control by heating alone is usually made possible at 90°F (32°C). However, it is very desirable to enter the controlled space, and a comfortable temperature is more satisfactory for that arrangement. Temperatures of 73.4°F (23°C) and 80°F (26.7°C) are in use and are satisfactory for this purpose. With cyclic control, the average test temperature may be obtained from a sensitive thermometer in a mass of dry sand. The temperature of the chamber walls facing a specimen over water should not be cooler than the water to avoid condensation on the test specimen.

6.3 *Balance and Weights*—The balance shall be sensitive to a change smaller than 1 % of the weight change during the period when a steady state is considered to exist. The weights used shall be accurate to 1 % of the weight change during the steady-state period. For example: A 1-perm (5.7×10^{-11} $kg \cdot Pa^{-1} \cdot s^{-1} \cdot m^{-2}$) specimen 10 in. (254 mm) square at 80°F (26.7°C) passes 8.6 grains or 0.56 g/day. In 18 days of steady state, the transfer is 10 g. For this usage, the balance must have a sensitivity of 1 % of 10 g or 0.1 g and the weights must be accurate to 0.1 g. If, however, the balance has a sensitivity of 0.2 g or the weights are no better than 0.2 g, the requirements of this paragraph can be met by continuing the steady state for 36 days. An analytical balance that is much more sensitive will permit more rapid results on specimens below 1 perm (5.7×10^{-11} $kg \cdot Pa^{-1} \cdot s^{-1} \cdot m^{-2}$) when the assembled dish is not excessively heavy. A light wire sling may be substituted for the usual pan to accommodate a larger and heavier load.

7. Materials

7.1 *Desiccant and Water:*

7.1.1 For the Desiccant Method, anhydrous calcium chloride in the form of small lumps that will pass a No. 8 (2.36-mm) sieve, and free of fines that will pass a No. 30 (600-μm) sieve, shall be used (Note 2). It shall be dried at 400°F (200°C) before use.

NOTE 2—If $CaCl_2$ will react chemically on the specimen, an adsorbing desiccant such as silica gel, activated at 400°F (200°C), may be used; but the moisture gain by this desiccant during the test must be limited to 4 %.

7.1.2 For the Water Method, distilled water shall be used in the test dish.

7.2 *Sealant*—The sealant used for attaching the specimen to the dish, in order to be suitable for this purpose, must be highly resistant to the passage of water vapor (and water). It must not lose weight to, or gain weight from, the atmosphere in an amount, over the required period of time, that would affect the test result by more than 2 %. It must not affect the vapor pressure in a water-filled dish. Molten asphalt or wax is required for permeance tests below 4 perms (2.3×10^{-10} $kg \cdot Pa^{-1} \cdot s^{-1} \cdot m^{-1}$). Sealing methods are discussed in Appendix X2.

[6] Joy, F. A., and Wilson, H. G., "Standardization of the Dish Method for Measuring Water Vapor Transmissions," National Research Council of Canada, Research Paper 279, January 1966, p. 263.

8. Sampling

8.1 The material shall be sampled in accordance with standard methods of sampling applicable to the material under test. The sample shall be of uniform thickness. If the material is of nonsymmetrical construction, the two faces shall be designated by distinguishing marks (for example, on a one-side-coated sample, "I" for the coated side and "II" for the uncoated side).

9. Test Specimens

9.1 Test specimens shall be representative of the material tested. When a product is designed for use in only one position, three specimens shall be tested by the same method with the vapor flow in the designated direction. When the sides of a product are indistinguishable, three specimens shall be tested by the same method. When the sides of a product are different and either side may face the vapor source, four specimens shall be tested by the same method, two being tested with the vapor flow in each direction and so reported.

9.2 A slab, produced and used as a laminate (such as a foamed plastic with natural "skins") may be tested in the thickness of use. Alternatively, it may be sliced into two or more sheets, each being separately tested and so reported as provided in 9.4, provided also, that the "overlay upon the cup ledge" (6.1) of any laminate shall not exceed 1/8 in. (3 mm).

9.3 When the material as used has a pitted or textured surface, the tested thickness shall be that of use. When it is homogeneous, however, a thinner slice of the slab may be tested as provided in 9.4.

9.4 In either case (9.2 or 9.3), the tested overall thickness, if less than that of use, shall be at least five times the sum of the maximum pit depths in both its faces, and its tested permeance shall be not greater than 5 perms (3.3 metric perms).

9.5 The overall thickness of each specimen shall be measured at the center of each quadrant and the results averaged. Measurement of specimens of 0.125 in. or less in thickness shall be made to the nearest 0.0001 in. Measurement of specimens greater than 0.125 in. in thickness shall be made to the nearest 0.001 in.

9.6 When testing any material with a permeance less than 0.05 perms or when testing a low permeance material that may be expected to lose or gain weight throughout the test (because of evaporation or oxidation), it is strongly recommended that an additional specimen, or "dummy," be tested exactly like the others, except that no desiccant or water is put in the dish. Failure to use this dummy specimen to establish modified dish weights may significantly increase the time required to complete the test. Because time to reach equilibrium of water permeance increases as the square of thickness, thick, particularly hygroscopic, materials may take as long as 60 days to reach equilibrium conditions.

10. Attachment of Specimen to Test Dish

10.1 Attach the specimen to the dish by sealing (and clamping if desired) in such a manner that the dish mouth defines the area of the specimen exposed to the vapor pressure in the dish. If necessary, mask the specimen top surface, exposed to conditioned air so that its exposure duplicates the mouth shape and size and is directly above it. A template is recommended for locating the mask. Thoroughly seal the edges of the specimen to prevent the passage of vapor into, or out of, or around the specimen edges or any portion thereof. The same assurance must apply to any part of the specimen faces outside their defined areas. Suggested methods of attachment are described in Appendix X2.

NOTE 4—In order to minimize the risk of condensation on the interior surface of the sample when it is placed in the chamber, the temperature of the water prior to preparation of the test specimen should be within ±2°F (±1.1°C) of the test condition.

11. Procedure for Desiccant Method

11.1 Fill the test dish with desiccant within 1/4 in. (6 mm) of the specimen. Leave enough space so that shaking of the dish, which must be done at each weighing, will mix the desiccant.

11.2 Attach the specimen to the dish (see 10.1) and place it in the controlled chamber, specimen up, weighing it at once. (This weight may be helpful to an understanding of the initial moisture in the specimen.)

11.3 Weigh the dish assembly periodically, often enough to provide eight or ten data points during the test. A data point is the weight at a particular time. The time that the weight is made should be recorded to a precision of approximately 1 % of the time span between successive weighing. Thus, if weighings are made every hour, record the time to the nearest 30 s; if recordings are made every day, a time to the nearest 15 min would be allowed. At first the weight may change rapidly; later a steady state will be reached where the rate of change is substantially constant. Weighings should be accomplished without removal of the test dishes from the controlled atmosphere, but if removal is prescribed necessary, the time the specimens are kept at different conditions, temperature or relative humidity, or both, should be kept to a minimum. When results of water vapor transmission are expected to be less than .05 perm, a dummy specimen is strongly recommended. Such a dummy specimen should be attached to an empty cup in the normal manner. The environment effects of temperature variation and buoyancy variability due to barometric pressure fluctuation can be arithmetically tared out of the weighing values. This precaution permits earlier and more reliable achievement of equilibrium conditions. Analyze the results as prescribed in 13.1.

11.4 Terminate the test or change the desiccant before the water added to the desiccant exceeds 10 % of its starting weight (Notes 1 and 4). This limit cannot be exactly determined and judgement is required. The desiccant gain may be more or less than the dish weight-gain when the moisture content of the specimen has changed.

NOTE 5—The WVT of some materials (especially wood) may depend on the ambient relative humidity immediately before the test. An apparent hysteresis results in higher WVT if the prior relative humidity was above the test condition and vice versa. It is therefore recommended that specimens of wood and paper products be conditioned to constant weight in a 50 % relative humidity atmosphere before they are tested. Some specimens may be advantageously preconditioned to minimize the moisture that the specimen will give up to the desiccant. This applies when the specimen is likely to have high moisture content or when it is coated on the top (vapor source) side.

12. Procedure for Water Method

12.1 Fill the test dish with distilled water to a level 3/4 ± 1/4 in. (19 ± 6 mm) from the specimen. The air space thus allowed has a small vapor resistance, but it is necessary in order to reduce the risk of water touching the specimen when the dish is handled. Such contact invalidates a test on some materials such as paper, wood, or other hygroscopic materials. The water depth shall be not less than ⅛ in. (3 mm) to ensure coverage of the dish bottom throughout the test. However, if the dish is of glass, its bottom must be visibly covered at all times but no specific depth is required. Water surges may be reduced in placing a grid of light noncorroding material in the dish to break the water surface. This grid shall be at least ¼ in. (6 mm) below the specimen, and it shall not reduce the water surface by more than 10 % (Note 5).

NOTE 6—For the Water Method, baking the empty dish and promptly coating its mouth with sealant before assembly is recommended. The water may be added most conveniently after the specimen is attached, through a small sealable hole in the dish above the water line.

12.2 Attach the specimen to the dish (see 10.1). Some specimens are likely to warp and break the seal during the test. The risk is reduced by preconditioning the specimen, and by clamping it to the dish ledge (if one is provided).

12.3 Weigh the dish assembly and place it in the controlled chamber on a true horizontal surface. Follow the procedure given in 11.3. If the test specimen cannot tolerate condensation on the surface, the dish assembly shall not be exposed to a temperature that differs by more than 5°F (2.8°C) from the control atmosphere to minimize the risk of condensation on the specimen. When results of water vapor transmission are expected to be less than .05 perm, a dummy specimen is strongly recommended. Such a dummy specimen should be attached to an empty cup in the normal manner. The environment effects of temperature variation and buoyancy variability due to barometric pressure fluctuation can be arithmetically tared out of the weighing values. This precaution permits earlier and more reliable achievement of equilibrium conditions. Analyze the results as prescribed in 13.1.

12.4 Where water is expected to be in contact with the barrier in service, proceed as in 11.3 except place the dish in an inverted position. The dish must be sufficiently level so that water covers the inner surface of the specimen despite any distortion of the specimen due to the weight of the water. With highly permeable specimens it is especially important to locate the test dish so that air circulates over the exposed surface at the specified velocity. The test dishes may be placed on the balance in the upright position for weighing, but the period during which the wetted surface of the specimen is not covered with water must be kept to a minimum.

13. Calculation and Analysis of Results

13.1 The results of the rate of water vapor transmission may be determined either graphically or numerically.

13.1.1 *Dummy Specimen*—If a dummy specimen has been used to compensate for variability in test conditions, due to temperature or barometric pressure, or both, the daily recorded weights should be adjusted by calculating the weight change from initial to time of weighing. This adjustment is made by reversing the direction of the dummy's weight change, relative to its initial weight, and modifying all the appropriate specimen weight(s) recorded at this time.

13.1.2 *Graphic Analysis*—Plot the weight, modified by the dummy specimen when used, against elapsed time, and inscribe a curve which tends to become straight. Judgment here is required and numerous points are helpful. When a straight line adequately fits the plot of at least six properly spaced points (periodic weight changes matching, or exceeding 20 % of the multiple of 100 times the scale sensitivity), a nominally steady state is assumed, and the slope of the straight line is the rate of water vapor transmission.

13.1.3 *Numerical Analysis*—A mathematical least squares regression analysis of the weight, modified by the dummy specimen when used, as a function of time will give the rate of water vapor transmission. An uncertainty, or standard deviation of this rate, can also be calculated to define the confidence band. For very low permeability materials, this method can be used to determine the results after 30 to 60 days when using an analytical balance, with a sensitivity of ±1 mg, even if the weight change does not meet the 100 times the sensitivity requirement of 6.3. Specimens analyzed in this manner must be clearly identified in the report.

13.2 Calculate the water vapor transmission, WVT, and permeance as follows:

13.2.1 *Water Vapor Transmission:*

$$\mathrm{WVT} = G/tA = (G/t)/A$$

where:

In inch-pound units:

G = weight change, grains (from the straight line),
t = time during which G occurred, h,
G/t = slope of the straight line, grains/h,
A = test area (cup mouth area), ft^2, and
WVT = rate of water vapor transmission, grains/h·ft^2.

In metric units:

G = weight change (from the straight line), g,
t = time, h,
G/t = slope of the straight line, g/h,
A = test area (cup mouth area), m^2, and
WVT = rate of water vapor transmission, g/h·m^2.

13.2.2 *Permeance:*

$$\text{Permeance} = \mathrm{WVT}/\Delta p = \mathrm{WVT}/S(R_1 - R_2)$$

where:

In inch-pound units:

Δp = vapor pressure difference, in. Hg,
S = saturation vapor pressure at test temperature, in. Hg,
R_1 = relative humidity at the source expressed as a fraction (the test chamber for desiccant method; in the dish for water method), and
R_2 = relative humidity at the vapor sink expressed as a fraction

In metric units:

Δp = vapor pressure difference, mm Hg (1.333×10^2 Pa),
S = saturation vapor pressure at test temperature, mm Hg (1.333×10^2 Pa),
R_1 = relative humidity at the source expressed as a fraction (the test chamber for desiccant method; in the dish for water method), and
R_2 = relative humidity at the vapor sink expressed as a fraction.

13.2.3 In the controlled chamber the relative humidity and temperature are the average values actually measured during the test and (unless continuously recorded) these measurements shall be made as frequently as the weight measurements. In the dish the relative humidity is nominally 0 % for the desiccant and 100 % for the water. These values are usually within 3 % relative humidity of the actual relative humidity for specimens below 4 perms (2.3×10^{-7} $g \cdot Pa^{-1} \cdot s^{-1} \cdot m^{-1}$) when the required conditions are maintained (no more than 10 % moisture in $CaCl_2$ and no more than 1 in. (25 mm) air space above water).

13.3 Only when the test specimen is homogeneous (not laminated) and not less than ½ in. (12.5 mm) thick, calculate its average permeability (perm in.) (metric perm-cm) as follows:

$$\text{Average permeability} = \text{permeance} \times \text{thickness}$$

NOTE 7: *Example*—In a desiccant test that ran 288 h (12 days) on an exposed area of 100 in.^{2} (0.0645 m^2), it was found that the rate of gain was substantially constant after 48 h and during the subsequent 240 h, the weight gain was 12 g. The controlled chamber conditions were measured at 89.0°F (31.7°C) and 49 % relative humidity.

Required: WVT and permeance

Calculation (inch-pound units):

$$G/t = 12\text{ g} \times \frac{15.43\text{ grains}}{\text{g}} \div 240\text{ h} = 0.771\text{ grains/h},$$

$$A = 100\text{ in.}^2 \times \frac{1\text{ ft}^2}{144\text{ in.}^2} = 0.695\text{ ft}^2,$$

S = 1.378 in. Hg (from standard references tables),

R_1 = 49 % (in chamber),

R_2 = 0 % (vapor sink), and

WVT = 0.771 grains/h ÷ 0.694 ft^2 = 1.11 grains/ft$^2 \cdot$h.

$$\text{Permeance} = WVT/\Delta P = WVT/S(R_1 - R_2) = 1.11\text{ grains/ft}^2\cdot\text{h} \div 1.378\text{ in. Hg }(0.49 - 0) = 1.64\text{ grains/ft}^2\cdot\text{h}\cdot\text{in. Hg} = 1.64\text{ perms}$$

Calculation (metric units):

G/t = 12 g/240 h = 0.05 g/h,

A = 0.0645 m^2,

S = 35 mm Hg (from reference tables),

= 35 mm Hg × 1.333×10^2 Pa/mm Hg = 46.66×10^2 Pa,

R_1 = 49 % (in chamber),

R_2 = 0 % (vapor sink), and

WVT = 0.05 g/h ÷ 0.0645 m^2 = 0.775 g/h·m^2.

$$\text{Permeance} = WVT/\Delta P - WVT/S(R_1 - R_2) = 0.775\text{ g/h}\cdot\text{m}^2 \times 1\text{ h/3600s} \div 46.66 \times 10^2\text{ Pa} \times (0.49 - 0) = 9.42 \times 10^{-8}\text{ g/Pa}\cdot\text{s}\cdot\text{m}^2$$

13.4 Metric units and conversion factor are given in Table 1.

14. Report

14.1 The report shall include the following:

14.1.1 Identification of the material tested, including its thickness.

14.1.2 Test method used (desiccant or water).

14.1.3 Test temperature.

14.1.4 Relative humidity in the test chamber.

14.1.5 Permeance of each specimen in perms (to two significant figures).

14.1.6 The side of each specimen on which the higher vapor pressure was applied. (The sides shall be distinguished as "side A" and "side B" when there is no obvious difference between them. When there is an obvious difference, this difference shall also be stated, such as "side A waxed" and "side B unwaxed.").

14.1.7 The average permeance of all specimens tested in each position.

14.1.8 The permeability of each specimen (as limited by 13.3), and the average permeability of all specimens tested.

14.1.9 Include a portion of the plot indicating the section of the curve used to calculate permeability.

14.1.10 State design of cup and type or composition of sealant.

15. Precision and Bias[7]

15.1 *Precision*—Table 2 is based on an interlaboratory test conducted in 1988. Four materials were tested using the dessicant method and the water method in triplicate. Fifteen

[7] Supporting data have been filed at ASTM Headquarters. Request RR: C-16-1014.

TABLE 2 Precision Results from Interlaboratory Testing

For Desiccant Method at 73.4°F:

Material	Thickness, in.	WVT (mean), perm	Repeatability[A]			Reproducibility[A]		
			S	*CV* %	*LSD*	*S*	*CV* %	*LSD*
A	0.001	0.606	0.0166	2.70	0.047	0.098	15.0	0.278
B	0.0055	0.0129	0.0028	22.1	0.008	0.0055	42.6	0.016
C	0.5	0.0613	0.0044	7.22	0.012	0.0185	30.6	0.052
D	1.0	0.783	0.0259	3.30	0.073	0.0613	7.8	0.174

For Water Method at 73.4°F:

Material	Thickness, in.	WVT (mean), perm	Repeatability[A]			Reproducibility[A]		
			S	*CV* %	*LSD*	*S*	*CV* %	*LSD*
A	0.001	0.715	0.0134	1.95	0.039	0.156	21.9	0.44
B	0.0055	0.0157	0.0022	13.8	0.0062	0.0021	19.4	0.006
C	0.5	0.097	0.0055	5.7	0.016	0.0195	20.9	0.055
D	1.0	1.04	0.0192	1.8	0.054	0.217	20.9	0.62

[A] For this data,

S = standard deviation,

CV = percent coefficient of variation ($S \times 100$/mean), and

LSD = least significant difference between two individual test results based on a 95 % confidence level = $2\sqrt{2}S$.

laboratories contributed data, with full results secured from four laboratories.

15.1.1 Test results were analyzed using Practice E 691.

15.2 *Bias*—This test method has no bias because water vapor transmission of materials is defined in terms of this test method.

APPENDIXES

(Nonmandatory Information)

X1. STANDARD TEST CONDITIONS

X1.1 Standard test conditons that have been useful are:

X1.1.1 *Procedure A*—Desiccant Method at 73.4°F (23°C).

X1.1.2 *Procedure B*—Water Method at 73.4°F (23°C).

X1.1.3 *Procedure BW*—Inverted Water Method at 73.4°F (23°C).

X1.1.4 *Procedure C*—Desiccant Method at 90°F (32.2°C).

X1.1.5 *Procedure D*—Water Method at 90°F (32.2°C).

X1.1.6 *Procedure E*—Desiccant Method at 100°F (37.8°C).

X2. CUP DESIGN AND SEALING METHODS

X2.1 An ideal sealing material has the following properties:

X2.1.1 Impermeability to water in either vapor or liquid form.

X2.1.2 No gain or loss of weight from or to the test chamber (evaporation, oxidation, hygroscopicity, and water solubility being undersirable).

X2.1.3 Good adhesion to any specimen and to the dish (even when wet).

X2.1.4 Complete conformity to a rough surface.

X2.1.5 Compatibility with the specimen and no excessive penetration into it.

X2.1.6 Strength or pliability (or both).

X2.1.7 Easy handleability (including desirable viscosity and thermal of molten sealant).

X2.1.8 Satisfactory sealants possess these properties in varying degrees and the choice is a compromise, with more tolerance in items at the beginning of this list for the sake of those at the latter part of the list when the requirements of 7.2 are met (Note A2). Molten asphalt or wax is required for permeance tests below 4 perms (2.6 metric perms). Tests to determine sealant behavior should include:

X2.1.8.1 An impervious specimen (metal) normally sealed to the dish and so tested, and

X2.1.8.2 The seal normally assembled to an empty dish with no specimen and so tested.

X2.2 The following materials are recommended for general use when the test specimen will not be affected by the temperature of the sealant:

X2.2.1 Asphalt, 180 to 200°F (82 to 93°C) softening point, meeting the requirements of Specification D 449, Type C. Apply by pouring.

X2.2.2 Beeswax and rosin (equal weights). A temperature of 275°F (135°C) is desirable for brush application. Pour at lower temperature.

X2.2.3 Microcrystalline wax[8] (60 %), mixed with refined crystalline paraffin wax (40 %).

X2.3 The materials listed in X2.3.1 are recommended for particular uses such as those shown in Figure X2.1. The suggested procedure described in X2.3.2 applies to an 11-3/8-in. (289-mm) square specimen if its permeance exceeds 4 perms (2.6 metric perms) (limited by evaporation of sealants).

X2.3.1 *Materials:*

X2.3.1.1 Aluminum foil, 0.005 in. (0.125 mm) minimum thickness.

X2.3.1.2 Tape, meeting the requirements of Specification D 2301, vinyl chloride plastic pressure-sensitive, electrical insulating tape.

X2.3.1.3 Cement, contact bond, preferably rubber base.

X2.3.2 *Procedure:*

X2.3.2.1 *Step 1*—Seal aluminum foil around edges of specimen, leaving a 100-in.^{2} (0.0654-m^2) exposed test area on each side. Use contact bond cement as directed by the manufacturer.

X2.3.2.2 *Step 2*—Spread sealant on inside of rim and ledge. Place desiccant (dry), or water and surge control material (wet) in pan. Press specimen in place. Avoid squeezing compound into the test area.

X2.3.2.3 *Step 3*—Coat outside of rim and bottom of ledge with contact bond cement, and place foil strips from edge of template, around rim, and bottom of ledge.

X2.4 A method of using hot asphalt, as applied to a 10-in. (254-mm) square-mouth dish with ledge and rim, is as follows:

X2.4.1 *Apparatus:*

X2.4.1.1 *Template*—A square frame of brass or steel, 3/16 in. (5 mm) thick and 3/4 in. (19 mm) deep. The 3/16-in. (5-mm) thickness is tapered to zero at the bottom of the frame where it will touch the test specimen and maintain a 10-in. (254-mm) square test area.

X2.4.1.2 *Sealant*—Asphalt (see X2.2.1 used at the proper pouring consistency of 375 to 450°F (179 to 232°C).

X2.4.1.3 *Melting Pot* for the asphalt, electrically heated, with one dimension greater than 11 3/8 in. (289 mm).

X2.4.1.4 *Small Ladle* for pouring.

X2.4.2 *Procedure*—Mark the 11 3/8-in. (289-mm) square

[8] Grade Nos. 2305 or 2310 of the Mobil Oil Corp., or their equivalent, have been found satisfactory for this purpose.

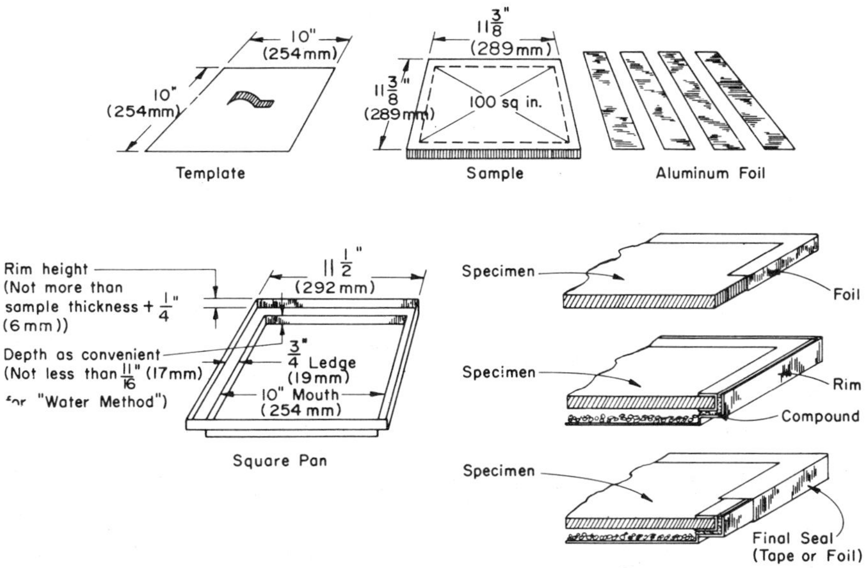

FIG. X2.1 Apparatus for Water Vapor Transmission Tests of Large Thick Specimens

specimen with a line at an equal distance from each edge, so that the area enclosed by the lines is as nearly as possible a 10-in. (254-mm) square. The template may be used for marking. Dip each edge of the specimen in molten asphalt up to the line, so that the test area is defined and all edges are coated with a heavy layer of asphalt. Place the specimen over the pan containing water or desiccant. Lightly oil the template or coat with petroleum jelly on its outer side, and place on the specimen. Pour molten asphalt into the space between the template and the rim of the pan. After the asphalt has cooled for a few minutes, the template should be easily removable.

X2.5 Hot wax may be applied like asphalt. It may also be applied (freely) with a small brush. Its lower working temperature may be advantageous when a specimen contains moisture.

X2.6 Several designs for dishes with supporting rings and flanges are shown in Fig. X2.2. Various modifications of these designs may be made provided that the principle of prevention of edge leakage by means of a complete seal is retained. The dishes may be constructed of any rigid, impermeable, corrosion-resistant material, provided that they can be accommodated on the available analytical balance. A lightweight metal, such as aluminum or one of its alloys, is generally used for larger-size dishes. In some cases when an aluminum dish is employed and moisture is allowed to condense on its surface, there may be appreciable oxidation of the aluminum with a resulting gain in weight. Any gain in weight will ordinarily depend on the previous history of the dish and the cleanness of the surface. An empty dish carried through the test procedure as a control will help to determine whether any error may be expected from this cause. When aluminum dishes are used for the water methods, a pressure may develop inside the assembly during a test due to corrosion. This can cause seal failure or otherwise affect the result. Where this is a problem, it can be overcome by providing inside the dish a protective coating of baked-on epoxy resin or similar material. Dishes with flanges or rings that project from the inner walls of the dish are to be avoided, as such projections influence the diffusion of the water vapor. The depth of the dish for the water procedures is such that there is a 0.80 ± 0.20 in. (20 ± 5-mm) distance between the water surface and the under surface of the specimen, with a water depth of about 0.20 in. (5 mm).

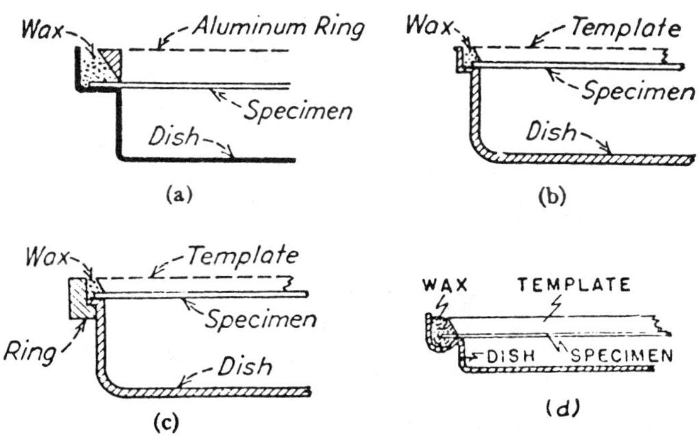

FIG. X2.2 Several Types of Dishes for Water Vapor Transmission Tests of Materials in Sheet Form

X2.6.1 For the desiccant-in-dish procedures, the dishes need not be as deep as those required for the water-in-dish procedures. The desiccant is within 1/4 in. (6 mm) of the under surface, and a minimum depth of only 1/2 in. (12 mm) of desiccant is required.

X2.6.2 The dishes shown in Fig. X2.2 require a molten seal.

X2.6.3 A template such as is shown in Fig. X2.3 is usually used for defining the test area and effecting the wax seal. It consists of a circular metal dish 1/8 in. (3.18 mm) or more in thickness with the edge beveled to an angle of about 45°. The diameter of the bottom (smaller) face of the template is approximately equal to, but not greater than, the diameter of the effective opening of the dish in contact with the specimen. Small guides may be attached to the template to center it automatically on the test specimen. A small hole

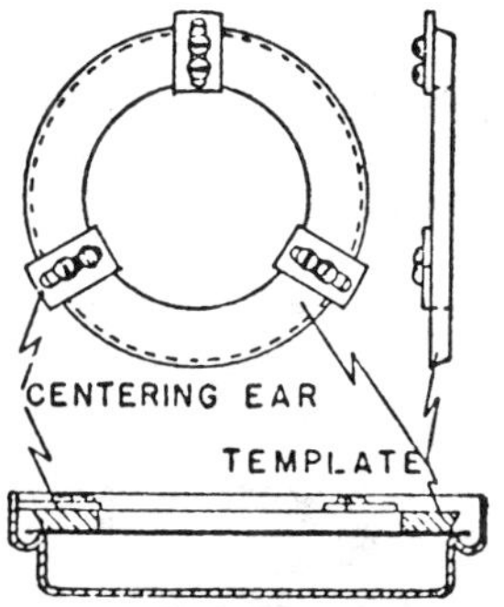

FIG. X2.3 Template Suitable for Use in Making the Wax Seals on Test Dishes

through the template to admit air, and petrolatum applied to the beveled edge of the template facilitate its removal after sealing the test specimen to the dish. In use, the template is placed over the test specimen and when it is carefully centered with the dish opening, molten wax is flowed into the annular space surrounding the beveled edge of the template. As soon as the wax has solidified, the template is removed from the sheet with a twisting motion. The outside flange of the dish should be high enough to extend over the top of the specimen, thus allowing the wax to completely envelop the edge.

X2.6.4 Gasketed types of seals are also in use on appropriately designed dishes. These simplify the mounting of the specimen, but must be used with caution, since the possibility of edge leakage is greater with gasketed seals than with wax seals. Gasketed seals are not permitted for the measurement of permeance less than 0.044 perms (2.5 $ng \cdot Pa^{-1} \cdot s^{-1} \cdot m^2$). As a further precaution when gasketed seals are used instead of preferred sealants, a blank test run is suggested using glass or metal as a dummy specimen.

X2.6.5 A suitable weighing cover consists of a circular disk of aluminum 1⁄32 to 3⁄32 in. (0.8 to 2.4 mm) in thickness provided with a suitable knob in the center for lifting. The cover fits over the test specimen when assembled and makes contact with the inside beveled surface of the wax seal at, or just above, the plane of the specimen. The cover is free of sharp edges which might remove the wax and is numbered or otherwise identified to facilitate its exclusive use with the same dish.

Designation: G 26 – 90

Standard Practice for Operating Light-Exposure Apparatus (Xenon-Arc Type) With and Without Water for Exposure of Nonmetallic Materials[1]

This standard is issued under the fixed designation G 26; the number immediately following the designation indicates the year of original adoption or, in the case of revision, the year of last revision. A number in parentheses indicates the year of last reapproval. A superscript epsilon (ε) indicates an editorial change since the last revision or reapproval.

1. Scope

1.1 This practice covers the basic principles and operating procedure for water- or light-exposure apparatus, or both, employing a xenon-arc light source.

NOTE 1—This practice combines the practices previously referred to as G 26 and G 27. Practice G 27, for Operating Xenon-Arc Type Apparatus for Light Exposure of Nonmetallic Materials[2], has been discontinued since it is now covered in this edition of G 26 under Test Methods C and D.

1.2 This practice does not specify the exposure conditions best suited for the material to be tested, but is limited to the method of obtaining, measuring, and controlling the conditions and procedures of the exposure. Sample preparation, test conditions, and evaluation of results are covered in ASTM methods or specifications for specific materials.

1.3 This practice includes four test methods:

1.3.1 *Test Method 1*—Continuous exposure to light and intermittent exposure to water spray.

1.3.2 *Test Method 2*—Alternate exposure to light and darkness and intermittent exposure to water spray.

1.3.3 *Test Method 3*—Continuous exposure to light without water spray. Exposure conditions are characteristic of those specified by AATCC Test Method 16E 1976.

1.3.4 *Test Method 4*—Alternate exposure to light and darkness without water spray. Exposure conditions characteristic of those natural conditions experienced when exposing are in accordance with Practice G 24.

1.4 *This practice may involve hazardous operations and equipment. This practice does not purport to address all of the safety problems associated with its use. It is the responsibility of the user of this practice to establish appropriate safety and health practices and determine the applicability of regulatory limitations prior to use.*

1.5 The values stated in SI units are to be regarded as the standard. The inch-pound equivalents of the SI units may be approximate.

2. Referenced Documents

2.1 *ASTM Standard:*

G 24 Practice for Conducting Natural Light Exposures Under Glass[2]

2.2 *AATCC Standard:*

Method 16E 1976 Colorfastness to Light, Water-Cooled Xenon-Arc Lamp Continuous Light[3]

2.3 *CIE Standard:*

No. 20 Recommendations for the Integrated Irradiance and the Spectral Distribution of Simulated Solar Radiation for Testing Purposes[4]

2.4 *ISO Standard:*

Gray Scale

3. Significance and Use

3.1 Several types of apparatus with different exposure conditions are available for use. No single operating procedure for light-exposure apparatus with and without water can be specified as a direct simulation of natural exposure. This practice does not imply expressly or otherwise an accelerated weathering test.

3.2 Since the natural environment varies with respect to time, geography, and topography, it may be expected that the effects of natural exposure will vary accordingly. All materials are not affected equally by the same environment. Results obtained by use of this practice should not be represented as equivalent to those of any natural weathering test until the degree of quantitative correlation has been established for the material in question.

3.3 Variations in results may be expected among instruments of different types or when operating conditions among similar type instruments vary within the accepted limits of this practice. Therefore, no reference should be made to results from use of this practice unless accompanied by the report form shown in Fig. 1 or unless otherwise specified in a referenced procedure.

4. Apparatus

4.1 *Water-Cooled Type:*[5]

4.1.1 The apparatus employed should utilize a water-cooled xenon-arc lamp as the source of radiation and should be one of the following general types, or their equivalent:

4.1.1.1 *Type A*—An exposure apparatus in which the source of radiant energy shall be a water-cooled xenon-arc vertically located at the central axis of either a 508-mm (20-in.) diameter vertical specimen rack, or of a 648-mm (25.5-in.) diameter inclined rack. Means shall be provided for automatic programming of temperature and cycles.

[1] This practice is under the jurisdiction of ASTM Committee G-3 on Durability of Nonmetallic Materials and is the direct responsibility of Subcommittee G03.03 on Simulated and Controlled Environmental Tests.

Current edition approved Jan. 26, 1990. Published March 1990. Originally published as E 239 – 64 T. Redesignated G 26 in 1970. Last previous edition G 26 – 88$^{\epsilon 1}$.

[2] *Annual Book of ASTM Standards,* Vol 14.02.

[3] Available from The Secretary, American Association of Textile Chemists and Colorists, P.O. Box 12215, Research Triangle Park, NC 27709.

[4] Available from Secretary, U.S. National Committee, CIE, National Institute of Standards and Technology, Gaithersburg, MD 20899.

[5] Available from Atlas Electric Devices Co., 4114 N. Ravenswood Ave., Chicago, IL 60613.

Laboratory ______________________________
Material ______________________________
G 26 Test Method No. ______________________________
Reference Standard Used: ______________________________
Other ASTM Test No. ______________ Method No. ______________
Exposure Apparatus: ASTM Type ______________________________
Mfr. Model ______________ Serial No. ______________
Light Source: Water-Cooled ____W, Air-Cooled ____W
Method Used to Regulate Wattage to Lamp ______________________________
Optical Filters: Type ______________, Age ______________ h
Spectral Irradiance: ______________W/m^2 at ______________nm
Irradiation Units: ______________kJ/m^2 How Monitored: ______________
Elapsed Exposure Time: ________h
Exposure Conditions: Program ________h
Light ________min Dark ________min

Black-Panel Temperature	___°C (°F)	
Air Temperature	___°C (°F)	
Dry Bulb Temperature	___°C (°F)	___°C (°F)
Relative Humidity	___%	___%
Specimen Water Spray	___min cycle	___min cycle
Specimen Nonspray	___min cycle	___min cycle

Specimen Spray Water Type: ______________________________
Specimen Spray Nozzle Type: Mfg. Designation ______________________________
Specimen Relocation Procedure During Exposure: ______________________________

Identify properties to be determined on test specimens and identify test procedures or methods used for property measurement.

Operator/Date: ______________________________
Supervisor/Date: ______________________________
Company: ______________________________

FIG. 1 Report Form

Means shall be provided for adjustment of relative humidity. The specimen rack shall rotate at 1 ± 0.1 rpm.

NOTE 2—In the commercial descriptions of the four types, the term "cycle" is defined as each time interval of light, darkness, and water spray that is specified differently in accordance with the different testing methods.

4.1.1.2 *Type AH*—The exposure apparatus shall be identical to Type A except it shall have automatic humidity control.

4.1.1.3 *Type B*—An exposure apparatus in which the source of radiant energy shall be a water-cooled xenon-arc vertically located at the central axis of a 960-mm (37.75-in.) diameter specimen rack. Means shall be provided for automatic programming of temperature and cycles. Means may be provided for adjustment of humidity. The specimen rack shall rotate at 1 ± 0.1 rpm.

4.1.1.4 *Type BH*—The exposure apparatus shall be similar to Type B except it shall have automatic humidity control.

4.1.2 The xenon-arcs employed shall be of the "long arc" water-cooled type operated through suitable reactance transformers and electrical equipment from a 50 or 60-Hz power supply. They shall employ cylindrical inner and outer optical filters to direct the flow of cooling water and to simulate a designed spectral energy distribution.

4.1.2.1 For the purpose of this practice, the xenon-arc lamp shall consist of a quartz xenon burner tube and one of the following optical filter combinations:[6,7]

(*a*) Borosilicate glass inner and outer optical filter to simulate the spectral power distribution (SPD) of natural daylight throughout the actinic region.

(*b*) Infrared absorbing glass inner optical filter with quartz outer optical filter to simulate the SPD of natural daylight (300 to 1000 nm).

(*c*) Borosilicate glass inner optical filter with soda lime glass outer optical filter to simulate the SPD of natural daylight (actinic wavelengths) filtered through window glass.

(*d*) Infrared absorbing glass inner and outer optical filter to simulate the SPD of natural daylight (310 to 1000 nm) filtered through window glass.

(*e*) Quartz inner and outer optical filter to approximate sunlight intensities unfiltered by the earth's atmosphere.[6]

4.1.2.2 To prevent loss in levels of irradiance due to excessive solarization and to prevent possible breakage caused by stresses in optical filters exposed to high-intensity UV energy, inner optical filters[8] shall be replaced periodically. Suggested intervals are 300 and 400 h for Types A and B, respectively. The outer optical filter shall be replaced after 1500 and 2000 h for Types A and B, respectively.

4.1.3 Distilled or deionized water should be recirculated past the burner at a flow rate sufficient to remove excess heat. Passing water through a cartridge demineralizer installed in the recirculation line just ahead of the lamp minimizes contamination of the quartz envelope of the burner. A heat-exchange unit should be used to cool the recirculated lamp water.

4.1.4 Since xenon-arc lamps, like all gas discharge lamps, will have a progressive drop in radiation output with continued use, and since optical filters will change in their

[6] Norton, Kiuntke, and Connor, *Canadian Textile Journal,* CTJOA, May 1969.

[7] Corning 7740 Pyrex® (borosilicate), Corning 4601 IR absorbing visible transmitting, Kimble R6 soda lime glass tubing, and General Electric GA-204 quartz tubing of proper dimensions have been found suitable for use as xenon-arc optical filters. Available from Atlas Electric Devices Co., 4114 N. Ravenswood Ave., Chicago, IL 60613.

[8] Kishii, T., and Ooka, K., *Symposium on Coloured Glasses,* International Comm. on Glass of Czech. Sci. & Tech. Soc. of the Silicate Industry, 1967.

transmission characteristics, provision shall be made in the apparatus (automatically or manually) for progressively increasing the wattage of the lamp to minimize changes in the intensity of the radiation at the face of the sample. The greatest change in both the xenon burner tube and optical filters occurs in the first 20 h of use. For this reason, burners and optical filters preaged by the instrument manufacturer are recommended for critical testing when a more rapidly changing rate of intensity is undesirable. For routine testing on a comparative basis, the 20 h of preaging may be omitted.

4.1.5 Many nonmetallic materials are selectively absorbing. The energy that a molecule absorbs depends upon the wavelength of the incident radiation.[9] It is desirable to monitor the level of irradiance of the photochemically effective wavelengths if intercomparisons are to be made among samples not exposed simultaneously. Samples may then be exposed to known amounts of irradiation, the time integral of irradiance.

4.1.5.1 Where radiometers capable of monitoring discrete portions of a continuous spectrum are available, exposure to a mutually agreed-upon level of irradiation may be specified as the exposure interval in place of a time interval.

(*a*) When using the optical filter combination described in 4.1.2.1 (*a*), the suggested minimum spectral irradiance levels are:

.2 $W/m^2/nm$ band at 320 nm
.35 $W/m^2/nm$ band at 340 nm
.5 $W/m^2/nm$ band at 380 nm
.75 $W/m^2/nm$ band at 420 nm

(*b*) When using the optical filter combination described in 4.1.2.1 (*c*), the suggested minimum spectral irradiance levels are:

.2 $W/m^2/nm$ band at 340 nm
.45 $W/m^2/nm$ band at 380 nm
.7 $W/m^2/nm$ band at 420 nm

(*c*) Operating at the suggested minimum levels approximates the average daily solar irradiance under ideal conditions. It is about half of the maximum values for total irradiance on a horizontal plane when the sun is at 90° altitude as reported in CIE No. 20. When irradiance is monitored and periodic manual adjustment of wattage is made to compensate for changes of intensity, no lamp assembly should be used that cannot maintain the minimum level of irradiance within a 10 % tolerance at the monitored wavelength. Irradiation expressed as joules per square metre is the product of irradiance × exposure time in seconds.

4.1.5.2 Xenon-arc type equipment not having a radiometer shall be operated with periodic increases in wattage to minimize any drift in levels of irradiation. Such intervals shall be established by the parties concerned or follow the suggested schedule below in 4.1.5.2 (*a*).

NOTE 3—The use of suggested wattage steps does not imply that irradiance will be maintained at levels equivalent to those obtained when employing a light monitor. They are intended as a guide to minimize the reduction of UV intensity with maximum lamp longevity when using either Test Method A or C for 2.5, 6, or 6.5-kW lamps, respectively.

(*a*) The following are the suggested wattage settings for each exposure interval of a 2500-W xenon burner tube based upon the average performance of xenon burner tubes with borosilicate filters:

Time, h	Power, W
0 to 20	1600
20 to 200	1800
200 to 400	2000
400 to 600	2250
600 to 800	2500
800 to 1000	2600
1000 to 1200	2800
1200 to 1500	3000

(*b*) The following are the suggested wattage settings for each exposure interval of a 6000-W xenon burner tube based upon the performance of xenon burner tubes with borosilicate filters in Test Method A:

Time, h	Power, W
0 to 20	min
20 to 100	4750
100 to 250	5000
250 to 400	5250
400 to 550	5500
550 to 700	6000
700 to 850	6500
850 to 1000	7000

(*c*) The following are the suggested wattage settings for each exposure interval of a 6500-W xenon burner tube based upon the average performance of xenon burner tubes with borosilicate filters in Test Method A:

Time, h	Power, W
0 to 20	min
20 to 100	5500
100 to 200	6000
200 to 500	6200
500 to 1000	6500
1000 to 1500	7000
1500 to 2000	7500
2000 and over	8500

4.1.6 Specimens should be mounted on a rack rotating about the lamp at a distance such that the location of each specimen assures that the irradiance incident on its surface does not vary by more than 5 % from the average. Figure 4 illustrates a rack meeting this condition.

4.1.7 Any apparatus that does not comply with the condition of 4.1.6 shall not have a variation in irradiance exceeding 10 % of the average, and may require specimen rearrangement during exposure, in accordance with the test procedure, in order to minimize the variability in radiant exposure.

4.1.8 Testing temperatures should be measured and regulated on the basis of a black panel thermometer unit mounted on the specimen rack so that the face of the unit is in the same relative position and is subjected to the same influences as the test specimens. The black panel thermometer unit should consist of a stainless steel panel about 70 by 50 by 0.95 mm (2.750 by 5.875 by 0.0375 in.), to which a stainless steel bimetallic dial-type thermometer is mechanically fastened. This thermometer should have a stem approximately 4 mm (0.16 in.) in diameter with a 44.5-mm (1.75-in.) dial. The sensitive portion extending about 38 mm (1.5 in.) from the end of the stem should be located in the center of the panel approximately 64 mm (2.5 in.) from the top and approximately 48 mm (1.875 in.) from the bottom

[9] Hawkins, W.L., ed., "Polymer Stabilization," *Wiley-Interscience*, 1972. p. 166.

of the panel. The face of the panel with the thermometer stem attached should be finished with a baked-on black infrared-absorbing coating having good resistance to light.

4.1.8.1 A thermocouple or resistance bulb thermometer mounted at the center face of the black panel, which provides temperature values equivalent to the dial thermometer, may be substituted.

4.1.9 A blower unit in the base of the apparatus should provide a flow of air through the test chamber and over the test specimens. Control of specimen and black-panel temperature should be accomplished by thermostatic control of the temperature of the constant volume of air from the blower. Black-panel temperatures should be read through the window in the test chamber door without opening the door.

4.1.10 Apparatus operated as a light- and water-exposure test shall be equipped with a specimen spray unit as illustrated in Figs. 2 and 3. All components of the specimen spray unit should be fabricated from stainless steel, plastic, or other material that does not contaminate the water. Apparatus operated only as a light-exposure test are not required to have the specimen spray unit or, if present, it should not be used.

4.1.11 Types A, AH, BH, and some Type B apparatus are equipped with an electrically operated vaporizing unit for adding moisture to the air as it passes through the conditioning chamber in the base section prior to its entry into the test chamber of the apparatus.

4.1.11.1 In Types A and B, the vaporizing unit, when manually turned on, may operate continuously while the apparatus is in operation. Type A and some Type B apparatus may be programmed to operate the vaporizer during a dark cycle only. The temperature of the water, with which the vaporizer is supplied, is not controlled. In Type B apparatus not supplied with a vaporizing unit, the relative humidity within the test chamber is governed by evaporation of water in the bottom of the chamber and from water emitted through the specimen spray unit.

4.1.11.2 In Types AH and BH apparatus, operation of the vaporizer is controlled automatically by a wet-bulb thermostat. The temperature of the water supplied to the vaporizer is regulated automatically by thermostatically-actuated electric immersion heaters. Control automatically shifts between two separate sets of thermostats as the arc lamp is turned on and off by the program control unit.

4.1.12 In Types A, B, AH, and BH apparatus containing vaporizing units, relative humidity is determined from wet- and dry-bulb thermometers located in the air stream at the vaporizer's point of exit from the test chamber.

4.1.12.1 Determine the relative humidity for Type B apparatus not equipped with a vaporizing unit from wet- and dry-bulb thermometers mounted in holders on the specimen rack so that their sensitive portions are in the same relative position as the face of the test specimen but shielded from the radiation.

4.1.12.2 Any apparatus with a vaporizing unit operated with the unit and immersion heaters turned off will provide essentially the same conditions of relative humidity as are produced in Type B apparatus without a vaporizing unit.

4.1.13 The apparatus shall include means for measuring the following:

4.1.13.1 Wattage of the xenon-arc lamp,

4.1.13.2 Irradiance at the specimen rack,

4.1.13.3 Black-panel temperature,

4.1.13.4 Dry-bulb temperature (test chamber),

4.1.13.5 Wet-bulb temperature (test chamber),

4.1.13.6 Exposure interval, and

4.1.13.7 Water spray pressure when applicable.

4.1.14 Where specified, the apparatus shall include means for regulating or controlling the following:

4.1.14.1 Wattage of the xenon lamp,

4.1.14.2 Irradiance or irradiation, or both, at the specimen rack,

4.1.14.3 Temperature (test chamber),

4.1.14.4 Relative humidity (test chamber),

4.1.14.5 Light-dark-spray-humidity-temperature cycles, and

4.1.14.6 Water spray pressure when applicable.

4.2 *Air-Cooled Type:*[10]

4.2.1 The apparatus employed shall use an air-cooled xenon-arc lamp as the source of radiation and should be one of the following types, or their equivalent:

4.2.1.1 *Type C*—Air-cooled xenon-arc apparatus, 1500 W, 158-mm (6.2-in.) diameter specimen rack, with automatic programming of cycles and humidity. The specimen rack should rotate at 5.2 ± 0.1 rpm.

4.2.1.2 *Type D*—Air-cooled xenon-arc apparatus, 4500 W, 360-mm (14.2-in.) diameter specimen rack, with automatic programming of temperature, cycles, and humidity. The specimen rack should rotate at 3.7 ± 0.1 rpm.

4.2.1.3 *Type E*—Air-cooled xenon-arc apparatus, three lamps operating simultaneously at 4500 W each, 610-mm (24.1-in.) diameter specimen rack, with automatic programming of temperature, cycles, and humidity. The specimen rack should rotate at 2.0 ± 0.1 rpm.

4.2.2 The xenon arcs employed in Types C and D should be of the medium-pressure, air-cooled type and one or more optical filters shall be fitted between the light source and the samples to filter out undesired wavelengths of radiation. A combination of optical filters recommended by the manufacturer shall be used to simulate (*1*) sunlight in the open, (*2*) sunlight behind window glass, or (*3*) any other simulated solar irradiation condition desired.

4.2.3 All gas discharge lamps have a progressive drop in radiant output with continued use. For Types C and D, it is recommended to exchange the IR-optical filters for new ones after 3500 h of use. Answering the purpose, not all seven IR-optical filters of the filter lantern should be exchanged at the same time but each should be exchanged after it has been running for 3500 h. For Type E which has three lamps, one lamp should be replaced every 500 hours to minimize the overall drop in radiant output. For Type E, the three partial preaged filters should be cleaned monthly and replaced after five years of normal usage.

4.2.4 Specimen holders should rotate around the arc, describing a cylindrical surface, so that specimens may face, or be turned opposite to, the arc. No part of the specimens shall be above or below the ends of the arc. Repositioning of specimens in upper, center, or lower positions can improve the uniformity of intensity of the exposure.

---

[10] Available from DSET Labs, Box 1850, Black Canyon Stage 1, Phoenix, AZ 85027.

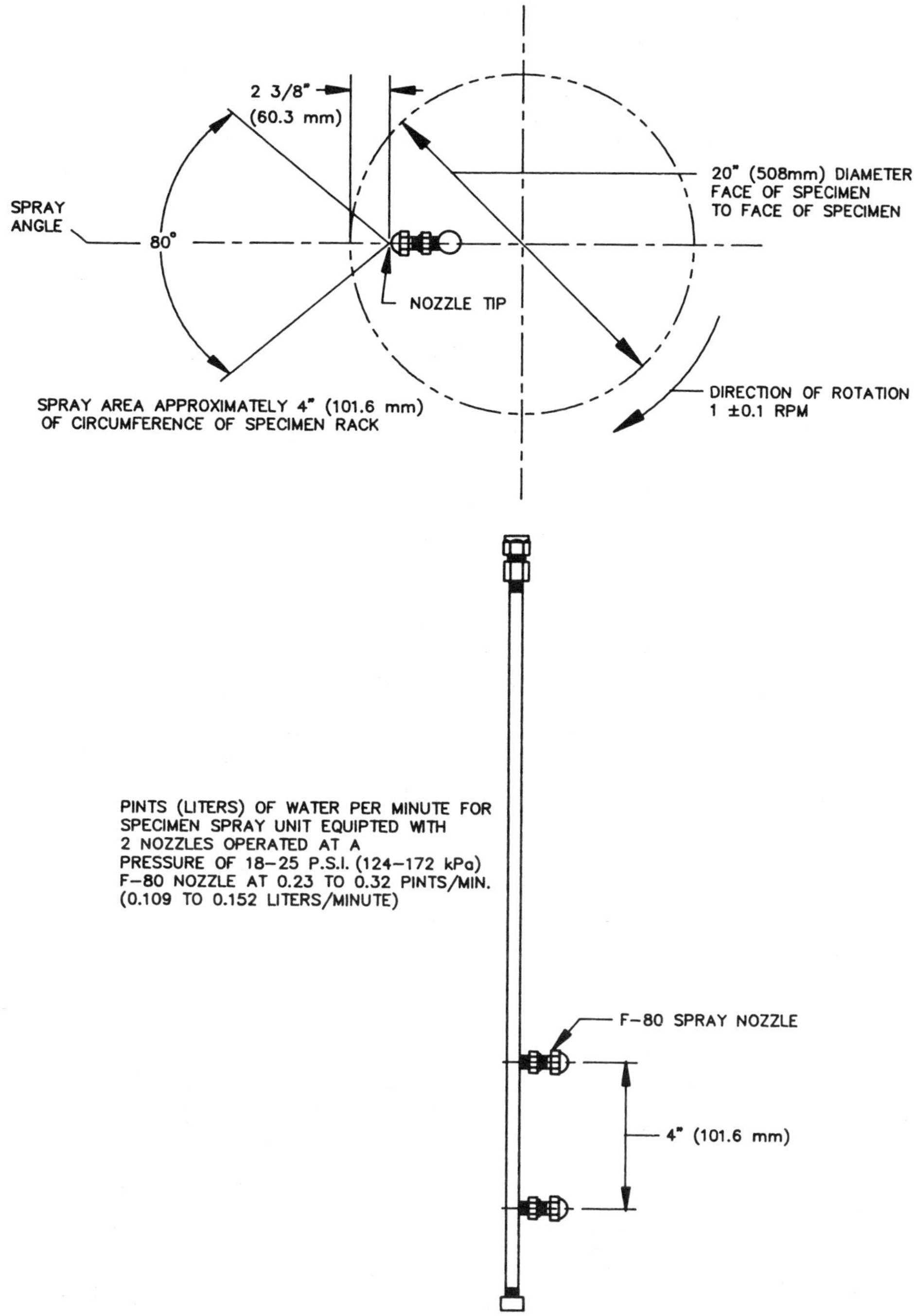

Metric Equivalents

2⅜ in.	60.325 mm
4 in.	101.6 mm
20 in.	508 mm
18 psi	124 kPa
25 psi	172 kPa
0.23 pt	0.109 liter
0.32 pt	0.152 liter

FIG. 2 Specimen Spray Arrangement for Type A and AH Apparatus

4.2.5 Testing temperatures should be measured and regulated on the basis of a black-panel thermometer unit that is mounted so that the face of the unit is in the same relative position and is subjected to the same influences as the test specimens. The black-panel thermometer unit should consist of a stainless steel panel approximately 48 by 200 by 1 mm (1.9 by 8.0 by 0.04 in.), to which a stainless steel bimetallic dial-type thermometer is mechanically fastened. The temper-

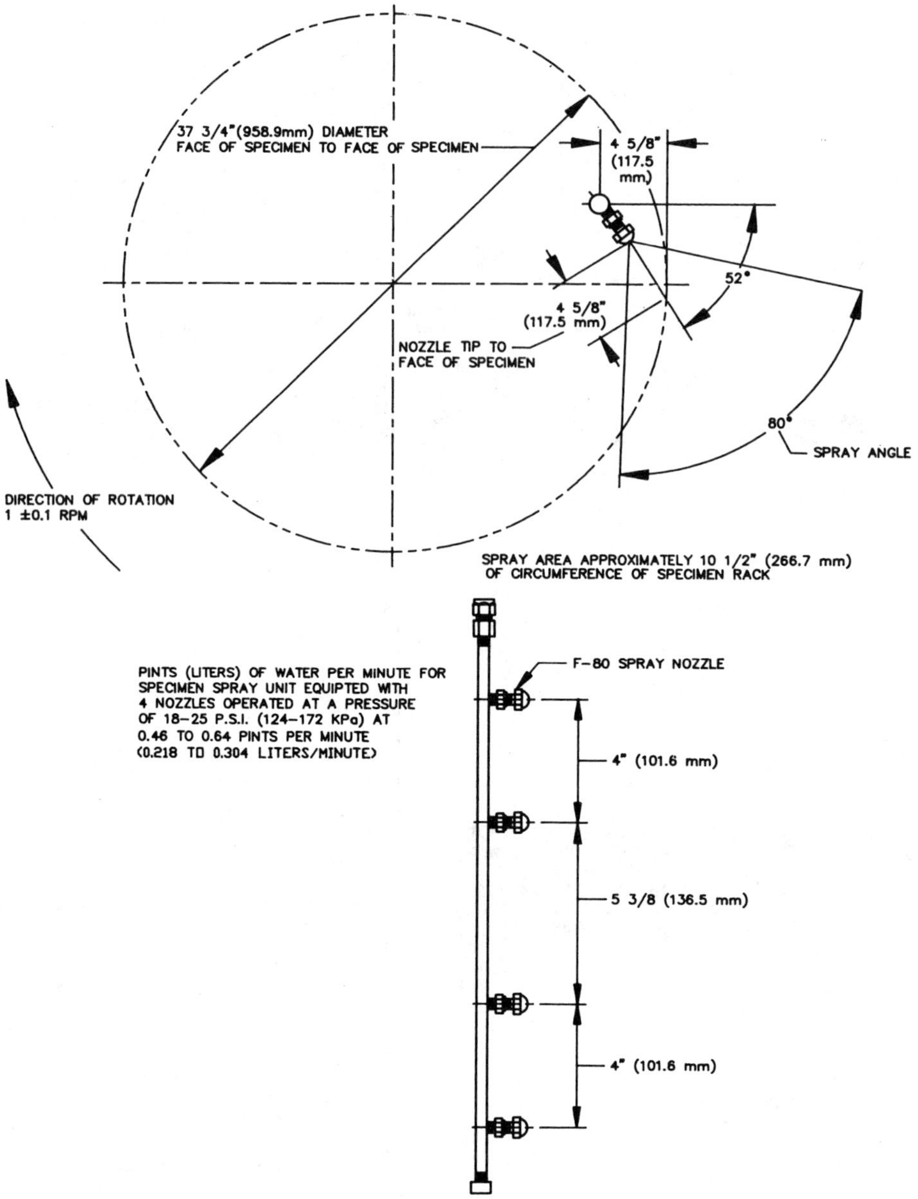

Metric Equivalents

4 in.	101.6 mm	18 psi	124 kPa
4⅝ in.	117.48 mm	25 psi	172 kPa
5⅜ in.	136.52 mm	0.46 pt	0.218 liter
10½ in.	266.7 mm	0.64 pt	0.304 liter
37¾ in.	958.85 mm		

FIG. 3 Specimen Spray Arrangement for Type B and BH Apparatus

ature at the center of the panel should be sensed by the thermometer. The face of the panel should be finished with a baked-on black glossy enamel having good stability to light.

4.2.6 A blower unit in the base of the apparatus shall provide a flow of air through the test chamber and over the test specimens. Control of the specimen and the black-panel temperature should be accomplished by thermostatic control of the temperature of the constant volume of air from the blower. Black-panel temperature should be read through the window in the test chamber door without opening the door.

4.2.7 The apparatus should be equipped with a system to spray the specimens uniformly with water. This system should be of stainless steel, plastic, or other material that does not react with or contaminate the water employed.

4.2.8 Relative humidity in the test chamber should be measured and controlled by a contact hygrometer. Water should be vaporized and diffused to enrich the air with moisture and produce the required humidity.

4.2.9 The apparatus shall include equipment necessary for measuring and controlling the same parameters as listed for water-cooled xenon arcs.

5. General Procedure

5.1 Check to be sure the apparatus is operating properly at the start of each test. Check the lamp operation at 100-h intervals to be sure the burner tube and optical filters are clean and that they have not exceeded the maximum recommended period of use.

5.2 Program the instrument to operate in the continuous light-on mode without water spray. Fill the specimen rack with blanks and the black-panel thermometer. Operate the instrument in this mode while regulating the chamber dry-bulb temperature to provide the desired black-panel temperature of 63 ± 3°C (145 ± 5°F).

5.3 When the chamber dry-bulb temperature has been regulated, adjust instruments with automatic humidity control to the desired relative humidity (Types C, D, and E) or to the chamber wet-bulb temperature (Types A and B) at the level that will ensure the desired relative humidity.

5.4 In Types A, B, C, D, and E apparatus, use the type and number of spray nozzles (Figs. 2 and 3) recommended by the manufacturer and operate at a pressure of 124 to 172 kPa (18 to 25 psi), measured at the nozzle unless otherwise recommended. The water shall have a pH of 6.0 to 8.0, contain less than 20 ppm solids, and leave no deposit or stain on the specimens after continued exposure in the apparatus. The temperature of the water should be 16 ± 5°C (60.8 ± 9°F), and recirculation of the water shall not be permitted unless the recirculated water meets the above requirements. Pass the specimens through the spray once in each minute or revolution of the rack during the spray cycle. Allow the water to strike the test specimens in the form of a fine spray equally distributed over the specimens.

5.5 The temperature of the rack spray water should be sufficiently low to reduce the specimen temperature below the dewpoint when the specimen is continuously sprayed during a dark cycle.

5.6 *Reference Standards:*

5.6.1 *AATCC Blue Wool Lightfastness Standards.*[3]

NOTE 4—The eight standards L-2 to L-9 have been made by blending wool dyed with the fugitive dye Erio Chrome Azurole B and wool dyed with the fast dye Indigosol Blue AGG in different proportions. The blended wools were spun into yarns and the yarns woven into cloths. Each standard is approximately twice as fast as the next lower numbered standard. It has been found that when new lots of the standards are produced, the amounts of the dyes required and the properties of the two wools in the blends are often different from those used originally. The dyeing strengths and blending proportions would therefore be misleading, and they are intentionally not specified.

5.6.2 Any standard sample established by mutual agreement between the purchaser and the supplier.

5.6.3 *ISO Gray Scale for Assessing Change in Color.*[11]

5.7 When the material to be evaluated is suspected of being very sensitive to variations in radiant exposure, and the apparatus does not meet the requirement of 4.1.6, it is the responsibility of the concerned parties to mutually agree upon a rearrangement schedule for specimens as part of the exposure program in order to minimize any effect that variations in irradiance might have as a function of specimen position.

TEST METHOD A—CONTINUOUS EXPOSURE TO LIGHT AND INTERMITTENT EXPOSURE TO WATER SPRAY

6. Procedure

6.1 Apparatus shall be Type A, B, AH, BH, C, D, or E light-exposure device.

6.2 Program the instrument for continuous light and intermittent water spray in accordance with the manufacturer's instructions. Choice of the program selected will be by mutual agreement among the interested parties. Historical convention has established a cycle of 102 min of light followed by a cycle of 18 min of light and water spray as a commonly accepted program that permits the attainment of the maximum black-panel temperature during the light-only portion of the cycle.

6.3 In Types AH and BH apparatus, adjust dry- and wet-bulb temperature controls, humidifier, and immersion heater controls to maintain the desired conditions as specified in Section 5. In Types C, D, and E, adjust the chamber temperature and humidity controls to maintain desired conditions as specified in Section 5.

6.4 In Types A and B with humidifier, humidity may be adjusted but not controlled.

6.4.1 Type B instruments without a humidifier must operate with whatever humidity occurs from the evaporation of water in the bottom of the test chamber and from specimen spray.

TEST METHOD B—ALTERNATE EXPOSURE TO LIGHT AND DARKNESS AND INTERMITTENT EXPOSURE TO WATER SPRAY

7. Procedure

7.1 Apparatus shall be Type A, B, AH, BH, or E light-exposure device.

7.1.1 Types AH, BH, and E apparatus with automatic humidity controls may be operated on alternate light and dark cycles.

7.1.1.1 Operation during the light-on cycle shall be as described in Section 6.

7.1.1.2 Separate controls for temperatures and humidification may be adjusted during the dark cycle for automatic control as the program alternates from light to dark. In Type BH and E apparatus, a rack spray to cool the specimens by wetting the unexposed back surface can result in development of condensation on the exposed specimen surface during the dark cycle.

7.1.2 Types A, B, AH, BH, and E may be programmed to operate on alternate light and dark cycles without control of relative humidity.

TEST METHOD C—CONTINUOUS EXPOSURE TO LIGHT WITHOUT WATER SPRAY

8. Procedure

8.1 Apparatus shall be Types A, B, AH, BH, C, D, or E light-exposure devices programmed for continuous light only, in accordance with the manufacturer's instructions. Optical filters C or D described in 4.1.2.1 should be used.

---

[11] Reference Draft ISO Recommendation No. 177, Document ISO/TC-38 (Secretariat 102) 186. Scales with instructions for use are available from ISO member bodies in the various countries of the world. In the USA they may be obtained from The Secretary, American Association of Textile Chemists and Colorists, P.O. Box 12215, Research Triangle Park, NC 27709.

8.1.1 Adjust the controls on the apparatus so that the black-panel temperature is 63 ± 3°C (145 ± 5°F) and the relative humidity is 30 ± 5 % when the xenon lamp is operated at a wattage that provides 20 AATCC Fading Units of exposure in about 15 to 25 clock hours. Check at regular intervals and, when necessary, readjust the controls to maintain the specified values for these variables. Target values for black panel temperature, 20 AATCC Fading Units, and relative humidity should be 63°C, 20 clock hours, and 30 %, respectively.

NOTE 5—20 AATCC Fading Units is equivalent to a 1.5 ± 0.2 ANLAB(40) unit change in Blue Wool L-4 (#4 on the ISO Gray Scale). Irradiation of about 110 kJ/m^2 (1.5 W/m^2·20-h exposure) at 420 nm has been found sufficient to produce the specified condition.

8.1.2 Expose the material to be tested as determined by mutual agreement among the concerned parties or, when not otherwise specified, in accordance with one of the following:

8.1.2.1 *Versus One AATCC Blue Wool Lightfastness Standard*—Expose the test specimen and any mutually agreed-upon AATCC Blue Wool Lightfastness Standard until the selected standard between its exposed and unexposed portion exhibits a color change equal to Step 4 of the ISO Gray Scale. Report the results by any mutually agreed-upon method of measuring change in the test specimen.

8.1.2.2 *Versus Set of AATCC Blue Wool Lightfastness Standards*—Expose the test specimen and a set of AATCC Blue Wool Lightfastness Standards until a maximum permissible amount of change, as determined by any specified or mutually agreed-upon method of measurement, occurs in the test specimen. Assign it a classification number equal to that of the numbered standard that most nearly exhibits a change in color equal to Step 4 on the ISO Gray Scale between its exposed and unexposed portion. Where one standard shows a color change greater than Step 4 and the next higher number standard shows less than a Step 4 color change, an intermediate or half-grade rating may be used.

8.1.2.3 *Versus Other Standard Sample*—Expose the test specimen and any mutually agreed-upon standard sample until either, by any specified or mutually agreed-upon method of measurement, shows an agreed-upon amount of change. Report the results on the basis of a comparison of the specimen with the standard sample.

TEST METHOD D—ALTERNATE EXPOSURE TO LIGHT AND DARKNESS WITHOUT WATER SPRAY

9. Procedure

9.1 Apparatus shall be Types A, B, AH, BH, or E programmed to predetermined cycles of light and darkness by turning off the xenon arc.

NOTE 6—Type B instruments not equipped with a vaporizing unit do not meet the requirements of Test Method D. In Type A and Type B apparatus with intermittent blower operation during the light-off cycle and the atomizer in continuous operation, the test chamber air temperature will drop to that of room air less the heat removed by vaporation of water from wicks and atomizer.

9.1.1 Unless otherwise specified, exposure conditions shall be identical to Test Method B except that during the light-on cycle the equilibrium condition for relative humidity shall be 35 ± 5 % and during periods of darkness equilibrium conditions shall be 90 ± 5 % relative humidity at a dry-bulb temperature of 35 ± 3°C (95 ± 5°F).

10. Report

10.1 Report the following information:

10.1.1 Type and model of exposure device,

10.1.2 Type of light source and wattage,

10.1.3 Type and age of filters,

10.1.4 Spectral irradiance at sample location, W/m^2·nm (see Note 9),

10.1.5 Irradiation, kJ/m^2,

10.1.6 Elapsed exposure time, h,

10.1.7 Light- or dark-water-humidity program employed,

10.1.8 Operating black-panel temperature,

10.1.9 Operating relative humidity,

10.1.10 Type of spray water,

10.1.11 Type of spray nozzle, and

10.1.12 Specimen relocation procedure.

NOTE 7—When direct measurement of irradiation and spectral irradiance cannot be made, data supplied by the manufacturer shall be substituted. Irradiation in kilojoules per square metre may be calculated by multiplying the product of the irradiance in watts per square metre and the exposure time in hours by the factor 3.6.

FIG. 4 Three-Tiered Specimen Rack

Designation: G 53 – 88

Standard Practice for Operating Light- and Water-Exposure Apparatus (Fluorescent UV-Condensation Type) for Exposure of Nonmetallic Materials[1]

This standard is issued under the fixed designation G 53; the number immediately following the designation indicates the year of original adoption or, in the case of revision, the year of last revision. A number in parentheses indicates the year of last reapproval. A superscript epsilon (ε) indicates an editorial change since the last revision or reapproval.

This standard has been approved for use by agencies of the Department of Defense. Consult the DoD Index of Specifications and Standards for the specific year of issue which has been adopted by the Department of Defense.

1. Scope

1.1 This practice covers the basic principles and operating procedures for using fluorescent ultraviolet (UV) and condensation apparatus to simulate the deterioration caused by sunlight and water as rain or dew.

1.2 This practice is limited to the method of obtaining, measuring, and controlling the conditions and procedures of exposure. It does not specify the exposure conditions best suited for the material to be tested. Specimen preparation and evaluation of the results are covered in ASTM test methods or specifications for specific materials.

1.3 The values stated in SI units are to be regarded as the standard.

1.4 *This standard may involve hazardous materials, operations, and equipment. This standard does not purport to address all of the safety problems associated with its use. It is the responsibility of the user of this standard to establish appropriate safety and health practices and determine the applicability of regulatory limitations prior to use.*

2. Referenced Documents

2.1 *ASTM Standards:*

E 220 Method for Calibration of Thermocouples by Comparison Techniques[2]

G 7 Practice for Atmospheric Environmental Exposure Testing of Nonmetallic Materials[3]

2.2 *CIE Standard:*

No. 20 Recommendations for the Integrated Irradiance and the Spectral Distribution of Simulated Solar Radiation for Testing Purposes[4]

3. Description of Terms Specific to This Standard

3.1 *irradiance*—the radiation incident on a surface expressed in W/m^2. Irradiance is the total of the incident radiation at *all* wavelengths. Forty watt fluorescent lamps of the UV-B and UV-A types generate similar amounts of irradiance. However, since this irradiance is distributed at different wavelengths, the photochemical effects caused by these different lamps vary greatly. Therefore, irradiance should not be used to compare UV light sources.

3.2 *spectral irradiance*—distribution of irradiance as a function of wavelength. It is expressed in W/m^2 per wavelength band. The spectral irradiance of sunlight is often shown as W/m^2 per 10 nm band. The spectral irradiance of fluorescent UV lamps should be shown in bands 1 or 2 nm wide. Spectral irradiance is the proper method for comparing sources with different energy distributions.

3.3 *spectral energy distribution (SED)*—general term for the characterization of the amount of radiation present at each wavelength. SEDs can be expressed by power in watts, irradiance in W/m^2, or energy in joules. The shape of the SED would be identical in all of these units. Fluorescent lamps are frequently described by relative SEDs which show the amount of radiation at each wavelength as a percentage of the amount of radiation at the peak wavelength. Fig. 1 is a relative SED.

3.4 *ultraviolet regions*—CIE Publication No. 20 (1972) divides the ultraviolet spectrum into three regions: UV-A, radiation in wavelengths between 315 nm and 400 nm; UV-B, radiation in wavelengths between 280 nm and 315 nm; and UV-C, radiation in wavelengths shorter than 280 nm.

3.5 *fluorescent UV lamp*—lamp in which radiation at 254 nm from a low-pressure mercury arc is transformed to longer wavelength UV by a phosphor. The spectral energy distribution of a fluorescent UV lamp is determined by the emission

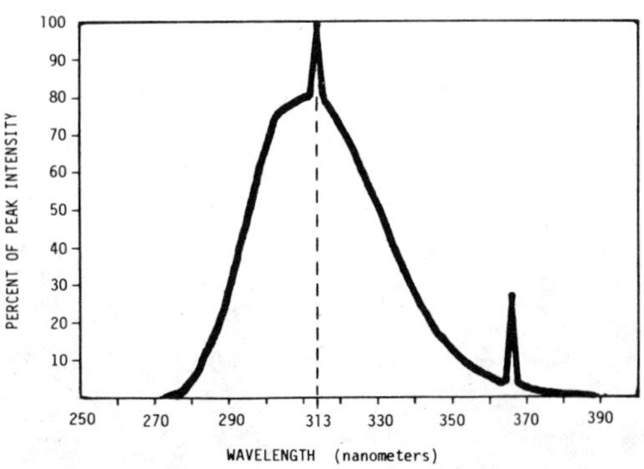

FIG. 1 Typical Relative SED of UV-B Lamp

---

[1] This practice is under the jurisdiction of ASTM Committee G-3 on Durability of Nonmetallic Materials, and is the direct responsibility of Subcommittee G03.03 on Simulated and Controlled Environmental Tests.

Current edition approved July 29, 1988. Published September 1988. Originally published as G 53 – 77. Last previous edition G 53 – 84.

[2] *Annual Book of ASTM Standards*, Vol 14.03.

[3] *Annual Book of ASTM Standards*, Vol 14.02.

[4] Available from Secretary, U.S. National Committee, CIE, National Bureau of Standards, Gaithersburg, MD 20899.

spectrum of the phosphor and the UV transmission of the glass tube.

4. Summary of Practice

4.1 Specimens are alternately exposed to ultraviolet light alone and to condensation alone in a repetitive cycle.

4.2 The UV source is an array of fluorescent lamps, with lamp emission concentrated in the UV range.

4.3 Condensation is produced by exposing the test surface to a heated, saturated mixture of air and water vapor, while the reverse side of the test specimen is exposed to the cooling influence of ambient room air.

4.4 The exposure condition may be varied by selection of: the fluorescent UV lamp; the timing of the UV and condensation exposure; the temperature of UV exposure; and the temperature of condensation exposure.

5. Significance and Use

5.1 The use of the apparatus under this practice is intended to simulate the deterioration caused by water as rain or dew and the ultraviolet energy in sunlight. It is not intended to simulate the deterioration caused by localized weather phenomena, such as atmospheric pollution, biological attack, and salt water exposure.

5.2 Variation in results may be expected when operating conditions are varied within the accepted limits of this practice. Therefore, no reference shall be made to results from use of this practice unless accompanied by a report detailing the specific operating conditions in conformance with Section 11.

5.3 Any report correlating results from use of this practice to results from a period of natural weathering shall specify in detail the conditions of natural exposure, since sunlight and water effects upon materials exposed to the weather will vary from year to year and also vary with location, latitude, time of year, temperature, proximity to water sources, etc.

NOTE 1—Practice G 7 lists the information required to describe a particular condition of outdoor exposure.

5.4 Correlations established and reported in conformance with 5.2 and 5.3 shall not be extrapolated to other test conditions permitted by this practice, to other conditions of natural exposure, or to materials other than those tested.

6. Apparatus[5]

6.1 *Test Chamber,* (see Fig. 2) constructed of corrosion-resistant materials enclosing eight fluorescent UV lamps, a heated water pan, test specimen racks, and provisions for controlling and indicating operating times and temperatures.

6.2 *Lamps,* rapid start, medium bipin fluorescent UV lamps with a length of 1220 mm, and a nominal rating of 40 W when operated from a ballast providing a controlled current of 430 mA at 102V.

6.2.1 Unless otherwise specified, the lamps shall be UV-B lamps with a peak emission at 313 nm and a spectral energy distribution as shown in Fig. 1.

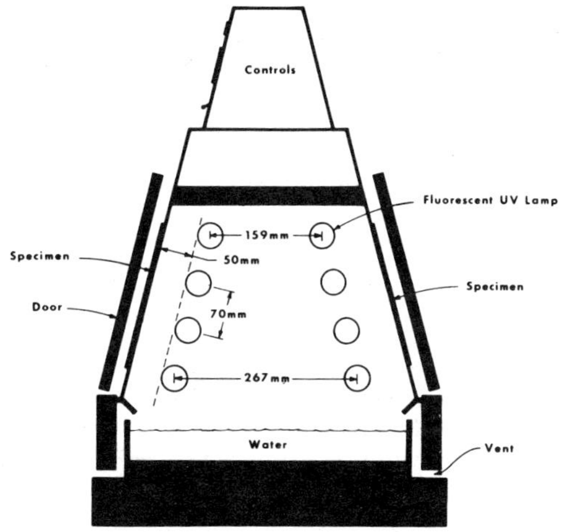

FIG. 2 Apparatus Schematic Cross Section

6.2.2 Other fluorescent UV lamps meeting the size and electrical characteristics in 6.2 may be used, provided that the lamp and spectral energy distribution are reported in conformance with Section 11.

6.3 *Lamp Spacing and Arrangement*—The lamps shall be mounted in two banks of four lamps each as shown in Fig. 2. The lamps in each bank shall be mounted parallel in a flat plane on 70-mm centers.

6.4 *Specimen Mounting and Arrangement*—The test specimens shall be mounted in stationary racks with the plane of the test surface parallel to the plane of the lamps at a distance of 50 mm from the nearest surface of the lamps, as shown in Fig. 2.

6.4.1 The test specimens shall be exposed within an area 210 mm in height by 900 mm wide on each side of the apparatus located as shown in Fig. 4.

NOTE 2—It is possible to mount specimens above, below, and beside the 210 by 900 mm area, but specimens so mounted will be exposed to lower UV intensities.

6.5 *Condensation Mechanism*—Water vapor shall be generated by heating a water pan extending under the entire sample area and containing a minimum water depth of 25 mm. Specimen racks and the test specimens themselves shall constitute the side walls of the chamber. The back side of the

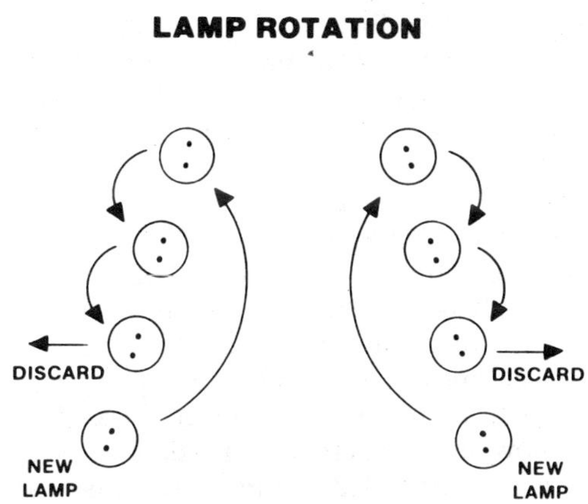

FIG. 3 Lamp Rotation

[5] Apparatus and lamps from Q-Panel Co., 26200 First St., Cleveland, OH 44145, and from Atlas Electric Devices Co., 4114 N. Ravenswood Ave., Chicago, IL 60613, have been found satisfactory.

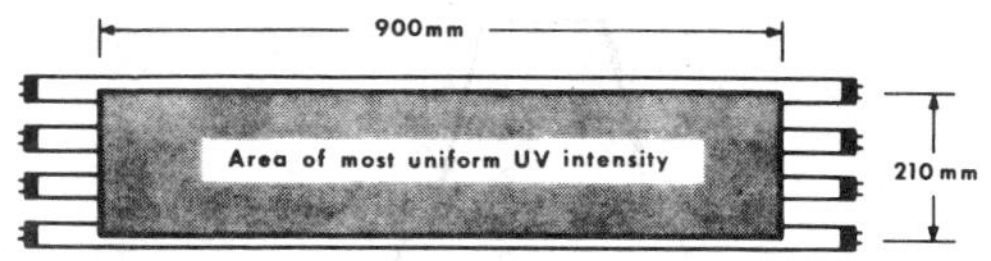

FIG. 4 Limits of Area of Uniform Intensity

specimens shall be exposed to cooling effects of ambient room air. The resulting heat transfer causes water to condense on the test surface.

6.5.1 The specimens shall be arranged so that condensate runs off the test surface by gravity and is replaced by fresh condensate in a continuous process. Vents along the bottom of the test chamber shall be provided to permit an exchange of ambient air and water vapor to prevent oxygen depletion of the condensate.

6.6 *Water Supply,* with an automatic control to regulate the level in the water pan shall be provided. Distilled, deionized, or potable tap water are equally acceptable for purposes of the test, since the condensation process itself distills water onto the test surface.

6.7 *Cycle Timer,* a continuously operating cycle time, for programming the selected cycle of UV periods and condensation periods.

6.7.1 Hour meters shall be provided to record total time of operation and total time of UV exposure.

6.8 *Specimen Temperature Measurement:*

6.8.1 Specimen temperature shall be measured by a thermometer with a remote sensor attached to a black aluminum panel 75 by 100 by 2.5 mm thick. The thermometer shall be precise to ±1°C through a range from 30° to 80°C. The indicator dial shall be located outside the test chamber.

6.8.2 The black aluminum panel with the thermometer sensor shall be positioned in the center of the exposure area so that the sensor is subject to the same conditions as the specimens.

6.9 *Specimen Temperature Control:*

6.9.1 During UV exposure, the selected equilibrium temperature shall be maintained within ±3°C by supplying heated air to the test chamber.

6.9.2 During condensation exposure, the selected equilibrium temperature shall be maintained within ±3°C by heating the water in the water pan.

6.9.3 The UV and condensation temperature controls shall be independent of each other.

6.9.4 Doors shall be located on the room air side of the specimen racks to act as insulation during the UV exposure and to minimize drafts. Such doors shall not interfere with the room air cooling of the specimen during the condensation exposure.

6.10 *Test Chamber Location:*

6.10.1 The apparatus shall be located in an area maintained at a temperature between 20°C and 30°C. The room temperature shall be measured by thermometers mounted on interior walls or column approximately 1500 mm above the floor level and at least 300 mm from any heated apparatus. Three or more thermometers located at various points will show any temperature variation in the area.

6.10.2 It is recommended that the apparatus be located at least 300 mm from walls or other apparatus. Nearby heat sources, such as ovens or heated test apparatus, should be avoided or shielded, because such heat sources can reduce the cooling required for condensation.

6.10.3 The room where the apparatus is located shall be ventilated to remove the heat and moisture produced and to maintain the temperatures specified in 6.10.1. Two to four air changes per hour will normally provide sufficient ventilation.

7. Test Specimens

7.1 Replicate specimens are desirable to provide a record of degradation at different time intervals. Retention of an unexposed specimen is recommended as it is difficult to mask a specimen to prevent exposure to condensation.

7.2 For specimens of insulating materials, such as wood, plastic, or porous laminates, maximum specimen thickness should be 20 mm to allow adequate heat transfer for condensation.

7.3 To provide rigidity, flexible specimens may be attached to a backing panel made of aluminum or other noncorrosive heat conductive material.

7.4 Cut edges of coated steel specimens shall be protected so that rust does not contaminate the test surface.

7.5 Holes in specimens and any openings larger than 1 mm around irregularly shaped specimens shall be sealed to prevent loss of water vapor. Porous specimens, such as textiles or wood shall be backed with a vapor barrier such as metal or plastic.

8. Calibration and Standardization

8.1 The thermometer or thermocouple which indicates test temperature shall be calibrated by immersing the sensing element and a liquid-in-glass thermometer in water heated to approximately 70°C and comparing the two temperatures as in Method E 220.

8.2 The formation of condensation may be observed by using clear glass or plastic blanks in the specimen holders or rack. One large sheet of plastic may also be used for this purpose. The condensate that forms on a given area during a period of condensation may be collected and then measured.

8.3 The fading and other changes caused by the fluorescent UV lamps may be observed by exposing lightfastness standards in the apparatus using UV alone without condensation.

8.3.1 The AATCC Blue Wool Lightfastness reference materials[6] may be used to measure the changes caused by UV. The L2 Blue Wool has been found satisfactory for evaluating fluorescent UV lamps.

8.3.2 Other lightfastness reference materials may be used by agreement.

8.4 Reference standards for calibrating the operation of the apparatus in alternate exposure to UV light and condensation may be prepared from painted metal, plastics, or other materials. Painted metal specimens produced by the coil-coating process have been found suitable for such reference standards.

9. Procedure

9.1 Mount the test specimens in the specimen racks with

[6] Available from American Association of Textile Chemists and Colorists, P.O. Box 12215, Research Triangle Park, NC 27709.

the test surfaces facing the lamp. When the test specimens do not completely fill the racks, fill the empty spaces with blank panels to maintain the test conditions within the chamber.

9.2 Program the selected test conditions. Operate continuously, repeating the cycle, except for servicing the instrument and inspection of specimens.

9.3 Various test conditions may be used. If no conditions are specified, the following cycle and temperatures are suggested: 4 h UV at 60°C, 4 h Condensation at 50°C.

NOTE 3—Prior versions of this practice recommended a condensation temperature of 40°C in 9.3. When operating in room conditions that do not comply with those set forth in 6.10, a 40°C condensation temperature can result in inadequate condensation. Therefore, a 50°C condensation temperature is now suggested.

9.3.1 Any test temperature that can be maintained within the limits specified in 6.9.1 and 6.9.2 may be used. UV test temperature of 50°C, 60°C, and 70°C are widely used. A condensation test temperature of 50°C is commonly used.

9.3.2 The following time cycles are widely used: 4 h UV/4 h CON, and 8 h UV/4 hr CON. Use UV and condensation periods of at least 2-h duration to allow sufficient time to reach equilibrium.

9.3.3 The severity of the UV exposure is influenced by test temperatures and time cycles. Photochemical reaction begins as soon as the UV lamps are turned on. The rate of UV degradation is proportional to the time of UV exposure or the temperature of UV exposure, or both. UV exposures at temperatures higher than those expected in the service environment can cause abnormal thermal degradation.

9.3.4 Water reactions during condensation exposure are affected by the permeability of the specimen and require time to initiate.The rate of water degradation is increased by increased temperature. However, long hot condensation exposures can cause abnormal degradation. Four-hour condensation exposures are often used for paints on metals, while condensation exposures of 20 h duration may be used on wood.

9.4 *Maintenance*—Periodic maintenance is required to maintain uniform UV and condensation exposure conditions.

9.4.1 After 400 to 450 h of lamp operating time, replace one lamp in each bank of lamps, and rotate the other lamps as shown in Fig. 3. This procedure provides a useful lamp life of 1600 to 1800 h.

9.4.2 Drain water and clean water pan when conducting lamp replacement and rotation procedure. Scum on the top of the water can inhibit water vaporization.

9.5 *Sample Rotation*—The samples' positions within the test area should be rotated on a regular basis in order to minimize any effects from horizontal or vertical variations in UV or temperature. The sample rotation sequence should provide that all samples spend approximately the same amount of exposure time near the center of the exposure area and near the edges of the exposure area. The following rotation sequence is recommended:

9.5.1 *Horizontal Rotation*—Once a day rotate the extreme left hand and extreme right hand samples into the center as shown in Fig. 6.

9.5.2 *Vertical Rotation*—Rotate the samples vertically so that each sample spends the same amount of exposure time in each vertical position in the sample holder. For instance, if two samples are stacked vertically in each sample holder, then the top and bottom sample should switch places halfway through the test method. If four samples are stacked vertically, then the samples should be repositioned vertically three times during the test method.

9.5.3 Other sample rotation sequences may be used by agreement.

9.6 Inspect specimens daily in tests with a length of 1 week or less. Tests for longer periods should be inspected weekly.

9.6.1 A permanent record of degradation at various times may be obtained by using three replicate specimens and

REPORT FORM ASTM PRACTICE G- 53 FLUORESCENT–UV / CONDENSATION TEST

TEST No.:________ Date: ________ By: ________

APPARATUS Type and Model				TEST TIME ______ hrs.
LAMP	MANUFACTURER	TYPE	PEAK EMISSION ______ nm	LOW CUT-OFF (1% of peak) ______ nm
CYCLE	______ hrs UV at ______ °C. / ______ hrs COND at ______ °C.			
SPECIAL TEST CONDITIONS				
START DATE:	END DATE:	TIMER HRS. START	TIMER HRS. STOP	LAMP SERVICE at ______ hrs.
SPECIMEN SIZE______				Rotated ☐ Yes ☐ No
MATERIAL				
PROPERTIES TO BE MEASURED			MEASURING PROCEDURE (ASTM Spec., etc.)	
1.				
2.				
3.				

FIG. 5 Sample Report Form

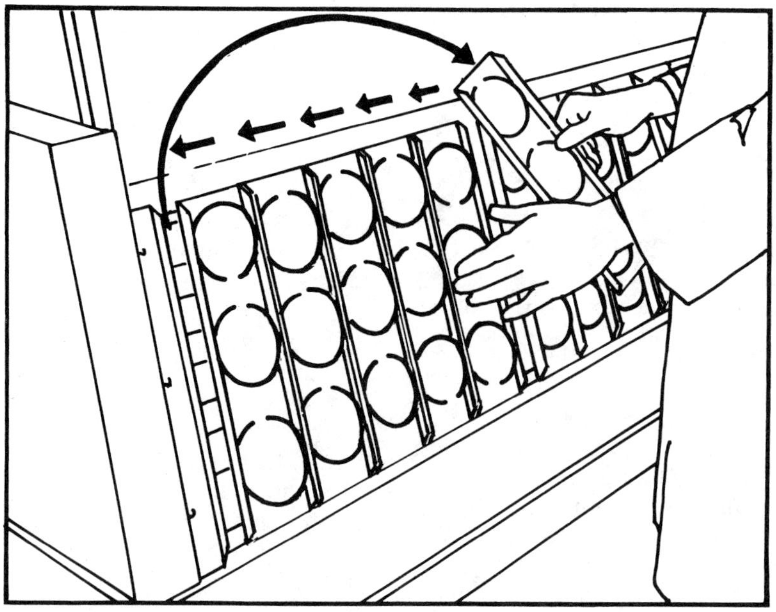

FIG. 6 Daily Horizontal Sample Rotation

removing the replicates sequentially at intervals equal to one third of the test length.

9.7 Conclude the test when either a mutually agreed-upon number of total test hours or a mutually agreed-upon change has occurred in the test specimen or a standard test specimen. For convenience in inspecting and concluding tests during normal working hours, the following test durations are recommended:

96 h = 4 days	672 h = 4 weeks
168 h = 1 week	1008 h = 6 weeks
336 h = 2 weeks	1512 h = 9 weeks
504 h = 3 weeks	2016 h = 12 weeks

10. Interpretation of Results

10.1 This practice is intended to simulate the deterioration caused by natural weathering. However, the rate of degradation in natural weathering varies from year to year. Laboratory accelerated weathering is more consistent in rate. Therefore, in comparing natural and laboratory weathering one should not attempt to predict the number of hours of laboratory exposure that might equal a year of natural weathering. Even if the details of the natural exposure are fully reported as required by 5.3, any time to time comparison would apply only to a given year at a particular site.

10.2 It is possible to correlate the deterioration resulting from a period of natural weathering, such as 1 year, to the deterioration resulting from a period of laboratory accelerated weathering. Spearman rank correlation, which is discussed in most basic statistical texts, has proved to be a useful method for this. A minimum of seven variables must be ranked by both test methods to obtain statistical validity from Spearman rank correlation. Ranking 10 to 15 variables by both methods will improve the reliability of such correlations.

10.3 It is frequently not possible to predict how many hours or weeks of accelerated testing will yield the best correlation to some period of outdoor testing. Therefore, laboratory tests should be measured periodically, such as weekly or biweekly, so that the correlation coefficient may be calculated at several points during the accelerated test.

11. Report

11.1 The report shall include the following:

11.1.1 Manufacturer and model of fluorescent UV/condensation apparatus,

11.1.2 Manufacturer's designation for the fluorescent UV lamp, the wavelength (nm) at which peak emission occurs, and the short wavelength at which 1 % of peak emission occurs (for example, FS-40 313/280 nm),

11.1.3 Cycle of UV exposure time and temperature, condensation time and temperature (for example 4 h UV/60°C, 4 h CON/50°C),

11.1.4 Total exposure time,

11.1.5 Special conditions of test, such as rotation of test specimens, and

11.2 A sample report form is shown in Fig. 5.

TAPPI TEST METHODS

CAUTION:
This method may require the use of some chemicals which may present serious health hazards to humans. Procedures for the handling of such substances are set forth on Material Safety Data Sheets which must be developed by all manufacturers and importers of potentially hazardous chemicals and maintained by all distributors of potentially hazardous chemicals. Prior to the use of this test method, the user should determine whether any of the chemicals to be used are potentially hazardous and, if so, should follow strictly the procedures specified by both the manufacturer, as well as state and federal authorities, for safe use of these chemicals.

T 208 om-89

TENTATIVE STANDARD – 1932
OFFICIAL STANDARD – 1934
REVISED – 1978
OFFICIAL TEST METHOD – 1984
REVISED – 1989

Moisture in wood, pulp, paper, and paperboard by toluene distillation

1. Scope and significance

1.1 This method can be applied to wood chips, sawdust, groundwood, pulp, paper, paperboard, and non-wood fibers. It is especially useful for determination of moisture content in materials containing volatile substances such as turpentine and resins in wood and groundwood or easily melting waxes and varnishes in paper and paperboard. These substances, by volatilization or melting, interfere with moisture determination by oven-drying methods.

1.2 In addition, water cannot be expelled completely on oven drying of cellulosic materials. Therefore, the moisture content, as determined by this method of distillation, is a better measure of the true water content than by oven-drying procedures for some materials.

2. Summary

The material to be tested for moisture content is boiled in toluene. The resultant water vapor is condensed, collected in a distilling receiver, and measured.

3. Apparatus

3.1 *Distillation apparatus* (Fig. 1), consisting of:

3.1.1 *Boiling flask*, 500 to 2000 mL.

3.1.2 *Distilling receiver*, 5 to 25 mL; Stark and Dean or other type.

3.1.3 *Condenser*, Liebig type; jacket length 500 to 600 mm.

3.1.4 *Drying tube* with a desiccant such as alumina, silica gel, or anhydrous calcium chloride or sulfate.

WARNING: Do not use magnesium perchlorate ("Anhydrone") as a desiccant because it may form an explosive mixture in contact with toluene vapors.

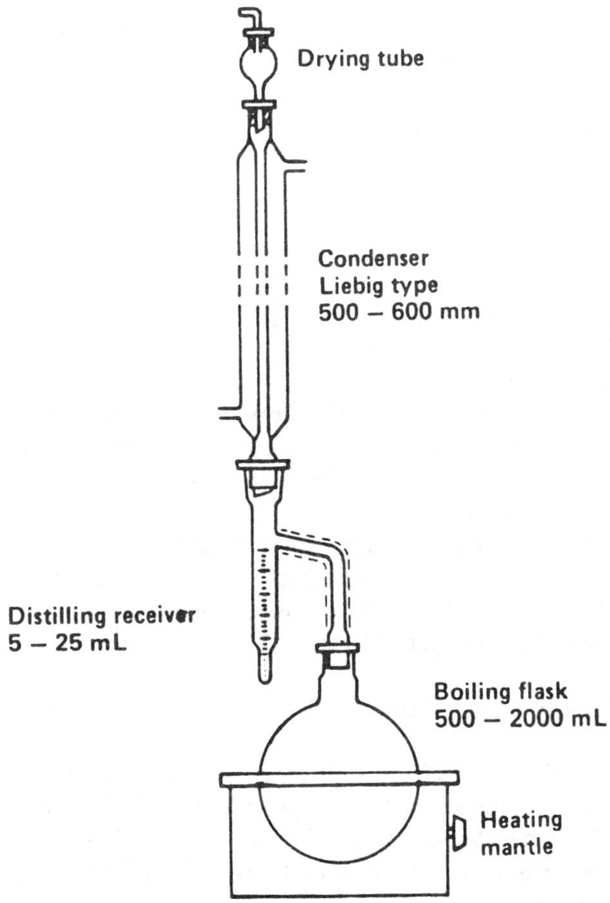

Fig. 1. Distillation apparatus.

Approved by the Chemical Properties Committee of the Process and Product Quality Division
TAPPI

3.1.5 *Electric heating mantle*, adjustable heat, designed for distillation of flammable solvents.

3.1.5.1 Size of the boiling flask and the distilling receiver used depends on the nature, size, and moisture content of a test specimen. In general, large boiling flasks should be used on bulky materials or materials of higher moisture content, such as wood chips and paper. Small distilling receivers with fine graduations are used on materials low in moisture content.

3.1.5.2 The flask, receiver, and inner tube of the condenser must be cleaned thoroughly with a dichromate-sulfuric acid cleaning solution, rinsed with water, and dried in an oven before use.

3.1.5.3 The distilling receiver shall be calibrated by adding first a few milliliters of toluene and then certain volumes of water from a calibrated pipet or buret.

NOTE 1: This calibration is necessary because of a different shape of water meniscus under toluene.

3.1.5.4 The arm of the distilling receiver shall be covered with an insulating material such as glass fiber tape in order to facilitate distillation.

3.1.5.5 Set up the distillation apparatus in a well-ventilated place, preferably in a fume cupboard and behind a safety shield.

3.2 *Copper or brass wire*, about 700 mm long, with a loop formed at one end to fit the inner tube of the condenser.

3.3 *Calibrated buret or pipets.*

4. Reagent

4.1 *Toluene (toluol)*, $C_6H_5CH_3$, reagent grade or purified, meeting ACS specification. Before using, prepare extracted toluene as follows:

4.1.1 Add about 50 mL of water to 1 L of toluene, shake for 5 min and allow to settle for about 15 min. Decant the toluene to a distillation flask and reflux in the distillation apparatus for about 60 min.

4.1.2 The toluene may be used repeatedly until it contains an excess of soluble substances extracted from the samples. Then redistill the toluene and retain the portion boiling within a range of 110° to 112°C.

5. Safety precaution

5.1 Toluene is a flammable organic solvent and must be handled with proper care. The procedures of boiling and distillation must be carried out behind a safety shield in a well-ventilated place, preferably in a fume hood, and away from an open flame.

5.2 Toluene is a toxic chemical substance, hazardous to health. Contact of toluene with skin and inhalation of toluene vapors must be avoided.

6. Sample

Obtain a representative sample of about 1000 g of wood or pulp, or about 500 of paper or paperboard, in accordance with TAPPI T 257 "Sampling and Preparing Wood for Analysis," TAPPI T 210 "Sampling and Testing Wood Pulp Shipments for Moisture," or TAPPI T 400 "Sampling and Accepting a Single Lot of Paper, Paperboard, Fiberboard, or Related Product." Place the sample immediately in an air-tight container to avoid changes in moisture content.

7. Test specimen

Weigh out rapidly, to the nearest 0.1 g, a test specimen according to the moisture content and nature of the test material. Record the weight of the test specimen as W. The amount of water in the test specimen shall not exceed the capacity of the distilling receiver. A suitable test specimen for wood and pulp generally ranges from 100 to 300 g; a suitable test specimen for paper and paperboard generally ranges from 50 to 100 g.

8. Procedure

8.1 Place the test specimen immediately after weighing in a boiling flask and add about 500 mL of toluene. (It is not necessary to cover the test specimen fully with toluene.) The last 20 to 25 mL of toluene shall be added through the condenser in order to wet the surface of the inner tube. Add a few glass beads to the flask.

8.2 Begin distillation and adjust the heat so that the water vapor condenses in the condenser tube and the rate of reflux is 3 to 4 drops of water per second.

8.3 Continue distillation until the condensation of water ceases and the level of water remains constant in the distilling receiver for a period of 15 min.

NOTE 2: Water drops adhering to the walls of condenser tube indicate insufficient cleaning of glassware. The drops could be dislodged by means of a looped wire, passed down from the top of condenser during the course of distillation.

8.4 Turn off the heat and allow the receiver to cool to room temperature. Read the volume of water collected.

9. Calculation and report

Calculate moisture content (B) in the test specimen as follows:

$$B, \% = \frac{A \times 100}{W}$$

where

A = volume of water collected in the receiver, mL

W = weight of the test specimen, g

Report the moisture content in the sample to the nearest 0.1%.

10. Precision

10.1 Repeatability: 0.2 mL as volume of water collected, as found in several laboratories and stated in CPPA G.1. On 100-g test specimen, for example, repeatability is thus 0.2% in terms of moisture content.

10.2 Reproducibility and comparability: not known.

11. Additional information

11.1 Effective date of issue: February 1, 1989.

11.2 Related methods: Canadian G.1 (1969); British B.S. 3433:1961.

Your comments and suggestions on this procedure are earnestly requested and should be sent to the TAPPI Test Methods Administrator. ■

OFFICIAL STANDARD - 1926
REVISED - 1968
REVISED - 1979
OFFICIAL TEST METHOD - 1983
REVISED - 1988
CORRECTION - 1991

Grammage of paper and paperboard (weight per unit area)

1. Scope

1.1 In the United States the customary or commercial term for expressing the "weight" per unit area (more properly "mass per unit area") of paper has been "basis weight," "ream weight," or "substance". These are defined as the mass in pounds of a ream of a given sheet size and number of sheets (usually 500 sheets, occasionally 480 sheets). In most other countries the mass per unit area is expressed in grams per square meter, g/m². The French term for mass per unit area, "grammage," is recommended by ISO Committee TC 6 on Paper for use in English as well as in French because of its convenience and clear relationship to g/m².

1.2 The mass per unit area of paperboard has been expressed in the customary system as the mass in pounds per thousand square feet, and in the metric system as grams per square meter (g/m²).

1.3 The SI metric units, in which grammage (mass per unit area) is expressed in g/m², are the preferred units for TAPPI Test Methods.

2. Summary

The area of several sheets of the paper or paperboard is determined from linear measurements and the mass (commonly called "weight") is determined by weighing. The grammage is calculated from the ratio of the mass to the area after conversion to metric units when necessary.

3. Significance

Most paper is bought and sold in accordance with its mass per unit area, and therefore the grammage has great significance both to the consumer and the producer in defining price. The values of many physical properties such as bursting strength, thickness, and bulk are interpreted and specified with regard to grammage.

4. Apparatus[1]

4.1 *Weighing device*, readable and accurate to within 0.25% of the applied load. When in use, the weighing device shall be shielded from air currents.

4.1.1 Special sheet-weighing devices designed to weigh test specimens of a given size may be used, provided that the above conditions are fulfilled and that the total area of each test specimen (see 6.3) in a single weighing is not less than 500 cm² (80 in²).

4.1.2 For a test specimen of smaller total area, an analytical balance is essential to obtain the required accuracy.

4.2 *Cutting device*, such as a "four square" cutter, or other device for ensuring parallelism of the opposite edges, normally capable of repeatedly cutting out test specimens whose area, in at least 95 instances out of 100, falls within ±0.5% of a known area, as determined by the method specified in 5.2.1.

4.2.1 When a template is used for preparing test specimens, the specimen shall be cut accurately as described in 4.2 with a sharp knife on a hard surface, such as a hardwood board.

4.3 *Scale*, a steel rule graduated in 0.5 mm ($^1/_{50}$ or $^1/_{64}$ in) increments, capable of measuring the dimensions of the test specimen to an accuracy of 0.2%.

5. Calibration

5.1 *Checking of weighing device*

5.1.1 Check the accuracy of the weighing device frequently (5.3) by applying accurately measured masses with both increasing and decreasing loads. If a sheet-weighing device is used, it must be properly calibrated to the required accuracy (4.1).

[1]Names of suppliers of equipment and materials may be available from the TAPPI Information Resources Administrator.

Approved by the Physical Properties Committee of the Process and Product Quality Division
TAPPI

5.1.2 Frictional effects shall be sufficiently minimal and the zero reading shall be sufficiently correct so as to attain the required weighing accuracy.

5.2 *Checking of cutting device*

5.2.1 Frequently (5.3) check the specimen area cut by using the scale (4.3) to measure 20 test specimens. The cutting accuracy specified in 4.2 is attained when the standard deviation of the individual areas is less than 0.25% of the mean area, in which case this mean area may be used for calculating grammage in subsequent tests.

5.3 *Frequency of checking*

5.3.1 The frequency of checking in 5.1 and 5.2.1 should be based on experience. A new device should be checked several times before being put into use. Then, if in continuous use, it should be checked twice daily until stability is established, then weekly, monthly, or less frequently as indicated by its stability, unless moved or unduly disturbed. Because of wear, the cutting device may require more frequent checking than the weighing device.

6. Sampling

6.1 Obtain the sample of the paper or paperboard to be tested in accordance with T 400 "Sampling and Accepting a Single Lot of Paper, Paperboard, Fiberboard, or Related Product."

6.2 From each test unit of the sample, cut a sufficient number of representative sheets for a total area per test unit of at least 5000 cm² (800 in²). The dimensions and hence the area of each sheet will depend on the sheet cutting device available. If a template is used, a minimum dimension of 20.0 x 25.0 cm (8 x 10 in) is recommended.

6.3 Select the number of sheets composing a test specimen (total area of each test specimen at least 500 cm²) to give the required weighing accuracy.

NOTE 1: As a consequence of the above rules which allow for flexibility in the selection of sheet weighing and cutting devices, the number of test specimens representing a test unit and the number of sheets per test specimen will depend on the choice of weighing and cutting devices, as illustrated by the following examples:

(a) If for a particular sample of paper, the sheet weighing device requires the weighing together of two sheets at least 20.0 x 25.0 cm to obtain the requisite weighing accuracy, then each test specimen will consist of two sheets at least 20.0 x 25.0 cm, and five such test specimens (each a pair of sheets) will be needed to obtain the required total area of 5000 cm².

(b) If the available precise cutting device gives a sheet size of 4.00 x 4.00 in (approximately 10 x 10 cm), then for the same paper and sheet weighing device used above, each of the needed five test specimens would require 10 sheets.

NOTE 2: For some uses, grammage variation across the web or roll is of interest; e.g., the original wide roll (or web) may be slit into narrow rolls, each 7.5 to 10 cm (3 to 4 in) wide for medical packaging. A "tailing-off" of the grammage near the edge of the web could be unacceptable. In such cases, test specimens should be taken from selected positions across the web and the grammage reported separately for each position.

7. Conditioning

7.1 Condition and test the test specimens in an atmosphere in accordance with T 402 "Standard Conditioning and Testing Atmospheres for Paper, Board, Pulp Handsheets, and Related Products," unless grammage "as received" is required.

7.2 The precision of this work is such that hysteresis has important effects; therefore, precondition the test specimens so that the equilibrium moisture content is approached from the drier state.

8. Procedure

8.1 If the cutting accuracy specified in 4.2 is attained (see 5.2.1), use the mean area obtained in 5.2.1; otherwise, determine the area of each test specimen to within 0.3%, i.e., measure the dimensions of each sheet to within 0.2%.

8.2 If the special sheet-weighing device is used, be sure that the dimensions of the sheets are within 0.2% of the required size; otherwise, make a correction in accordance with the following formula:

$$G = \frac{A'}{A \times G'}$$

where

G = corrected indicated mass per unit area

G' = indicated mass per unit area of the test specimen

A' = area of the test specimen for which the device is calibrated

A = area of the weighed test specimen

8.3 Weigh each test specimen to within 0.25%.

9. Calculations

9.1 From the measurements, calculate the mass per unit area of each test specimen. If M is the mass and A is the area of the test specimen in the units of measurements, then the grammage in grams per square meter (g/m²) may be calculated by the formula:

$$G = \frac{K \times M}{A}$$

where K is the conversion factor given in Table 1.

9.1.1 If the ream weight in pounds (BW) has been determined for a customary trade size of 500 sheets, the grammage in g/m² may be calculated as follows:

$$G = P \times \text{BW}$$

where P is the conversion factor given in Table 2.

9.2 Calculate the mean of the results for each test unit, and express in g/m² to three significant figures.

9.3 To convert grammage (G) to mass in pounds (BW) of custom trade sizes for 500-sheet reams or, for paperboard, 1000 ft², the following formula may be used:

$$BW = Q \times G$$

where Q is the conversion factor given in Table 2. Express the results in pounds to three significant figures.

Table 1. Conversion factors

Units of measurement: Mass (M)	Area (A)	Conversion factors K
Gram	cm²	10,000
Gram	in²	1,550
Indicated weight (lb) for 500-sheet ream	cm²	9,070
Indicated weight (lb) for 500-sheet ream	in²	1,406

Table 2. Ream weight conversion factors

Kind of paper	Trade or basic size	Trade size area, ft²	P (BW to G)	Q (G to BW)
Paperboard	1000 ft²	1000.0	4.882	0.205
Writing and printing	17 x 22 - 500	1298.6	3.760	0.266
Blotting	19 x 24 - 500	1583.3	3.084	0.324
Cover	20 x 26 - 500	1805.6	2.704	0.370
Carbon paper	20 x 30 - 500	2083.3	2.344	0.427
Cardboard	22 x 28 - 500	2138.9	2.283	0.438
Bristol and tag	22½ x 28½ - 500	2226.6	2.193	0.456
Binder's board	25¼ x 30¼ - 500	2652.1	1.841	0.543
Index	25½ x 30½ - 500	2700.5	1.808	0.553
News, wrapping, tissue, paperboard, bag paper	24 x 36 - 500 (3000 ft²)	3000.0	1.627	0.614
Tissue	24 x 36 - 480	2880.0	1.695	0.590
Book	25 x 38 - 500	3298.6	1.480	0.676
Newsboard	26 x 38 - 500	3430.6	1.423	0.703
Former TAPPI standard size	25 x 40 - 500	3472.2	1.406	0.711

10. Report

10.1 Report the grammage in g/m² to three significant figures for each test unit.

10.2 If the test units have been taken from more than one position across a roll or sheet and information on grammage variation is required (e.g., Note 2), report the average for each position separately.

10.3 If desired, the results may also be reported in pounds for customary trade sizes.

10.4 If the total area of a test unit is less than that specified, state the actual total area tested.

11. Precision

11.1 Repeatability of test results within a laboratory = 0.94%

11.2 Reproducibility of test results between laboratories = 2.84%

11.3 The above values of precision are averages calculated from 18 reports of the TAPPI/NBS Collaborative Reference Program for Paper. Each test result is an average for approximately 5000 cm² of paper or paperboard.

12. Additional information

12.1 Effective date of issue: April 19, 1988.

12.2 Table 2 is for 500-sheet reams except as noted. If the trade size uses 480-sheet reams, then the factor P is multiplied by 50/48 or 1.042 for conversion and factor Q is multiplied by 48/50 or 0.960.

12.3 The U.S. Government Printing Office has discontinued using twice the trade size (i.e., 1000 sheets instead of 500) and is now following trade practice.

12.4 To within less than 0.5%, mass per unit area in g/m² is numerically equivalent to basis weight in pounds (35 x 40-500), so that by adjusting the size of the template most ream weight scales may be used to read directly in g/m².

12.5 Related method: ISO 536.

12.6 This method was revised in 1968 and 1979. The latter revision adopted the ISO terminology "grammage." In 1982, Table 2 was corrected and expanded, Note 2 was added. The 1991 correction deleted a reference to a withdrawn TI Sheet.

Your comments and suggestions on this procedure are earnestly requested and should be sent to the TAPPI Technical Divisions Administrator. ■

OFFICIAL STANDARD – 1926
REVISED – 1968
REVISED – 1976
OFFICIAL TEST METHOD – 1983
REVISED – 1984
REVISED – 1989

Thickness (caliper) of paper, paperboard, and combined board

1. Scope

1.1 This method describes the procedure for measuring single-sheet thickness and variations in single sheet thickness of paper, paperboard, and combined board except for creped papers. The term "combined board" encompasses both corrugated and solid fiberboard.

1.2 A modification of this method suitable for thin sheets under about 0.04 to 0.05 mm is described in Note 3.

1.3 Another method for measuring the thickness of paper is TAPPI T 500 "Book Bulk and Bulking Number of Paper," which describes a procedure for measuring the overall thickness of a stack of book paper under pressure of 250 kPa (35 psi).

2. Summary

The method involves measuring the thickness of a single sheet of paper, paperboard, or combined board by the use of an automatically operated micrometer when a specified static load is applied for a minimum specified time.[1]

3. Significance

Thickness is an important property of paper, paperboard, or combined board, and variations in thickness are also important especially for papers and paperboards used for mechanical purposes. The test is useful for research work, routine control, design of end-use products, and for acceptance testing for conformance to specifications. Thickness of combined board is one of the characteristics that importantly affects flexural stiffness.

4. Definition

Thickness or caliper of paper, paperboard, or combined board as measured by this method is defined as the perpendicular distance between the two principal surfaces of the paper or paperboard under prescribed conditions.

5. Apparatus[2]

5.1 *Automatically-operated micrometer*[1], a dead-weight type instrument, provided with:

5.1.1 A flat ground circular movable face (hereafter called the pressure foot), having an area of 200 ± 5 mm^2 (equivalent to about 0.31 ± 0.01 in.^{2}) and corresponding to a diameter of 16 ± 0.2 mm (0.63 ± 0.01 in.).

5.1.2 A flat ground circular fixed face (hereafter called the anvil) of such size that it is in contact with the whole area of the pressure foot in the zero position.

5.1.3 Surfaces of the pressure foot and anvil parallel to within 0.001 mm (0.00005 in.) and pressure foot movement on an axis perpendicular to the anvil.

5.1.4 Pressure foot lowering speed of 0.8 ± 0.1 mm (0.03 ± 0.004 in.)/s (see also Additional Information).

NOTE 1: This may be checked by measuring the eccentricity and speed of rotation of the cam operating the pressure foot.

5.1.5 Pressure foot, when lowered, exerting steady pressure on the specimen of 50 ± 2 kPa (approximately 7.3 ± 0.3 psi) for 3 ± 1 s.

5.1.6 Readings (5.1.6.1 or 5.1.6.2 as applicable)

5.1.6.1 Dial, graduated to 0.002 mm (0.0001 in.) or less, readable to at least 0.001 mm (0.00005 in.), or 1% of paper thickness, whichever is larger. If the large indicating hand of the dial is required to revolve more than once to cover the capacity of the gage, equip the dial with a revolution counter to indicate the number of complete revolutions of the indicating hand.

5.1.6.2 Digital display with indication to at least 0.001 mm (0.00005 in.), or 1% of paper thickness, whichever is larger.

5.1.7 Capability of repeating readings to within 0.001 mm (0.00005 in.) at zero settings or on a steel gage block.

[1]See Appendix A.1 for manually operated micrometer
[2]Names of suppliers of equipment and materials may be available from the TAPPI Information Resources Administrator.

Approved by the Physical Properties Committee of the Process and Product Quality Division
TAPPI

5.1.8 Accuracy within 0.001 mm (0.00005 in.) or 1% of paper thickness, whichever is larger.

6. Calibration

6.1 *Accuracy of dial indications.* Using standard steel gage whose thickness is known to within 0.0005 mm (0.00002 in.), check the accuracy of the dial reading over the required range. Prepare a calibration curve or table if necessary.

NOTE 2: In the use of steel gage, a nondeformable material, it must be remembered that the value corresponding to the thickest portion of the gage, rather than the average thickness, should be used in calibration of the micrometer.

6.2 *Parallelism of the faces.* Use a uniform diameter wire of any size up to half the operational opening of the faces, e.g., .020 inch diameter. Place the wire alternately on the left side, right side, front side, and back side approximately one-eighth inch from each respective edge of the foot and note the readings. Adjust anvil so that all readings are within .00005 inch of one another.

6.3 *Pressure between the faces.* Measure the force required to just prevent the movement of the pressure foot from a reading corresponding to about the average paper thickness tested, to a lower reading. For example, where the pressure foot projects through the top of the apparatus, attach a fine copper wire to it and, by means of a balance or a calibrated spring, measure the force necessary to prevent the closing of the foot. Alternatively, suspend from the balance a stirrup made of a flat metal plate having a hole larger than the diameter of the micrometer foot, covered at the bottom of the plate by a thin metal disk of about the average thickness of the paper to be measured.

7. Sampling

Obtain a sample in accordance with TAPPI T 400 "Sampling and Accepting a Single Lot of Paper, Paperboard, Fiberboard, or Related Product."

8. Test specimens

From each test unit of the sample, cut ten specimens, each specimen consisting of one thickness of paper (i.e., a single sheet) at least 40 mm (1.57 in.) (preferably 80 mm for combined board) in its shortest dimension and at least 120 mm (approximately 4.75 in.) in its longest dimension. Cut the specimen strips so that the longer dimension is across the machine direction. (Determine the machine direction in accordance with TAPPI T 409 "Machine Direction of Paper and Paperboard.")

NOTE 3: Some paper manufacturers and users measure the thickness of thin papers using a specimen that contains 4, 8, or 10 sheets of paper. Usually, but not always, if a micrometer is used to measure the thickness of thin sheets under about 0.04 to 0.05 mm, the error in reading the instrument is likely to make a significant difference in the results. If a pile of 10 sheets is measured and the result divided by 10, there will also be an error due to the "piling effect," but the latter error will, in general, be smaller than the former. It is therefore customary to measure the thickness of untreated paper used for electrical insulation under 0.051 mm (0.0020 in.) in thickness using a pile of 10 sheets, and the thickness of fine tissue grades, such as carbonizing, foil mounting, waxing and printing tissues under 0.041 mm (0.0016 in.), using a pile of 4 or 8 sheets.

9. Conditioning

Precondition on the dry side, condition, and test all specimens in accordance with TAPPI T 402 "Standard Conditioning and Testing Atmospheres for Paper, Board, Pulp Handsheets, and Related Products."

10. Procedure

10.1 Before using the micrometer, make sure that the pressure foot and anvil surfaces are clean, that the calibration of the instrument has been verified and a calibration curve has been prepared, if necessary, and that the instrument is mounted on a solid level surface free from noticeable vibration. Place the specimen on the anvil in such a position that all points on the peripheries of the contact surfaces are at least 6 mm (approximately 0.25 in.), or 12 mm for combined board, from the edges of the specimen. Measure the thickness of each specimen at five nonoverlapping regular intervals in a line at right angles to the machine direction of the paper. If only an estimate of lot or test unit average is required, i.e., variations of thickness within a sheet are of no interest, then only two readings per specimen need be made.

10.2 Take each reading near the end of the dwell time. If necessary, apply the calibration corrections to the readings.

NOTE 4: If the thickness of noticeably compressible papers is being measured, it is particulary important that the rate of fall of the pressure foot and the dwell time be within the specified limits. (See 5.1.4)

11. Report

For each test unit of the sample report the overall average (50 or 20 readings) in millimeters (or inches) to the nearest 0.001 mm (0.00004 in.). Also, when specified, report the minimum and maximum thickness obtained for each specimen to the nearest 0.002 mm (0.00008 in.). For coarse paperboards and combined boards, report average to 0.01 mm (0.0004 in.). If the procedure described in Note 3 is used, it should be so indicated.

12. Precision

12.1 Repeatability (within a laboratory) = 1.25%.

12.2 Reproducibility (between laboratories) = 5.50% in accordance with the definitions of these terms in TAPPI T 1206 "Precision Statement for Test Methods."

12.3 The above precision was calculated from data taken from 24 reports of the Collaborative Reference Program for paper.

13. Additional Information

13.1 Effective date of issue: June 15, 1989.

13.2 For a method in which the pressure foot is lowered at a speed negligible in comparison with the speed in this method, test results have been reported to be as much as 3% higher, depending on the type of paper.

13.3 Related methods: APPITA P 426 "Thickness of Paper and Paperboard," Technical Association of the Australian and New Zealand Pulp and Paper Industry, Parkville, Australia; BS 3983 "Methods for the Determination of the Thickness and Bulk of Paper" British Standards Institution, London, England; CPPA D 4 "Thickness and Density of Paper and Paperboard," Canadian Pulp and Paper Association, Montreal, Canada; DIN-Norm 53105, Zellcheming Germany; ISO 534 "Paper and Board Determination of the Thickness of Single Sheets (and Method of Calculation of the Apparent Density of Board)," (see Appendix A.4) International Organization for Standardization, Geneva, Switzerland; SCAN P7 "Thickness, Density and Bulk of Paper and Paperboard," Scandinavian Pulp, Paper and Board Testing Committee, Stockholm, Sweden; ISO 3034 "Corrugated Fiberboard: Determination of Thickness."

Appendix

Instructions for using manually operated instrument to obtain approximately same results as with motor operated instrument

A.1 *Apparatus.* Manually operated, dead-weight actuated, dial-type micrometer conforming in general to the requirements for the automatically operated instrument.

A.2 *Procedure.* Raise the foot until the dial reads .030 inches, and then lower to Zero over a one second period. Practice until the lowering rate becomes relatively uniform. As an alternative raise the foot to .060 inch and practice lowering to Zero in a period of two seconds. When actually measuring a specimen be sure to take the reading within 3±1 second after the foot comes to rest.

A.3 *Precision and accuracy.* Because of the greater influence of the operator, the manually operated instrument has shown less agreement among laboratories than the automatic instrument. Properly and skillfully operated, the accuracy of the manually operated instrument should be as good as that of the automatic instrument except possibly on highly compressible papers for which differences of 3% have been reported.

A.4 *Difference from ISO procedure.* The most recent draft of the International Standard (ISO 534-1980) lists the preferred pressure exerted by the foot to be 100 kPa and, as an alternative, the 50 kPa employed in this TAPPI procedure.

Your comments and suggestions on this procedure are earnestly requested and should be sent to the TAPPI Test Methods Administrator. ■

T 414 om-88

OFFICIAL STANDARD – 1926
TENTATIVE STANDARD – 1964
OFFICIAL TEST METHOD – 1982
REVISED – 1988

Internal tearing resistance of paper (Elmendorf-type method)

1. Scope

1.1 This method measures the force perpendicular to the plane of the paper required to tear multiple sheets of paper through a specified distance after the tear has been started using an Elmendorf-type tearing tester. The measured results can be used to calculate the approximate tearing resistance of a single sheet. In the case of tearing a single sheet of paper, the tearing resistance is measured directly.

1.2 It is not suitable for determining the cross-directional tearing resistance of highly directional boards and papers. These materials are covered by TAPPI T 496 "Cross Directional Internal Tearing Resistance of Paperboard."

1.3 For edge-tear resistance see TAPPI T 470 "Edge Tearing Resistance of Paper (Finch Method)."

2. Summary

One or more sheets of the sample material are torn together through a fixed distance by means of the pendulum of an Elmendorf-type tearing tester. The work done in tearing is measured by the loss in potential energy of the pendulum. The instrument scale is calibrated to indicate the average force exerted when a certain number of plies are torn together (work done divided by the total distance torn).

3. Significant test variables

3.1 Several Elmendorf-type tearing testers are available and in use throughout the world, principally those of Australian, British, German, Swedish, and United States manufacture. In addition, testing practices also vary, as is reflected in the related methods for these countries or others listed in 10.3. Instruments and practices in use vary in at least three major respects:

3.1.1 One difference is in the design of the pendulum sector. The oldest model without deep cutout permitted the specimen to come in contact with the sector during the test and gave values significantly higher than those obtained using the newer models *with* a deep cutout (see Fig. 1) which eliminate this undesirable friction. The magnitude of the difference in value obtained using different styles of the instrument de-

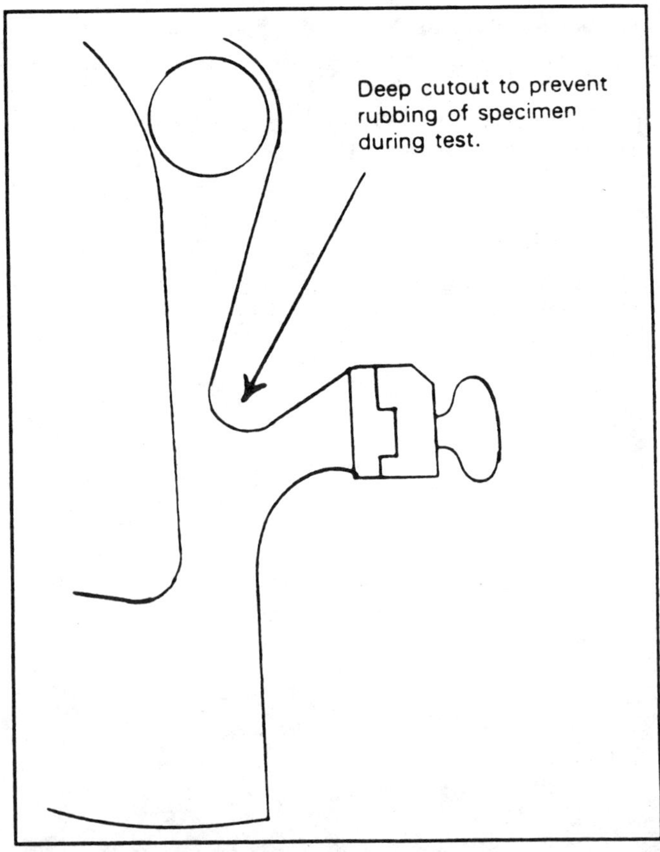

Fig. 1. Newer testing model with deep cutout.

Approved by the Physical Properties Committee of the Process and Product Quality Division
TAPPI

Reprinted by permission of the Technical Association of the Pulp and Paper Industry, P.O. Box 105113, Atlanta, GA 30348

scribed in this test method varies as a function of instrument and with different types and grammages of paper. The instrument having the oldest-style sector does not meet the requirements of this test method. With a few materials, test values have been observed (*1*) to be as much as 10% greater with the oldest-style sector.

3.1.2 The second difference is in the design of the specimen clamps which, together with the structural characteristics of the paper which govern the nature of the tear with respect to its splitting tendencies during the test, can have an appreciable influence on the mode of tearing and may result in significant differences (*2*). The procedure described in Section 7.3 reduces this effect. The clamp designs used by some manufacturers may vary even for their own models. Instruments are available with pneumatically activated grips as well. When available, their use minimizes variations due to differences in clamping pressures exerted by manually tightened grips.

3.1.3 The third difference results from a combined variation in testers and testing practices. As measured tearing resistance increases or decreases for different types of paper, it may become so large or so small as to be outside the practical range of the instrument. This problem may be overcome in one of two ways; the number of sample sheets tested at one time may be changed, or the mass of the instrument pendulum may be changed either by addition of augmenting weights or by replacement of the entire pendulum with one of a different known mass. The tearing length must never be varied in an effort to alter the pendulum capacity.

3.2 The foregoing, together with other lesser differences in design details between instruments or testing practices, preclude specifying a tearing instrument and method that would give essentially the same test results when using Elmendorf instruments of different design and manufacture. Even for one specific model, some procedural variables such as the number of plies torn may alter the test values calculated on a single sheet basis substantially. Hence, by necessity, this reference method must be arbitrary and is limited to the described procedure used with instruments conforming to all of the requirements specified under Section 4.

4. Apparatus

4.1 Elmendorf tearing tester (*3–5*), with a cut-out as shown in Fig. 1, which prevents the specimen from coming in contact with the pendulum sector during the test, and having the following elements:

4.1.1 A stationary clamp; a movable clamp carried on a pendulum formed by a sector of a circle free to swing on a ball bearing; a knife mounted on a stationary post for starting the tear; means for leveling the instrument; means for holding the pendulum in a raised position and for releasing it instantaneously; and means for registering the maximum arc through which the pendulum swings when released.

4.1.1.1 The registering means may consist of a graduated scale mounted on the pendulum, a pointer mounted on the same axis as the pendulum with constant friction just sufficient to stop the pointer at the highest point reached by the swing of the sector, and an adjustable pointer stop for setting the zero of the instrument.

4.1.1.2 The pointer and scale may be replaced by a digital readout unit which gives readings of equivalent accuracy and precision (*6*).

4.1.2 With the pendulum in its initial position ready for a test, the clamps are separated by an interval of 2.8 ± 0.3 mm and are so aligned that the specimen clamped in them lies in a plane parallel to the axis of the pendulum, the plane making an angle of 27.5 ± 0.5° with the perpendicular line joining the axis and the horizontal line formed by the top edges of the clamping jaws. The distance between the axis and the top edges of the clamping jaws is 103.0 ± 0.1 mm. The clamping surface in each jaw is at least 25 mm wide and 15.9 ± 0.1 mm deep.

NOTE 1: In the past, it has been the practice for instruments commonly available in the United States to be equipped with 36 ± 1 mm wide jaws. Instruments currently available, however, may be equipped with jaws as narrow as 25 mm. Testing has shown that the effect of jaw width on test results is statistically insignificant. It is recommended, however, that the test specimen length be adjusted to match jaw width. See Note 6.

4.1.3 The instrument measures the energy (work done) used by the pendulum in tearing the test specimen. In order to convert to average tearing force, the energy must be divided by the total distance through which the force is applied. This division may be accomplished by the electronics in digital read-out instruments so that the read-out is directly in grams-force or in millinewtons (SI unit of force). For pointer and scale instruments, the scale may be in millinewtons or in grams-force for a specified number of plies; i.e., when the specified number of plies are torn together, the scale reading gives the average tearing resistance (force) of a single ply.

4.1.4 Instruments of several capacities e.g., about 2000, 4000, 8000, 16,000, 32,000 mN (200, 400, 800, 1600, 3200 gf) and perhaps others are available, with the several capacities being achieved by individual instruments, interchangeable pendulum sectors, or augmenting weights. The instrument recognized as "standard" for this method has a capacity of 1600 gf (SI equivalent 15.7 N); i.e., it has a pendulum sector of such mass and mass distribution that its 0 to 100 scale is direct reading in grams-force per ply when 16 plies are torn together. For a 16-ply test specimen, the tearing distance $K = 16 \times 4.3$ cm (tearing distance per ply) x 2 = 137.6 cm, the factor 2 being included since in tearing a given length the force is applied twice the distance. Likewise, for a 16-ply test specimen, the tearing energy per ply for a scale reading of 100 would then be 100 gf x 137.6 cm or 13760 gf·cm (SI equivalent 1349.4 mJ). For some of the instruments of different capacities where different numbers of plies are required, or when the number of plies tested using the "standard" instrument differs from 16, different values of *K* and/or the tearing energy per ply may be calculated, using the above calculation as a model.

4.1.5 In the "standard" instrument, the zero reading on the scale is at about 70° from the center line (i.e., the vertical balance line when the pendulum hangs freely), the 100 reading is at about 21° from the center line, and a vertical force of 1057.3 ± 2.0 gf (SI equivalent 10.369 ± 0.020 N) applied at 22.000 ± 0.005 cm from the pendulum axis is required to hold the pendulum sector at 90° from its freely hanging position. Other tearing instruments will require vertical forces that are factors of 2 greater or smaller than 1057.3 gf and, if calibrated in millinewtons, the zero reading would remain at 70° and the 1000 reading would be at about 19° (or the 981 reading at about 21°).

4.1.6 The cutting knife for the test specimen is centered between the clamps and adjusted in height so that the tearing distance is 43.0 ± 0.2 mm; i.e., the distance between the end of the slit made by the knife and the upper edge of the specimen is 43.0 ± 0.2 mm when the lower edge of the 63.0-mm wide specimen rests against the bottom of the clamp.

4.2 *Specimen cutter*, to insure parallel specimens 63.0 ± 0.15-mm wide with sharp and clean edges. For this purpose it is desirable to use the type having two hardened and ground base shears, twin knives tensioned against the base shears, and a hold-down mechanism.

5. Calibration and adjustment

5.1 *Verification of scale*

5.1.1 Once the scale has been verified, it is unnecessary to repeat this step, provided the tester is kept in adjustment and no parts become changed or perceptibly worn. The scale may be verified either by the potential energy method or by the method which uses the check weights obtainable from the manufacturer. The potential energy method is relatively time-consuming and complicated. The check weight method is relatively simple.

5.1.2 *Potential energy method.* The procedure (7) for verification is as follows: Anchor and level the tester as later described. Clamp a known weight (in grams) W to the radial edge of the sector beneath the jaws, the center of gravity of the weight (including means of attaching) having been previously marked by a punched dot on the face of the weight that is to be toward the front of the instrument. Close the jaw of the clamp in the sector. Raise and set the sector as for tearing a sheet and, by means of a surface gauge or cathetometer, measure in centimeters, to the nearest 0.01 cm, the height H of the center of gravity of the weight above a fixed horizontal surface. Then release the sector, allow it to swing and note the pointer reading. Without touching the pointer, raise the sector until the edge of the pointer just meets with its stop, in which position again determine the height h of the center of gravity of the weight above the fixed surface.

5.1.2.1 The work done is $W(h\text{-}H)$ gf·cm. For the standard 1600 gf tester, the pointer reading should be $W(h\text{-}H)/K$, where K = 137.6 cm. For other instruments graduated for grams-force of greater or lesser capacity, the reading will be factors of 2 greater or smaller. If graduated for millinewtons, the additional factor 9.81 must be applied.

5.1.2.2 One or more weights may be clamped on the edge of the sector for each calibration point, the work done in raising each weight being calculated and added together.

5.1.2.3 If the deviations of the indicated readings are greater than one-half division, the instrument should be returned to the manufacturer for repair and adjustment.

5.1.2.4 Calibration weights may or may not be available from the manufacturer of the instrument for use in calibration.

5.1.3 *Verification of Scale – Check Weight Method.* Use Check Weights calibrated for suitable scale values (i.e., 20%, 50% and 80% of pendulum capacity.) Different Check Weights are needed for each pendulum capacity. These weights should be so constructed that each weight can be inserted in the clamps by the procedure used for a test specimen.

5.1.3.1 With the pendulum in the raised position, open the clamp of the pendulum. Slide the weight into position and fasten it securely into the clamp. The body of the weight must be beneath the clamp. Depress the pendulum stop, thus releasing the pendulum. Hold down the stop until after the pendulum swing is completed, and catch the pendulum on the return swing. Read the indicating device to the nearest division.

5.1.3.2 Repeat this procedure with each of the check weights.

5.2 *Adjustment of tearing distance*

5.2.1 To check the 43.0-mm tearing distance, apply a small amount of graphite (from an ordinary pencil), to the cutting knife so that when the cut is made, some of the graphite transfers to the paper. This serves to contrast the cut from the uncut portion of the paper and facilitates the measurement. Make this measurement with a vernier caliper with a depth gauge or a good quality steel rule, readable to 0.2 mm or better under magnification. An alternative procedure is to use a go, no-go gauge, which may be available from the manufacturer of the instrument.

5.3 *Adjustment of instrument for operation*

5.3.1 *Pendulum notching.* Sometimes, as a result of frequent use, a notch is worn in the pendulum sector at the point of contact with the sector stop, giving a jerky release of the pendulum. If this happens, either repair the sector by cutting out and replacing the worn edge, or adjust the height of the stop to the very lowest point of the sector edge. In this case, recheck the calibration of the scale.

5.3.2 *Clamp alignment and knife condition.* Rest the pendulum sector against its stop, and check the alignment of the clamps. Adjust the pendulum stop if necessary. Verify by visual check that the knife is centered between the clamps, and adjust if necessary. Check the sharpness of the knife. A dull knife will result in a square notch near the top of the cut with the paper pushed out. If necessary, sharpen the knife with a rough stone; a rough edge is better than a sharp, smooth edge. Check the tearing distance and adjust the height of the knife if necessary. *Do not* change the dimensions of the specimen to adjust the tearing distance.

5.3.3 *Instrument mounting.* Support the instrument on a table so rigid that there will be no perceptible movement of the table or instrument during the swing of the pendulum. Any movement of the instrument base during the swinging of the pendulum may be a significant source of error.

NOTE 2: Threaded bolt holes are usually provided in the base of the instrument and may be used to secure the instrument to the table. An alternative procedure is to place the instrument on a guide which ensures that the instrument always has the same position on the table. Such a guide may be available from the manufacturer of the instrument.

5.3.4 *Instrument leveling*. Level the instrument so that, with the sector free, the line on the sector indicating the vertical from the point of suspension is bisected by the edge of the pendulum stop mechanism.

5.3.5 *Pendulum friction (older instruments)*. Draw a pencil line on the stop-mechanism 25 mm to the right of the edge of the sector stop. Raise the sector to its initial position and set the pointer against its stop. On releasing the sector and holding the sector stop down, the sector should make at least 20 complete oscillations before the edge of the section which engages the stop no longer passes to the left of the pencil line. Otherwise, clean, oil, and adjust the bearing.

5.3.6 *Pendulum friction (newer instruments)*. In recent years, a new type of frictionless bearing made of synthetic material has been used. This bearing will not necessarily allow the pendulum sector to make 20 complete oscillations as the older one did. This does not mean that there is excess friction in the pendulum swing. These newer bearings should not be oiled. Consult the instructions supplied with the instrument for guidance.

5.3.7 *Pointer zero reading*. Operate the leveled instrument several times with nothing in the jaws, the movable jaw being closed. If zero is not registered, the pointer stop should be adjusted until the zero reading is obtained. Do not change the level to adjust the zero.

5.3.8 *Pointer friction*. Set the pointer at the zero reading on the scale before releasing the sector, and after release see that the pointer is pushed not less than 2.5 mm nor more than 4.0 mm beyond the zero. If the pointer friction does not cause it to lie between these two distances, remove the pointer, wipe the bearing clean, and apply a trace of good clock oil to the groove of the bearing, adjust the spring tension or make other adjustments to achieve the specified friction. Reassemble, readjust the zero setting, and recheck the pointer friction.

6. Sampling and test specimens

6.1 Sample the paper in accordance with T 400 "Sampling and Accepting a Single Lot of Paper, Paperboard, Fiberboard, or Related Product." From each test unit of the sample, prepare 10 representative specimens in each principal direction of the paper, unless a test in only one direction is required. For each specimen keep the wire sides of all the plies facing the same way.

NOTE 3: It has been found (8) that there is usually no advantage in testing more than 10 specimens of a homogeneous test unit of the sample.

6.2 Cut each ply for a test specimen at least 53 mm long by 63.0 ± 0.15 mm wide, taking all the plies to be torn together from a single sheet, or if this does not provide sufficient material, from adjacent sheets of a unit.

NOTE 4: The correct length of the test specimen to be used in making a test, measured in millimeters, is equal to the distance between the outermost edges of each of the instrument's jaws (± 2 mm). For the instrument described in 4.1.2, that distance is at least 2 x 25 mm (the minimum width for each jaw face) plus 2.8 mm (the distance between the clamps) or at least 53 mm. In the United States, the majority of instruments have jaws 36 ± 1 mm wide. In such cases, a test specimen length of 76 ± 2.0 mm, specified in previous versions of this method, continues to be the correct length.

6.3 Determine from a preliminary test or the product specification how many plies are needed to make up a specimen so that, when torn together on the instrument having a 15.7-N (1600-gf) capacity, they give an instrument scale reading nearest 40% of full scale.

NOTE 5: The work done in tearing a number of sheets of paper includes a certain amount of work to bend the paper continuously as it is torn to provide for the rubbing of the torn edges of the specimen together and to lift the paper. The number of plies torn at one time and their size can affect the test result with some papers. Empirical requirements for both the apparatus and the method are therefore necessary to keep the additional work not used for tearing to a definite quantity. For this reason, in making comparisons between two or more sets of paper of the same type and grammage, use the same number of plies for each set.

6.4 If a single-ply test specimen gives a reading higher than 75 on the standard 1600 gf instrument (75% of full scale on other instruments), use the next higher capacity instruments with one ply or, if necessary, a still higher capacity instrument.

NOTE 6: For weaker papers, the standard 1600-gf instrument may require that 16 or more plies be torn together under the procedure specified in 6.3. For these papers, and provided lower capacity instruments are available, the number of plies may be restricted to four and the next lower capacity instrument may be used whenever the reading falls below 20% of full scale. The ISO 1974 procedure provides for testing four-ply specimens with multiple pendulum instruments. If this alternative procedure is used, so state in the report.

7. Procedure

7.1 Precondition, condition, and test the specimens in accordance with T 402 "Standard Conditioning and Testing Atmospheres for Paper, Board, Pulp Handsheets, and Related Products."

7.2 Raise the pendulum sector to its initial position and set the pointer against its stop. Center the specimen in the clamps with the bottom edge carefully set against the stops. Securely clamp the specimen using approximately the same pressure on both clamps. Make the initial slit. Depress the pendulum stop quickly as far as it will go to release the pendulum. Hold down the stop until after the tear is completed and catch the pendulum on the return swing without disturbing the position of the pointer.

7.3 Make only one test per specimen, each specimen consisting of the same number of plies. Make tests alternately with the wire sides of all the plies of a specimen facing the pendulum and with the wire sides of all the plies away from the pendulum. Make certain that the specimen leans toward and not away from the pendulum by gently creasing the specimen at the clamp if necessary, but in doing so avoid affecting the relative humidity of the test area (9).

7.4 Record the scale readings to the nearest half division; also record the number of plies used in the specimens.

7.5 If the line of tear fails to pass through the top edge of the specimen but deviates to one side, note and report this, but do not use the reading so obtained. If more than one-third of the tests exhibit this behavior, this method should not be used for the material concerned. If the sheets split extensively when being torn, this also should be reported.

7.6 Calculate the average tearing force in millinewtons and, if desired, in grams-force required to tear a single ply as follows:

7.6.1 If the standard 1600-gf instrument with 0-100 scale is used:

Average tearing force, mN = (16 x 9.81 x average scale reading) / number of plies

Average tearing force, gf = (16 x average scale reading) / number of piles

7.6.2 If an instrument of different grams-force capacity with 0–100 scale is used:

Average tearing force, mN =
(16 x 9.81 x avg. scale reading x gf-capacity) / (number of plies x 1600 gf)

Average tearing force, gf =
(16 x average scale reading x gf-capacity) / (number of plies x 1600 gf)

7.6.3 If an instrument has an SI metric scale (e.g., 0–1000 graduations):

Average tearing force, mN =
(16 x avg. scale reading x capacity, N) / (number of plies x 15.7 N)

Average tearing force, gf =
(16 x avg. scale reading x capacity, N) / (9.81 x number of plies x 15.7 N)

7.6.4 If an instrument has a direct-reading scale (e.g., digital read-out) that directly gives the force per ply when preset for the number of plies:

Average tearing force, mN = scale reading if directly in millinewtons, *or*
= 9.81 x scale reading if in grams-force

Average tearing force, gf = scale reading / 9.81, if scale is in millinewtons, *or*
= scale reading if directly in grams-force

NOTE 7: Previously, a standard reference material (NBS Standard Sample No. 704) was available for use with this method (*10*). Instructions accompanying the material described a procedure to correct values when using this standard. Currently, this standard reference material has been exhausted and will not be replaced.

8. Report

8.1 Report results with the tear parallel with the machine direction as resistance to internal tearing in the machine direction and those with the tear perpendicular to the machine direction as resistance to internal tearing in the cross direction.

8.2 For each principal direction, report the average, maximum, and minimum of accepted test values of the force required to tear a single ply to three significant figures.

8.3 For a complete report, state the number of plies torn at one time; the number and value of any rejected readings and reasons for their rejection; if an augmenting weight was used; and the make and model number of the instrument used.

9. Precision

9.1 On the basis of studies made in accordance with T 1200 "Interlaboratory Evaluation of Test Methods Used with Paper and Board Products," the precision of test results representing the average of ten readings, has been found to be as follows:

9.1.1 Repeatability (within a laboratory) = 4.2%

9.1.2 Reproducibility (between laboratories) = 12.5%

9.1.3 Comparability (between materials) = 6.9%, in accordance with the definition of these terms in T 1206 "Precision Statement for Test Methods." Previously, reproducibility could be reduced to 8.3% by using a reference material for standardizing the instruments; however, this material is no longer available (see Note 7).

9.2 In each of the above situations, two test results, each representing an average of ten readings, may be considered alike with a probability of 95% when the two results agree within the appropriate value shown above.

10. Additional information

10.1 Effective date of issue: November 30, 1987.

10.2 The principal changes made in this revision are: (a) discussion of the check weight method of scale verification; and (b) inclusion of the discussion on the new type of pendulum bearings.

10.3 Related methods: ASTM, D 689; Australian Appita, P400; British BS 4468; Canadian CPPA, D.9; ISO 1974; SCAN-P 11:73.

10.4 Other references related to this method are by Jones and Galley (*11*), Van den Akker, Wink, and Van Eperen (*12*), Sun, Wilson, and Bach (*13*), Lashof (*14*), and Swartout and Setterholm (*15*).

References

1. Cohen, W. E., and Watson, A. J., *Proc. Ausralian Pulp Paper Ind. Tech. Assn.* **3**: 212 (1949).
2. Wink, W. A., and Van Eperen, R. H., *Tappi* **46**: 323 (1963).
3. Elmendorf, A., "Strength Test for Paper," *Paper* **26**: 302 (1920).
4. Elmendorf, A. "The Principle of the Elmendorf Paper Tester," *Paper* **28**:(1921).
5. Institute of Paper Chemistry, "Tearing Strength of Paper, Part I," Instrumentation Studies XLVI, *Paper Trade J.* **118** (5):13 (1944).
6. Yarber, W. H. II, and Zdzieborski, J. H. George, *Tappi* **55**: 1064 (1972).
7. Clark, J. d'A., *Tech. Assn. Papers. Series XV* **1**:262 (1932); *Paper Trade J.* **94** (1): 33 (1932).
8. Lashof, T. W.. *Tappi* **45**: 656 (1962).
9. Balder, E. E., *Tappi* **63** (8):103 (1980).
10. "Association News and Events," *Tappi* **45** (4): 113A (1962).
11. Jones, H.W.H., and Galley, W., *Pulp Paper Mag. Can.* **53**(5): 116 (1952).
12. Van DenAkker, J. A., Wink, W. A., and Van Eperen, R. H., *Tappi* **50** (9): 46 (1967).
13. Sun, B.C., Wilson, J. W., and Bach, L., *Tappi* **59**(10): 106 (176).
14. Lashof, T. W., *Tappi* **47**: 445 (1964).
15. Swartout, J. T., and Setterholm, V. C., U.S. Forest Service Research Note FPL, 05 (May 1968).

Your comments and suggestions on this procedure are earnestly requested and should be sent to the TAPPI Technical Divisions Administrator. ■

TENTATIVE STANDARD – 1926
OFFICIAL STANDARD – 1927
SUGGESTED METHOD – 1968
OFFICIAL STANDARD – 1978
OFFICIAL TEST METHOD – 1983
REVISED – 1984
REVISED – 1989

Folding endurance of paper (Schopper type tester)

1. Scope

1.1 This method describes the use of the Schopper type of folding apparatus. It is suitable for papers having a thickness of 0.25 mm (0.010 in.) or less.

1.2 The procedure for the MIT type apparatus, which may be used for papers of any thickness, is described in TAPPI T 511 "Folding Endurance of Paper (MIT Tester)."

2. Significance

Folding endurance tests have been used for the estimation of the suitability of paper in use to withstand repeated bending, folding, and creasing.

3. Apparatus

3.1 *Folding tester* (Fig. 1 in the Appendix) consisting of the following:

3.1.1 Two horizontally opposed and movable *clamping jaws*, which hold a specimen 100 mm (4 in.) long under variable tension during the folding cycle, while a slotted folding blade, sliding back and forth between four creasing rollers, folds the paper at 105-125 double folds per minute. The clamps are supported from below on rollers and, while in motion, are freely suspended between tension springs. The folding blade is 0.50 ± 0.0125 mm (0.020 ± 0.0005 in.) thick, with a vertical folding slot 0.50 ± 0.0125 mm (0.020 ± 0.0005 in.) wide, the slot extending somewhat above and below the normal position of the test specimen. The vertical edges of the slot are cylindrical with radius equal to half the thickness of the blade. The four creasing rollers, each 6 mm (approximately $^1/_4$ in.) diameter and 18 mm (approximately $^3/_4$ in.) long, are arranged symmetrically about the midposition of the folding slot, and are provided with antifriction or jeweled bearings.

3.1.2 *Counter*, to register the number of double folds and to stop the instrument when the specimen breaks.

3.2 *Cutter*, to provide test strips, 15.0 ± 0.1 mm (0.590 ± 0.004 in.) wide, and with clean edges.

4. Adjustment and calibration

4.1 Test the clamps as follows: Insert a specimen in place and alternately apply and release the tension a number of times. Then, with tension released, note whether the specimen remains smooth and straight as originally inserted. Buckling or waviness indicates a faulty clamp which has allowed the specimen to slip.

4.2 Inspect the supporting rollers for worn surfaces and for bearing friction, and correct if necessary. Adjust the rollers so that they do not bind against the clamps in any position. With a feeler gage, check the four creasing rollers for parallelism and clearance. Also make sure that the two edges of the folding slot are parallel with each other and with the creasing rollers. The distance between the folding blade and the two creasing rollers on each side of it is required to be 0.38 ± 0.05 mm (0.015 ± 0.002 in.), and the width between the rollers of the space occupied by the unbent specimen should be approximately 0.5 mm (0.02 in.). As a final test of alignment, fold a specimen somewhat short of failure and inspect it for uniformity of wear along the crease. If the specimen seems weaker at one end of the crease than at the other, and the ends of the strip are satisfactorily clamped, this indicates faulty alignment of the rollers or the folding slot and will lead to low folding results.

4.3 Adjust the tension springs attached to the clamps against dead-weight loads, so that the tension on the specimen during a test is 7.60 ± 0.1 N (780 ± 10 gf) when the clamps are farthest apart (when the specimen is straight and free), and 9.81 ± 0.1 N (1000 + 10 gf) when they are nearest together. The adjustments are preferably made *in situ* with the aid of a calibrating device (*1*), as described (Fig. 1) in the Appendix.

NOTE 1: As stated in Section A.3.2, the minimum tension (7.60 ± 0.1N) is the most important value. If the difference between the maximum and minimum tension is not equal to the desirable 2.1 in., then the minimum tension should be adjusted to the 7.60 N

Approved by the Physical Properties Committee of the Process and Product Quality Division
TAPPI

Fig. 1. Schopper-type tester.

value and the maximum rechecked. If the maximum tension is not within 0.5N of the specified 9.81 N then the springs should be replaced.

4.4 Adjust and calibrate the instrument at least once a month if it is in continual use or immediately before a test.

5. Sampling and test specimens

5.1 Obtain a sample of the paper in accordance with TAPPI T 400 "Sampling and Accepting a Single Lot of Paper, Paperboard, Fiberboard, or Related Product." From each test unit obtain 10 specimens, cut in each principal direction of the paper, having a width of 15.0 ± 0.1 mm (0.590 ± 0.004 in.) and a length of 100 mm (4 in.). Select specimens that are free from wrinkles or blemishes not inherent in the paper and be sure that the area where the folding is to take place does not contain any portion of a watermark and appears to be of average opacity.

5.2 Handle each specimen by an end and do not touch it with the hands in the region in which it is to be folded.

6. Procedure

6.1 Precondition, condition, and test the paper in accordance with TAPPI T 402 "Standard Conditioning and Testing Atmospheres for Paper, Board, Pulp Handsheets, and Related Products."

NOTE 2: Because of the extreme sensitivity of this test to the moisture content of the specimen, it is important that the moisture equilibrium be approached from a drier state as specified in T 402.

6.2 Lock the vertical slot of the reciprocating blade is in its central position. Without touching the center of the specimen, place it in the slot and fasten the ends firmly and squarely in the jaws so that the surface of the specimen lies wholly within one plane. Apply the tension and fold the specimen until it breaks. Record the number of double folds made before fracture.

7. Report

7.1 Folding endurance is defined as the logarithm (to the base 10) of the number of double folds.

7.1.1 For each test unit record the number of double folds for the machine and cross directions. Convert the raw data to the logarithm (base 10) by using common log tables. Calculate the mean of the logs and report as $\log_{10}$ Schopper type folding endurance to three significant figures. Include the number of specimens tested and the maximum and minimum $\log_{10}$ of the fold number obtained in each direction.

7.2 *Alternative Report.* For each test unit, report the average test result for the machine and cross directions of the paper separately, as Schopper type tester folding number (double folds), to two significant figures. Include the number of specimens tested and the maximum and the minimum values obtained in each direction, and the temperature and relative humidity under which sample was tested.

7.3 Tests made on strips having their lengths in the machine direction are designated as "machine direction," and, correspondingly, those having their lengths at right angles to the machine direction are designated as "cross direction."

8. Precision

8.1 Repeatability = 15%;

8.2 Comparability = 34%;

8.3 Reproducibility = 40%; according to the definition of these terms in TAPPI T 1206 "Precision Statement for Test Methods."

NOTE 3: Precision statement is based on folding number.

NOTE 4: Folding endurance numbers are very susceptible to small errors in the adjustment and calibration of the instrument and in the relative humidity of the test room. Limits as much as 2 $^1/_2$ times those shown may be expected if the instructions of this method are not followed meticulously (2-5).

9. Additional information

9.1 Effective date of issue: October 1, 1989.

9.2 The 1978 revision specified the minimum and maximum tension in SI units, with tolerances, and also gave a tolerance in the distance between the folding blade and the two creasing rollers. A statement of significance was also added.

9.3 The use of a suction fan to cool the folding blade has been found useful in minimizing variations in results.

9.4 Related methods: British BS 4419 "Methods for Determination of Folding Endurance of Paper"; German, DIN VZ and PCIV/12; French, AFNOR Q03-001; ISO 5626 "Determination of Folding Endurance."

References

1. Carson, F. T., and Snyder, L. W., "Calibration and Adjustments of Schopper Folding Tester," Bureau of Standards Technologic Paper No. 375 (1929).
2. Brecht, W., and Korner, L., "The Accuracy in Testing of Paper Properties," *Das Papier* **5**: 155 (1951).
3. Reitz, L. K., and Sillay, F. J., "Application of Statistical Methods to Paper Testing Procedures," *Paper Trade J.* **126** (17): 54 (1948).
4. Kahlson, T., and Martensson, B., "The Reason for Variations in Folding Endurance Values," *Paperi ja Puu* **46** (10): 581 (1964).
5. Brecht, W., and Wesp, A., "A New Method for Testing Folding Strength," *Das Papier* **6**: 443 (1952).

Appendix

Instructions for use of the calibrating device

A.1 *Apparatus* (see Fig. 1).

A.1.1 *Wire-pulling frame.*

A.1.2 *Blocks,* two, with protruding limit guides (one for 1000 gf extension, the other for 780 gf extension).

A.1.3 *Upright,* with free-turning pulley.

A.1.4 *Weights,* 1000-g and 780-g.

A.1.5 *Brackets,* for use with the upright.

A.2 *Mounting*

A.2.1 Screw the upright brackets to each side of the wooden base so that when the upright with the free-turning pulley is placed in the bracket cutout, the center of the pulley is exactly in line with the center of the barrels, the center of the jaws, the center of the bearing block, and the movement of the jaws.

A.2.2 Place the wire-pulling frame over one of the barrel assemblies and central bearing blocks. Grip the flat metal tang of the frame in the jaw of the other assembly so that the top of the tang is flush with the top of the jaw.

A.2.3 With the flywheel of the instrument in its locked position, place the block with protruding limit guides on the sliding mechanism, so that the sliding arm fits into the cutout of the block and that the protruding limit guides at the top of the block extend out over the wire-pulling frame. Insert the upright with free-turning pulley in its bracket. Hook the string attached to the weight to the center of the end section of the wire-pulling frame and pass the string over the pulley to hold the weight vertically.

A.3 *Operation*

A.3.1 There are two sets of limit guides: one to indicate the relaxed position of the jaws and the second fixed to indicate the extended limit of the jaws. With the first set in position, pull the barrel mechanism to its outward position and gently release the suspended 1000-g weight. When the spring of the jaw is correctly calibrated, the jaw will then barely touch the edge of the protruding limit guide. If the jaw does not reach the protruding limit guide or if it touches and is stopped by it, loosen the small Allen setscrew on the rear side of the barrel and turn the knurled knob one way or the other, until contact is just barely made.

A.3.2 Next, remove the 1000-g weight and place the limit guide for the relaxed position. Apply the 780-g weight; the jaw should then barely touch the extended limit guide. By turning the knob on the barrel assembly, a delicate adjustment should result in an agreement between the spri..g tension and the extension of the jaws in both the relaxed and the extended positions.

NOTE 5: The adjustment made with the 780-g weight is more important than with the 1000-g weight.

A.3.3 Apply the same procedure for the opposite jaw after reversing the position of the wire-pulling frame and the upright with the free-turning pulley.

Your comments and suggestions are earnestly requested and should be sent to the TAPPI Technical Divisions Administrator. ■

T 441 om-90

TENTATIVE STANDARD – 1937
OFFICIAL STANDARD – 1960
REVISED – 1977
OFFICIAL TEST METHOD – 1984
REVISED – 1990

Water absorptiveness of sized (non-bibulous) paper and paperboard (Cobb test)

1. Scope

1.1 This method describes a procedure for determining the quantity of water absorbed by nonbibulous paper and paperboard in a specified time under standardized conditions. It is based on studies by Cobb and Lowe (*1*), Cobb (2) and other investigators (*3, 4*).

1.2 Water absorptiveness is a function of various characteristics of paper or board such as sizing, porosity, etc. This method is generally applicable to sized paper and paperboard, but it is not recommended as a sizing test for writing paper.

1.3 For testing unsized and absorbent paper or paperboard, see T 432 "Water Absorbency of Bibulous Paper" and T 492 "Water Absorption of Paperboard (Non-Bibulous)" (water drop test); for slack sized paper and paperboard, see T 433 "Water Resistance of Sized Paper and Paperboard (Dry-Indicator Method)."

2. Apparatus

2.1 *Water absorption apparatus*[1], to permit one side of the specimen to be wetted uniformly at the moment the soaking period begins, and to allow controlled rapid removal of the water from the specimen at the end of the test period. The specimen holder (*3*) as shown in Fig. 1 comprises a metal ring with a machined lower face, 11.28 ± 0.02 cm inside diameter (corresponding to a cross-sectional area of 100 cm^2), 2.5 cm high and about 0.6 cm thick, clamped to a flat base plate about 15 x 15 cm with a metal cross bar 17 x 2.5 x 0.6 cm and two wing nuts on a pair of studs. The cross bar has a hole at one end and a slot at the other to facilitate assembly and use. On the base plate is a rubber mat, larger than the outside dimensions of the ring, on which the specimen is clamped.

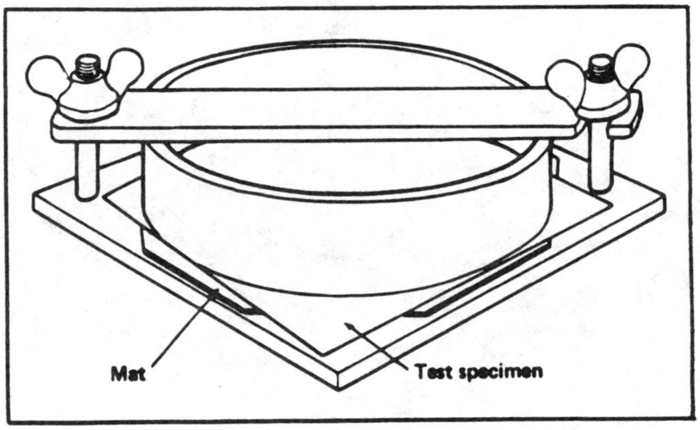

Fig. 1. Water absorption apparatus.

NOTE 1: The metal parts should preferably be of brass or other corrosion-resistant material.

NOTE 2: Several versions of the water absoption apparatus are now available. Although the basic procedure for performing the test is not changed, the clamping mechanism is. This may affect the speed at which the sample can be removed from the apparatus.

2.2 *Metal roller*, solid brass having a smooth face 20 cm wide and weighing 10.0 ± 0.5 kg.

2.3 *Timer*, stopwatch or electric timer reading in seconds.

2.4 *Graduated cylinder*, 100 mL.

2.5 *Balance*, with an accuracy of 0.01 g, or better.

[1]Names of suppliers of testing equipment and materials may be available from TAPPI Information Resources Center.

Approved by the Physical Properties Committee of the Process and Product Quality Division
TAPPI

3. Materials

3.1 *Blotting paper*[1], sheets of blotting paper, 200-250 g/m², with a capillary rise of 50-100 mm of water (mean of MD + CD) as measured by the Klemm method (see Appendix). Normally, the blotter specified in T 205 "Forming Handsheets for Physical Tests of Pulp" will meet this requirement.

3.2 *Water*, distilled or deionized.

NOTE 3: The temperature of the water is important and must be maintained at 23 ± 1°C. Renew the water for each determination.

4. Sampling and test specimens

Obtain a sample of the paper in accordance with T 400 "Sampling and Accepting a Single Lot of Paper, Paperboard, Fiberboard, or Related Product." From each test unit, cut specimens to a size slightly greater than the outside dimensions of the ring of the apparatus, i.e., squares 12.5 x 12.5 cm. The specimens should be free from folds, wrinkles, or other blemishes not commonly inherent in the paper. For hard-sized papers (absorbing less than 100 g of water/m²), use 10 specimens per test unit. For soft-sized papers (absorbing more than 100 g/m²), use 20 specimens.

5. Procedure

5.1 Condition the specimens in an atmosphere in accordance with T 402 "Standard Conditioning and Testing Atmospheres for Paper, Pulp Handsheets, and Related Products."

5.2 Weigh each specimen to the nearest 0.01 g.

5.3 Test half the specimens with the wire side up, the other half with the felt side up.

5.4 Place a dry rubber mat on the metal plate and lay a weighed specimen on it. After wiping the metal ring perfectly dry, place it upon the specimen, and fasten it firmly enough in place with the crossbar to prevent any leakage between the ring and the specimen. For reporting, the test side is the one that is in contact with the water during the test.

NOTE 4: For materials where leakage between the ring and the upper surface of the test specimen may occur during the test, a soft elastic non-absorbent gasket may be used. The gasket should have the same internal dimensions as the ring.

5.5 Pour 100 mL of water into the ring as rapidly as possible thus giving a head of 1 ± 0.1 cm. Start the stopwatch immediately. At 15 ± 5 s before the expiration of the predetermined test period, usually 120 s (see 8.3, 8.4, and 8.5), pour the water quickly from the ring, taking great care not to drop any of the water upon the outside portion of the specimen. Promptly loosen the wing nuts, swing the crossbar out of the way while holding the ring in position by pressing it down with one hand. Carefully, but quickly, remove the ring and place the specimen with its wetted side up on a sheet of blotting paper resting on a flat rigid surface.

5.6 Exactly at the end of the predetermined test period, place a second sheet of blotting paper on top of the specimen and remove the surplus water by moving the hand roller once back and once forward over the pad without exerting any additional pressure on the roller. Specimens which contain an excess of surplus water after blotting, as shown by glossy areas on the surface, should be rejected and the test repeated, decanting the water from the ring sooner. Fold the specimen with the wetted area inside. Immediately reweigh it to the nearest 0.01 g.

5.7 Subtract the conditioned weight of the specimen from its final weight, and multiply by 100 the gain in weight in grams to obtain the weight of water absorbed in grams per square meter.

NOTE 5: If any liquid has passed through the sheet to the rubber mat, the test is not acceptable. When this occurs, either shorten the time to 60 s, or staple two or more specimen sheets together outside the test area. In such a case, the calculated test area remains that of the inside ring. For hardsized papers, a longer period (e.g., 300 s) may be found advisable.

6. Report

Report the absorptiveness, calculated as the average weight of the water absorbed in grams per square meter, for the wire and the felt side of the paper separately. If desired, state also the maximum and minimum values for each side.

NOTE 6: Unless otherwise stated in the report, it is assumed that an exposure period of 120 s on a single-sheet thickness has been employed. Such conditions are suitable for most well-sized papers.

7. Precision

Based on limited data for hard sized papers and for means of five specimens on the same side, the repeatability is within 8% and the reproducibility between laboratories is within 10%, as defined by TAPPI T 1206 "Precision Statement for Test Methods."

8. Additional information

8.1 Effective date of issue: April 20, 1990.

8.2 The standard test area is 100 cm². If the available specimens are too small, a proportionately smaller test area may be used providing that the volume of water is reduced to provide a pressure head of 1 ± 0.1 cm. The change in area should be noted in the report.

8.3 This test is designed for nonbibulous papers and paperboards, but satisfactory results have been obtained with up to about 20 sheets of highly absorbent paper stapled together and tested for 60 instead of 120 s.

8.4 For very hard-sized or specially treated papers, the test may be extended to periods up to 18 h, to increase the sensitivity.

8.5 For layers of absorbent papers, the quantity of water absorbed is almost proportional to the time of exposure. For well-sized papers, the quantity is approximately proportional to the square root of this time.

8.6 An effect of natural aging has been noticed in many papers. For example, results on cupstock papers, aged a few weeks, are usually about 2 g/m² less than on papers tested immediately after being made.

8.7 The Cobb test may also be suitable with other (water base) solutions, such as dilute lactic acid and hot coffee for food board and cup stock and possibly ink for writing and printing papers.

8.8 Weighing the wetted specimens can be facilitated by using tared water-vapor-proof containers which will eliminate evaporation losses. Metal cans or polyethylene bags about 11 cm x 23 cm x 0.0076 cm with wire closure have been reported to be satisfactory for this purpose.

8.9 For specimens that have long penetration times or have surfaces that are difficult to wet, substitute the distilled or deionized water with water having a known concentration of wetting agent. With the results also indicate the exposure time, total head, wetting agent and concentration.

8.10 This method was revised in 1942, 1945, 1958, 1960, 1963, 1969, 1977, and 1990. The 1969 revision incorporated standardization of the test area at 100 cm² and the pressure head depth at 1.0 ± 0.1 cm, standardization of the blotting method with specification of the blotters and rollers used, and specification of the water as distilled or demineralized. The 1977 revisions were editorial. This 1990 revision makes mention of the use of alternate clamping mechanisms.

8.11 Related methods: ASTM D 3285, Canadian CPPA F.2; British Standard 2644; British PPMA PT-15; Australian APPITA P 411, Scandinavian SCAN P-12; ISO Standard 535-1976.

References

1. Cobb, R. M., and Lowe, D.V., "A Sizing Test and Sizing Theory," *Tech. Assoc. Papers* **17**:213 (1934); *Paper Trade J.* **98**(12):43 (1934).
2. Cobb, R.M., "What I Do Not Know About Sizing and Capillary Flow," *Tech. Assoc. Papers* **18**:209 (1935); *Paper Trade J.* **100**(16):42 (1935).
3. Egy, W.L., "The Cobb Sizing Test," *Pulp Paper Mag. Can.* **38**(7):499 (1937).
4. Codwise, P.W., "Water Resistance of Paper," *Tech. Assoc. Papers* **19**:231 (1936); *Paper Trade J.* **102**(3):39 (1936).

Appendix
Measurement of blotter absorptiveness (Klemm method)

Make the test on samples conditioned according to T 402. Cut parallel specimen strips about 15 mm wide and at least 200 mm long in both machine and cross directions. Immerse the specimen strips suspended vertically to a depth of 10 mm in distilled or deionized water maintained at a temperature of 23 ±1°C. After an immersion time of 10 min, read the height in millimeters to which the water rises above the water level in the container.

Your comments and suggestions on this procedure are earnestly requested and should be sent to the TAPPI Technical Divisions Administrator. ■

T 803 om-88

TENTATIVE STANDARD – 1944
OFFICIAL STANDARD – 1967
OFFICIAL TEST METHOD – 1979
CLASSICAL METHOD – 1985
OFFICIAL METHOD – 1988

Puncture test of containerboard

1. Scope

This tester (Fig. 1) measures the energy required to puncture containerboard or corrugated board with a pyramidal point affixed to a pendulum arm. These energy units actually are made up of two major components: the energy to tear the material and the energy to bend it out of the way of the point.

Fig. 1. Puncture tester.

2. Significance

2.1 Damage in use, similar to that occurring in this test, may result from contact with solid objects, such as the end of a piece of lumber or the corner of a wooden box.

2.2 The test evaluates, in a composite fashion, some fabrication factors and properties of the materials.

NOTE 1: This test on single and doublewall corrugated board measures the energy to puncture the board and the energy to bend the board back to allow penetration of the point and arm through the test specimen. It is widely used as well on triplewall board. Here the clearance between the puncture point and the aperture in the loose plate 13.2 mm (0.5 in.) can be less than the thickness of the board. For some triplewall board then, the total energy measured combines the energy for puncture and bending with that necessary to crush the board.

3. Apparatus

3.1 *Puncture test apparatus* (see Fig. 1), consisting of the following:

3.1.1 *Pendulum*, incorporating an arm in the form of an arc of 90°.

3.1.2 *Puncture point*, affixed to the end of the arm and having the shape of a right angle triangular pyramid 25 ± 0.7 mm (1 in.) high, with leading edges rounded, base edges left sharp, and the point honed to 1.57 ± 0.05 mm (0.062 ± 0.002 in.) radius.

3.1.3 *Puncture point collar*, loose fitting around the base of the puncture point to avoid friction that would result in the punctured material rubbed against the puncture arm.

3.1.4 *Trigger*, or latch, to hold the pendulum in a horizontal position.

3.1.5 *Clamping jaws*, spring actuated, for rigidly holding the specimen to be tested in a horizontal plane.

3.1.6 *Pointer and scale*, to indicate the arc through which the pendulum travels after passing through the specimen. Scale graduations correspond to the weights used.

Approved by the Fiberboard Shipping Container Testing Committee of the Corrugated Containers Division
TAPPI

3.1.7 *Augmenting weights*, which when attached give scale values up to least 38.87 J (1300 units of energy) when the pendulum is in the horizontal position.

3.1.8 *Loose plate*, inserted between the clamping jaws when testing material to define the test area (means should be provided for securing the plate once it has been properly positioned). The loose plate is removed before checking the zero reading of the apparatus.

3.1.9 *Adjustable screw*, bearing against the pointer assembly during the measuring of half the pendulum swing, permitting adjustment of the full-scale position of the pointer.

NOTE 2: For testing lighter materials, the apparatus can be supplied with an attachment that allows the release of the pendulum at a 45° angle. An auxiliary low scale is also supplied which when attached to the standard scale plate provides additional calibration from 0-0.90 J (0-30 units) and allows reading to half a unit. Modified machines for heavier material are also available.

4. Adjustment and calibration

4.1 Place the apparatus on a rigid, heavy foundation of convenient height, and bolt it down so that it is level in both horizontal directions. The preferred mount is a separate concrete base. Adjust the apparatus with the entire clamping assembly unbolted and, with the newer instruments, swing the hinged assembly away from the path of the pendulum. With older instruments, completely remove the assembly from the machine.

4.2 *Pendulum friction*. Remove the weights and weight-support stud, put the puncture-point collar in place on the puncture head, rotate the pointer to a position below the zero mark, and then latch and release the pendulum. The pendulum should swing for a minimum of 8 minutes before coming to rest. If it does not, clean and lubricate the bearing.

4.3 *Zero reading*. Remove the weights and stud; remove the loose plate and fix the puncture-point collar on the puncture head. Allow the pendulum to hang vertically on its pivot shaft. Then with the adjusting screw, align the pointer exactly on full scale. With the pointer set to about 25 mm (1 in.) above zero on the scale, raise the pendulum until the latch engages. Release the pendulum. If the scale reading is not zero, loosen the four setscrews that position the latch tongue as required. Repeat this procedure until the scale reading is zero.

4.4 *Pointer friction*. Latch the bare pendulum, without the weights or weight stud and with the collar attached to the puncture head. Remove the clamping plate, rotate the pointer to the full-scale reading, release the pendulum, and note the position of the pointer at maximum travel. Then without moving the pointer, relatch the pendulum and release it, thus allowing it to swing and push the pointer further. Adjust the pointer friction by means of the screw in the stud projecting from the pointer hub, so that the additional movement is not more than two units. Recheck the friction with the pointer at mid-scale to guard against reducing the friction too much. If too much friction has been removed, the pointer will move or "bounce" below the position indicated in the second step.

4.5 *Punction point collar adjustment*. Place the collar in the same position each time on the end of the puncture point head. A dot of paint on one side of the collar to match a dot on the point base is an easy way to fix this position. An adjustable pin determines the force needed to dislodge the collar. This force is 1.0 ± 0.15 N (3.6 ± 0.5 oz) pull when the collar is in a horizontal position. The puncture head extends at least 0.38 mm (0.015 in.) beyond the outer edge of the collar when the collar is attached to the puncture point. Keep the flange angle at approximately 135°.

4.6 *Clamp load*. A force of 245 ± 22 N (55 ± 5 lb) applied downward from the middle of the lower plate should open the clamp 6 mm (1/4 in.). Tighten or loosen the wing nuts at the bottom of the spring tube to adjust this pressure if necessary. To check that the clamping is uniform over the test specimen, place a number of strips of the thinnest specimen to be tested between the jaws at various points. If any can be removed easily, change the jaw alignment until the load is uniform.

4.7 *Center of gravity*. If the center of gravity of the pendulum does not lie in the proper plane, addition of auxiliary weights will shift the pointer zero and necessitate the changing of both the pointer and the latch adjustments when changing scales. The alignment of the center of gravity may be most conveniently checked by removing the weights and stud from the pendulum, allowing the pendulum to come to its rest position (with collar attached at the puncture point), and gently rotating the pointer clockwise until it just touches its adjusting screw. Note the pointer position, attach the stud and weights to the pendulum, and repeat the above operations. If the center of gravity lies in the proper plane, no shift of the pointer will occur. If the pointer shifts, the center of gravity is incorrect. Correct it by shifting the balancing weight in or out on its screw. This weight is adjusted at the factory and then staked in place. It should never be moved unless it is certain that the center of gravity is incorrect. If adjustment appears to be required, it is good to first check the pendulum with all combinations of weights and by rotating the weights to make sure that they are concentric before returning the pendulum to the factory for adjustment.

4.8 *Loose plate*. Secure the loose plate beneath the upper clamping jaw in such a manner that a vertical projection of the lower edges of the puncture point is equidistant from the triangular edges of the loose plate opening when the base of the puncture head is 31.8 mm ($1^1/_4$ in.) below the upper clamping jaw.

4.9 *Puncture point*. Be sure that the three corners at the base of the puncture point are equidistant [within 1.5 mm (1/16 in.)] below the upper clamping jaw when the base of the puncture point is 31.8 mm ($1^1/_4$ in.) below the upper clamping jaw; that its apex does not deviate from a plane equidistant from the pendulum bearing by more than 1.5 mm (1/16 in.); and that its edge farthest from the axis of rotation is parallel to that axis. The condition of the puncture point is important.

important. It should be protected from abuse and damage, so that its surfaces remain smooth and free of nicks and scratches. Clean the surface of the point with a degreasing agent, such as 1-1-1-trichloroethane or methyl ethyl ketone, after each puncture is made on a wax-coated, asphalt-laminated, or other specially treated board which may transfer material to the point.

5. Sampling and test specimens

Sample the board in accordance with TAPPI T 400 "Sampling and Accepting a Single Lot of Paper, Paperboard, Fiberboard, or Related Product." From each test unit of the sample, select representative specimens at least 300 x 300 mm (12 x 12 in.); see Procedure for number of specimens depending on nature of material to be tested. Make all tests at least 95 mm (3³/₄ in.) from an edge, score line, or imperfection.

6. Conditioning

Precondition and condition the specimens and test in the atmosphere specified in TAPPI T 402 "Standard Conditioning and Testing Atmospheres for Paper, Board, Pulp Handsheets, and Related Products."

7. Procedure

7.1 Before making each test, place the puncture point collar on the base of the point and set the pointer about 25 mm (1 in.) above the expected reading. Raise the pendulum to the horizontal position and latch it.

7.2 On single wall, double wall, and triple wall corrugated and solid fiberboard, make four punctures which comprise a set. One set constitutes one test. The plane of the curved pendulum arm is used as the reference in relating the position of the specimens to the testing machine. Directions refer to the direction of the corrugations of corrugated board or grain direction of uncombined sheets or solid fiber. The orientation of the specimens for a set follows: (a) parallel, with one surface down; (b) parallel, with the other surface down; (c) perpendicular, with one surface down; and (d) perpendicular, with the other surface down.

7.3 On uncombined sheets or single-faced corrugated boards, make a set of two punctures per test with specimens oriented as follows: (a) parallel, with smooth surface down; and (b) perpendicular, with smooth surface down.

7.4 Clamp each specimen, then release the pendulum by pushing the latch handle to the left; after the pendulum has completed its swing, note the reading to the nearest one-half division on the proper scale. It is not advisable to perform a test giving a reading within the upper or lower quarter of the scale. In such a case, use another scale. The weights to be used for each scale are indicated on the scale plate of the apparatus. Follow the same procedure for testing lightweight material, but use the bare pendulum without additional weight or stud, release the pendulum from the 45° position, and use the auxiliary low scale.

8. Report

8.1 Report the average of at least two test sets in scale units of three significant figures. (Each unit is equal to 0.0299 joules.) The total tearing length of the head is 10.77 cm (4.24 in.).

9. Precision

9.1 The values of repeatability have been calculated for a test result each of which is the average of a set as defined in 7.2 (four individual tests). These values were based on work performed on a single instrument at the Institute of Paper Chemistry.

9.2 The repeatability as defined in TAPPI T 1206 "Precision Statement for Test Methods" is 5.25%.

9.3 The reproducibility of this method is not known.

10. Additional information

10.1 Effective date of issue: June 8, 1988.

10.2 Related methods: ISO 3036-1975.

10.3 Changes in this revision include:

10.3.1 Elimination of the procedure for measuring stiffness (this was rarely used).

10.3.2 A note regarding crushing of triplewall board.

10.3.3 Clarification of the description of the apparatus.

10.3.4 Explanation of how to treat pendulum friction.

10.3.5 Title and wording of Section 5.

10.3.6 Change in the precision statement.

10.3.7 Additional references to related methods.

References

1. Beach, R.L., "Puncture Testing of Box Board," *Paper Trade J.* **108** (5):30 (1939).
2. Wells, S.D., Kaplan, A.B., and Ayers, L.R., "Tests on Fiber Containers—The Beach Puncture Tester," *Paper Trade J.* **116**(13):29 (1943).
3. Van Den Akker, J.A., and Wink, W.A., "Instrumentation Studies: LX. The Beach Puncture Tester," *Tappi* **33**(12):114A (1950).

Your comments and suggestions on this procedure are earnestly requested and should be sent to the TAPPI Test Methods Administrator. ■

T 808 om-86
SUGGESTED METHOD — 1965
OFFICIAL STANDARD — 1971
OFFICIAL TEST METHOD — 1981
REVISED — 1986

Flat crush test of corrugated board

1. Scope

The flat crush test (*1*) is a measure of the resistance of the flutes in corrugated board to a crushing force applied perpendicular to the surface of the board under prescribed conditions. The test is satisfactory for single-faced or single wall (double-faced) corrugated board, but not for double-wall or triple-wall corrugated board, because of lateral motion of the central facing or facings. This method is not intended for use in testing corrugated structures [for testing such structures see TAPPI Useful Method 810 "Leaning Flute Free (LFF) Flat Crush of Corrugated Board"].

2. Significance

2.1 Flat crush is a measure of the flute rigidity of corrugated board. A high flat crush value indicates a combination of good flute formation and at least adequate strength medium. Low flat crush can indicate a number of conditions. Among these are low strength medium, leaning flutes, and/or crushed flutes.

2.2 Care must be taken when measuring flat crush resistance of finished corrugated containers, to ensure that the method of measurement adequately simulates the reaction to crushing forces exerted on the container during actual use. This method generally does not simulate these conditions. Therefore, the correlation may be low between flat crush and the ability of a box to perform in the field.

2.3 Boxes made from crushed corrugated board may have low flat crush and may not perform well. Board with leaning flutes may have low flat crush, but this may not affect box performance. For leaning flutes to cause low flat crush, there must be lateral movement between the two liners during the tests. After the board is formed into a box, lateral movement between the two lines is restricted by the geometry of the box, and the low flat crush cannot become operative. Also, this test method does not take into account the effect of linerboard stiffness in resisting crushing forces.

3. Apparatus

3.1 *Compression machine*[1], consisting of upper and lower platens held rigidly parallel and free from lateral movement within 0.05 mm (0.002 in.). The platens must be parallel to within 1 part in 2000 when measured diagonally across the bare platens. To prevent slippage of the specimens, the platens are faced with crocus or fine emery cloth, free of ridges, by means of double-coated adhesive tape. The facing cloth should be changed after every 2000 tests. Means are provided for compressing a specimen placed between the two platens at a uniform rate.

3.1.1 The capacity of the tester may be 2000 N (approximately 500 lbf) or 5000 N (approximately 1000 lbf). The rate of applying the load is 111 ± 22 N/s (25 ± 5 lbf/s), and the pulley ratio should be such as to achieve this rate.

3.1.2 Means are provided for measuring and indicating the applied load within 5 N (approximately 1 lbf) and such that the indicator can be accurately calibrated with dead weights. The accuracy required is 0.5% or 2.5 N (approximately 0.5 lbf), whichever number is greater.

NOTE 1: Either a flexible beam or fixed platen type of compression tester may be used, but to achieve the precision noted in Section 6, conduct all tests on only one type of compression tester. Rigid support testers with vertical platen movement of not more than 0.150 mm (0.006 in.) should not be considered under this method.

NOTE 2: See Appendix for calibration procedures.

3.2 *Specimen cutter*, consisting of a device capable of cutting through the corrugated combined board structure without crushing areas at the cut edges. This may be circular, square, or rectangular.

3.3 *Stopwatch*, graduated to 0.2 s.

3.4 *Inside micrometer*, having an accuracy of 0.025 mm (0.001 in.).

4. Sampling and test specimens

4.1 From each test unit of a sample obtained in accordance with TAPPI T 400 "Sampling and Accepting a Single Lot of Paper, Paperboard, Fiberboard, or Related Product," select a minimum of 10 specimens free of abnormalities not representative of the combined board.

[1]Names of suppliers of testing equipment and materials for this method may be available from the TAPPI Information Resources Administrator.

Approved by the Fiberboard Shipping Container Testing Committee of the Corrugated Containers Division TAPPI

Each specimen should be cut, preferably in circular form, so that each is either 32.3 cm^2 (5.00 $in.^2$) or 64.5 cm^2 (10.00 $in.^2$) in area. All specimens should be cut at least 38 mm (1.5 in.) away from printed matter, scores, and die-cuts.

4.1.1 If the specimens are not circular, exercise special care to maintain the desired area accurately.

4.2 Avoid crushing areas at the cut edges, and, where possible, avoid fractional flute counts.

5. Procedure

5.1 Condition and test the board in an atmosphere in accordance with TAPPI T 402 "Standard Conditioning and Testing Atmospheres for Paper, Board, Pulp Handsheets, and Related Products."

5.1.1 After cutting to size, position the specimen centrally on the lower platen with the corrugations parallel to the front of the machine. Apply the crushing load to the specimen until the side walls of the corrugations collapse completely. Failure is defined as the maximum load sustained before complete collapse.

NOTE 3: Normally, a preliminary end point occurs when the tips of the corrugation flatten on one or both sides of the specimen. This should not be confused with the final end point when the corrugations collapse completely. If the collapse of the corrugations is so gradual that no such peak load is distinctly registered, note this fact.

6. Report

6.1 Report the flat crush test results, in kilopascals (kN/m^2) or pounds per square inch, as the average of ten determinations, to three significant figures (1 psi = 6.895 kPa).

6.2 Include for a complete report:

6.2.1 The standard deviation.

6.2.2 The total number of specimens tested.

6.2.3 The number of specimens exhibiting rolling failure.

6.2.4 The testing machines used, whether A, B, or C flute and the flute count.

NOTE 4: Rolling failure is defined as an irregular skewed pattern or leaning type of collapse of the flutes.

7. Precision

7.1 Repeatability (within a laboratory) = 5%.

7.2 Comparability (between materials) = 7%.

7.3 Reproducibility (between laboratories) = 10%.

7.4 These terms are in accordance with definitions in TAPPI T 1206 "Precision Statement for Test Methods."

7.5 These data were obtained in a round-robin among three laboratories, with six different testers, using 10 specimens of 64.5 cm^2 (10 $in.^2$) in area.

8. Additional information

8.1 Effective date of issue: March 24, 1986.

8.2 This method was first published in 1965 as a suggested method, and it became an official method in 1971.

8.3 Related methods: ASTM D 1225 "Flat Crush of Corrugated Fiberboard," American Society for Testing and Materials, Philadelphia, PA; APPITA P 429 "Flat Crush Resistance of Corrugated Board," Technical Association of the Australian and New Zealand Pulp and Paper Industry, Parkville, Australia; CPPA D.20 "Flat Crush Test of Corrugated Board," Canadian Pulp and Paper Association, Montreal, Canada; ISO 3035 "Determination of Flat Crush Resistance — Single-Faced and Single-Wall Corrugated Fiberboard," International Organization for Standardization, Geneva, Switzerland; SCAN P-32 "Flat Crush Resistance of Corrugated Fiberboard," Scandinavian Pulp, Paper, and Board Testing Committee, Stockholm, Sweden.

Appendix A. Calibration procedure for the flexible beam type of crush tester

A.1 *Lateral movement of the lower platen.* Remove the weighing beam and lower platen from the tester. Securely clamp the ends of the beam on a horizontal plane surface. Manually move the platen to its full extent towards the front and back and note the displacement on a dial micrometer. Repeat a similar procedure to determine the amount of motion parallel to the beam length.

A.2 *Platen parallelism.* With no load, this may be checked conveniently by means of the inside micrometer. Remove the facing cloth, separate the platens 25 to 50 mm (1 to 2 in.) and measure the clearance at each corner of the platens. For a 100 x 100 mm (4 x 4 in.) platen, the difference in clearance is required not to exceed 0.05 mm (0.002 in.) at the extreme edges.

NOTE 5: Determining the parallelism under conditions of normal loading is difficult. One suggested procedure is to load the platen through a centrally positioned metal ball and measure the platen edge clearances at various intensities of load over the normal testing range. Another procedure is to apply the load to a 6.0–6.5-mm (0.25-in.) thick piece of wood or metal with parallel faces, covered by a sheet of paper and with a soft carbon paper. A uniform carbon imprint should be obtained after slightly rotating the piece under pressure.

A.3 *Loading rate.* With the platens in contact, measure the time required to attain selected load levels, up to 90% of the capacity of the beam.

A.4 *Test specimen area.* As a precaution against eccentricity of the liners of a specimen or deflection or distortion of the knife blade used for cutting the specimen, check the size of the specimen on both sides by measuring its diameter at right angles.

A.5 *Calibration of the beam (dead weight method).* Remove the motor and top pulley and place the tester on a narrow table or bench. At right angles across the platen at its central point, extend a narrow rigid bar, arranged with a pan or hook at each end to support pairs of balanced dead weights which, with the added weight of the beam and their supports, do not exceed the calibrated capacity of the beam. Check the accuracy of the load-indicating mechanisms to within the prescribed limits, throughout the range of interest. If necessary, prepare a calibration curve.

NOTE 6: Calibration should be performed in ascending weight increments.

Appendix B. Calibration procedure of low-range, movable platen type of compression tester

B.1 Level the machine with a level placed on the bottom platen and clean out the dash pot. After checking the indicator for zero, place the high and then the low range weights in position. If the zero is off, reset it so that the high and low range zeros are the same. Then place a 446-N (100-lbf) load on the machine and check for any sticking.

NOTE 7: Sticking occurs when the load pointer arm comes to a sudden stop or does not return to the same reading after the weights are removed and replaced. If this occurs, check the bearings for cleanliness, the dash pot plunger for burrs, the dash pot for proper alignment, and the counterweight shaft for end play which should not be more than 0.10 divisions. Also, ascertain whether the load indicator arm is rubbing against the dial, and whether the guide bolts for the lower platen are too tight.

B.2 Add loads in 112-N (25-lbf) increments until 446 N (100 lbf) is added and note the readings. If the readings are off by more than 5 N (1 lbf) at 446 N (100 lbf), loosen two set screws in the pendulum arm, set the arm until its reading is 446 N (100 lbf), and retighten the set screws. Remove all the load and recheck to see that the zero and 446 N (100 lbf) readings are within the tolerance.

B.3 After the zero and 446 N (100 lbf) readings have been corrected, recheck and again note the readings at zero, 112 N (25 lbf), 223 N (50 lbf) 335 N (75 lbf), and 446 N (100 lbf).

NOTE 8: If the intermediate readings are off, proceed as follows: Place a 223-N (50-lbf) dead load on the platen. With two adjustable wrenches, bend the load pointer arm until the 223-N (50-lbf) reading is 223 ± 5N (50 ± 1 lbf). Recheck readings at zero, 112 N (25 lbf), 223 N (50 lbf) 335 N (75 lbf), and 446 N (100 lbf). Ensure that all the points are within the following tolerances: 0–90: ± 2.5 N (0–20: ± 0.5 lbf); 91–446 N (21–100: ± 1 lbf).

Reference

1. Nelson, H. G., and Maltenfort, G. G., report of TAPPI Flat Crush Study, *Tappi* **44** (9): 614 (1961).

Your comments and suggestions on this procedure are earnestly requested and should be sent to the TAPPI Technical Divisions Administrator. ■

T 810 om-85

SUGGESTED METHOD – 1966
OFFICIAL TEST METHOD – 1980
REVISED – 1985

Bursting strength of corrugated and solid fiberboard

1. Scope

This method describes the procedure for measuring the bursting strength of single wall and double wall corrugated and solid fiberboard. It is not designed to be used for the bursting strength of paper (TAPPI T 403 "Bursting Strength of Paper"), paperboard and linerboard (TAPPI T 807 "Bursting Strength of Paperboard and Linerboard"), or triple wall corrugated board.

2. Significance

The bursting strength of combined board is primarily an indication of the character of the materials used in manufacturing a fiberboard box and has value in this respect. On the other hand, it gives no direct information regarding the ultimate performance of design or construction of the finished container and correlates very poorly, if at all, with most of the performance values of the container. Bursting strength of combined board is a requirement of the various carrier regulations and federal specifications for shipping containers. The bursting strength of the component paperboard is of critical importance as a control test in the paperboard mill since the conformity of the finished container is generally controlled by the bursting strength of the paperboard. Triple-wall corrugated board cannot be tested suitably by the bursting method. Testing of double-wall board is of questionable accuracy since it is rarely possible to get sufficiently simultaneous bursts of the multiple facings. The test is simple and rapid to execute, but it must be recognized that it is subject to serious errors if instrument, diaphragm, and gages are not properly maintained or if improper procedures are used (*1*, *2*, *3*).

3. Apparatus

3.1 *Bursting tester*[1], consisting of the following.

3.1.1 Means for clamping the test specimen between two annular, plane surfaces having fine concentric tool marks to minimize slippage. The upper clamping platen (clamping ring) has a minimum diameter of 96 mm (3.75 in.), a minimum thickness of 9.5 mm (0.375 in.), and a circular opening of 31.50 ± 0.03 mm (1.240 ± 0.001 in.) diameter. The lower edge of the opening (side in contact with the board) has a 0.64-mm (0.025-in.) radius. The lower clamping surface (diaphragm plate) has a thickness of 5.56 ± 0.08 mm (0.219 ± 0.003 in.) with an opening 31.50 ± 0.03 mm (1.240 ± 0.001 in.) in diameter. The upper edge of the opening (in contact with the board) has a 0.4-mm ± 0.1 mm (0.016-in.) radius and the lower edge of the opening (in contact with the rubber diaphragm) has a radius of 3.1 ± 0.1 mm (0.004 in.) to prevent cutting the rubber when pressure is applied. The upper clamping ring is connected to the clamping mechanism through a swivel joint to facilitate an even clamping pressure. The openings in the two clamping plates are required to be concentric to within 0.13 mm (0.0051 in.) and their clamping faces flat and parallel (see T 807 Appendix A.1.1).

3.1.2 A molded (disk-shaped) diaphragm requiring a pressure of not less than 160 kPa nor more than 21 kPa (not less than 23 nor more than 30 psi) to distend it to a height of 9.53 mm (0.375 in.) above the diaphragm plate (see T 807, Appendix A.1.2.)

3.1.3 Means of forcing liquid into the pressure chamber below the diaphragm at a steady rate of 170 ± 16 mL/min. This pressure shall be generated by a motor-driven piston forcing a liquid (glycerin) into the pressure chamber of the apparatus (see T 807, Appendix A.1.3).

3.1.4 A Bourdon pressure gage of the maximum reading or the lazy hand type. The scale should have a radius of 50 mm (1.875 in.) with graduations extending over a minimum arc of 270° indicating bursting pressure in kPa or psi, with an accuracy of 0.5% of full scale, and have sufficient capacity so that all readings can be maintained in the middle half of the scale. In its operating posi-

[1] Names of suppliers of equipment and materials may be available from the TAPPI Information Resources Administrator.

Approved by the Fiberboard Shipping Container Testing Committee of the Corrugated Containers Division
TECHNICAL ASSOCIATION OF THE PULP AND PAPER INDUSTRY

Reprinted by permission of the Technical Association of the Pulp and Paper Industry, P.O. Box 105113, Atlanta, GA 30348.

tion, have the gage inclined between horizontal and not over 30° from the horizontal. When more than one gage is mounted on a single apparatus, only the gage on which the measurement is being made is open to the hydraulic system so as not to reduce the rate of distention of the sample.

3.1.5 As an alternate to 3.1.4, a pressure transducer with suitable signal processing circuitry to display the maximum bursting pressure may be used provided it gives comparable results. The pressure transducer should be connected to the tester in the same manner as the gage.

NOTE 1: Bursting strength tests should be made with the pressure transducer system to determine if the results obtained are comparable to those obtained with a properly calibrated Bourdon pressure gage.

3.1.5.1 Typical transducer ranges are shown in Table 1.

Table 1. Transducer/digital read-out characteristics

Range of bursting pressure		*Range of transducer*		*Accuracy*	
kPa	*psi*	*kPa*	*psi*	*kPa*	*psi*
5-500	2-100	0-500	0-100	2.5	0.5
10-1000	2-200	0-1000	0-200	5.0	1.0
30-3000	5-500	0-3000	0-500	15.0	2.5
30-6000	5-1000	0-6000	0-1000	25.0	5.0

4. Calibration

4.1 For calibration of apparatus see T 807, Appendix A.1.

4.2 Calibrate the gages frequently enough to ensure their accuracy (once a week if in constant use, or before using if they have been idle for a month or more). Calibrate the gages while inclined at the same angle as used during a test by means of a dead weight gage tester. During calibration, apply the pressure so that the rate of travel of the needle is similar to that when actually testing the board. If a gage is accidently subjected to a pressure beyond its capacity, recalibrate it before it is again used (see T 807, Appendix A.1.4).

5. Sampling and test specimens

5.1 *Solid fiberboard*

5.1.1 From each test unit obtained in accordance with TAPPI T 400 "Sampling and Accepting a Single Lot of Paper, Paperboard, Fiberboard, or Related Product," prepare five 300 × 300 mm (12 × 12 in.) specimens. If the dimensions of each sheet of the test unit are too small, then use specimens no less than 100 mm (4 in.) wide and of sufficient length or number to permit a total of 20 bursts.

5.2 *Corrugated board*

5.2.1 From each test unit obtained in accordance with T 400, prepare five 300 × 300 mm specimens. If size does not permit this, take specimens no less than 150 mm (6 in.) wide and of sufficient length or number to permit a total of 20 bursts.

6. Conditioning

Condition all specimens prior to testing and conduct tests in an atmosphere in accordance with TAPPI T 402 "Standard Conditioning and Testing Atmospheres for Paper, Board, Pulp Handsheets, and Related Products."

7. Procedure

7.1 *Solid fiberboard*

7.1.1 Insert the specimen between the clamping ring and diaphragm plate, then apply sufficient clamping pressure either manually or hydraulically to prevent slippage of the specimen during the test.

7.1.2 Apply the bursting pressure by forcing the piston forward until the diaphragm ruptures the specimen. Record the maximum registered by the gage and gently return the indicator of the gage to zero. Should slippage be indicated, discard the test value and increase the clamping pressure by appropriate increments.

7.1.3 Allot a minimum area of 100 × 100 mm (4 × 4 in.) for each burst to prevent the clamping areas from overlapping. Make an equal number of bursts from each side of the specimen. Arrange that no more than one burst from each side of the specimen falls in the same line of machine formation. Make no test on areas containing sand holes, creases, or obvious imperfections. Make a minimum of 6 bursts on each 300 × 300 mm (12 × 12 in.) specimen and a maximum of 10 bursts to determine the average bursting strength of the material tested.

7.2 *Corrugated board*

7.2.1 Insert the specimen between the clamping ring and the diaphragm plate. Apply sufficient clamping pressure, either manually or hydraulically, to prevent slippage of the specimen during the test. Apply the bursting pressure by forcing the piston forward until the diaphragm ruptures the specimen. Record the maximum registered by the gage and gently return the indicator of the gage to zero. Discard values obtained from bursts showing creases in the linerboard (indicating slippage) when computing the average.

NOTE 2: Clamping pressure is critical in the testing of corrugated board. It is difficult to define exactly because of the influence on the resistance to the clamping pressure of different flute structures. The pressure should be only great enough to prevent slippage of the facing during the burst but not great enough to completely crush the corrugations. Low clamping pressures which permit visible slippage to occur usually result in high test values; high clamping pressures which completely crush the corrugations usually result in low test values. Sometimes it is necessary to use maximum clamping pressure to eliminate double pops with double wall board.

7.2.2 On corrugated board a minimum area of 150 × 150 mm (6 × 6 in.) is required for each burst. A maximum of four bursts, two from each direction, is therefore made on each 900 cm^2 (1 ft^2) specimen. A margin of at least 25 mm (1 in.) is left between the periphery of the clamping ring and the edge of the specimen. Locate the bursts so that not more than one burst from each direction is made in line with the same corrugation. Make a minimum of 20 bursts.

8. Report

For each test unit report the average of the test determinations in kilopascals (or in lb/in.^{2} equivalent to kPa/6.89) to three significant figures.

NOTE 3: For purposes of determining compliance with carrier classification requirements, Uniform Freight Classification Rule 41 and National Motor Freight Classification Item 222 specify a minimum bursting test rather than an average of the test determinations. These rules state, in effect, that only one burst (out of the six prescribed) is permitted to fall below the minimum test required. Board failing to pass the foregoing will be accepted if, in a retest consisting of 24 bursts, not over 4 bursts fall below the minimum test required.

9. Precision

9.1 Repeatability (within a laboratory) = 5.7%

9.2 Reproducibility (between laboratories) = 13.5%

9.3 Comparability (between laboratories) = not known; in accordance with the definitions of these terms in TAPPI T 1206 "Precision Statement for Test Methods."

9.4 The above values were obtained using test results, each an average of 20 determinations. The interlaboratory study was conducted in accordance with TAPPI T 1200 "Interlaboratory Evaluation of Test Methods Used with Paper and Board Products" by the Fibre Box Association Technical Committee, 1971-2, among 12 laboratories on 6 different corrugated combinations.

10. Additional information

10.1 Effective date of issue: March 27, 1985.

10.2 This method has been written for the flat diaphragm type of instrument. An alternative instrument with a hat-shaped diaphragm may be used in a manner similar to that described, if prominently stated in the report. However, equivalency of results has not been established.

10.2.1 To change diaphragm on instruments using hat-shaped diaphragms, rotate the operating handle to start the motor pumping cycle and then immediately return the handle clockwise to its top. The machine will go through its pumping cycle and stop with the diaphragm in its retracted position. Unscrew the lower platen and remove the old diaphragm and fiber spacer washer and wipe the diaphragm seat dry. Insert a new diaphragm and place the fiber spacer washer on the diaphragm shoulder. Replace the lower platen and tighten it. It should be noted that turning the lower platen tighter does not clamp the diaphragm; if leakage occurs, a second fiber spacer washer may be required.

10.2.2 Liquid is replaced under the diaphragm by means of a special glycerin gun. A special bleeder tube, designed for the removal of air, is built into the chamber. Press the diaphragm toward the tube to remove the air, and when a slight excess of liquid is observed, remove the filler gun, disconnect the motor, and rotate the operating handle counterclockwise to the test position. It will be found that this raises the diaphragm slightly. While the operating handle is in the test position, bleed fluid out of the chamber until the diaphragm is level with the upper surface of the lower platen. Close the valve and rotate the operating level to the off position, replace the upper platen, and start the motor. The machine will automatically go through one last cycle and stop in position for a test.

10.2.3 The diameter of the upper clamping plate of the instrument using the hat-shaped diaphragm is 88.9 mm (3.50 in.). This results in a clamping area of approximately 5400 mm^2 (approximately 8.4 in.^{2}) compared to a diameter of 95.3 mm (3.75 in.) and an area of approximately 6500 mm^2 (approximately 10 in.^{2}) for the tester previously described.

10.3 Related methods: ASTM D 2738 (technically identical); Australian and New Zealand, APPITA P 438 (specifies clamping pressure.)

References

1. Institute of Paper Chemistry, "A Method for Determining the Bursting Strength (Mullen) of Paperboard and Paperboard Products," *Fiber Containers* (Feb. 1948).
2. McKee, R. C., Root, C. H., and Ayers, L. R., "Instrumental and Operational Variables Influencing Bursting Strength Results," *Fiber Containers* (May, June, and July 1948).
3. Pitman, G. A., "Problems of Entrapped Air in Mullen Testers," *Fiber Containers* (Nov. 1955).

Your comments and suggestions on this procedure are earnestly requested and should be sent to the TAPPI Test Methods Administrator. ■

Index

Selected ASTM Standards on Packaging

This index covers the standards and related material appearing in this volume only. The boldface references are to the ASTM designations; the standards appear in each volume in alphanumeric order. Proposed methods and specifications carry the index reference, "(Proposed)"; they appear, as does other ancillary material, in the Related Material (gray-edged) section in the back of the book. A triangle (Δ) preceding the ASTM designation denotes that Adjunct Material for the standard is published separately; the Adjunct No. is given in the standard. A Combined Index, covering the standards appearing in all volumes of the *1991 Annual Book of ASTM Standards*, is issued as Volume 00.01.

Alphabetization in the index is letter-for-letter, with no consideration to punctuation or word division. Initial prepositions of (indented) subentries are ignored for alphabetization.

In the preparation of indexes, every attempt has been made to index standards on three levels: (1) by main subject, using general and specific search terms; (2) by test methods or other significant sections of ASTM standards; and (3) by cross-references to locate main subject entry terms. (*See also* references are abbreviated as *Sa* and appear under main entry terms.) Specification F451, for Acrylic Bone Cements (Volume 13.01), illustrates ASTM's method of indexing.

INDEX TERMS FOR MAIN SUBJECT ENTRY

Adhesives—surgical implant materials
acrylic bone cements (for internal orthopedic prostheses), self-curing, spec., **F 451**

Orthopaedic medical devices—bone
acrylic bone cements (for internal orthopedic prostheses), self-curing, spec., **F 451**

INDEX TERMS FOR TESTS

Compressive strength
acrylic bone cements, test, **F 451**

Doughing time
acrylic bone cements, test, **F 451**

CROSS-REFERENCES

Bone—cement
See **Adhesives—surgical implant materials**

Fixation materials
Sa **Orthopaedic medical devices** (*headings*)

D

Peel adhesion
peel adhesion of pressure-sensitive tape at 180° angle, test, (for metric see D 3330M), **D 3330**

Performance—containers/packaging
gross leaks in heat-sealed flexible packages, test, **D 3078**
pallets/related structures employed in materials handling/shipping, tests, **D 1185**
performance testing of shipping containers/systems, practice, **D 4169**
random vibration testing of shipping containers, test, **D 4728**
testing hazardous materials packaging (intended for transportation), spec., **D 4919**

Performance-based closing practices
methods of closing/sealing/reinforcing of fiberboard shipping containers, practice, **D 1974**

Plastics (general)
low-density polyethylene films for general use/packaging applications, spec., **D 4635**
nonmetallic strapping and joining methods (for car/truck loading), spec., **D 3950**
water vapor transmission of materials, test, **E 96**

Plastics (general)—nylon (NP)
nonmetallic strapping and joining methods (for car/truck loading), spec., **D 3950**

Plastic sheet and film
tensile properties of thin plastic sheeting, test, **D 882**
water vapor transmission of materials, test, **E 96**

Plastics (thermoplastic)
operating light-exposure apparatus (xenon-arc type) with/without water for exposure of nonmetallic materials, practice, **G 26**

Plastics (thermosetting)
operating light-exposure apparatus (xenon-arc type) with/without water for exposure of nonmetallic materials, practice, **G 26**

Platforms
pallets/related structures employed in materials handling/shipping, tests, **D 1185**

Polyester plastics
nonmetallic strapping and joining methods (for car/truck loading), spec., **D 3950**

Polyethylene (PE) plastics
low-density polyethylene films for general use/packaging applications, spec., **D 4635**

Polyethylene (PE) plastics—sheet and film
corrugated/solid fiberboard sheet stock (container grade) and cut shapes, spec., **D 4727**

Polyolefin (PO) plastics
nonmetallic strapping and joining methods (for car/truck loading), spec., **D 3950**

Preservation
commercial packaging—minimum requirements for preservation/packing/packaging/marking, practice, **D 3951**

Pressure-sensitive tapes
adherence to linerboard of pressure-sensitive tape, at low temperature, test, (for metric see D 3889M), **D 3889**
adhesion of pressure-sensitive tape to fiberboard, at 90° angle and constant stress, test, (for metric see D 2860M), **D 2860**
bursting strength of pressure-sensitive tapes (used in closure of fiberboard containers), test, (for metric see D 3662M), **D 3662**
holding power of pressure-sensitive tapes, test, (for metric see D 3654M), **D 3654**
peel adhesion of pressure-sensitive tape at 180° angle, test, (for metric see D 3330M), **D 3330**
tensile strength/elongation of pressure-sensitive tapes, test, (for metric see D 3759M), **D 3759**

Pressure testing—packaging/containers
controlled horizontal impact test for shipping containers, methods, **D 4003**
leakage testing of empty containers, by vacuum method, test, **D 4991**

Pulp
moisture in wood (chils/sawdust/groundwood)/pulp/paper/paperboard, by toluene distillation, **T 208**

Puncture test
puncture test of containerboard, **T 803**

Q

Quality assurance
tensile strength/elongation of pressure-sensitive tapes, test, (for metric see D 3759M), **D 3759**

R

Radiation exposure
operating light-exposure apparatus (xenon-arc type) with/without water for exposure of nonmetallic materials, practice, **G 26**

Rail car switching impact test
controlled horizontal impact test for shipping containers, methods, **D 4003**

Railroad handling test
mechanical handling of unitized loads/large shipping cases/crates, test, Δ **D 1083**

Ream weight
grammage of paper/paperboard (weight per unit area), **T 410**

Reinforcing
methods of closing/sealing/reinforcing of fiberboard shipping containers, practice, **D 1974**

Resistance—compression
compression testing for shipping containers/components/unit loads, method, **D 642**

Resistance—impact
controlled horizontal impact test for shipping containers, methods, **D 4003**
pallets/related structures employed in materials handling/shipping, test, **D 1185**
shipping containers, loaded, boxes and crates, incline test, **D 880**

Resistance—leak
testing hazardous materials packaging (intended for transportation), spec., **D 4919**

Resistance—leakage
leakage testing of empty containers, by vacuum method, test, **D 4991**

Resistance—sliding friction
pallets/related structures employed in materials handling/shipping, test, **D 1185**

Resistance—sling hoist-lift
pallets/related structures employed in materials handling/shipping, test, **D 1185**

Resistance—slip
coefficient of static friction of corrugated/solid fiberboard, test, **D 4521**

Resistance—tear
internal tearing resistance of paper, by Elmendorf-type method, **T 414**

Resistance—vibration
pallets/related structures employed in materials handling/shipping, test, **D 1185**
shipping containers, loaded, test, **D 999**

Resistance—water
testing hazardous materials packaging (intended for transportation), spec., **D 4919**

T

V

W

X

1916 Race Street
Philadelphia, PA 19103
Phone: (215) 299-5454
FAX: (215) 977-9679

European Office
27-29 Knowl Piece, Wilbury Way
Hitchin, Herts SG4 OSX, England
Phone: 0462-437933
FAX: 0462-433678

1991 MEMBERSHIP APPLICATION

THIS APPLICATION IS FOR SOCIETY MEMBERSHIP ONLY.

EACH COMMITTEE REQUIRES A SEPARATE APPLICATION.

PLEASE SEND APPLICATIONS FOR THE FOLLOWING

COMMITTEE (S) ____________________

APPLICATION IS MADE FOR MEMBERSHIP IN ASTM:

SEE BENEFITS AND FEES ON REVERSE SIDE

☐ MEMBER — AN INDIVIDUAL, GOVERNMENT REPRESENTATIVE, OR AN INSTITUTION (EDUCATIONAL, PUBLIC LIBRARY, OR A SCIENTIFIC ENGINEERING, OR TECHNICAL NON-PROFIT SOCIETY) SUBSCRIBING TO THE PURPOSES OF THE SOCIETY PROVIDED IN THE CHARTER AND BYLAWS.

☐ ORGANIZATIONAL — AN INDIVIDUAL, BUSINESS, GOVERNMENTAL, RESEARCH, OR PROFESSIONAL ORGANIZATION, OR TRADE ASSOCIATION, OR SEPARATE FACILITY THEREOF SUBSCRIBING TO THE PURPOSES OF THE SOCIETY PROVIDED IN THE CHARTER AND BYLAWS.

PLEASE PRINT ALL INFORMATION CLEARLY. DO NOT EXCEED THE CAPACITY OF EACH LINE.

LAST NAME ____ FIRST NAME ____ INITIAL ____

COMPANY NAME ____

FACILITY ____

STREET ____

P.O. BOX ____

CITY ____ STATE ____ ZIP ____ – ____

PHONE ( ) — EXTENSION ____

FAX ( ) —

COUNTRY ____

JOB TITLE ____

PARENT COMPANY (IF DIFFERENT FROM ABOVE) ▶ ____________________

OFFICIAL REPRESENTATIVE (ORGANIZATIONAL MEMBERSHIP ONLY)

LAST NAME ____ FIRST NAME ____ INITIAL ____

TITLE ____

IF HOME ADDRESS IS TO BE USED FOR ALL MAILINGS & PUBLICATIONS, PLEASE COMPLETE BELOW (AFTER COMPLETING ABOVE).

ADDRESS ____

CITY ____ STATE ____ ZIP ____ – ____

MAIL TO: ASTM ATTN: MEMBER SERVICES

*** ALL CHECKS MADE PAYABLE TO ASTM IN U.S. FUNDS ON U.S. BANKS**

* OPTIONAL METHODS OF PAYMENT: ☐ AMERICAN EXPRESS ☐ MASTER CARD ☐ VISA

ACCOUNT NO. (ALL DIGTS) ____ EXPIRATION DATE ____ SIGNATURE ____

ONE FREE VOLUME — BOOK OF ASTM STANDARDS — YOUR CHOICE — VOLUME # ____

BENEFITS AND FEES FOR 1991

	Organizational	Member
Annual Fee	$350.00*	$50.00**
Participation on ASTM Technical Committees	One Official Rep. Only	Member Only
Annual Book of ASTM Standards	One Free Volume	One Free Volume
Complete Set of Book of ASTM Standards (Vols. 1-68)	$3,300.00 (Prepaid)	$3,700.00 (Prepaid)
Individual Book of Standards at Member Price	No Limit	None
Complete Sections at Member Price (Prepaid)	No Limit	No Limit
Other Publications at Member prices	No Limit	No Limit
Annual Subscription to *Standardization News*	1 Free	1 Free

* Of the $350 Administrative fee payment, $300 is tax deductible as a charitable contribution for federal tax purposes. The remaining $50 may be deducted as an ordinary and necessary business expense.

** Administrative fee payments made to ASTM are not tax deductible as charitable contributions for federal income tax purposes. They may be deductible as ordinary and necessary business expense.

MEMBERSHIP

- Membership year is 1 January to 31 December. Fees are payable in advance and are not prorated. Membership will become effective upon payment of fees.
- Organizations applying for membership should select one official representative qualified to exercise the rights and privileges of membership. This includes voting on Society Ballot, and if appropriate, on technical committee matters.
- Only the official representative may participate on committees under the organizational membership. All others must maintain individual memberships in order to participate on committees.
- The official representative for the organizational membership may be changed as needed by notifying ASTM Headquarters. Committee participation is not transferable. An application will have to be submitted by the new representative to the appropriate committees. A member not officially serving on ASTM committees may not vote in committee matters.
- For members serving on technical committees, their organizations may ask for this membership to be transferred without payment of additional fees, if the fees were paid by the company. However, no additional free volume of the Book of ASTM Standards will be allowed. Committee participation is not transferable. An application will have to be submitted by the new member to the appropriate committee.

PUBLICATIONS

- All publications must be shipped to the member.
- Shipping and handling costs will not be charged for prepaid orders shipped to the U.S. and Canada.

NOTE: Membership benefits and publication prices are subject to change.